P9-AFZ-461

METHODS IN

MICROBIOLOGY

Volume 23
Techniques for the Study of Mycorrhiza

Edited by

J. R. NORRIS

Reading, UK

D. J. READ

Department of Animal & Plant Sciences,
University of Sheffield, UK

A. K. VARMA

School of Life Sciences, Jawaharlal Nehru University
New Delhi, India

ACADEMIC PRESS
Harcourt Brace Jovanovich, Publishers
London San Diego New York Boston
Sydney Tokyo Toronto

ACADEMIC PRESS LIMITED
24–28 Oval Road
London NW1

United States Edition published by
ACADEMIC PRESS INC.
San Diego, CA 92101

Copyright © 1991 by
ACADEMIC PRESS LIMITED

Chapters 9 and 17 © 1991 US Government in the jurisdictional territory of the USA

All Rights Reserved
No part of this book may be reproduced in any form, by photostat, microfilm or any other means, without written permission from the publishers

British Library Cataloguing in Publication Data is available

ISBN 0-12-521523-1

Editorial and production services by Fisher Duncan
10 Barley Mow Passage, London W4 4PH, UK
Printed in Great Britain by St Edmundsbury Press,
Bury St Edmunds, Suffolk

CONTRIBUTORS

R. Agerer University of Munich, Institute for Systematic Botany, Menzingerstrasse 67, D-8000 München 19, Germany
I. Ahmad Centre for Plant Biotechnology, Department of Botany, University of Toronto, 25 Willcocks Street, Toronto, Ontario M5S 3B2, Canada
J. L. Armstrong US Environmental Protection Agency, Environmental Research Laboratory, Corvallis, OR 97333, USA
V. Barrett Department of Botany & Microbiology, Auburn University, 129 Funchess Hall, Auburn, Alabama 36849-5407, USA
B. Botton Université de Nancy 1, Laboratoire de Physiologie Végétale et Forestière, BP 239, 54506 Vandoeuvre les Nancy Cedex, France
J. W. G. Cairney Department of Soil Science, Waite Agricultural Research Institute, University of Adelaide, Glen Osmond, SA 5064, Australia
P. Chakravarty Department of Botany, University of Guelph, Guelph, Ontario N1G 2W1, Canada
M. Chalot Université de Nancy 1, Laboratoire de Physiologie Végétale et Forestière, BP 239, 54506 Vandoeuvre les Nancy Cedex, France
C. E. Cordell US Department of Agriculture, Forest Service, Southeastern Forest Experiment Station, Asheville, NC 28802, USA
P. Cudlín Institute of Landscape Ecology, Czechoslovak Academy of Sciences, NA Sadkach 7, 370 05 Ceské Budějovice, Czechoslovakia
R. K. Dixon Environmental Research Laboratory, Environmental Protection Agency, Corvallis, Oregon 97333, USA
D. M. Durall Natural Environment Research Council, Plant Mycorrhizal Unit, Department of Plant Sciences, Parks Road, Oxford OX1 3PF, UK
S. Egli Swiss Federal Institute of Forest, Snow and Landscape Research, CH-8903 Birmensdorf, Switzerland
Professor J. L. Harley†
H. Heinonen-Tanski Department of Environmental Sciences, University of Kuopio, PO Box 6, SF 70211 Kuopio, Finland
J. A. Hellebust Centre for Plant Biotechnology, Department of Botany, University of Toronto, 25 Willcocks Street, Toronto, Ontario M5S 3B2, Canada
T. Holopainen Department of Environmental Sciences, University of Kuopio, PO Box 6, SF 70211 Kuopio, Finland
I. Jakobsen Plant Biology Section, Environmental Science and Technology Department, Riso National Laboratory, DK-4000, Denmark
D. H. Jennings Department of Genetics and Microbiology, Life Sciences Building, The University of Liverpool, PO Box 147, Liverpool L69 3BX, UK
M. D. Jones Natural Environment Research Council, Plant Mycorrhizal Unit, Department of Plant Sciences, Parks Road, Oxford OX1 3PF, UK

†Deceased

I. Kälin Swiss Federal Institute of Forest, Snow and Landscape Research, CH-8903 Birmensdorf, Switzerland
I. Kottke Universität Tübingen, Institut für Botanik, Lehrstuhl Spezielle Botanik, Mykologie, Auf der Morgenstelle 1, D-7400 Tübingen, Germany
J. R. Leake Department of Animal and Plant Sciences, University of Sheffield, P.O. Box 601, Sheffield S10 2UQ, UK
P. A. Lemke Department of Botany & Microbiology, Auburn University, 129 Funchess Hall, Auburn, AL 36849-5407, USA
M. Madkour Institut für Mikrobiologie, Goerg-August-Universität Göttingen, D-3400 Göttingen, Germany
F. M. Martin Laboratoire de Microbiologie Forestière, Centre de Recherches Forestière de Nancy, Institut National de la Recherche Agronomique, 54280 Champenoux, France
D. H. Marx US Department of Agriculture, Forest Sciences Laboratory, Carlton Street, Athens, GA 30602, USA
F. Mayer Institut für Mikrobiologie, Georg-August-Universität Göttingen, D-3400 Göttingen, Germany
A. Nolte Institut für Mikrobiologie, Goerg-August-Universität Göttingen, D-3400 Göttingen, Germany
R. L. Peterson Department of Botany, University of Guelph, Guelph, Ontario N1G 2W1, Canada
D. J. Read Department of Animal and Plant Sciences, University of Sheffield, P.O. Box 601, Sheffield S10 2UQ, UK
J. L. Ruehle US Department of Agriculture, Forest Sciences Laboratory, Carlton Street, Athens, GA 30602, USA
P. T. Rygiewicz US Environmental Protection Agency, Environmental Research Laboratory, Corvallis, OR 97333, USA
P. B. Tinker National Environmental Research Council, Terrestrial and Freshwater Science Directorate, Polaris House, North Star Avenue, Swindon SN2 1EU, UK
A. Varma School of Life Sciences, Jawaharlal Nehru University, New Delhi 110067, India

PREFACE

It is now almost ten years since the appearance of the only previous volume devoted to the methodologies used in mycorrhizal research: *Methods and Principles of Mycorrhizal Research* (N. C. Schenck, ed.) American Phytopathological Society, St Pauls, Minnesota, 1982. These have been momentous years which have seen not only an explosive increase of interest in mycorrhiza as evidenced by the surge in number of published papers, but also, and most importantly, the welcome development of a greater awareness amongst plant physiologists, ecologists and microbiologists that the mycorrhizal condition is the norm rather than the exception in nature. Through the same period sophisticated new technologies (e.g. NMR, RFLP, DNA mediated transformation, immunocytochemistry), which are readily applicable to mycorrhizal research, have become widely available.

These changes have given rise to an urgent need for a comprehensive and up-to-date source book of techniques which can be routinely used in laboratories involved in mycorrhizal research to facilitate, and increase the effectiveness of, their research. The present volumes are designed to meet this need.

The volumes are divided so that the major emphasis of the first (Vol. 23) is placed upon a description of techniques applicable to ectomycorrhizal and ericoid systems, while that of the second (Vol. 24) is upon vesicular–arbuscular systems. Each contains a mix of papers, written by the leading authorities in their fields, which emphasise techniques, but which also include up-to-date reviews of particular topics e.g. carbon metabolism, nitrogen metabolism and applied aspects. The methods whereby, and the extent to which, mycorrhizal inoculation can be used to facilitate revegetation, as well as increases of agricultural or forest productivity are also described.

We envisage that over the next decade these volumes will provide essential information for all those initiating projects in this important subject area, and that at the same time they will facilitate considerable increases in effectiveness in established programmes of mycorrhizal research.

Inevitably in multi-authored volumes such as these the terminology used in the original submissions differed greatly between papers. Not the least of the difficulties was posed by the word ‘mycorrhiza’ itself. This appeared in various guises including ‘mycorrhiza’ ‘mycorrhizas’ and ‘mycorrhizae’. Faced with such problems, and with the need for internal consistency foremost in their minds, the editors elected to employ the word ‘mycorrhiza’ throughout, since it can be used to refer to both the singular and plural state and to describe ‘the mycorrhizal condition’. On etymological grounds, ‘mycorrhiza’ is preferable to ‘mycorrhizae’ because the latter involves the combination of two elements of Greek origin, ‘myco’ (fungus) and ‘rhiza’ (root), with a Latin ‘ae’ ending.

We hope that the need for internal consistency in the use of terminology will be recognised. For their part, the editors have striven to ensure that any minor changes of style arising from alteration of terminology are made in such a way as to have minimal impact on the essential meaning of the text.

October 1991

J. R. Norris
D. J. Read
A. K. Varma

CONTENTS

This book is dedicated to the memory of

Professor J. L. Harley

1

Introduction: The State of the Art

J. L. HARLEY†

I. Preliminary considerations

The study of mycorrhizal symbiosis may be said, after 100 years of investigation, to have come of age. We can now see the symbiotic associations of fungi with higher plants and with bryophytes in perspective and understand, broadly, their importance in the living world. It is essential, however, to be frank with ourselves and not to pretend to understanding in areas where there is still confusion and ignorance. This chapter provides only a quick survey of past work, but it also attempts to highlight our ignorance where it exists.

The study started with the work of Kamienski in 1881 on the chlorophyll-less *Monotropa* and of Frank in 1885 on forest trees. Indeed, it is important to realize that many properties of mycorrhizal

† Deceased

METHODS IN MICROBIOLOGY
VOLUME 23 ISBN 0-12-521523-1

Copyright © 1991 by Academic Press Limited
All rights of reproduction in any form reserved

symbiosis, confirmed by recent work, were adumbrated by the views expressed by the earliest workers. The section on *Historical Perspectives* which comprises the first 74 pages of the Proceedings of the 6th North American Conference on Mycorrhizae (1985), provides an historical account of the subject, with translations into English of early papers, and their scientific background. The relationship of mycorrhiza, lichens, and nitrogen-fixing symbioses was early recognized. We can now see that symbiotic associations of fungi with higher plants, bryophytes and algae are the most abundant and widespread of all symbioses, but it would perhaps be rash to conclude that they were therefore the most important. The corals, associations of algae and coelenterate animals, both in the physiological effects on their environment and in their reef- and island-building activities are, and have been for millions of years, of great global importance. In a different way the symbiotic nitrogen-fixing symbioses of *Rhizobium* and *Frankia* with green plants, harnessing as they do a direct supply of photosynthetic products to the microbial process of nitrogen fixation, are again of very great significance in natural ecosystems, as well as in plant production for human and animal use.

Although symbioses between fungi and green plants conform to relatively few different morphological types, they involve more than one quarter of the described species of fungi, including members of almost all large groups, with some 70% or more of the species of angiosperms, most or all species of gymnosperms, a large number of pteridophytes, a significant but yet unknown number of bryophytes and members of about 36 genera of algae and cyanobacteria of which only eight genera are common in lichens.

This summary of the occurrence of fungal symbioses with green plants is enough to convince us that they must be an important biological study, but one of considerable complexity involving many facets, taxonomy, physiology, genetics and ecology of the constituent organisms and their joint structures, together with their possible value in the understanding of plant life and ecology and their importance in the technologies of plant production.

Although in the study of mycorrhiza our aim is to understand the relationships between fungi and green plants which are mutualistic symbioses, we must realize that the results of work with corals, lichens and even damaging biotrophic parasites, like the rust fungi, may enlighten our problems, as has been shown by Smith, D. C. *et al*. (1979) and Smith, D. C. and Douglas (1987).

The mycorrhiza, composite organs of roots and fungi, formed on most species of angiosperms and gymnosperms and on some pteridophytes, have been classified under relatively few types, as shown in Table I. These types show variation within themselves, and from time to time new variants are described which depart from those previously recognized, but the classification has in general been very satisfactory and has led to the assumption that each category has most of its salient aspects of structure, physiology and biochemistry and functioning in common. In the mycorrhizal host*, the root systems usually consist of mycorrhiza and uncolonized roots. These latter are often represented only by new root apices and mature mother roots. At the same time mycorrhiza, like all subterranean organs, are surrounded by an entourage of other microorganisms—especially fungi and bacteria—which flourish on exudates and sloughed cells and tissues of their host organs. All these organisms, mycorrhizal fungi and other members of the populations of the rhizosphere and root or mycorrhizal surface, have significant effects on the physiological processes of their hosts, especially in terms of absorption, by virtue of their position. They exist in the interface of soil and root or mycorrhiza and so may alter the soil atmosphere and by absorption, release, or change of form or pH, the availability of chemical compounds in the root region. They may also inhibit or perhaps encourage soil-borne pathogens.

In spite of such complexities, the mycorrhizal fungi have dominant importance in the root region by virtue of their close relationship with the cells and tissues of their hosts, on the one hand, and with soil particles and soil solutions on the other.

Although mycorrhizal symbiosis is so common amongst angiosperms, certain genera and even families are usually or often non-mycorrhizal. This absence of mycorrhiza has been considered by Testa *et al.* (1987). It should be noted that a number of common beliefs in respect of the absence of mycorrhiza are not tenable as generalities. For instances, it is not true that all plants of wet places or that all annual plants are always free of fungi. The problems are much more complicated and may more likely turn on specificity and recognition of the symbionts, one to another.

II. The general features of mycorrhizal symbiosis

In the majority of mycorrhizal symbioses, the higher plant host is capable of active photosynthesis. In relatively few, such as in the so-called saprophytic plants and in the early growth of orchids and other

* The term "host" is used in this essay, as it so often is, as synonymous with "photobiont" where the host is a green plant.

TABLE I

The characteristics of the important kinds of mycorrhiza. The structural characters given relate to the mature state, not the developing or senescent states. Entries in brackets rare

	Kinds of mycorrhiza						
	Vesicular-arbuscular	Ectomycorrhiza	Ectendo-mycorrhiza	Arbutoid	Monotropoid	Ericoid	Orchid
Fungi septate	−	+	+	+	+	+	+
Fungi aseptate	+	(+)	−	−	−	−	−
Hyphae enter cells	+	−	+	+	+	+	+
Fungal sheath present	−	+	+ or −	+	+	−	−
Hartig net formed	−	+	+	+	+	−	−
Hyphal coils in cells	+	−	+	+	−	+	+
Haustoria dichotomous	+	−	−	−	−	−	−
Haustoria not dichotomous	−	−	−	−	+	−	+ or −
Vesicles in cells or tissues	+ (or −)	−	−	−	−	−	−
Achlorophylly	− (or +)	−	−	− (or +)	+	−	+
Fungal taxon	Phyco	Basidio Asco Phyco	Basidio Asco?	Basidio	Basidio	Asco (Basidio)	Basidio
Host taxon	Bryo Pterido Gymno Angio	Gymno Angio	Gymno Angio	Ericales	Monotropaceae	Ericales	Orchidaceae

plants with tiny seeds, the host is incapable of photosynthesis. In these plants the fungus supplies the whole symbiotic system with carbon and inorganic nutrients. In these special cases the plant is often considered to be parasitic on its fungus and indeed the type of fungus involved, one capable of obtaining carbon compounds from the soil or from living plants, is different. However, the structural relationship between the partners does not greatly differ in most cases from other associations, and the "saprophytes" and "partial saprophytes" are conveniently considered mycorrhizal.

In the general case, however, carbon compounds derived from photosynthesis are utilized by both the host and by the fungus. The latter absorbs soil-borne nutrients, part of which is translocated to the host. The system therefore consists of hyphae ramifying in the soil which connect with the root tissue or even penetrate the root cells. Nutrients synthesized or absorbed are, so to speak, shared between fungus and host. This general plan of functioning of mycorrhizal plants is well substantiated; what is needed and being actively sought is a greater knowledge of the physiological and biochemical processes and mechanisms of absorption and translocation and release of substances by the partners and the ecological input of the symbiosis.

Table I notes points on the structure of mycorrhiza. In vesicular-arbuscular mycorrhiza the fungal hyphae penetrate between and into the cells of the host. In the latter they form coils, branched haustoria (arbuscules) and perhaps vesicles. In ericoid mycorrhiza, once again, the cells are penetrated, and extensive hyphal coils develop within them. In ectomycorrhiza the hyphae penetrate between cells of the cortex of the root to form a branched structure called the Hartig net, as well as forming a compact sheath or mantle surrounding the rootlet. Other types with actively photosynthesizing hosts have combinations of Hartig net, intracellular penetration and perhaps fungal sheath also.

No matter what the gross structure, the context or interface between partners is not very dissimilar, so long as they are physiologically active. For instance, the penetrating hyphae and arbuscules of vesicular-arbuscular mycorrhiza are bounded by the hyphal plasmalemma and outside that, the hyphal wall. Outside again is an "interfacial" zone probably derived from host wall material. This is bounded by host plasmalemma. Hence no part of the fungus is "free" in the lumen of the cell but is bounded by interfacial zone and host plasmalemma. The interfacial zone contains membranaceous vesicles and fibrils. It seems to be a zone into which the host secretes the materials and enzymes to produce a cell wall, but this actually is prevented by the fungus (Dexheimer *et al.*, 1979).

A similar arrangement of hypha, interfacial zone and host is found in all types of mycorrhiza in which hyphae penetrate into the cells. In ectomycorrhiza the fungal hyphae penetrate between the cells and in some cases modify them. In any event there is never a juxtaposition of protoplast or plasmalemma but always a separating zone of walls or modified walls.

A system, such as this, of association of living fungus and functional host elements is present in the functional or mature state of all mycorrhiza. However, mycorrhizal roots have a developmental period which leads to the mature state which itself is followed by senescence. In vesicular-arbuscular and other mycorrhiza, the arbuscules themselves and the hyphal coils senesce and become encapsulated in an organized cell wall of host origin. It is as if the fungus no longer prevents the formation of the wall, i.e. it is unable to maintain the interfacial zone. Collated estimates of the longevity of arbuscules suggests that they have short lives of 4–15 days (see Harley and Smith, S.E., 1983). In ericoid mycorrhiza, however, the host cell may degenerate first and the hyphae then persist, active, for a time in the senescing cell. In ectomycorrhiza there may be a development of cambium in the inner cortex of the host which cuts off the cortex. As this happens, the fungus may penetrate the senescing cells and the mature structure breaks down.

Of course in most cases the apical growth of the root results in elongation of the axis in which mycorrhizal infection develops in the new tissue in advance of the old, and in the course of time it also senesces. A mycorrhizal root is a growing, developing and ageing axis and branches.

A problem which needs further work concerns the actual method of penetration of the tissues of the host by mycorrhizal hyphae. For most of them, in spite of much research, there is no evidence available that they produce enzymes capable of hydrolysing the polymers of the cell wall. Some few have been shown to produce restricted amounts of cellulase, pectinase and other hydrolytic enzymes, but it is certainly not a consistent property of theirs. Moreover, as Dexheimer *et al.* (1979) have shown, the hyphae of vesicular-arbuscular mycorrhiza do not actually penetrate the cell wall but cause it to invaginate and later to be converted into the interfacial zone. The same is true of other fungi which penetrate the cell lumina. Harley (1985) shortly reviewed this matter, pointing out that ectomycorrhizal fungi appear to penetrate the tissues in a zone behind the apex where the cell walls are forming and maturing. The view is indeed tenable that in all cases the hyphae interfere with the polymerizing enzymes forming the cell walls of the host so that penetration or invagination takes place. This matter needs much more investigation, for it should be appreciated that mycorrhizal fungi must have some peculiar property which enables them, and not the many other fungi of the root region, to penetrate and set up a relation with living roots. The actual structure of the interface so formed is all-important in the nutritional relationship between the partners. Gianinazzi-Pearson and her collaborators have studied and are studying the presence of enzymes such as ATPase at the membrane surface of arbuscules which doubtless have an important function in controlling movement of materials across the membrane.

III. Specificity of fungi and hosts

Although ericoid mycorrhiza are mainly restricted to one order, the Ericales, other kinds such as the ectomycorrhiza and vesicular-arbuscular mycorrhiza are not restricted in the same way to a taxonomic group. Vesicular-arbuscular mycorrhiza are found widely in pteridophytes, gymnosperms and angiosperms of all growth habits. Ectomycorrhiza are mainly restricted to arborescent plants of angiosperms and gymnosperms but not to particular taxonomic groups. Some of these can also develop vesicular-arbuscular mycorrhiza, e.g. some Rosaceae and some Salicaceae can be either ectomycorrhizal or vesicular-arbuscular mycorrhizal, or both.

Turning to actual specificity; species of host to species of fungus—this is extremely loose. Amongst vesicular-arbuscular fungi (Endogonaceae) most have been found to form functional mycorrhiza with many species of host and few are restricted. More is known of the ectomycorrhizal fungi, for they can be tested in culture. Very few are genus-specific, some are restricted to gymnosperms, some to a single family, but most are of very wide host range. These matters, which clearly have important implications for the efficiency of functioning and of ecological behaviour, are discussed by Harley and Smith, S. E. (1983) and will be considered briefly below. It should be noted that the low specificity poses difficulties in the understanding of the possible fungus-to-host recognition mechanism.

IV. Mycorrhiza and growth of the host

Comparisons of plants, with and without mycorrhiza, have been made, time without number, ever since mycorrhiza were discovered. Such comparisons have often shown that mycorrhizal plants grow faster in their early stages than do non-mycorrhizal ones. Only with herbaceous plants have experiments lasting the whole or even a significant part of the life-cycle been made. The measurements made have usually been of height, shoot, or whole plant weight (fresh and dry), root to shoot ratio, or similar simple measures of growth. Sometimes relative growth rates have been estimated and at times internal quantities or concentrations of soil-derived nutrients have also been estimated.

Increases of the rate of growth of mycorrhizal plants measured in these ways are usually observable when the potential for active photosynthesis is high (e.g. adequate light) and when there is a deficit or rather no excess of one or more essential nutrient such as phosphate or nitrogen source. Where increases are not observable they are often associated with very ready availability of nutrients or a low light supply. In addition, there are variations in the efficiency of the fungi which can form mycorrhiza, and a strain may be ineffective or inefficient in some environmental conditions. In some cases there may be little effect of mycorrhizal formation, or perhaps an actual decrease in growth rate of the host may be observed. In some cases, where growth experiments have been performed on a range of phosphate concentration in soil, a depression of growth in mycorrhizal plants may occur at a high nutrient level. The causes of this are discussed by Harley and Smith, S. E. (1983) with many references. They conclude that competition between the symbionts for carbohydrates is the most likely explanation and point out that at a given nutrient supply, reduced day-length reduces the growth rate of mycorrhizal plants more than that of non-mycorrhizal ones.

This kind of growth reduction in mycorrhizal plants should not be given any special emphasis, as for instance by dubbing it parasitism by the fungus. It actually seems to be a stage in the continuum of competition between the symbionts for carbon compounds where their production is inadequate to maintain the symbiotic system in full growth. What has been measured is the growth of the host only—that is reduced. It should be noted that no data are available to determine whether the growth of the fungus is also decreased.

This raises an important subject. In all types of mycorrhiza it has been observed that different fungal strains may colonize the host to different extents. They also may differ in the extent to which they affect growth rate. These variations may be correlated but are not necessarily so. Again, there is no corpus of data on the growth efficiency of fungal strains, e.g. the growth made per unit mass of carbohydrate consumed. Such estimates are needed, particularly to find out whether efficiency in respect of carbohydrate usage in culture is linked with efficiency as symbionts.

A further need is that in all symbiotic growth experiments of the kind under discussion, further details of structure and extent of hyphal connections with the soil should be made. These should include, where possible, frequency and extent of such structures as Hartig net and arbuscules.

An important philosophical point arises in respect of such experiments. It is a habit which we all mistakenly fall into at times, to talk of the beneficial or detrimental effects of mycorrhizal infection. It is fundamentally important to refrain from consideration of benefit in respect of mycorrhizal symbiosis, not only because in most experiments the fungal growth is not considered, but because it is essential to keep in mind what has actually been measured and to draw only those conclusions which can rationally be arrived at from those measurements. No conclusion of benefit, that is goodness or badness, can be made from scientific measurements. We cannot measure goodness or badness in biological terms, only in terms of human needs or desires. The concept of *survival value* is the nearest to a concept of goodness biologically speaking, and that is extremely difficult to estimate; it cannot be arrived at from the kind of experiment that we have been discussing.

V. Nutrient uptake by mycorrhizal roots

A. Genera! considerations

The factors which affect the rates of absorption of nutrients by mycorrhizal roots and their constituent fungi do not differ in any great respect from those affecting uptake by other plant organs or free-living fungi. Uptake of essential nutrients is linked with, and dependent upon, metabolic rate. Hence factors affecting metabolism are important, such as availability of respiratory and other substrates, oxygen supply, temperature, concentration and form of the nutrient and metabolic poisons or inhibitors. Where the nutrient, during or after absorption, is converted to other forms, or built into organic compounds, the supply of respiratory substrates is necessary not only for the energy expended in uptake but also for the synthesis.

Essentially the same macro- and micro-nutrients are as necessary for the growth of mycorrhizal plants as they are for others. However, research has concentrated particularly on the uptake of the two important macro-nutrients nitrogen and phosphate. The immediate source is the soil solution with which both hyphal surfaces and root surfaces are in contact. The hyphae typically ramify through the soil, producing for mycorrhizal plants a large exploited volume. This exploitation varies in importance, depending upon the nutrient considered, its concentration and mobility in the soil and the demand or rate of uptake by the plant.

As Nye and Tinker (1977) emphasized, the concentration of an absorbed nutrient tends to fall in the environs of the organ, root or hypha that is absorbing it. If the movement of that nutrient through the soil is slower than its rate of uptake, then its concentration at the absorbing surface will fall, so that a zone of deficit, or even a zone of zero concentration, will come to exist at the surface. In such a case the absorption rate of that nutrient is then controlled entirely by its rate of movement to the surface through the soil, and not by physiological factors. The hyphae emanating from a mycorrhizal root serve to cross the zone of deficit and maintain the rate of absorption. The maintenance of the absorbing area in contact with soil solution by hyphal growth is much more economic in terms of material and energy expenditure than by root growth (Harley, 1989).

The practical importance of this matter in respect of vesicular-arbuscular mycorrhiza was demonstrated experimentally by Sanders and Tinker (1973) in respect of phosphate uptake. It will be noted that this mechanism demands that the fungus be able to absorb the nutrient and translocate it to the host faster than it can move through the soil.

Of the essential nutrients phosphate is present in the soil solution in very low concentration and the greater part of the soil phosphate is attached to soil particles or in organic form (inositol phosphates). Phosphate has very low mobility. Ammonium is about ten times as mobile as phosphate in the soil but is required and absorbed in ten times greater quantity. These two macro-nutrients are therefore facilitated greatly in uptake by mycorrhizal infection.

Nitrate, in eutrophic moist warm soils, is present in quantity and ammonium is almost absent. Nitrate is very mobile and is therefore the main source of nitrogen in such soils, but it is low in quantity and almost absent in acid soils, especially those of high phenolic content such as acid forest soils and heathlands. In the eutrophic soils mycorrhiza is not essential to continued active nitrate uptake, but it is essential in acid soils where only ammonium or organic nitrogen compounds are available.

The mobility of potassium is similar to that of ammonium but it is required in lesser quantities. Calcium is usually present in excess except in very acid soils. Hence these two are less affected by mycorrhization than ammonium and phosphate.

A feature of some importance that complicates this somewhat simple picture is the possession by the surfaces of mycorrhizal fungi and ectomycorrhizal sheath surfaces of surface-attached or secreted enzymes. These include phosphatases capable of hydrolysis of organic and inorganic phosphates which may comprise up to about 80% of the total soil phosphate. Enzymes capable of hydrolysing proteins and polypeptides are also sometimes present. Experimental work on phosphate enzymes was done by Bartlett and Lewis (1973), Williamson and Alexander (1976), Calleja *et al.* (1980) and Mitchell and Read (1981). That on nitrogen compounds is discussed by Read *et al.* (1988).

The ecological importance of these enzymatic reactions at the mycorrhizal or hyphal surfaces has not been fully considered, but undoubtedly would seem to provide sources of nutrient in addition to the standing concentration of the soil solution in the immediate locality of a hypha or rootlet.

B. The absorption and physiology of phosphate

A very large number of experiments with mycorrhiza and mycorrhizal plants have emphasized the importance of mycorrhizal organs in the uptake of phosphate. As a result the increased growth of mycorrhizal plants, as compared with non-mycorrhizal ones, has often been ascribed solely to their increased phosphate uptake. Although phosphate is present in low concentration in the solution of all soils and is a very frequent nutrient which is deficient, this view is a warped one. It arises out of the ease with which phosphate may be studied using ^{32}p-labelled compounds and the early demonstration of the importance of mycorrhiza in phosphate physiology of the plant. Moreover, experiments with plants grown on eutrophic soils with good nitrogen availability overstressed the importance of phosphate uptake.

The source of phosphate is the soil solution, low in concentration even when perhaps locally increased by phosphatase action (mentioned above). Absorption dependent on respiratory activity results in its rapid assimilation into nucleotides and sugar phosphates, as is usual in plant material. Although there is considerable accumulation as inorganic orthophosphate in fungal hyphae and the sheath of ectomycorrhiza, much is converted into inorganic polyphosphates. These are accumulated in soluble and granular form. The granules are made evident particularly by staining techniques with metachromatic basic dyes, such as toluidine blue or methylene blue. It seems that the conspicuous nature of the granules may have caused them to be considered more important than they deserve to be, as compared with soluble polyphosphates.

Recent work has emphasized the quantitative importance of the soluble forms (Loughman and Ratcliffe, 1984). Strullu *et al*. (1983) put forward the view that only a small percentage of the phosphate in the fungal sheath of *Fagus* mycorrhiza was present as granules, although total polyphosphates were present in quantity.

Inorganic polyphosphates, soluble and granules, are common in micro-organisms, including bacteria, fungi and algae (see Kulaev, 1979). Hence they should not be considered a peculiarity of mycorrhizal fungi but as an important form of phosphate in their metabolism.

The phosphates in the molecule of polyphosphate are linked with "high energy" bonds (P—O—P) which release about 10 kcal or more on hydrolysis. This explains the fact that during the uptake of phosphate into beech mycorrhiza, Harley *et al.* (1954) observed that oxygen uptake and phosphate uptake were linearily linked under some conditions.

Polyphosphates may form linear chains of length a few to hundreds of phosphates of which the higher polymers are water-insoluble. Their formula is $M_{(n+2)}P_{(n)}O_{(2n+1)}$, where M is equivalent to a monovalent cation. Salts of divalent cations Ca^{2+} and Mg^{2+} are insoluble in water and it is therefore understandable that methods of fixation for electron microscopy (EM) where such ions are present may increase their content in the granules.

There have been few successful isolations of polyphosphate granules from living cells followed by chemical analysis. Kulaev (1979) gives examples. For instance, volutin granules of *Tetrahymena*, an alga, are reported to contain large amounts of calcium and magnesium and so could be regarded as calcium salts. They were extracted in conditions where calcium was not present in the extraction fluid. On the other hand, work recently reported, using EM, by Orlovich *et al.* (1989), has shown that high calcium content of polyphosphate granules of ectomycorrhizal fungi may be an artifact of fixation. Again, Lapeyrie *et al.* (1984) found that the presence of neither calcium nor magnesium in the culture medium of the ectomycorrhizal fungus *Pisolithus tinctorius* affected granule formation when these ions were supplied *before* phosphate was supplied. That is, during phosphate uptake neither divalent ion was present in the medium. It was assumed that a prior supply of Ca^{2+} or Mg^{2+} would suffice to encourage granule formation during the incubation with NaH_2PO_4 but it did not.

On the other hand, Strullu *et al.* (1982, 1983, 1986), working with mycorrhizal sheath tissue, showed that calcium and, to an unknown extent, potassium, were present in the granules, but that feeding KH_2PO_4 reduced the Ca/P ratio of the granules as granular phosphate increased. Moreover, the presence of Ca^{2+} or Mg^{2+} as chlorides in the solution of KH_2PO_4 from which phosphate was absorbed, increased considerably the rate of phosphate uptake and the number of granules formed, as compared with solutions containing KCl at the same concentration. The lowest P/Ca rate was observed in the granules in the presence of K^+ only and higher in the presence of Ca^{2+}. In spite of the fact that these analyses were performed in material fixed for EM (criticised by Orlovich *et al.*), these results could not have been obtained, it is believed, if Ca^{2+} was not an important ion in the granules.

The matter needs more consideration because both Lapeyrie *et al.* and Orlovich *et al.* used cultured fungal mycelium, as contrasted with the sheath tissue used by Strullu *et al*. This point is worth real consideration because others have been surprised at the small synthesis of polyphosphates in actively-growing mycelium (e.g. Martin *et al.*, 1983; Rollin *et al.*, 1984). It seems clear (see Kulaev, 1979) that polyphosphates intervene in synthetic metabolism and also active phosphorylation and ATP are needed for their synthesis. It is therefore probable that in actively-growing and synthesizing mycelium, the growth and synthetic processes will compete for ATP or other high energy phosphate compounds, and less polyphosphate will be accumulated; whereas in slow growing mycelium, on the other hand, more soluble and granular polyphosphate will be formed. Clearly here is a subject demanding detailed future research.

Movement of phosphate from fungus to host was first experimentally demonstrated by Melin and Nilsson (1950) and has since been further widely observed in experiments with whole plants. Harley and McCready (1952) demonstrated that during phosphate uptake by ectomycorrhiza a large proportion was stored in the sheath. At the same time a proportion passed to the host through the sheath. This mostly did not mix with the stored phosphate in the sheath but passed by a route in the protoplast (Harley *et al.*, 1954). It was also shown that phosphate stored in the sheath might be mobilised and transported to the host by a mechanism dependent on oxygen supply, temperature and the absence of metabolic inhibitors, in conditions where the external availability of phosphate was poor (Harley and Brierley, 1954, 1955). The actual compound passing into the host was identified by Harley and Loughman (1963) as inorganic orthophosphate which was subsequently metabolized into nucleotides and sugar phosphates in the host.

C. Absorption and metabolism of nitrogen compounds

Most of the information in this section concerns ectomycorrhiza and ericoid mycorrhiza, for there has been most interest in the effect of these on nitrogen uptake.

Although it was one of the original assumptions of Frank that mycorrhizas of trees aided uptake of nitrogenous compounds from the soil, this aspect of their functioning received little proper research until recently. It is true that the "nitrogen theory of mycorrhiza" was widely held but without much advance. Hatch (1937) especially was the proponent of the "salt uptake theory" which suggested that mycorrhiza aided uptake of all or any macro-nutrients from the soil. He and others working with young pines showed that mycorrhizal plants absorbed N, P and K more readily than non-mycorrhizal ones. However, research into the mechanisms of nitrogen absorption fell behind as work with radiotracers emphasized the importance of phosphate uptake.

In eutrophic agricultural soils, work with plants having vesicular-arbuscular mycorrhiza showed them to have no greater potential for absorbing nitrogen than their uninfected counterparts. The explanation for this doubtless lies in the fact that nitrate is plentiful in such soils, for breakdown of organic compounds leads to nitrification of ammonium by bacteria and rapid nitrate formation. By contrast, as Frank showed long ago, acid woodland soil is deficient of nitrate, and nitrogen is present in them in ammonium and organic forms.

Research with ectomycorrhizal fungi in culture has shown that many may not be able to utilize nitrate, although there is much variation in this regard, even within a single species (Laiho, 1970; Lundeberg, 1970). Pearson and Read (1975) have shown that the fungi of ericoid mycorrhiza *Hymenoscyphus* (*Pezizella*) *ericae* can utilize both ammonium and nitrate as well as organic nitrogen compounds (see also Read *et al.*, 1988). In any event, it is clear that ectomycorrhizal and ericoid mycorrhiza have the capacity to absorb and metabolize ammonium compounds and other simple amino compounds (Melin, 1925; Melin and Nilsson, 1953; Carrodus, 1966; Sangwanit and Bledsoe, 1985; Abuzinadah and Read, 1986a, b). A review of this matter is that of Read *et al.*, 1988.

Nitrate, when absorbed, is reduced by nitrate reductase and the ammonium so formed or directly absorbed is assimilated via the glutamine synthetase and glutamate synthase, although the glutamate dehydrogenase pathway has also been identified in some material (Smith, S. E. *et al.*, 1985; Martin *et al.*, 1986, 1987). This process of assimilation to produce glutamine requires the provision of carbon skeletons derived from keto acid of the tricarboxylic acid cycle. The maintenance of their concentration requires anaplerotic CO_2 fixation, which has been shown to occur (Harley, 1964; Carrodus, 1966).

The actual compounds transferred from the fungus to the host are essentially unknown. The hypothesis has been put forward by Carrodus and others that glutamine, as the most abundant and first-formed amino compound, is a probable candidate. Indeed, Reid and Lewis (in Lewis, 1976) give some experimental evidence that this suggestion is true, but the matter needs more investigation.

Although mycorrhizal fungi, like all other fungi, have no potential whatever to fix gaseous nitrogen themselves, they are commonly a third symbiont in most nitrogen-fixing symbioses—host, bacterium, mycorrhizal fungus. This was first emphasized long ago by Asai (1943, 1944) who showed that many species of Leguminosae, symbiotic with *Rhizobium*, formed more and larger nodules and grew larger if also mycorrhizally infected. Later this was demonstrated in more detail by others, on *Alnus* and *Frankia* and ectomycorrhiza; also with *Rhizobium* and vesicular-arbuscular mycorrhiza in Legumes (e.g. Daft and Elgiahmi, 1974). Since then there has been further work on such associations (see Daft *et al.*, 1988; Gardner, 1986).

The explanation of the stimulation of growth and fixation of nitrogen by mycorrhization lies, most probably, in the facility that mycorrhiza confers for the absorption of phosphate. Phosphate is especially necessary for the process of nitrogen fixation by the associated micro-organism as well as for the growth of the host.

D. Carbon physiology of mycorrhizal symbioses

Green mycorrhizal hosts provide, by photosynthesis, all or most of the carbon compounds of their mycorrhizal system. By contrast, the chlorophyll-free hosts, e.g. Monotropaceae, some Orchidaceae as adults, all Orchidaceae as seedlings and the so-called saprophytes of other families, derive all their carbon compounds by the activity of their fungi. Indeed there are broadly two kinds of mycorrhizal fungi, with respect to carbon economy: those of the green hosts have no or very limited capability of digesting and absorbing soil-borne carbon polymers, cellulose, pectins, lignin, etc. On the other hand, the fungi of chlorophyll-less plants may derive carbon by digesting such carbon polymers and translocating the products through their mycelium and releasing them to their hosts. These species of fungi belong to well-known ligninoclastic or celluloclastic species. Others are parasitic on other green plants or mycorrhizal with them, redistributing, as it were, carbon compounds by parasitizing another living source (see Harley, 1973; Harley and Smith, 1983).

It is the mycorrhizas of green hosts which are of interest to us here.

Although the earliest workers on mycorrhizas assumed that the movement of photosynthates to the fungus did occur, the first direct demonstration of it was due to Melin and Nilsson (1967), who showed, using $^{14}CO_2$, that the products of photosynthesis by young pine plants were translocated through the host and into and through the mycelium of the fungal symbiont. Nelson (1964) later demonstrated that mycorrhizal infection altered the distribution of photosynthate in young pine so that a greater quantity was passed to mycorrhizal root systems than to non-mycorrhizal ones. This distortion or re-orientation of translocation towards a fungal-infected part parallels that which occurs in other biotrophic fungal infections, such as those of rust fungi. A discussion of this matter is given by Smith, D.C. *et al.* (1979).

A graphic experimental demonstration of this behaviour was provided by Lewis and Harley (1965), using excised mycorrhiza. They showed that if [^{14}C] sucrose was applied to the cut axis, [^{14}C]carbohydrate was translocated to the tip and accumulated particularly in the fungal sheath in the form of the fungal carbohydrates trehalose, mannitol and glycogen, carbohydrates not readily used by the host tissues. These examples emphasize a similar pattern to rust fungi and have indeed been accepted widely as the most likely behaviour in all mycorrhiza. However, caution is necessary, because it does not seem so easily applicable to vesicular-arbuscular mycorrhiza (see Smith, S.E. and Gianinazzi-Pearson, 1988), although lipids, glycogen granules, polyols and trehalose have at times and to some extent been found in their hyphae. It is clear that more investigation is needed, especially into the substances which become labelled in vesicular-arbuscular fungi and into the actual carbonaceous substances passing from the host and the mechanism of movement.

Roots of plants in pure culture have been shown to release carbon compounds into their surroundings, and, in the seedling stage, a large proportion of that photosynthesized may be lost as exudates and sloughed cells. It is possible that such a leak is quantitatively sufficient to provide for transfer to a mycorrhizal fungus. However, this is doubted by most observers, and a mechanism of generating leakiness has been sought. Indeed, simple calculations of the quantity of carbon necessary for the maintenance of ectomycorrhizal fungi lead to the conclusion that something greater than 10% of photosynthates is consumed by the fungi of adult trees (see below).

A mechanism was suggested by the experiments of Wedding and Harley (1976) that mannitol, a fungal carbohydrate which is able to penetrate host tissue, inhibited differentially the carbohydrate enzymes of the host so that glucose was produced in excess and might leak to the fungus. However, Jirgis *et al.* (1986) were later unable to repeat Wedding and Harley's experimental findings, so that their contention needs further examination. The subject of the release of carbohydrates and indeed the physiological interaction between symbionts needs much more research, which must be combined with further and more detailed biochemical work upon them (see Martin *et al.*, 1987; Smith, S.E. and Gianinazzi-Pearson, 1988).

In respect to the movement of carbon compounds from the host to the fungus and the reverse movement of phosphate to the host, the view has been put forward that these processes may be linked. In summary, the view is that monosaccharides, such as glucose, released from the root are absorbed by the fungus and phosphorylated by means of ATP or polyphosphates or a combination of both. The sugar phosphates react to form trehalose, glycogen or mannitol, by well-known pathways and inorganic phosphate is released. In this way the active concentration of phosphate at the interface is increased and the active concentration of monosaccharide decreased, so facilitating reciprocal movement. Quantitative estimates of phosphate and carbon moving are needed to justify these suggestions.

VI. Translocation in mycorrhizal systems

The mechanisms of translocation in plants and fungi are far from satisfactorily known. However, phloem translocation in higher plants is understood in the broadest outline and will not be discussed further here. Since substances absorbed by hyphae from the soil find their way to the interface with the host tissue and so into the host, translocation within the hypha takes place. What needs to be ascertained is the route through the hypha, what is translocated, how fast in mass per unit length per unit time, and factors affecting the process. The subject is discussed by Harley and Smith, S.E. (1983). Their essay gives, amongst other things, a list of substances whose movement has been studied in mycorrhizal hyphae together with their flux rates, when estimated, and full references.

Translocation in orchid fungi was first investigated by S.E. Smith (1967), using orchid seedlings as receptors and the hyphae of species of *Rhizoctonia* as translocators. Carbohydrate absorbed by the fungus from the culture medium appeared in the hyphae as trehalose, and in some strains as mannitol also. In the seedling tissue these fungal sugars diminished and sucrose was formed, i.e. the reverse of the behaviour in ectomycorrhiza.

Translocation over considerable distances in centimetres, of phosphate in particular, has been shown to occur in vesicular-arbuscular, ericoid, orchid and ectomycorrhizal fungi. Estimates of quantity per unit length per unit time for individual hyphae have been made, but they are often subject to some doubt, in particular instances because they depend on accurate estimates of the number of hyphae connecting the root or seedling to the source.

The estimates made are usually of movement of a single substance in one direction; however, bidirectional translocation—movement of different substances simultaneously in opposite directions—must be envisaged and limits severely the kind of mechanism which can operate. Its occurrence eliminates the possibility that translocation is, in general, operated by mass flow of fluid in the hyphae, such as might arise from transpiration of water by the host symbiont, although it may contribute to the movement in one direction. The fact that translocation is temperature-sensitive might suggest that it is metabolically operated, but it might also depend upon the rate of uptake or loss, both of which are temperature-dependent.

Cytoplasmic streaming, observable in hyphae, is a possible mechanism, dependent on metabolic activity, including temperature. It might act as a mixing mechanism, reinforcing diffusional movement so that substances are swept from the point of their entry into the hyphae and arrive at their point of removal as a result of the streaming movement of the fluid contents, which is in considerable measure a random movement. The net effect of this would of course be bidirectional translocation.

With respect to the translocation of phosphate, the suggestion has been made that granules of polyphosphate might be translocated in the hyphae and this might help to explain the considerable rates of phosphate movement (Cox and Tinker, 1976). In a similar way it has been suggested that glycogen granules might be the vehicle of carbohydrate movement in fungi. There are considerable difficulties in these suggestions, first because of the existence of pits and plasmodesmata through which the granules could not pass. Again, utilization of the substances translocated depends on their enzymic hydrolysis. The enzymes concerned must either be in the hyphal fluid (when the granules would be subject to hydrolysis all the time) or attached at the point of discharge of the substances. In the latter case, the random meeting of enzyme and granular substrate would be statistically improbable compared with molecules as translocated subjects.

A compromise to these difficulties might be the translocation of polyphosphate in soluble short chain lengths and carbohydrates similarly in short chains rather than granules. However, the whole subject needs much more original thought and active experiment.

VII. Recognition between the symbionts

The very fact that in the soil a root is surrounded by many species of fungi and only certain of these form composite mycorrhizal organs with it, demonstrates that there must be some method of recognition between potential symbionts. This method of recognition cannot solely involve the production of common substances like simple carbohydrates, an earlier suggestion for this process, for such common substances can be utilized by many soil organisms.

A fact that complicates and tends to confuse our thinking about recognition is the remarkable absence of close specificity between the symbionts. There is however some specialization. For instance, Ericales have their own peculiar brand of mycorrhiza not found, to any extent, outside that order, with a few species of fungi which do not consort with other plants. The same is true in great measure of the Orchidaceae. On the other hand ectomycorrhiza formed by large numbers of coniferous and angiospermous trees and a few pteridophytes and very few herbs, with any of many species of basidiomycetes and ascomycetes, may also hold fungal species in common with arbutoid and monotropoid mycorrhizal plants. In neither the ectomycorrhizal plants nor those forming vesicular-arbuscular mycorrhiza is there a single example of a species of fungus restricted to a single species of host. There are cases of a single genus being restricted to a single family or genus of host, but most are of wide specificity.

The subject of recognition has been discussed briefly (Harley and Smith, 1983), but one must admit that up to date no great enlightenment of its problems has occurred. Comparisons with recognition mechanisms of host and biotrophic parasite are not particularly helpful, for the relationships in those cases are specific and genetically determined, e.g. by gene-for-gene specificity and the surface reactions very selective to compatibles.

VIII. The importance of mycorrhizal infection in ecosystems

In 1971, I put forward the view that cycling of carbon in ecosystems had two main routes. In the first route the fall of litter (e.g. leaf litter) was followed by its digestion and incorporation into the soil by soil organisms and the release of CO_2 by their activity. The second route involved the translocation of carbon to the root systems of photosynthesizing plants where it was used in the maintenance and respiration of roots and their associated rhizosphere organisms and mycorrhizal fungi. In that paper and in a later one (1973) I made estimates by various means, based on our own and others' observations, of the expenditure of an ectomycorrhizal tree on its mycorrhizal fungus, and concluded that *at least* 10% of the photosynthetic product was so expended. Later work by Fogel and Hunt (1979) and Vogt *et al.* (1982) confirmed this order of expenditure on the mycorrhizal fungi and indeed suggested that 10% was too low.

This level of expenditure on the fungal symbiont emphasizes a salient significance of mycorrhiza formation. In the soil the main carbonaceous substances are resistant polymers; in addition, high concentrations of phenolic compounds are present in acid woodland and heathland soil. It requires, therefore, specialized fungi to utilize soil carbon compounds. By contrast, in the mycorrhizal fungi we have a group for which a supply of carbon compounds is continuous and usually plentiful from the photosynthesizing host. The mycorrhizal fungus is therefore in a particularly advantageous position to absorb nutrients from the soil, using that ready supply. Of course, destructive parasites have, for the limited time of their host's survival, a similar potential, but not the continued potential such as the mycorrhizal symbiont experiences. It is not remarkable, therefore, that mycorrhizal fungi of green hosts have no, or only limited, powers of utilizing complex carbon compounds.

In 1973 Lewis and I both quite separately put forward the view that the non-specificity of mycorrhizal fungi might lead to an ecological complication, in that a common mycelium might connect two or more hosts of the same or different species. Hence there could be a movement of carbon compounds or other substances between them. This might have repercussions on our conception of root competition and especially on the growth of young seedlings or other plants below the canopy. Even if they did not receive carbon from the dominant plants via their common mycelium, they might be spared much of the possible carbon drain by being attached to a mycelium fed with carbon by dominant plants.

A further examination of these matters has now been undertaken experimentally by a number of workers, particularly D. J. Read and his colleagues. A good account of the work which is fundamental to our concepts of ecosystem structure and our understanding of the place of mycorrhiza in ecosystems is given by Read *et al.* (1985) and it is also discussed in more general terms by Harley and Smith (1983).

IX. Comment

It is clearly impossible to cover the whole "state of the art" relevant to mycorrhizal work in one short introductory chapter. Here it has been only possible to discuss, briefly, many of the facets requiring more research. The list of references given, although selective, includes several works of reference and reviews that should help to heal the imperfections of the account.

References

Abuzindah, R. A. and Read, D. J. (1986). *New Phytol.* **103**, 481–493; also 507–514.
Asai, T. (1943). *Jap. J. Bot.* **12**, 359–436.
Asai, T. (1944). *Jap. J. Bot.* **13**, 463–485.
Bartlett, E. M. and Lewis, D. H. (1973). *Soil Biol. Biochem.* **5**, 249–257.
Calleja, M., Mousain, D., Lacouvreur, B. and D'Anzac, J. (1980). *Physiol. Veg.* **18**, 489–504.
Carrodus, B. B. (1966). *New Phytol.* **65**, 358–371.
Carrodus, B. B. (1967). *New Phytol.* **66**, 1–4.
Cox, G. and Tinker, P. B. (1976). *New Phytol.* **77**, 371–378.
Daft, M. J. and Elgiahmi, (1974). *New Phytol.* **83**, 1139–1147.
Daft, M. J., Clelland, D. M. and Gardiner, I. C. (1988). *Proc. Roy. Soc. Edin.* **85b**, 283–298.
Dexheimer, J., Gianinazzi, S. and Gianinazzi-Pearson, V. (1979). *Z. Pfl. Physiol.* **92**, 191–207.
Fogel, R. and Hunt, G. (1979). *Can. J. For. Res.* **9**, 245–256.
Gardner, I. C. (1986). *Mircen J.* **2**, 147–160.
Harley, J. L. (1964). *New Phytol.* **63**, 203–208.
Harley, J. L. (1971). *J. Ecol.* **59**, 653–668.
Harley, J. L. (1973). *J. Nat. Sci. Council, Sri Lanka* **1**, 31–48.
Harley, J. L. (1985). *Proc. Indian, Acad. Sci.* (Plant Sci.) **94**, 99–109.
Harley, J. L. (1989). *Mycol. Res.* **92**, 129–139.
Harley, J. L. and Brierley, J. K. (1954). *New Phytol.* **53**, 240–252.
Harley, J. L. and Brierley, J. K. (1955). *New Phytol.* **54**, 296–302.
Harley, J. L. and Loughman, B. C. (1963). *New Phytol.* **62**, 350–359.
Harley, J. L. and McCready, C. C. (1952). *New Phytol.* **51**, 56–64.
Harley, J. L. and Smith, S. E. (1983). *Mycorrhizal Symbiosis*. Academic Press, London and New York. 483 pp.
Harley, J. L., McCready, C. C. and Brierley, J. K. (1954). *New Phytol.* **53**, 92–98.
Hatch, A. B. (1937). *Black Rock For. Bull.* **6**, 1–168.
Jirgis, R., Ramstedt, M. and Söderhäll, K. (1986). *New Phytol.* **102**, 285–291.
Kulaev, I. S. (1979). *The Biochemistry of Inorganic Polyphosphates* (R. F. Brookes, trans.). John Wiley and Sons, Chichester and New York. 255 pp.
Laiho, O. (1970). *Acta For. Fenn.* **106**, 1–65.

Lapeyrie, F. F., Chilvers, G. A. and Douglas, P. A. (1984). *New Phytol.* **98**, 345–360.
Lewis, D. H. (1973). In *Taxonomy and Ecolog.* (V. H. Heywood, ed.), pp. 151–172.
Lewis, D. H. (1976). In *Perspectives in Experimental Biology* (N. Sutherland, ed.), Vol. 2, pp. 207–219. Pergamon Press, Oxford.
Lewis, D. H. and Harley, J. L. (1965). *New Phytol.* **64**, 256–269.
Lundeberg, G. (1970). *Stud. For. Suec.* **79**, 1–95.
Loughman, B. C. and Ratcliffe, R. G. (1984). *Adv. Plant Nutr.* **1**, 241–283.
Martin, F., Canet, D., Rolin, D., Marchal, J. P. and Larker, F. (1983). *Plant and Soil* **71**, 469–476.
Martin, F., Stewart, G. R., Genetet, L. and Le Tacon, F. (1986). *New Phytol.* **102**, 85–94.
Martin, F., Ramstedt, and M Söderäll, K. (1987). *Biochemie* **69**, 569–581.
Melin, E. (1925). *Untersuchungen über die bedeutung der Baummykorrhiza*, pp. 1–152. C. Fischer, Jena.
Melin, E. and Nilsson, H. (1950). *Physio. Planta.* **3**, 88–92.
Melin, E. and Nilsson, H. (1953). *Nature* **171**, 134.
Melin, E. and Nilsson, H. (1967). *Svensk. Bot. Tidsk.* **51**, 166–186.
Mitchell, D. T. and Read, D. J. (1981). *Trans. Br. Mycol. Soc.* **7**, 195–198.
Nelson, C. D. (1964). In *Formation of Wood in Forest Trees* (M. H. Zimmerman, ed.) pp. 235–257. Maria Mooes Cabot Foundation, New York.
Nye, P. H. and Tinker, P. B. (1977). *Solute Movement in the Soil Root System*, pp. 1–342. Blackwell Scientific Publications, Oxford.
Orlovich, D. A., Ashford, E. A. and Cox, G. C. (1989). In *Plant–Microbe Interface Structure and Function* (P. A. McGee, S. E. Smith, and F. A. Smith, eds). Report for *Aust. J. Plant Physiol.* **16**, 107–115.
Pearson, V. and Read, D. J. (1975). *Trans. Br. Mycol. Soc.* **94**, 1–7.
Proceedings of 6th North American Conference on Mycorrhizae (1985). Historical Perspectives, pp. 1–74. Forest Research Laboratory, Corvallis.
Read, D. J., Francis, R. and Finlay, R. D. (1985). In *Ecological Interactions in Soil* (A. H. Fitter, ed.) pp. 193–217. Blackwell Scientific, Oxford.
Read, D. J., Leake, J. R. and Langdale, A. R. (1988). *Nitrogen, Phosphorus and Sulphur Utilization by Fungi*. Symposium of the British Mycological Society (L. Boddy, R. Marchant, and D. J. Read, eds), pp. 181–182.
Rollin, D., Le Tacon, F. and Larker, F. (1984). *New Phytol.* **98** 335–343.
Sanders, F. E. and Tinker, P. B. (1973). *Pesticide Sci.* **4**, 385–395.
Sangwanit, V. and Bledsoe, C. (1985). In *Proceedings of the 6th North American Conference on Mycorrhizae*, p. 346. Forest Research Laboratory, Corvallis.
Smith, D. C. and Douglas, A. E. (1987). *The Biology of Symbiosis*. Edward Arnold, London. 302 pp.
Smith, D. C., Muscatine, L. and Lewis, D. (1979). *Biol. Rev.* **44**, 17–90.
Smith, S. E. (1967). *New Phytol.* **65**, 488–499.
Smith, S. E. and Gianinazzi-Pearson, V. (1988). *Ann. Rev. Plant Physiol. Plant Mol. Biol.* **39**, 221–244.
Smith, S. E., St. John B., Smith, F. A. and Nicholas, D. J. D. (1985). *New Phytol.* **99**, 211–227.
Strullus, D.-G., Harley, J. L., Gourret, J. P. and Garrec, J. P. (1982). *New Phytol.* **92**, 417–423.

Strullus, D.-G., Harley, J. L., Gourret, J. P. and Garrec, J. P. (1983). *New Phytol*. **94**, 89–94.
Strullu, D.-G., Grellier, B., Garrec, J. P., McCready, C. C. and Harley, J. L. (1986). *New Phytol*. **103**, 403–416.
Testa, M., Smith, S. E. and Smith, F. A. (1987). *Can. J. Bot.* **65**, 419–431.
Vogt, K. A., Grier, C. C., Meier, C. E. and Edmunds, R. L. (1982). *Ecology* **63**, 370–380.
Wedding, R. J. and Harley, J. L. (1976). *New Phytol*. **72**, 675–688.
Williamson, B. and Alexander, I. (1976). *Soil Biol. Biochem.* **7**, 195–198.

2

Characterization of Ectomycorrhiza

REINHARD AGERER

University of Munich, Institute for Systematic Botany, Menzingerstrasse 67, D-8000 München 19, Germany

---

METHODS IN MICROBIOLOGY
VOLUME 23 ISBN 0-12-521523-1

Copyright © 1991 by Academic Press Limited
All rights of reproduction in any form reserved

I. Introduction

Studies of ectomycorrhiza can contribute to elucidation of fungal relationships (Agerer *et al.*, 1990; Agerer, 1991a). Ectomycorrhiza structures consist of fungal tissues, the arrangement and organization of which can be used to describe fungal species in the same way as any other taxonomically suitable feature. Their characteristics are well conserved (Agerer *et al.*, 1990). As ectomycorrhiza formed by different fungi are structurally distinct (Frank, 1885; Chilvers, 1968; Voiry, 1981; Agerer, 1987–1990), especially with respect to their rhizomorphs (Ogawa, 1981; Agerer *et al.*, 1990), differences between them in nutrition and growth-promoting efficiencies can be expected. Some studies have already confirmed this assumption (Duddridge *et al.*, 1980; Read, 1984; Chu-Chou and Grace, 1985; Kammerbauer *et al.*, 1989). Thus anatomical studies of ectomycorrhiza can provide important information with respect to their physiological capabilities.

It has also been shown that there can be species-specific differences in ability to colonize roots, depending upon age of individual trees or stands (e.g. Fleming, 1983). It is also known that application of fertilizer (Alexander and Fairley, 1983; Brand and Agerer, 1988) or natural flooding events (Stenström, 1990) can influence ectomycorrhizal structure. Furthermore, it has been shown that some ectomycorrhizal fungi can grow their hyphae within the rhizomorphs and ectomycorrhiza of other fungi (Agerer, 1990a, 1991a), suggesting that these ectomycorrhizal fungi can influence each other with respect to plant nutrition and with respect to their fruit body formation.

A large number of physiological, ecological and taxonomic studies on ectomycorrhiza have indicated that different species of ectomycorrhiza behave differently in several respects. For this reason alone, comprehensive characterizations of ectomycorrhizal structures are required urgently, so that recognition of naturally occurring types can be achieved.

As early as the end of the last century some authors had characterized ectomycorrhiza to a high standard. Very informative drawings of ectomycorrhiza of *Castanea sativa* were published by Gibelli (1883) and of *Fagus sylvatica* and *Carpinus betulus* by Frank (1885). These authors made detailed drawings of the surface views of plectenchymatous and pseudoparenchymatous ectomycorrhizal mantles. For some time after this it was apparently forgotten that such surface views and other plan views of mantles show important features. It was not until the 1960s that such techniques were used again by Chilvers (1968), Zak (1969) and Fontana (1962). Other researchers made tangential sections through mantles (e.g. Haug *et al.*, 1986). Without doubt, however, plan views of different mantle layers are more informative and are easier to prepare and to interpret than are tangential sections. However, if the mantles are very dark, tangential sections are appropriate.

Early studies of the internal organization of rhizomorphs of ectomycorrhiza and of ectomycorrhizal fruit bodies were made by Bommer (1896), Masui (1926a,b) and Melin (1927). Compilations of different types are given by Ogawa (1981) and Agerer *et al.* (1986, 1990). The first comprehensive introductions to methods for the characterization of ectomycorrhiza were provided by Trappe (1965) and Zak (1973). Only brief remarks by Wilcox (1982) on the characterization of ectomycorrhiza are included in the more recent book, *Methods and Principles of Mycorrhizal Research*, edited by Schenck (1982). As ectomycorrhizal characterization has progressed considerably over the last 10 years, the updated compilation given in this chapter is intended to facilitate anatomical research on ectomycorrhiza.

II. Samples

Soil samples containing the mycorrhizal roots to be examined are taken either with the intention of identifying the specimens by tracing rhizomorphal or hyphal connections to fruit bodies (identification probes) or with the intention of characterizing as many ectomycorrhiza as possible in order to obtain a picture of the "species" composition of ectomycorrhiza within the soil ("blind" probes). In both cases it is necessary to use methods that ensure that the ectomycorrhiza retain their natural features.

A comprehensive description of procedures required for sampling and isolating ectomycorrhiza can be found in the *Colour Atlas of Ectomycorrhizae* (Agerer, 1987–1990). Only a few additional remarks are necessary. The steps necessary for characterization of ectomycorrhiza starting from taking soil cores are shown in Fig. 1.

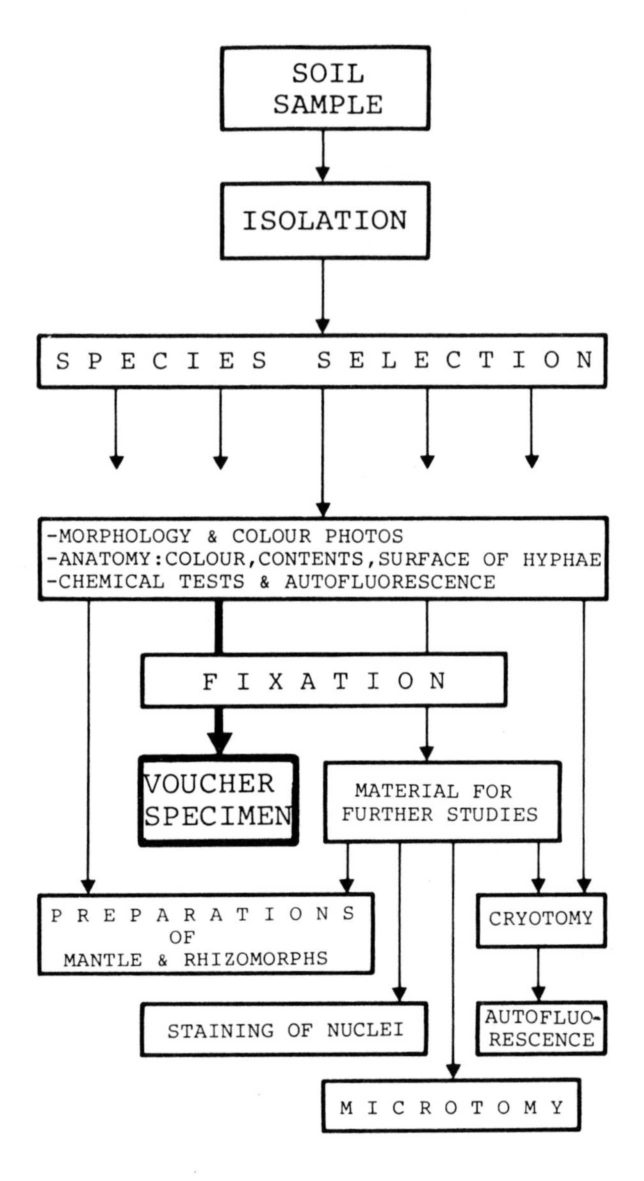

Fig. 1. Scheme showing the different steps of characterization of ectomycorrhiza, starting with soil samples and examination of fresh material, and finishing with preservation of voucher collections and different kinds of anatomical studies.

A. Collection of soil samples (Fig. 2)

A sharp, robust knife with a long blade should be used to take soil cores to a depth of *c*. 10 cm. A whetstone is an appropriate device to sharpen the knife if necessary, to prevent pulling of the roots through the soil. This ensures that neither ectomycorrhiza nor hyphal connections between stipe and ectomycorrhiza are damaged or disrupted. Before identification probes are taken the stipe of the fruit body should be carefully cut; in the case of species with dark stipes, pins with coloured heads should be used to mark the stipe base, otherwise its position may be lost. The soil cores should be wrapped in aluminium foil to prevent crumbling. They can then be stored in a refrigerator for up to two weeks.

B. Designations

All samples of ectomycorrhiza must be given special designations (collection numbers, see Section VI.A) which allow referral to all observations from one ectomycorrhizal collection. Each ectomycorrhizal sample isolated from a single soil core is given an individual collection number.

C. Washing procedure

The soil cores must be completely immersed in water until they are saturated before attempting to isolate the ectomycorrhiza (Fig. 3). Only the finest instruments should be used (Fig. 2B). During washing the soil must be covered by water to reduce the vigour necessary to uncover and clean the ectomycorrhiza and also to allow the soil particles to float away (Fig. 3). Moderate magnification (6×, 12×), a black background and lamps of daylight quality are necessary; this is to get a real impression of the ectomycorrhizal colours from the beginning of the washing procedure. A white background is dazzling and masks the natural colours. The water should be changed several times.

Cleaned ectomycorrhiza must not be stored in water because in some species hyphae can continue to grow, and, in addition, colour changes can occur.

D. Confirmation of identification

A great deal of patience is required to reveal intact connections between stipe base and ectomycorrhiza (Fig. 2D), but in some cases large

1
2
4
5
A
11
11
2
2
3
C
9
7
8
7
6
6
10
B
11
2
3
D

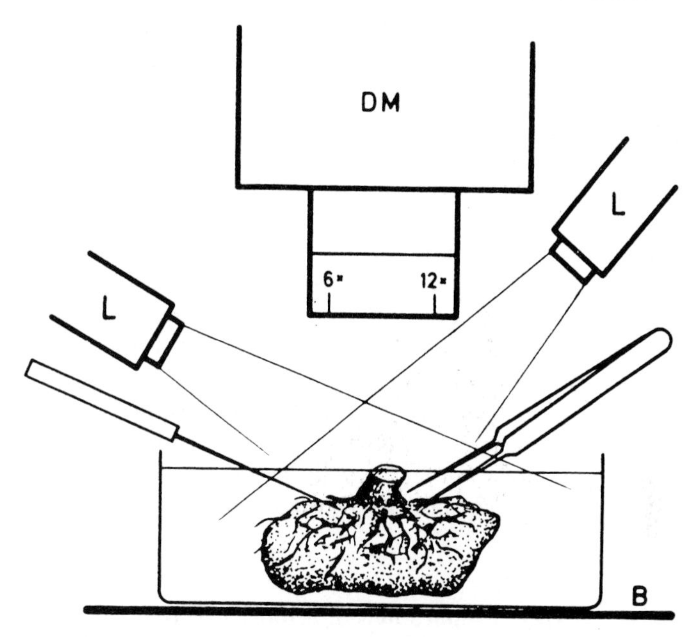

Fig. 3. Apparatus for examination of ectomycorrhiza (B, black background; DM, dissecting microscope; L, lamps of daylight quality). For further explanation, see text.

numbers of ectomycorrhiza can be detected, connected repeatedly with the stipe of the fruit body (Fig. 2C). In identification probes, those ectomycorrhiza which are directly attached to the stipe, or are enveloped by stipe tissue, must not be regarded as belonging to that fruit body because foreign ectomycorrhiza can be overgrown by the fungus.

Only ectomycorrhiza in which, to which or from which unequivocal hyphal or rhizomorph connections can be traced should be regarded as identified. In addition, the rhizomorphs and hyphae of both stipe base and ectomycorrhiza should be compared microscopically. Not all fungal relationships are amenable to this method of hyphal tracing. This procedure is very difficult in some genera, e.g. *Lactarius*, *Russula*,

Fig. 2. Samples and first steps of preparation. (A) Taking identification probes. (B) Tools for isolation of ectomycorrhiza. (C) Several ectomycorrhiza connected with two stipe bases. (D) First uncovered ectomycorrhiza near stipe base. Key: 1, fruit body; 2, cut stipe base; 3, ectomycorrhiza; 4, knife; 5, whetstone; 6, fine paint brushes; 7, fine needles; 8, thin pipette; 9, fine forceps; 10, small pair of scissors; 11, pin, marking the stipe base. Bar, 5 mm.

Hygrophorus and *Inocybe*. However, it works well with species of the genera *Cortinarius*, *Dermocybe*, *Sarcodon*, *Tricholoma* and *Xerocomus*. Sometimes several attempts are necessary, but there are some cases where success has never been achieved.

All ectomycorrhiza regarded as being of one species and one sample must be checked for homogeneity. In order to do this a few tips should be carefully compared both morphologically and anatomically using rhizomorph and mantle preparations.

If agar cultures can be obtained from isolates of fruit bodies and ectomycorrhiza, then further confirmation of the identification can be achieved by comparing these cultures. If cultures are only available from fruit bodies, then synthesis tests under at least semi-natural conditions (Read *et al.*, 1985; Brand and Agerer, 1986) should result in the production of ectomycorrhiza of similar structure with respect to mantle and rhizomorph organization and hyphal features. Voucher specimens of each homogeneous sample should be stored (see Section VI.A).

III. Preparations of ectomycorrhiza

A. Morphological characterization, colour photography and autofluorescence

Morphological characterization should always be carried out on freshly isolated material, using lamps of daylight quality and a black background (see Section II.C). The checklist and comments given in the *Colour Atlas of Ectomycorrhizae* (Agerer, 1987–1990) can be used as a guideline. The best record of colours is achieved by taking photographs.

Although colour charts are used for descriptions of fungi and have also been used for ectomycorrhiza (Ingleby *et al.*, 1990), for comparative purposes it is easier to refer to published colour photographs (Agerer, 1987–1990).

The apparatus for taking photographs of ectomycorrhiza used in this laboratory is shown in Fig. 4. The ectomycorrhiza are photographed in water to allow emanating hyphae and rhizomorphs to spread out into approximately their natural position. The ectomycorrhizal system can be held in forceps which in turn are held by a clamp. If a clamp with a universal joint is used, the mycorrhizal system can be oriented in any direction. The ectomycorrhiza should be completely submerged in water, otherwise light reflections can occur. For best results they should be held approximately midway between the black background and the water surface in a Petri dish filled to a depth of *c*. 4 cm with water. By

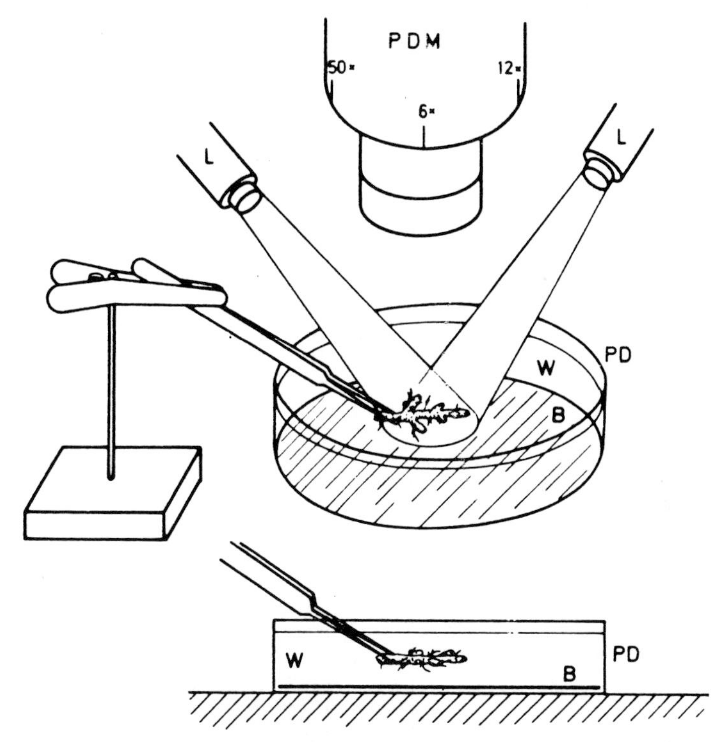

Fig. 4. Apparatus for photography of ectomycorrhiza (B, black background; L, lamps of daylight quality; PD, large Petri dish; PDM, photo-dissecting microscope; W, stored or degassed water). For further explanation, see text.

this means shadows can be avoided and very small particles floating on the water surface are out of focus. The use of stored or degassed water reduces the number of air bubbles formed on the ectomycorrhiza. The black background can be irradiated by the lamps to give a lighter background (grey) when photographing dark ectomycorrhiza. The main reason for using lamps of daylight quality and an appropriate daylight film is that there must not be any difference between the actual visual impression of ectomycorrhizal colours and that recorded on the colour slides.

B. Anatomical characterization

Anatomical characterization of ectomycorrhiza can best be performed using Normarski's interference contrast microscopy. It shows definite mantle or rhizomorph layers without interference from hyphal layers lying beneath or above the focused level. Only very dark mantles and rhizomorphs need to be sectioned.

Anatomical features of mantle and rhizomorphs can be drawn and/or photographed. Drawings have the great advantage that the hyphal arrangement within a plectenchyma can be shown in three dimensions (Fig. 5), whereas photographs show only a single layer in two dimensions (Fig. 6). In some cases photographs taken consecutively from the mantle surface to the inner layer can offer some additional information on the hyphal arrangement (Fig. 6A–D). The same is true of rhizo-

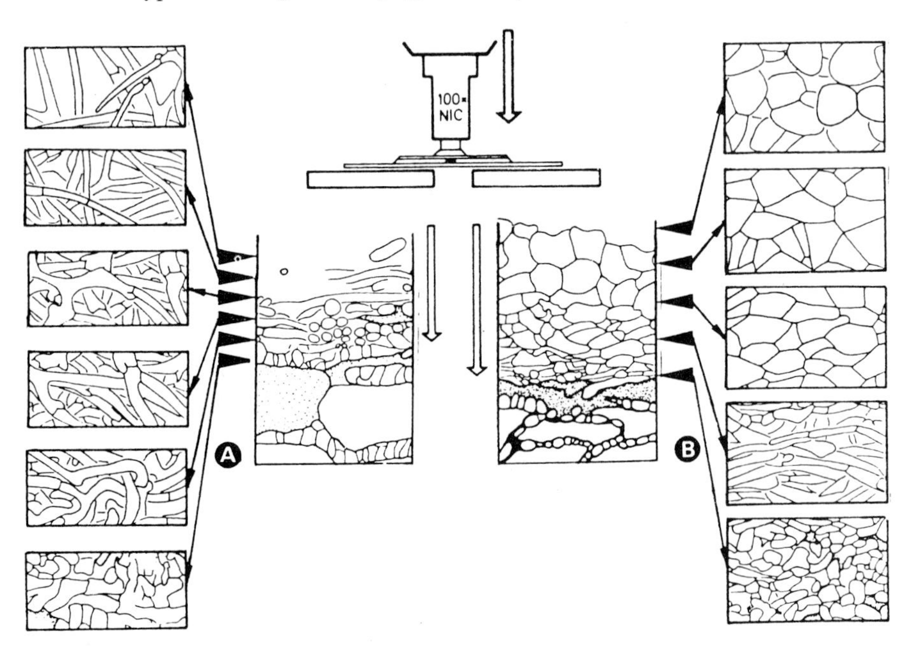

Fig. 5. Construction of consecutive plan views of differently organized ectomycorrhizal mantles. (A) plectenchymatous mantle in *Thelephora terrestris*; (B) pseudoparenchymatous mantle in *Piceirhiza nigra*. NIC, Normarski's interference contrast microscopy (after Weiss, 1990).

Fig. 6. (A)–(D): Plan views of different mantle layers, photographed in each case from the same position. From surface (A) to the inner side of the mantle (D): (A) an apical part of a cystidium on a loose layer of inflated hyphae; (B) inflated hyphae more densely arranged; (C) compact pseudoparenchymatous layer of hyphae; (D) dense plectenchyma. (E)–(G): Plan views of a rhizomorph, photographed in each case from the same position, from a surface view (E) to the middle of the rhizomorph (G). Using Normarski's interference contrast microscopy, bar represents 10 μm. (A–D: from Herb FB 13. 10. 87 in M; E–G: from Herb. RA 11554 in M.)

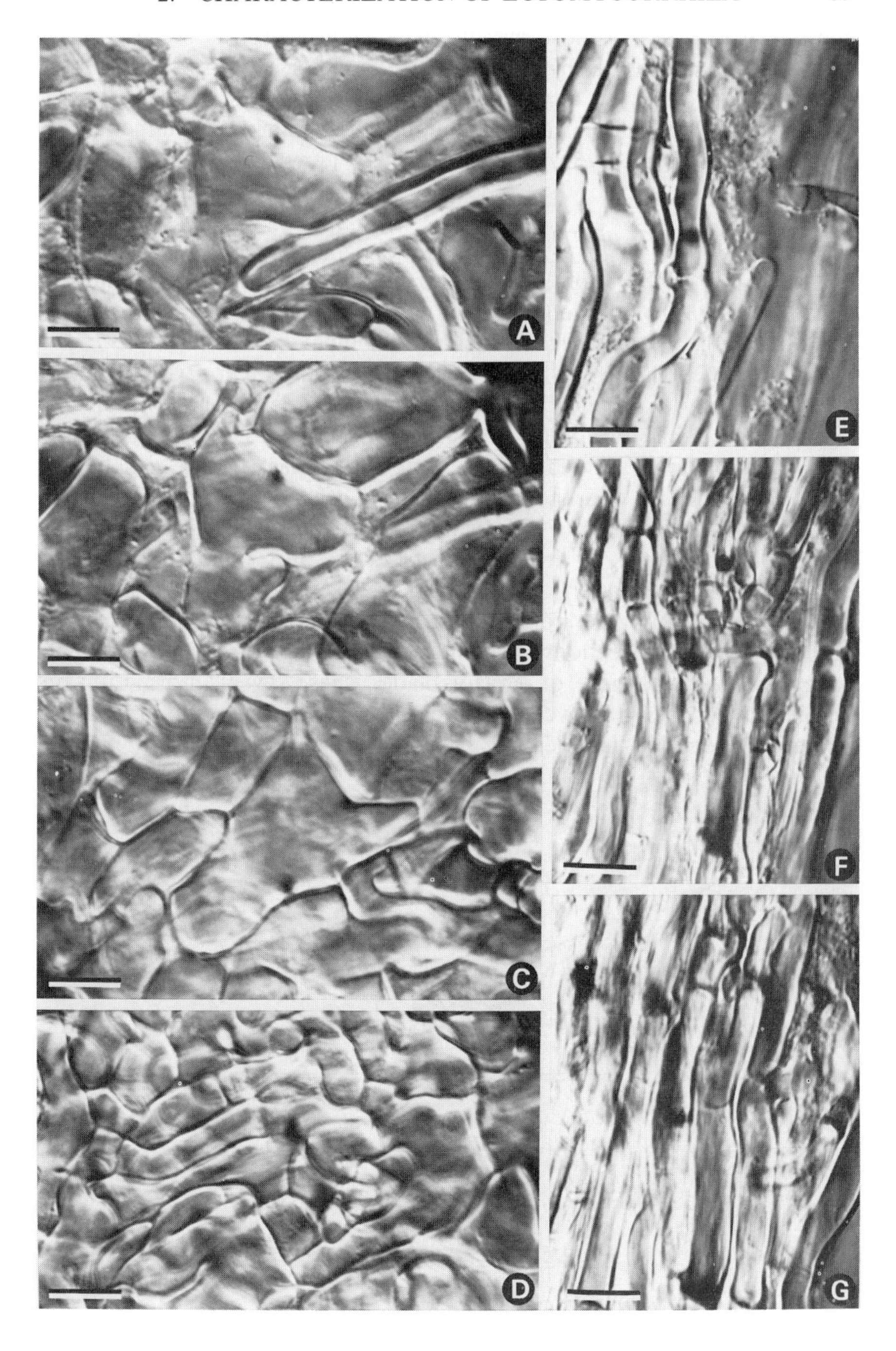
A
B
C
D
E
F
G

morphs (Fig. 6E–G). Sections through ectomycorrhiza, however, can best be photographed in normal phase contrast (cf. Section III.B.4 below).

Guidelines for anatomical examination, checklists and comments are given in the *Colour Atlas of Ectomycorrhizae* (Agerer, 1987–1990).

1. Preparations for plan views

Whenever possible, mantle and rhizomorph preparations should be made from freshly isolated material. Fixed ectomycorrhiza can also be used but unfixed structures are easier to handle.

An ectomycorrhizal tip should be held with fine forceps within a drop of water lying on an objective slide (Fig. 7). Mantle preparations should be prepared while observing the structure through a dissecting microscope at a magnification of 25 or 50 ×. A compact mantle can be slit with a fine needle and the mantle peeled off. In the case of loosely woven mantles, fragments only can be detached. The rest of the root should be removed or it can be dissected into minute pieces for microscopical observation of the cortical cells. Care must be taken that the root fragments are not too thick. Rhizomorphs can also be detached. Tangential sections of mantles near a rhizomorph base can be taken with a small piece of a razor blade. These reveal information on the transitional zone between mantle and rhizomorph. The very tips of the ectomycorrhizal mantles can best be studied either after pulling off or after cutting off the mantle cap.

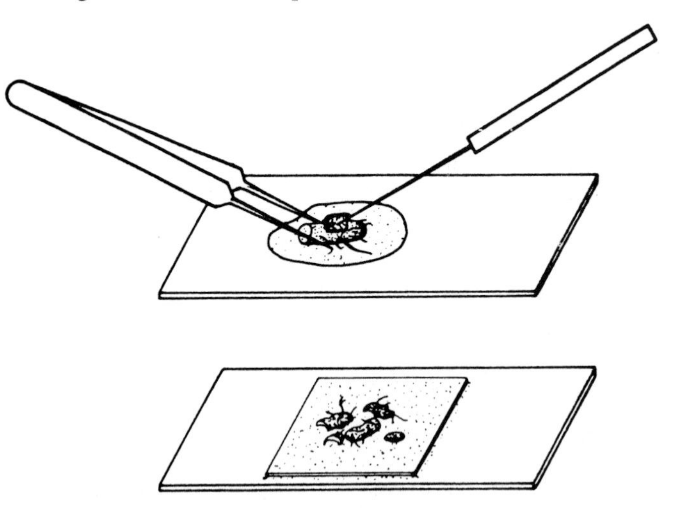

Fig. 7. Making mantle preparations. For explanation, see text.

After features have been checked in the fresh condition (Agerer, 1987–1990), the mantle preparations should be studied in lactic acid, as this medium consistently increases the contrast for microscopical studies. Specimens can be stored as permanent slides after adding sufficient lactic acid and final sealing. Focusing from the outer surface to the inner surface can often reveal different structures and hyphal layers (Figs 5, 6A–D).

If transmitted light only is available, some mantle fragments must be oriented with their outer surface upwards and some with their outer surface downwards, so that the mantle can be studied from both sides. For transmitted light microscopy staining with cotton blue in lactic acid or toluidine blue is recommended (Ingleby *et al.*, 1990). Ectomycorrhizal preparations to be examined by interference contrast need not be stained.

2. *Crystals and warts*

Crystals associated with ectomycorrhiza often consist of calcium oxalate (Malajczuk and Cromack, 1982). In contrast to warts they should be readily dissolved by HCl. After dissolution some warts may remain, their positions corresponding to those of the crystals. These structures have been observed several times in non-mycorrhizal fungi (Agerer, 1980).

3. *Dolipores and Woronin bodies*

Dolipore-like structures can best be observed using phase contrast. They appear as hyaline hemispherical structures on both sides of the centre of the septa. Woronin bodies appear as roundish granules which are strongly light reflecting and can best be distinguished from lipid droplets by application of KOH, which dissolves Woronin bodies but leaves lipid droplets unchanged (Baral, 1991).

4. *Sections of ectomycorrhiza*

The decision, whether to section ectomycorrhiza with a cryotome or after embedding with a normal microtome, is dependent upon the structure of the ectomycorrhiza and the information wanted from the study. The type of microscope which is used (transmitted light, phase contrast, interference contrast) influences the required thickness of the sections, and also determines whether staining is necessary.

In general, when ectomycorrhiza possess a rather loose mantle cryotomy is disadvantageous and should be avoided, because after the transfer of sections of such ectomycorrhiza in a drop of mounting medium the outer mantle parts can float away. Thus needed information about emanating hyphae, cystidia and outer mantle layers can be lost. If only the structure of the Hartig net is to be investigated, then this method can be used without disadvantage. Ectomycorrhiza with pseudoparenchymatous or densely woven mantles are suitable for sectioning by cryotome. Thicker sections (15–20 μm) are appropriate for identification and for autofluorescence studies of the mantle.

If phase contrast is to be used, sections have to be very thin (2–4 μm). They need not be stained because the contrast allows the different cellular structures to be distinguished. Sections for phase contrast studies should always be made of embedded material. Interference contrast microscopy needs somewhat thicker sections (4–7 μm), otherwise the contrast is too weak. If transmitted light microscopy only is available, then the sections can also be thicker but should be stained. Often cotton blue in lactic acid or toluidine blue in water is sufficient (Agerer, 1986a; Ingleby *et al.*, 1990). Complex staining procedures of several stains applied consecutively (e.g. after Delafield: Gerlach, 1969; or quadruple stains: Wilcox, 1982) are not necessary.

The descriptions of cryotomy and microtomy of embedded material given in Tables I and II and Fig. 8 are very detailed. This is necessary for beginners and in our experience has been helpful for students and for those attending workshops. Care should always be taken that the original designations of the ectomycorrhiza are not lost; this is especially important when ectomycorrhiza must be transferred several times.

Cryotomy and microtomy of embedded material are frequently used methods in some laboratories. Each group develops its own procedures. Thus the procedures given in Tables I and II are only intended to describe the methods successfully used by our group. We have found that, for best results, ectomycorrhiza of Angiosperms should, whenever possible, be cryosectioned only as fresh material, whereas those of Gymnosperms are mostly better cryosectioned after fixation. Historesin and histo-moulds from LKB are used for embedding.

5. *Autofluorescence*

Reasonably thick cryotome sections (10–15 μm) should be used for checking autofluorescence of sectioned ectomycorrhiza. Such sections fluoresce more brightly than thinner ones. For the same reason lactic acid should be used as mounting medium. The most informative filter is

TABLE I
Cryotomy of ectomycorrhiza

Preparation of ectomycorrhiza
Fresh ectomycorrhiza
- Very fresh material is suitable for cryotome sections in demineralized water. Such sections are free of cytoplasmic contents and give a brilliant picture of mantle structures.
- Cut preferably straight or only slightly curved ectomycorrhiza under a dissecting microscope into pieces *c*. 3–4 mm long, include tips.
- Place them in small flasks which contain demineralized water.

Ectomycorrhiza fixed in FEA
- Fix ectomycorrhiza in FEA at for least 12 h, using only fresh and turgescent material.
- Cut preferably straight or only slightly curved ectomycorrhiza under a dissecting microscope into pieces *c*. 3–4 mm long, include tips.
- Place these pieces into small flasks which contain a 2% glycerine water solution.
- Leave for 2–3 h, shaking gently several times.

Sections of ectomycorrhiza
Preparation of the cryotome
- Use "type C" knives.
- Choose the temperature (for 2% glycerine water solution −25 °C is recommended; for fresh ectomycorrhiza use water and −30 °C). A test section should be made because ectomycorrhiza can behave differently.

Section
- Put 2–3 drops of glycerine and water solution (demineralized water if fresh ectomycorrhiza are to be cut) on the section table of the cryotome; the amount is dependent on the dimensions of the ectomycorrhiza.
- Orientate the ectomycorrhiza in the freezing drop (horizontally for longitudinal sections, vertically for cross-sections).
- Allow the drop to freeze completely.
- Choose the thickness of the sections.
- Elevate the section table until the knife touches the frozen drop.
- Put 1–2 drops of mounting medium (lactic acid or cotton blue in lactic acid) on a microscope slide.
- Make the sections and observe the procedure through a dissecting microscope.
- Remove the sections with a fine paint brush (which has been previously dipped in 2% glycerine water solution) from the knife and put them immediately in the drop of mounting medium (do not put more than five sections in one drop).
- Orientate the sections with a needle under a dissecting microscope in the middle of the drop; flatten the longitudinal sections if they have become distorted.
- Cover with a cover slip and soak away the surplus medium with a piece of filter paper.

- Label the preparation unequivocally.
- Seal the preparation with Entellan (use a fume hood for health reasons).
- Allow the preparation to dry overnight.
- Be sure that the immersion oil does not mix with Entellan; clean the preparation after the use of immersion oil because it can solubilize Entellan.

TABLE II

Microtomy of embedded ectomycorrhiza

Fixation

- The ectomycorrhiza should be fixed for at least 12 h in FEA; only fresh and turgescent material should be used.
- Cut straight or only slightly bent ectomycorrhiza under a dissecting microscope into pieces 3–4 mm long. Include the tips of the ectomycorrhiza.

Dehydration and infiltration

- Four steps of ethanol dehydration, 20%, 40%, 70% and 80%, each of 30 min, are required.
- The infiltration solution should then be prepared by mixing 50 ml basic resin with 1 bag of activator.
- Infiltration should be performed in two steps: (a) ethanol 96% : infiltration solution = 1 : 1 (at least 6 h); and (b) pure infiltration solution (at least 24 h).

Embedding and polymerization

- Mix the embedding medium (15 ml infiltration solution) and add 1 ml (= 20 drops) of the hardener; mix thoroughly.
- Fill 0.06 ml embedding medium in the lower part of the histo-moulds with an automatic pipette.
- Transfer the ectomycorrhiza into small dishes which contain some infiltration solution.
- Take the ectomycorrhiza with fine forceps, soak away the surplus infiltration solution with a piece of filter paper, and put the ectomycorrhiza singly into the holes of the histo-moulds.
- Enough time is now available (20–30 min) to orientate the ectomycorrhiza under a dissecting microscope, horizontally for longitudinal sections, or with the tip downwards for cross-sections; bear in mind that the viscosity of the embedding medium becomes greater with time; after 20–30 min the ectomycorrhiza must not be moved any more.
- Allow the medium to polymerize at room temperature (best overnight).
- Make new embedding solution.
- Lay the holders in the holes of the histo-moulds.
- Fill 1 ml of the newly mixed embedding medium with an automatic pipette through the holes of the holders.
- Allow the medium to polymerize at room temperature (at least 12 h).
- Take out the holders with the embedded ectomycorrhiza.

Sectioning

- Cut the historesin block into a pyramid shape.
- Take a hard metal knife (d-grinding) and fix the block in the microtome holder.
- Add a large volume of demineralized water to pre-cleaned microscope slides.
- Remove the sections with a fine needle from the knife and put them serially arranged into the water.
- Let the water dry and the sections adhere onto the microscope slides in a warm place (60 °C, at least overnight).
- Add some drops of Entellan (use a fume hood for health reasons) to the microscope slide and lay a cover slip carefully, slowly and at first obliquely on the drops.
- Let the preparation dry at least overnight; be sure that immersion oil does not come into contact with the Entellan during microscopical studies.

---

the ultraviolet filter, but blue and green filters can also be applied for detection of differences in wavelength. Both mantle and rhizomorphs can show differently autofluorescing layers.

C. Ultrastructure

Methods of preparing ectomycorrhiza for scanning electron microscopy can be found in Massicotte *et al.* (1986, 1987). For transmission electron microscopy the reader is referred to Berndt *et al.* (1990), Haug and Oberwinkler (1987) or Massicotte *et al.* (1985).

D. Chemical tests

Chemical tests on whole ectomycorrhiza can best be carried out within shallow cavities on a porcelain plate (Fig. 9A). First some drops of water should be placed in all cavities followed by the ectomycorrhiza. One cavity is used for ectomycorrhiza which are left untreated for reference. After removing the water with a piece of filter paper the reagents (Table III) can be added separately to each cavity. Colour changes should be checked under a dissecting microscope. Zak (1969) recommended 5 min as the optimal observation time.

Chemical tests on mantle preparations and rhizomorphs can be carried out directly on microscope slides (Fig. 9B). After peeling off the fragments, water should be removed with a piece of filter paper and the desired reagent (Table III) added to the almost dry fragments. The preparation is then covered by a cover-slip and observed in transmitted

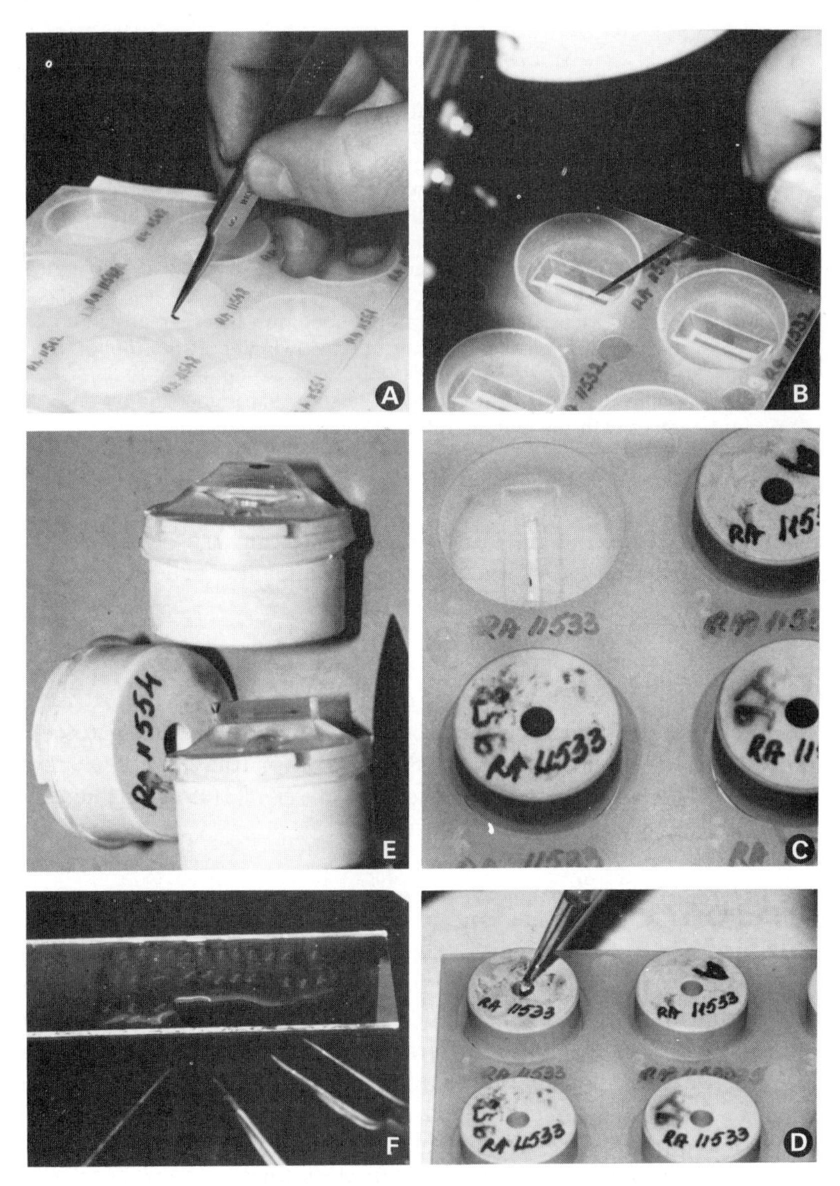

Fig. 8. Some steps in the procedure for embedding ectomycorrhiza in histo-resin. (A) Transfer of ectomycorrhiza from infiltration solution with forceps into the histo-moulds containing embedding medium in the lower parts of the cavities. (B) Orientation of the ectomycorrhiza in the lower parts of the histo-mould cavities with a needle under a dissecting microscope. (C) Ectomycorrhiza already embedded, holders in part put into the upper large cavities. (D)

light. In a further step the surplus reagent should be removed by drawing water through the preparation and a final observation should be made. One untreated preparation should be used for purposes of comparison. Care must be taken that the same light conditions are always used. Furthermore, it is important always to use similarly aged ectomycorrhiza, because aged mantles can give misleading colour changes. For further comments, see Agerer (1987–1990).

TABLE III

Recipes for some of the more important reagents

- Cotton blue in lactic acid (Erb and Matheis, 1983): solution of 0.05 g cotton blue in 30 ml lactic acid 85–90%.
- Ethanol: 70%.
- FEA (= FAA): formaldehyde : ethanol 70% : acetic acid = 5 : 90 : 5.
- $FeSO_4$ (Meixner, 1975): solution of 1 g Iron(II)-sulphate in 10 ml distilled water, finally some drops of conc. H_2SO_4 should be added.
- Guaiacol (Erb and Matheis, 1983): solution of 1 g Guaiac resin in 6 ml ethanol 70%, solution is stable only for *c*. 1 year.
- KOH: 15% aqueous solution (w/v).
- Lactic acid: 85%.
- Melzer's reagent (Moser, 1978): solution of 0.5 g iodine, 1.5 g KI, 20 ml distilled water, and 20 ml chloral hydrate.
- NH_4OH (Singer, 1986): concentrated solution.
- Pyrogallol (Singer, 1986): 5% aqueous solution.
- Sudan III: dissolve in a water bath 1 g Sudan III in 500 ml ethanol 96%, and add finally 500 ml glycerin.
- Sulfovanillin: for this reagent a few crystals of vanillin should be placed close to the cover slip on one side of the preparation. After addition of one drop of conc. H_2SO_4 a piece of filter paper should be laid on the other side of the cover slip (as, in Fig. 9B). Vanillin is dissolved in the acid and the fresh reagent is drawn through the preparation.
- Toluidine blue in water (Ingleby *et al.*, 1990): 1% (w/v) aqueous solution.

(*Fig. 8 continued*)
Holders laid into the cavities, newly made embedding medium being filled through the central hole of the holders. (E) Embedding procedure finished; holder beneath shows the cubic historesin block which contains the ectomycorrhiza, the historesin block above already cut to a pyramid. (F) Microscope slide with a layer of distilled water with sections floating on it, some tools. (The numbers on the histo-moulds and holders are the designation numbers of the ectomycorrhiza.)

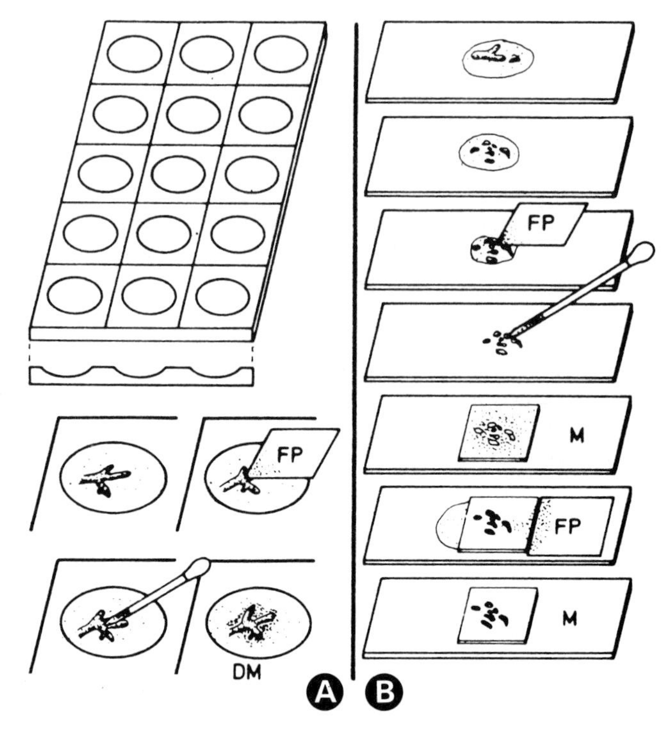

Fig. 9. Application of chemical reagents to test specific colour reactions of ectomycorrhiza. (A) Procedure for whole ectomycorrhiza; (B) procedure for mantle fragments (DM, dissecting microscope study; FP, filter paper; M, microscope study). For further explanations, see text.

E. Staining of nuclei and siderophilous granules

A staining procedure using aceto-carmine (KE) for detection of siderophilous granules in basidia of fruit bodies was first described by Clemençon (1978). This method can also be applied to stain nuclei and siderophilous granules of ectomycorrhiza (1986a). Hoyer's mounting medium is prepared after Cunningham (1972) (see Table IV).

Rhizomorphs with heavily encrusted or very thick-walled hyphae are difficult to stain. Staining of nuclei with aceto-carmine is appropriate for ectomycorrhiza with strong autofluorescence, otherwise fluorochrome stains can also be used (but not for detection of siderophilous granules) (see Table V).

During staining the ectomycorrhiza must be transferred several times, thus care should be taken that their designations are also transferred.

TABLE IV
Staining of nuclei and siderophilous granules

Preparation of the ectomycorrhiza
Both fresh ectomycorrhiza and specimens fixed in FEA can be used.

Mordanting procedure
Pick up the ectomycorrhiza with a needle or fine forceps, remove the surplus water or FEA with a piece of filter paper and put them for at least 1 h (or up to 2 days) in the mordanting solution (FBV) in small flasks (e.g. pyrex tubes).

Staining procedure
For the staining procedure *c*. 1/3 ml of KE should be placed in small pyrex tubes of 1–3 ml capacity (Eppendorf tubes).

- Retrieve the ectomycorrhiza from the mordanting solution, remove the surplus FBV with a piece of filter paper, and put them into the KE in the Eppendorf tubes; these must be closed, their caps having a small hole to prevent bursting.
- They should then be boiled for *c*. 5 min.

Preparation of microscope slides

- Pour the contents of the Eppendorf tubes onto a filter paper; remove the ectomycorrhiza with a fine needle from the filter paper or from the wall of the Eppendorf tube.
- Transfer the ectomycorrhiza directly into a drop of Hoyer's Mounting Medium (HMM) lying on a microscope slide. Peel off mantle fragments and rhizomorphs under a dissecting microscope and cover with a cover slip.
- If the mantle fragments are stained too intensely, the ectomycorrhiza can first be transferred for partial decolouration into acetic acid 50% or chloral hydrate for a short period.
- The HMM dries and the slides become permanent.

Recipes
Mordanting solution (FBV)

- The following chemicals should be mixed: 5 ml ferric chloride (a 10% solution in acetic acid 50%), 5 ml copper acetate (a 10% solution in acetic acid 50%), 5 ml picric acid (saturated in distilled water), 5 ml formaldehyde (saturated in distilled water).
- 1 ml lead acetate (1% solution in acetic acid 50%) should be added drop by drop while stirring constantly.
- The final solution is stable for years.

Aceto-carmine (KE)

- Boil a few grams of carmine under reflux with 200–300 ml of acetic acid 50% for 2–3 h.
- Filter next day, and the solution is ready for use.
- The solution is stable for several years.

Continued

Table IV continued

Hoyer's Mounting Medium (HMM)

- Chemicals: 15 g gum Arabic or gum guaiacol, 100 g chloral hydrate, 10 g glycerin, 25 ml water.
- Add to the water a crystal of chloral hydrate the size of a small pea to prevent bacterial growth upon the gum.
- Soak the gum in the water for about 24 h.
- Add the rest of the 100 g chloral hydrate and let the solution stand until all material is dissolved; this might take several days.
- Add the glycerin.
- The solution is now ready for use and is stable for several years.

TABLE V

Staining of nuclei with fluorochromes

Preparation and staining

Thin-walled, unpigmented living hyphae (very fresh ectomycorrhiza, e.g. of *Cortinarius* spp., *Laccaria* spp.) allow the staining of nuclei with fluorochromes (bisbenimide, acridine-orange, etc.); autofluorescence of the ectomycorrhiza must not be strong.

- Make mantle and rhizomorph preparations directly in a drop of fluorochrome (1 mg in 10 ml distilled water) lying on a microscope slide.
- Stain for at least 2 min.

Microscopy

- The slides must be studied under a fluorescence microscope with magnification 1000 ×.
- Use different filters (UV, green, blue) to find the best fluorescence contrast.
- If the background is stained too strongly, pass water through the preparation with the aid of filter paper.

Colouration

- Acridine orange stains nuclei and cell walls of vital hyphae green and dead hyphae orange.
- Bisbenzimide stains cell walls, nuclei and dolipores bright blue.

IV. Characteristics

A large number of features are available with which to characterize ectomycorrhiza (Table VI). The first features which can be seen after isolation from soil concern the gross morphology. Subsequently mantle and rhizomorph preparations can easily be made. Emanating hyphae and cystidia can be checked on the same preparations.

TABLE VI
Important characteristics

Morphological characteristics of the ectomycorrhiza
Shape and dimensions
Features of surface seen at moderate magnification (25–50 ×)
Colour (different old parts of specimens)
Rhizomorphs (frequency, occurrence, colour, characteristics of the margin)
Mantle anatomy in plan views including features of hyphae
Surface views
Plan views of different layers
Plan views of ectomycorrhizal tips
Characteristics of emanating elements
Rhizomorphs (internal organization, anastomoses)
Cystidia
Emanating hyphae
Sections of ectomycorrhiza
Thickness and organization of the mantle (including the tip)
Hartig net in section and in plan view, depth
Shape and dimensions of cortical cells
Chemical tests, autofluorescence, nuclei, siderophily

See also checklists and explanation given in Agerer (1987–1990) and Agerer (1987).

This is not intended as a complete overview for the characteristics or a compilation of literature. Only some examples are quoted which are suitable as examples of possible features of ectomycorrhiza.

A. Morphology

The morphology of ectomycorrhiza can vary depending on the fungal relationship and the tree genus. A compilation of morphological types and guidelines for measurements and descriptions are available (Agerer, 1987–1990).

As is widely known, pine ,ectomycorrhiza are usually forked and the dichotomous ramification can be repeated several times, sometimes resulting in dense clusters of ectomycorrhiza, the so-called coralloid systems. When these coralloid systems are enveloped by a layer of hyphae then tubercle-like systems are formed. Both types occur on the genus *Pinus*. The form of the ectomycorrhiza may also be influenced by the type of fungus. For example, in association with *Pinus cembra*, *Suillus plorans* forms tubercles, whereas *Suillus sibiricus* forms only coralloid systems with this host (Treu, 1990a–c). The same can be found in *Pinus sylvestris*: coralloid systems are formed by *Suillus bovinus*, while tubercle-like systems are formed by *Suillus variegatus* (Agerer, 1990a). Unramified or infrequently ramified ectomycorrhiza are formed by *Cenococcum* on the latter tree species (Uhl, 1988a). Members of the genera *Dermocybe* and *Cortinarius* (Uhl and Agerer, 1987; Uhl, 1988a) can change the dichotomous forms into irregularly shaped systems.

Coralloid and tubercle-like systems can also be found in genera other than *Pinus*. Coralloid systems with a loose hyphal envelope similar to tuberculate ectomycorrhiza are formed by *Leccinum* spp. on *Betula* (Müller and Agerer, 1990a). Tuberculate ectomycorrhiza are also found on *Pseudotsuga* (Trappe, 1965; Zak, 1971). The tuberculate ectomycorrhiza found by Zak (1971) were formed by *Rhizopogon vinicolor*. Masui (1926a) reported tuberculate ectomycorrhiza on *Quercus pausidendata*.

Monopodial systems of ectomycorrhiza are common in most tree genera and can be formed by different fungal relationships. Pinnate or pyramidal systems may be dependent on the growth conditions or on the tree genera involved, and probably to a lesser extent on the fungus. Pinnate systems occur preferentially in layered litter horizons, or on roots with diarch steles. Some fungal relationships are again able to cause irregularly shaped pinnate or pyramidal systems. This can be found in the genera *Dermocybe* and *Cortinarius* (e.g. *Cortinarius obtusus*: Agerer and Gronbach, 1988).

The unramified ends of ectomycorrhizal systems can be straight, bent, tortuous or beaded (Agerer, 1987–1990).

Some ectomycorrhiza can show hypertrophy. This situation appears to arise when other fungi, possibly of a non-mutualistic nature, grow within the ectomycorrhizal roots. Such hypertrophy has been found in some *Lactarius* spp. (Brand, 1991), in *Russula ochroleuca* (Brand, 1991), in *Boletinus cavipes* (R. Agerer, unpubl. res.) and in *Piceirhiza gelatinosa* (Haug, 1989).

The ectomycorrhiza can present different surface features. Several types are shown by Agerer (1987–1990). Smooth ectomycorrhiza with scanty or no rhizomorphs seem to be characteristic of the genera *Lactarius* and *Russula* but they can also be found in other genera (e.g. *Boletus*, *Xerocomus*). Cottony ectomycorrhiza occur in the genus *Hebeloma*, and ectomycorrhiza with abundant rhizomorphs can be formed by many different fungal relationships (for further examples, see Agerer, 1987–1990).

B. Mantle in plan views

The most informative anatomical features are in the ectomycorrhizal mantle as seen in plan views (Agerer *et al.*, 1990). Of special importance is the surface view of the mantle, i.e. the plan view of the outermost mantle layer. Two basic types can be distinguished: (1) plectenchymatous mantles, in which the hyphae can be recognized individually (Fig. 10A–I); and (2) pseudoparenchymatous mantles, in which the individual hyphae that have formed these mantle cells cannot be distinguished because they have been enlarged and have lost their original form (Fig. 10K–P), thus resembling a true parenchyma.

Nine main types of plectenchymatous mantles can actually be distinguished; some examples are compiled in Table VII to facilitate the search for species with such mantles. The nine main types are:

(1) Plectenchymatous mantles with a net-like arrangement of hyphal bundles (Fig. 10A).
(2) Plectenchymatous mantles where such a hyphal pattern is lacking and in which the hyphae grow in a rather irregular manner over the mantle surface often preferentially arranged in longitudinal directions (Fig. 10B).

TABLE VII
Examples of the different mantle types

A[a]	*Xerocomus chrysenteron* *Boletus edulis* *Leccinum scabrum*	B	*Dermocybe cinnamomea* *Lactarius deliciosus* *Hebeloma edurum*	C	*Piceirhiza gelatinosa* *Lactarius porninsis* *Elaphomyces muricatus*
D	*Thelephora terrestris* *Gomphidius glutinosus* *Tuber puberulum*	E	*Rhizopogon luteolus*	F	*Boletinus cavipes* *Suillus plorans* *Suillus flavus*
G	*Cenococcum geophilum* *Pinirhiza spinulosa*	H	*Russula xerampelina* *Lactarius alpinus*	I	*Lactarius picinus* *Lactarius fuliginosus*
K	*Piceirhiza nigra*	L	*Fagirhiza fusca*	M	*Lactarius vellereus*
N	*Russula laricina*	O	*Russula ochroleuca*	P	*Lactarius subdulcis*

[a]Letters A-P refer to Fig. 10.
Compiled from: Agerer (1986b, 1987c, 1991a), Brand and Agerer (1986), Gronbach and Agerer (1986), Biaschke (1987), Gronbach (1988), Uhl (1988a), Agerer and Weiss (1989), Brand (1989, 1991), Müller and Agerer (1990a) and Treu (1990a).

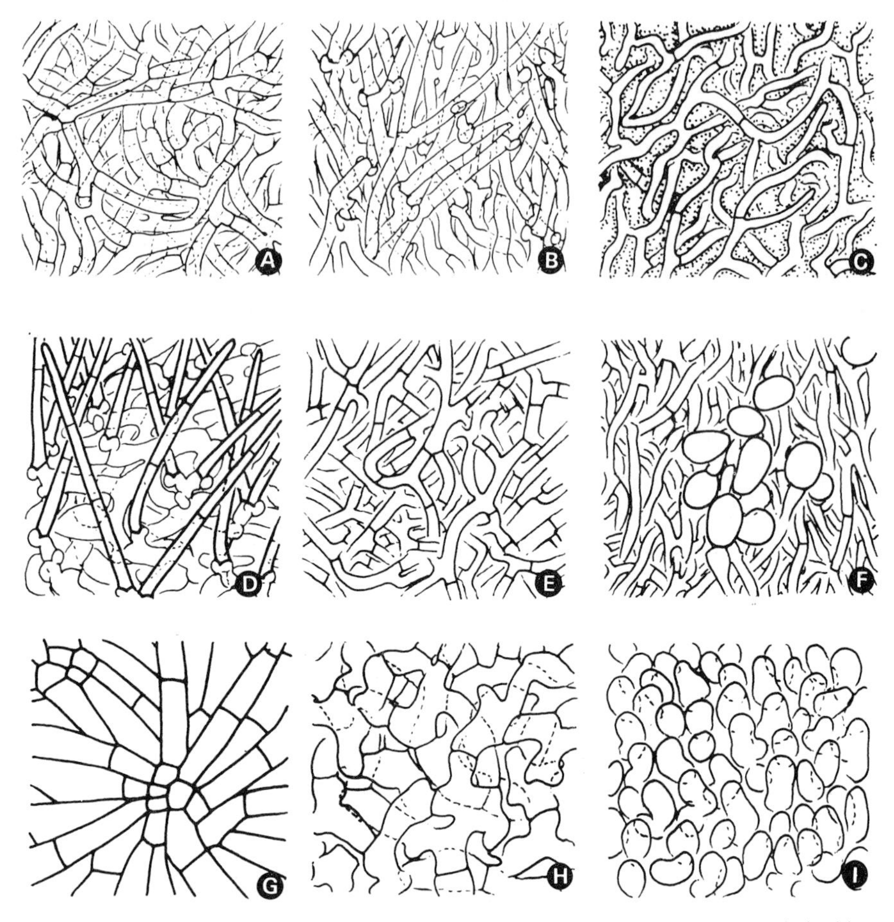

Fig. 10. Schematic drawings of different mantle types in surface view; (A)–(I), plectenchymatous mantles (after Agerer *et al.*, 1990). For explanation, see text.

(3) Plectenchymatous mantles with a gelatinous matrix. The hyphae are embedded in a gelatinous substance originating from the hyphal walls (Fig. 10C).
(4) Plectenchymatous mantles, with hyphae arranged in a net-like formation and bearing prominent cystidia (Fig. 10D).
(5) Plectenchymatous mantles with net-like hyphal arrangement produced by multiply-branched, squarrose hyphae (Fig. 10E).
(6) Plectenchymatous mantles with patches of roundish cells lying on the mantle consisting of otherwise normally shaped hyphae (Fig. 10F).

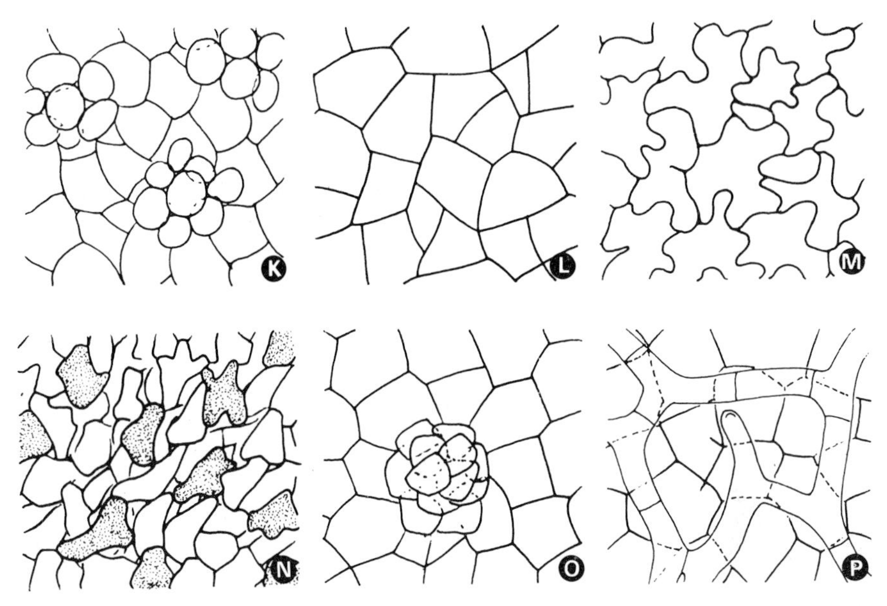

Fig. 10 (cont.). (K)–(P) Pseudoparenchymatous mantles (after Agerer *et al.*, 1990). For explanation, see text.

(7) Plectenchymatous mantles with hyphae in star-like arrangements which are tightly glued together (Fig. 10G).
(8) A transitional type between the plectenchymatous and pseudoparenchymatous mantle, in which irregularly shaped hyphae form a coarse net (Fig. 10H).
(9) Plectenchymatous mantles with approximately perpendicularly protruding, stout and often slightly curved hyphal end-cells, which are filled with oily droplets (Fig. 10I).

Six main types of pseudoparenchymatous mantles are known at present (Fig. 10K–P):

(1) Pseudoparenchymatous mantles composed of angular cells, bearing mounds of roundish cells (Fig. 10K).
(2) Pseudoparenchymatous mantles with angular cells (Fig. 10L).
(3) Pseudoparenchymatous mantles with epidermoid cells which have the appearance of cells of a leaf epidermis (Fig. 10M).
(4) Pseudoparenchymatous mantles with some cells containing oil droplets, stainable in sulfovanillin (Fig. 10N).
(5) Pseudoparenchymatous mantles with angular cells bearing mounds of flattened cells (Fig. 10O).

(6) Pseudoparenchymatous mantles with angular cells bearing a delicate hyphal net (Fig. 10P).

The hyphae below the outer mantle surface often change shape gradually and the inner mantle layers are mostly composed of normal hyphae, which form a loose or a dense plectenchymatous layer (Figs 5, 6D). Only occasionally is a pseudoparenchymatous layer present (*Fagirhiza arachnoidea*: Brand, 1991). Laticifers, characteristic of *Lactarius* ectomycorrhiza (Peyronel, 1934; Agerer, 1986b; Brand and Agerer, 1986; Agerer *et al.*, 1990; Ingleby *et al.*, 1990), can be found mostly within inner mantle layers. Species-specific differences occur regarding ramification, septa, dimensions and colour of contents (Peyronel, 1934; Agerer, 1986b; Gronbach, 1988; Brand, 1991).

C. Rhizomorphs

Rhizomorphs can also be subdivided into several types (Fig. 11). Some species showing different types of structure are listed in Table VIII. Ogawa (1981) arranged ectomycorrhizal rhizomorphs with respect to their internal organization, as did Agerer *et al.* (1990). The term rhizomorph is used here in a broad sense (Agerer, 1987–1990); Cairney *et al.*, 1991). The different rhizomorph types illustrated in Fig. 11 are:

(1) Undifferentiated rhizomorphs possessing rather loosely woven hyphae of uniform diameter and several hyphae which grow out of the margin (Fig. 11A). The hyphae may be held together by several different mechanisms, e.g. anastomoses or intertwining.
(2) Undifferentiated rhizomorphs with rather smooth margins; the hyphae are compactly arranged and are of uniform diameter (Fig. 11B).
(3) Slightly differentiated rhizomorphs with somewhat enlarged central hyphae (Fig. 11C).
(4) Differentiated rhizomorphs possessing thicker hyphae, which seem to be randomly distributed (Fig. 11D).
(5) Highly differentiated rhizomorphs with a central core of very thick hyphae (Figs 11E, F). The septa of the central hyphae may all be present (Fig. 11E) or partially or completely dissolved (Fig. 11F).

Rhizomorphs can show dimorphism. Those connecting fruit bodies with ectomycorrhiza can be wholly enveloped relative to those growing

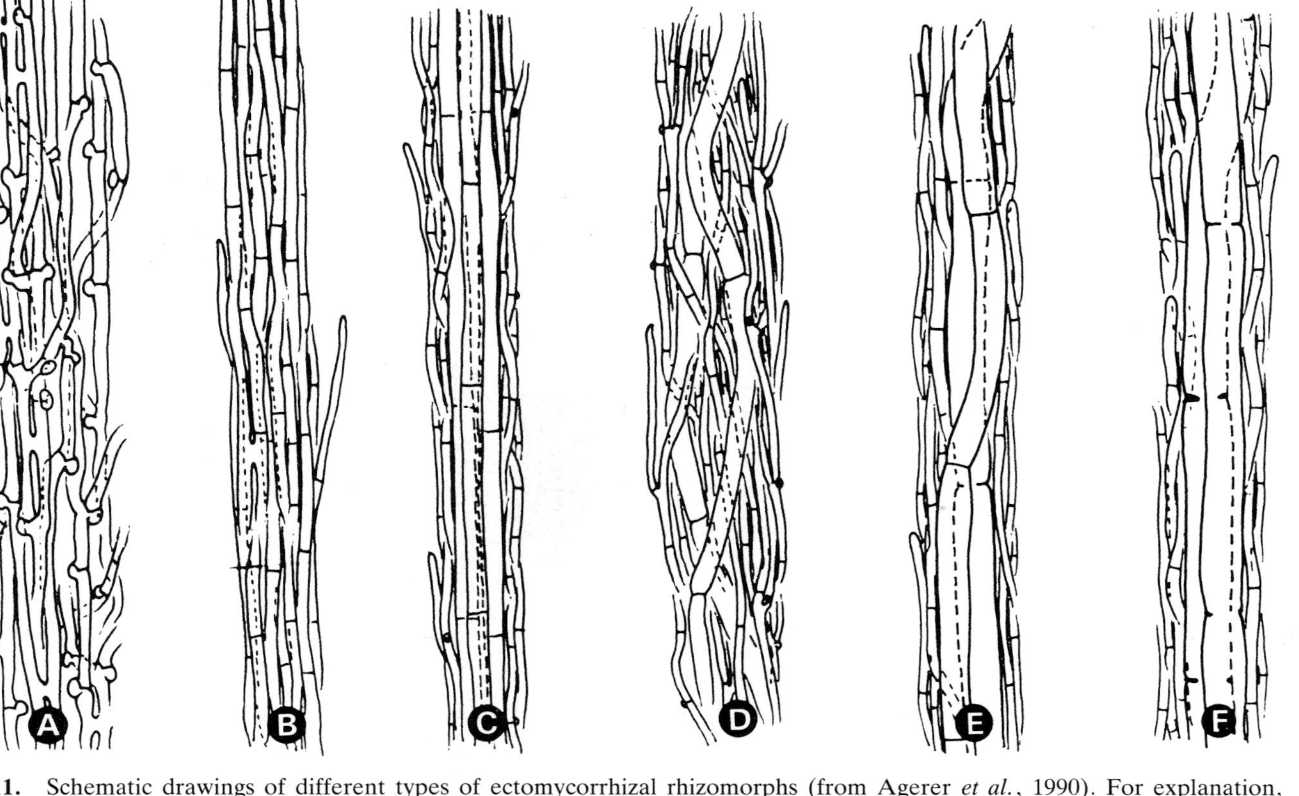

Fig. 11. Schematic drawings of different types of ectomycorrhizal rhizomorphs (from Agerer *et al.*, 1990). For explanation, see text.

TABLE VIII
Examples of different rhizomorph organization

A[a]	*Cortinarius obtusus, Dermocybe cinnamomea, Tricholoma sulfureum*
B	*Laccaria amethystina, Lactarius deterrimus, Lactarius vellereus*
C	*Gomphidius glutinosus, Thelephora terrestris*
D	*Cortinarius hercynicus, Cortinarius variecolor*
E	*Tricholoma saponaceum*
F	*Leccinum scabrum, Paxillus involutus, Scleroderma citrinum, Suillus bovinus, Xerocomus chrysenteron*

[a]Letters A-F refer to Fig. 11.
Compiled from: Bommer (1896), Duddridge *et al.*, (1980), Agerer (1986a, 1987b,c,d, 1988a,b,c, 1991a), Brand and Agerer (1986), Massicotte *et al.* (1986), Agerer and Weiss (1989), Brand (1989).

from the ectomycorrhiza into the soil (*Thelephora terrestris*: Schramm, 1966; *Dermocybe* spp.: Uhl and Agerer, 1987).

Some additional valuable features can be observed by examination of the rhizomorph ontogeny. The individual hyphae of young rhizomorphs must be investigated during their formation and growth from the point of origin. Amongst the most important characteristics of a number of rhizomorph types are backwardly growing hyphal branches (Fig. 13A), anastomoses (Fig. 12), and reversely oriented septa of clamps (Fig. 13B); see also Agerer (1987–1990, 1988b, 1990b). Some fungi lack for example backwards ramifications (e.g. *Laccaria amethystina*: Raidl and R. Agerer, in prep.).

D. Emanating hyphae and cystidia

As in the case of the hyphae in rhizomorphs, the hyphae emanating from the sheath can also form anastomoses. Several kinds of hyphal connections can be found (Fig. 12). Neighbouring hyphae can either be connected by rather long hyphal bridges (Fig. 12D–F) or the anastomosis can be very short (Fig. 12A–C). They can be open (Fig. 12A,D), closed by simple septa (Fig. 12B,E) or by clamps (Fig. 12C,F). Formation "B" is called a contact-septum, formation "C" a contact-clamp. Contact clamps are characteristic of ectomycorrhiza of the genus *Dermocybe* (Agerer *et al.*, 1990). Sometimes dolipore-like structures or presumed Woronin bodies may be present (see below).

Emanating hyphae can have various shapes. In addition to their colour, the thickness of their walls, the presence or absence of clamps

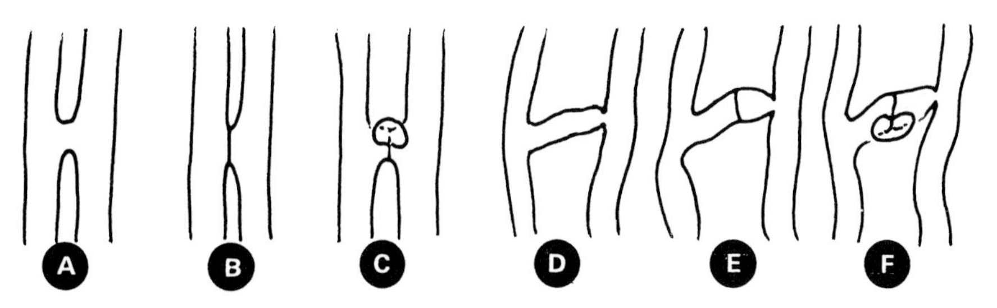

Fig. 12. Different types of anastomoses (from Agerer *et al.*, 1990). For explanation, see text.

and the presence of intrahyphal hyphae (Fig. 13C) are all characteristics of general value. Some such hyphae can completely lack septa, for example in ectomycorrhiza of *Endogone* (*E. flammicorona*: Chu-Chou and Grace, 1983).

Some ectomycorrhiza have emanating hyphae with rough surfaces (*Amphinema byssoides*: Fassi and De Vecchi, 1962); some possess crystals (*Piloderma croceum*: e.g. Brand, 1991). Some crystals can

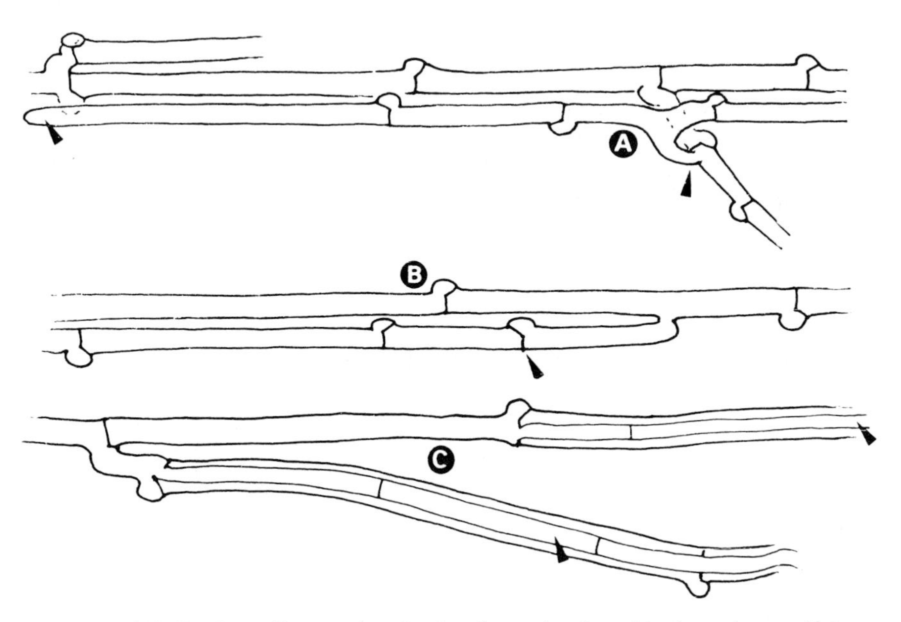

Fig. 13. (A) Backwardly growing hypha (arrow) of a side branch, ramifying into one branch growing backwards (left, arrow) and one forwards (right). (B) Anastomosis of a hypha with formation of a backwardly oriented clamp (arrow). (C) Intrahyphal hyphae.

strongly reflect light when viewed with Normarski interference contrast microscopy. These so-called "Suillus-crystaloids" are characteristic of *Suillus* spp. and of *Boletinus cavipes* (Treu, 1990a).

Although cystidia on ectomycorrhizal mantles (Fig. 14) are not very common, they provide valuable features for characterization of ectomycorrhiza as well as, in some instances, information on the fungal relationships from which these ectomycorrhiza are formed. Thick-walled, awl-shaped cystidia (Fig. 14A) without a basal clamp occur on ectomycorrhiza of the genus *Tuber* (Müller and Agerer, 1990b). Similarly shaped cystidia, but possessing a basal clamp, are characteristic of *Thelephora terrestris* (Schramm, 1966; Agerer and Weiss, 1989) and of *Gomphidius* spp. (Agerer, 1991a). *Chroogomphus* spp. are often furnished with strongly curved, thick-walled cystidia (Fig. 14K) (Agerer, 1990a). Dark brown ectomycorrhiza, which frequently occur on a wide range of host trees, very often possess cystidia with a basal inflation, either with a straight (Fig. 14B) or a curved appendage (Fig. 14C); examples are seen in *Piceirhiza obscura* (Gronbach, 1988) and *Pinirhiza*

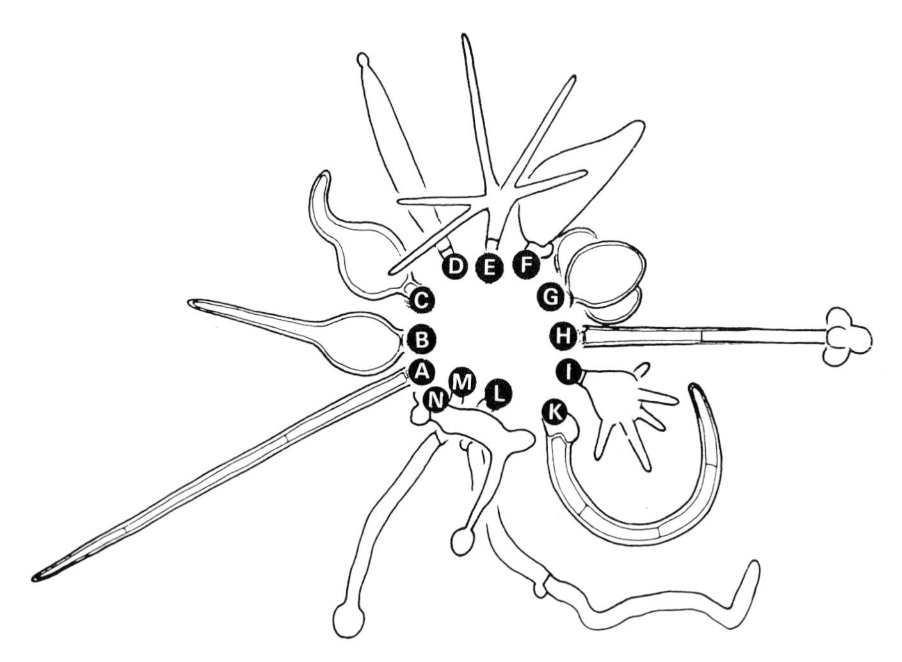

Fig. 14. Different types of cystidia of ectomycorrhiza (compiled after Dominick, 1959; Fontana, 1962; Peyronel, 1963; Chilvers, 1968; Jülich, 1985; Agerer *et al.*, 1986; Gronbach, 1988; Agerer, 1990a, 1991a; Berg, 1990; Müller and Agerer, 1990b; Brand, 1991). For explanation, see text.

spinulosa (Uhl, 1988a). Slender clavate cystidia with a small globule at the distal end (Fig. 14D) (Agerer *et al.*, 1986) are characteristic of some *Russula* ectomycorrhiza (Agerer *et al.*, 1990; Taylor and Alexander, 1989). Clavate cystidia (Fig. 14F) are known from *Paxillus involutus*, while spherical cystidia (Fig. 14G) have been observed on several occasions, e.g. in *Chroogomphus* (Agerer, 1990a), *Gomphidius* (Agerer, 1991a) and in *Suillus* spp. (Treu, 1990a), as well as in some unidentified ectomycorrhiza (*Piceirhiza nigra*: Gronbach, 1988). Slender cystidia with an apical lobed extension (Fig. 14H) are characteristic of the ectomycorrhiza of *Riessia* (Jülich, 1985). Other kinds of cystidia are known from different unidentified ectomycorrhiza (Fig. 14E,I,M,N). In some cases it might be difficult to distinguish between cystidia and short emanating hyphae (Fig. 14L), as transitional types can be found.

E. Chlamydospores

Recently, prominent chlamydospores have been found on ectomycorrhiza isolated from soil (Agerer, 1990a, 1991a). *Gomphidius roseus* forms oval, sometimes irregularly shaped, thick-walled, yellowish chlamydospores within the ectomycorrhizal mantle of *Suillus bovinus* (Agerer, 1990a, 1991a). Other types of chlamydospores are known from ectomycorrhiza of the *Thelephoraceae* (Agerer, 1991a). Brown star-like chlamydospores with thick walls and elongate warts are typical of *Sarcodon imbricatus* (R. Agerer, in prep.); brownish, thick-walled, roundish, and smooth chlamydospores are characteristic of *Bankera fuligineo-alba* (Danielson, 1984a) and *Hydnellum peckii* (R. Agerer, in prep.). The formation of chlamydospores in cultures of *Elaphomyces muricatus* was reported by Miller and Miller (1984).

F. Sections of ectomycorrhiza

For several reasons median longitudinal sections of ectomycorrhiza are more informative than cross-sections. Firstly, longitudinal sections include the ectomycorrhizal mantle from the tip (the youngest part) to basal regions, where the mantle hyphae have their mature shape. It is also not easy to conclude from cross-sections the portion at which the ectomycorrhiza has been cut. Secondly, longitudinal sections clearly show where the Hartig net initiation zone lies; and where it is not fully established. Cross-sections of this zone could be interpreted as representing reduced penetration by the Hartig net. Thirdly, if the cortical cells are obliquely oriented, it might well be that two neighbouring cortical cells are cut in cross-section, thus indicating incorrectly a Hartig net of two cortical rows depth, when only a single layer occurs. Fourthly, the shape of the cortical cells enveloped by a Hartig net is more characteristic in longitudinal section than in cross-section (Uhl and Agerer, 1987). Because of the greater importance of longitudinal sections, the following explanations refer only to such sections.

The mantle in section mostly merely confirms the structures already seen in plan views; in only a few species does the mantle section provide additional information. One example is *Lactarius picinus*, in which a cross-section shows three distinct mantle layers (Agerer, 1986b).

The Hartig net can comprise either only one row of cortical cells (Fig. 15A,B), as seems to be the case in Angiospermous trees (Godbout and Fortin, 1983), or it can protrude more deeply towards the endodermis (Fig. 15C,D), as is characteristic of Gymnosperms (Godbout and Fortin, 1983). In most genera of ectomycorrhizal Angiosperms, e.g. *Alnus*, *Betula*, *Corylus* and *Populus* (Godbout and Fortin, 1983, 1985), the Hartig net is restricted to the anticlinal walls of the cortex cells (Fig. 15A). This is called a paraepidermal Hartig net (Godbout and Fortin, 1983). In others (e.g. *Fagus*: Brand and Agerer, 1986, 1988) it com-

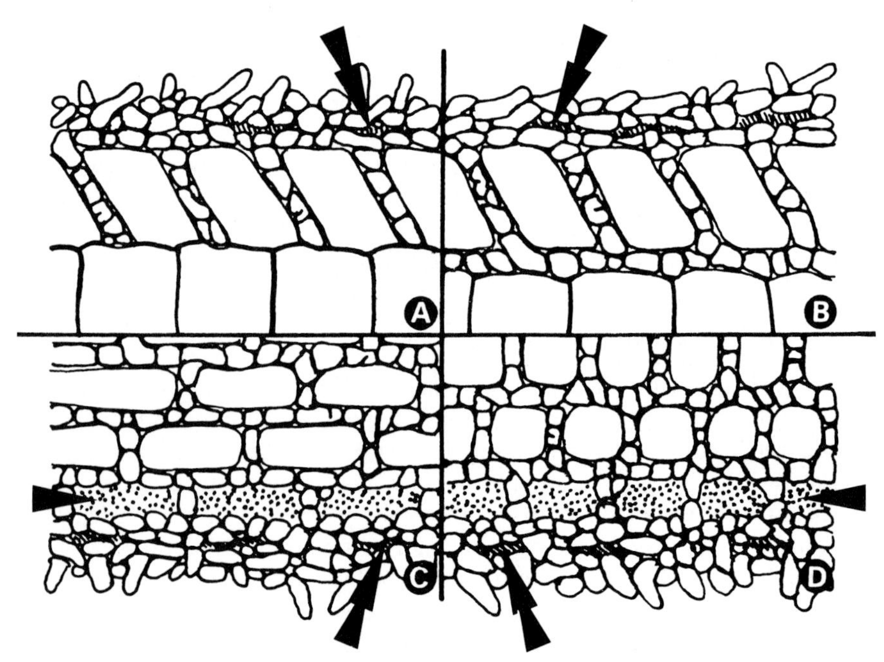

Fig. 15. (A)–(D). Features of the Hartig net (arrows, tannin cells; double arrows, residues of root cap cells). For explanation, see text.

pletely envelopes the cortex cells (Fig. 15B), a situation described as periepidermal (Godbout and Fortin, 1983).

A further difference between Gymnosperms and Angiosperms concerns the formation of tannin cells. Although brownish substances are present in all ectomycorrhiza in the inner mantle layers, which are residues of root cap cells, cells filled with tannins as far as is known can only be found in ectomycorrhiza of Gymnosperms (compare Fig. 15A,B with 15C,D).

In addition, a feature characteristic of the tree genera involved, rather than of the fungi, is the shape of cortical cells in median longitudinal sections. The ratios between tangential and radial dimensions of the cortex cells which are enveloped by a Hartig net (CCq: Agerer, 1987–1990; elongation ratio: Godbout and Fortin, 1983), reveal for Angiosperms values considerably smaller than 1 (Fig. 15A,B), for *Pinus*

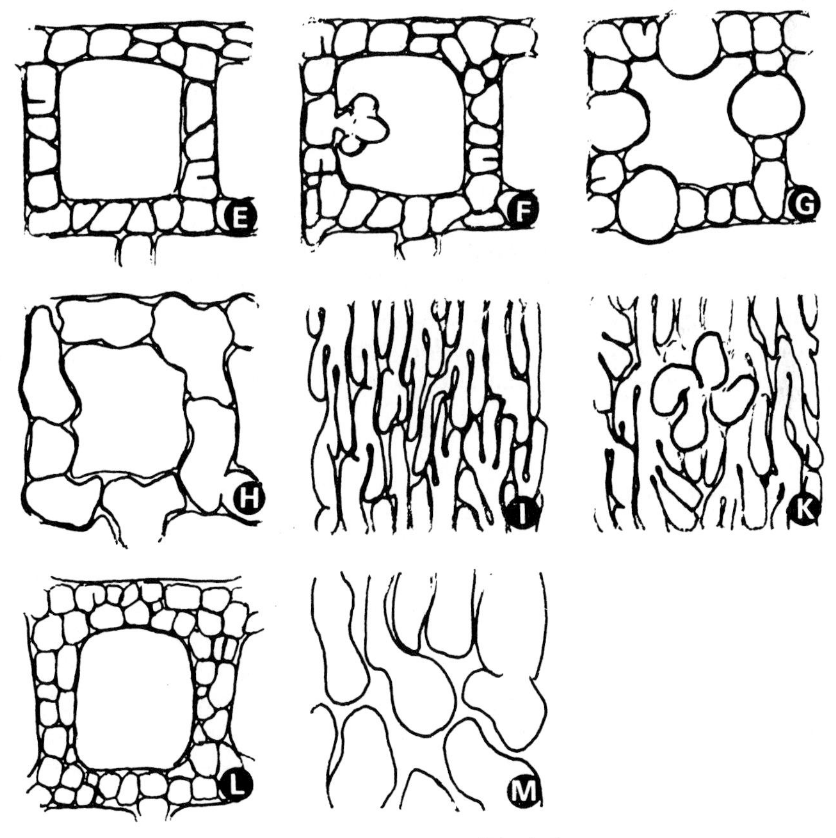

Fig. 15 (cont.). (E)–(M).

approximately 1 (Fig. 15D), and for *Picea abies*, *Larix decidua* and *Pseudotsuga menziesii* more than 2.5 (Fig. 15C) (Godbout and Fortin, 1985; Uhl and Agerer, 1987; Pillukat and R. Agerer, unpub. res.).

The structure of the Hartig net can often provide characteristics specific for the fungi involved. The most common type of Hartig net is composed of a single row of hyphae between the cortex cells (Fig. 15E–K, M), but occasionally several hyphal rows exist (Fig. 15L), as in *Piceirhiza gelatinosa* (Gronbach and Agerer, 1986). However, Hartig nets between tannin cells can sometimes consist of more than one row of hyphal cells and in addition are often thicker than those between the cortical cells. The Hartig net in section is usually of constant width (Fig. 15E). A characteristic of *Leccinum* spp., however, is a beaded Hartig net (Fig. 15C). This is described by Godbout and Fortin (1985) and Müller and Agerer (1990a). A rather coarse Hartig net (Fig. 15H) seems characteristic of some ascomycetous ectomycorrhiza (Danielson, 1984b) and was also found for a *Lactarius* sp. (Brand and Agerer, 1986).

The Hartig net is in plan view mostly of the palmetti type (Fig. 15I,K); occasionally there are rather broad, infrequently lobed, loosely arranged hyphae (Fig. 15M), as seen in *Lactarius vellereus* (Brand and Agerer, 1988).

Very occasionally haustoria-like intrusions may be formed (Fig. 15F) e.g. *Leccinum scabrum* (Müller and Agerer, 1990a); *Russula emetica*, *R. mairei* and *R. nana* (Brand, 1991). Occasionally apparently non-mutualistic fungi can form haustoria within a host which is ectomycorrhizal. Usually it is difficult to discern the affiliations of these fungi (*Lactarius* spp.: Brand, 1991; *Suillus* spp.: Agerer, 1990a, 1991a). If haustoria are present, they can also be seen in plan views of the Hartig net (Fig. 15K).

G. Autofluorescence

Both Trappe (1967) and Zak (1973) discuss autofluorescence of sections of ectomycorrhiza and of whole specimens and both conclude that autofluorescence is very valuable for characterization of ectomycorrhiza. It is now known to be a valuable character for some additional species including *Cortinarius venetus* (Agerer, 1987b) and *Dermocybe semisanguinea* (Uhl, 1988a). It appears that ectomycorrhiza of species with strongly autofluorescing fruit bodies also fluoresce strongly.

Sections of ectomycorrhiza can display mantle layers with different kinds of autofluorescence (e.g. *Lactarius vellereus*: Brand and Agerer, 1986; *Rhizopogon luteolus*: Uhl, 1988b; and *Russula ochroleuca*: Agerer, 1986a).

H. Ultrastructure

The ultrastructure organization of ectomycorrhizal fungi has been shown recently to provide informative characters, among which are details of the structure of septal pores, of the matrix material between hyphae and of deposits on cell walls (Haug and Oberwinkler, 1987).

Based upon septal pore structures, Haug and Oberwinkler (1987) could recognize several ectomycorrhiza as being formed by ascomycetous or basidiomycetous fungi. Ascomycetes could be identified by their Woronin bodies, and basidiomycetes by their dolipores. In addition, a continuous parenthesome suggested that a few of the investigated ectomycorrhiza were probably formed by heterobasidiomycetes. Most of the ectomycorrhiza, however, possessed perforated parenthesomes. In the hyphae of some ectomycorrhiza near the centre of the septa, structures similar to dolipores can be seen using phase contrast (e.g. *Dermocybe palustris*: Uhl and Agerer, 1987; *Cortinarius obtusus*: Gronbach, 1988). Woronin bodies can be seen as highly refractive granules near the septa (Danielson, 1982; e.g. *Fagirhiza tubulosa*: Brand, 1991).

Matrix material between mantle hyphae may be electron transparent or electron dense, but both kinds of matrix can occur in the same ectomycorrhiza (Haug and Oberwinkler, 1987). Strullu (1985) suggested that on account of their matrix material, electron dense ascomycetous mycorrhiza could be distinguished from their basidiomycetes counterparts, which are electron transparent. However, Haug and Oberwinkler (1987) showed that an electron dense matrix can also be found in basidiomycetous ectomycorrhiza, and Berndt *et al.* (1990) observed an electron transparent matrix in ascomycetous ectomycorrhiza. The previously assumed difference between Ascomycetes and Basidiomycetes with respect to their wall layers (two layers or multilayered: Kreger Van-Rij and Veenhuis, 1971) seems not to be an appropriate criterion with which to distinguish these fungal relationships (Berndt *et al.*, 1990).

Minute globules of secreted material can be detected on the outer mantle hyphae or on emanating hyphae by transmission electron microscopy (Haug *et al.*, 1986).

Massicotte *et al.* (1987) reviewed the value of scanning electron microscopy for the characterization of ectomycorrhiza. One advantage of this approach is that it provides the opportunity for topographical research of mantle hyphae. However, the hyphal arrangements are equally well shown by Normarski interference contrast and this easy, rapid and inexpensive method reveals additional information on the structure and arrangement of septa. The main advantage of scanning electron microscopy (SEM) seems to be the possibility of investigating the shape of crystals and the structure of warty hyphae (e.g. *Pisolithus tinctorius*: Massicotte *et al.*, 1987).

I. Nuclei and siderophilous granules

Staining of nuclei has been carried out when examining clampless ectomycorrhiza to determine whether they have been formed by Basidiomycetes or Ascomycetes. Basidiomycetes are expected to possess nuclear pairs within their cells, whereas Ascomycetes should have solitary nuclei; *Tuber* spp., however, are known to have pairs of nuclei (Bonfante-Fasolo, 1973).

Pairs of nuclei could often be found at variable distances from one another, but a close association seems to be characteristic of some species (*Tylopilus felleus*, *Amanita muscaria*: Uhl, 1988a). Species with contact-clamps sometimes show an increased number of nuclei per cell (e.g. *Dermocybe palustris*: Uhl and Agerer, 1987). Cells with up to ten nuclei have been detected in *Russula ochroleuca* ectomycorrhiza (Gronbach, 1988; Berg, 1990). Using transmission electron microscopy Kottke and Oberwinkler (1986) found multinucleate palmetti of the Hartig net, suggesting that the larger number of nuclei is a response, specific to the Hartig net, associated with its greater physiological activity. Cells with single nuclei are occasionally found in *Lactarius badiosanguineus* (Treu, 1990a) and in *Piceirhiza nigra* (Gronbach, 1988). Interestingly, the solitary nuclei showed greater diameters after staining with aceto-carmine (3–4 μm) than did pairs (mostly 1.5–2.5 μm), perhaps indicating that they are polyploid. Other species consistently displayed one small nucleus per cell. This is true of *Piceirhiza glutinosaesimilis* (Berg, 1990) a feature which could even be used to distinguish this species from *Piceirhiza glutinosa* (Gronbach, 1988), in which two and more nuclei are characteristically present. Further evidence of a taxonomic value of these nuclei characteristics has still to be obtained.

Siderophily of ectomycorrhizal hyphae is a poorly understood feature. After staining with aceto-carmine, siderophilous granules have a black colouration. They can be very minute and densely arranged (*Russula xerampelina*: Agerer, 1986b) or larger and rather infrequent within the cells (*Cortinarius variecolor*: Agerer, 1988a). Uhl (1988a) found siderophilous granules only in *Tricholoma* and *Dermocybe* spp.

J. Colour changes with chemicals

A large number of different chemicals are used as taxonomic aids in mycology (Meixner, 1975; Singer, 1986). Because macrochemical reactions can help in the identification of fruit bodies, it was assumed that ectomycorrhiza, too, would show specific reactions (Trappe, 1967; Zak, 1973; Agerer, 1986a). However, only a few chemicals have so far been successfully applied to ectomycorrhiza.

Perhaps the most important chemical is Melzer's reagent. Parts of some ectomycorrhiza stain blue in this reagent, producing what is called an amyloid reaction (*Chroogomphus*: Miller, 1983; Agerer, 1990a; *Endogone lactiflua*: Walker, 1985; *Gomphidius*: Agerer, 1991a; *Thelephora terrestris*: Agerer, 1991a). Some hyphae of *Gomphidius* spp. are stained slightly brownish (Agerer, 1991a), a so-called dextrinoid reaction. Furthermore, this reagent is able to reveal what are possibly parasitic hyphae of the genera *Chroogomphus* and *Gomphidius* within ectomycorrhiza of *Suillus* and *Rhizopogon* spp. (Agerer, 1990a, 1991a). Sulfovanillin can be used to stain the laticifers of *Lactarius* ectomycorrhiza red or blue or almost black (Peyronel, 1934; Agerer, 1986b; Brand and Agerer, 1986; Agerer *et al.*, 1990) and the outer mantle cells of *Lactarius* spp. of the section Plinthogali (e.g. *Lactarius picinus*: Agerer *et al.*, 1990) or the special cells of the mantle surface of some *Russula* spp. (e.g. *Russula emetica*: Agerer *et al.*, 1990; Brand, 1991). The whole mantle can also be stained a distinctly violet colour with sulfovanillin (e.g. *Hygrophorus lucorum*: Treu, 1990a). Toluidine blue is utilized by Ingleby *et al.* (1990) because it displays metachromatic reactions of the hyphal walls. They can become purple, violet or pink. Potassium hydroxide (KOH) stains some ectomycorrhiza brown (e.g. *Tricholoma flavobrunneum*: Uhl, 1988a) or brownish red (*Leccinum* spp.: Müller and Agerer, 1990a). The cell mounds of *Russula ochroleuca* stain red (Brand, 1991). The colour change, at least in *Leccinum* and *Russula ochroleuca*, is reversible by acids (Müller and Agerer, 1990a; Brand 1991). Further species turn to violet (e.g. *Cortinarius cinnabarinus*: Brand, 1991). Ethanol is useful as an extractant of pigments from ectomycorrhiza (e.g. *Dermocybe*: Thoen, 1977).

Other reagents cause irregular reactions or have only been used infrequently: Sudan III stained some cystidial contents red (*Fagirhiza globulifera*: Brand, 1991); guaiacol stained ectomycorrhiza of *Tricholoma vaccinium* blue (Agerer, 1987d); ectomycorrhizal mantles of *Tuber macrosporum* are stained brown in $FeSO_4$ (Giovanenetti and Fontana, 1981); ectomycorrhiza of *Lactarius obscuratus* were stained violet by pyrogallol and green by concentrated NH_4OH (Froidevaux, 1973); and the addition of weak acids (lactic acid) to *Leccinum* ectomycorrhiza, which had turned blue after lying for some time in water, can change the blue colour to red (Müller and Agerer, 1990a).

There is a need to continue the search for additional chemicals which can cause colour changes. In spite of some disappointing results up to now with some chemicals (Agerer, 1987b), the above-mentioned results show that chemical reagents can contribute to the successful characterization of ectomycorrhiza.

K. Ecology

That fruit bodies often prefer certain soil types has already been shown by several authors (e.g. Bohus, 1973; Tyler, 1985). Thus it can be suggested that their ectomycorrhiza also have preferences for particular soil conditions.

Since the investigations of Egli (1981) it has been known that different ectomycorrhiza may prefer particular soil horizons. Similar results were found subsequently by Haug *et al.* (1986) and Alexander *et al.* (1990). Brand (1989) has already described in detail the distribution of *Xerocomus chrysenteron* ectomycorrhiza and extended such description to several other species (Brand, 1991), indicating that such ecological features are also important for characterization.

V. General value of ectomycorrhizal characteristics

Generally, the features now available for characterization of ectomycorrhiza can be arranged in a sequence from those that are most important (Group 1) to those of seemingly lesser importance (Group 4).

The question of the ease with which the different characteristics can be determined must also be considered.

A. Group 1 features

The most informative features are the mantles in surface view and the anatomy of rhizomorphs. In addition, the structures of the emanating hyphae and occasionally of the cystidia must be examined. These anatomical features can be investigated quickly and relatively easily and—depending upon the length of the ectomycorrhizal root in question—only a quarter or a half of it may be necessary for study. Thus these features can be used for routine determination of ectomycorrhizal structure. The anatomical features are complemented by morphological characteristics, especially by the colour and the surface of the ectomycorrhizal sheath as seen at intermediate magnification. As the host trees can influence the extent of ramification of rootlets the tree genus, at least, should be known. Modern descriptions or keys for determination of mycorrhiza are mostly based on these characteristics. Some chemicals (e.g. Melzer's reagent, KOH and sulfovanillin) should immediately be applied for identification of freshly isolated ectomycorrhiza because they often reveal important features. But for comprehensive descriptions of new or lesser known species a more comprehensive set should be used. Analysis of autofluorescence of the whole ectomycorrhiza should also be included because of the possibly conclusive information it may provide.

B. Group 2 features

As the occurrence of ectomycorrhiza depends upon soil characteristics, their ecology with respect to soil types and preference for soil layers should have already been considered when taking the soil cores and during the isolation procedure.

C. Group 3 features

This group comprises anatomical studies of sections, which should preferably be cut longitudinally. They can be obtained quickly using a cryotome for studies of the Hartig net. Comprehensive descriptions require time-consuming embedded sections. The same is true of ultrastructural studies, but the latter can also provide very important information concerning septal pores, matrix and hyphal surface.

D. Group 4 features

Of apparently lesser importance are studies using stains to determine the number of nuclei per hyphal cell, their distribution and their dimensions, as well as the occurrence of siderophilous granules. If transmission electron microscopy is available, this technique can help to determine whether the ectomycorrhiza are formed by Ascomycetes or Basidiomycetes. Siderophily of ectomycorrhizal hyphae is not understood sufficiently for it to be useful in identification.

E. Additional methods

Some methods which are either infrequently used, or are as yet not fully developed, may provide additional information in future. There are a number of potentially useful sophisticated modern methods available: chromatography of pigments, DNA/RNA researches, investigation of protein patterns, and immunological methods. Each might contribute important results to different problems. However, they probably will not become routine methods and cannot be applied independently from comprehensive anatomical investigations. Comprehensive anatomical characterization of ectomycorrhiza will still be urgently needed for some decades. Certainly, relative to our knowledge of fruit bodies of ectomycorrhizal fungi or of lichens, our understanding of the anatomical features of ectomycorrhiza is very weak.

VI. Voucher specimens

The importance of depositing voucher specimens has been stressed by Trappe and Schenck (1982) for the ectomycorrhiza and for fruit bodies of fungi in general by Ammirati (1979) and Trappe (1977).

Many reports on ectomycorrhizal research suffer from a lack of voucher specimens. Often the descriptions of ectomycorrhiza are too cursory to provide a complete picture of the characterized ectomycorrhiza. Even the most comprehensive descriptions now available often need further expansion. The deposition of voucher specimens is not only necessary for taxonomic studies; it is also imperative when physiological, chemical, cultural and ecological studies, and synthesis experiments are to be performed. Scientific researches without quotation and deposition of voucher specimens are almost worthless, because there is no opportunity to reconsider the results. Ammirati (1979) states: "Voucher collections can be used by future researchers to compare their specimens and results with those of earlier workers, but, if they are not retained, accurate comparisons are impossible".

A. Preservation of voucher specimens

Preservation of the ectomycorrhizal specimens should be achieved in three ways.

(1) Representative examples of the ectomycorrhiza are fixed in FAA (see Section II.D). These voucher specimens should include homogeneous material; a mixture of different species must be avoided (see Section II.D). Small air-tight flasks containing the ectomycorrhiza can be kept at room temperature indefinitely. Other workers (Ingleby *et al.*, 1990) prefer 2% glutaraldehyde, but these specimens must be stored at 4 °C.

(2) Colour slides which show the ectomycorrhiza in natural colours (see Section III.A)—preferably at four different magnifications—should be stored.
(3) All microscope slides (of mantles, rhizomorphs, sections etc.) should also be stored.

All materials (colour slides, microscope slides, fixed ectomycorrhiza) should be labelled unequivocally, e.g. by a collection number. All material belonging to the same specimen should have the same designation and at least the fixed material should be deposited in an official herbarium, where it can be made available to other researchers for reference. If the ectomycorrhiza have been identified by tracing hyphal connections to fruit bodies, then these fruit bodies must also be preserved and supplied with the same collection number as the associated ectomycorrhiza. The fruit bodies must be dried at *c.* 45–50°C and stored in a dry environment (Ammirati, 1979). If fruit bodies are used to make cultures, these specimens must also be preserved. The labels of the voucher specimens should bear the following information:

- Name of species (of ectomycorrhiza or fungus).
- Record of the date and locality (exact geographical position, species of trees in the vicinity, host species of the ectomycorrhiza and, if possible, soil type and soil layer).
- Name of the collector(s).
- Collection number, which preferably starts with the collector's initials (to simplify reference and retrieval).

B. Voucher collections

Voucher collections of ectomycorrhiza have an additional value. From a number of collections, a single reference specimen (Agerer, 1987b) should be selected for each species of ectomycorrhiza, regardless of whether the identity of the fungus is known or not. That collection should be quoted as the reference specimen on which the descriptions were mainly based, and which is typical in all features of the comprehensively described ectomycorrhiza. Reference specimens of ectomycorrhiza have the same significance as type specimens of plants and fungal fruit bodies. They should always be referred to if difficulties arise in the examination of the species. This is especially important for unidentified ectomycorrhiza. A few examples of reference specimens are given below:

Piceirhiza gelatinosa (on *Picea abies*): reference specimen (cf. Gronbach and Agerer, 1986): Deutschland, Bayern, Lkr. Aichach, Höglwald between Odelzhausen and Mering, Plot A1 23.10.1984, Herb. E. Gronbach (in M).

Russula ochroleuca (on *Picea abies*): reference specimen of *Russula ochroleuca* on *Picea* (cf. Agerer, 1986b): Deutschland, Bayern, between Odelzhausen and Mering, in Höglwald, 19.9.1984, Herb. RA 10705 (in M).

Russula ochroleuca (on *Fagus sylvatica*): reference specimen of *Russula ochroleuca* on *Fagus* (cf. Brand, 1991): Deutschland, Baden-Württemberg, Staatswald Rittnert bei Karlsruhe, Buchenwald mit lehmiger Parabraunerde über Löß; fruit body in Herb. F. Brand FB237, ectomycorrhiza FBM51 (in M).

C. Deposition of voucher collections

A list of official herbaria (including addresses and the commonly used abbreviations) where voucher collections can be preserved is published in the *Index Herbariorum* (Holmgren *et al.*, 1981). The directors of the herbaria should be asked in advance if material kept in fixative is acceptable.

Trappe and Schenck (1982) offered to preserve all Endogonales specimens in the "Oregon State Mycological Herbarium". Schenck and Perez (1987) reported that an "International Culture Collection of VA Mycorrhiza Fungi (INVAM)" was established in Gainesville, Florida. It is of great advantage to concentrate the deposition of important collections (new species, specimens used for comprehensive descriptions) in a few commonly known and accessible herbaria, so that extended searches for voucher specimens are not necessary if ectomycorrhiza are to be compared. At the "Botanische Staatssammlung München" a large number of voucher specimens of ectomycorrhiza are already preserved. This herbarium has offered to keep collections of ectomycorrhiza from all over the world, to establish the "International Collection of Ectomycorrhizae München" (ICEM). Researchers on ectomycorrhiza are kindly asked to send voucher collections to: The Director, Botanische Staatssammlung München, Menzingerstrasse 67, D-8000 München 19, Germany.

D. Naming of unidentified ectomycorrhiza

As has been previously reported, unidentified ectomycorrhiza require binary names in the same way as do fruit bodies of fungi (Gronbach and

Agerer, 1986; Agerer, 1987–1990; Weiss and Agerer, 1988), but only when they are described comprehensively and in detail. The name is made up of the genus name of the host tree and a characterizing epithet, for example *Piceirhiza gelatinosa* (Gronbach and Agerer, 1986) or *Laricirhiza alpina* (Treu, 1990a). Those names identify an ectomycorrhiza of the genus *Picea* (or *Larix*); "gelatinosa" refers to the conspicuously gelatinous mantle, while "alpina" indicates that it was found in the Alps. The genus name of the tree is sufficient to include in the name because tree species-specific ectomycorrhiza are formed only very infrequently. When the identity of the fungus is discovered, the artificial name of the ectomycorrhiza can be replaced by the name of the fungus.

These names are not intended to classify ectomycorrhiza in the sense of the *Botanical Code of Nomenclature*, therefore a Latin diagnosis is not necessary.

Such names are much more informative and more easily organized, and retrieved than the special abbreviations commonly used up to now by different authors in dissimilar ways, for example: Chilvers (1968), Eucalypt Mycorrhiza Type 1 or Eucalypt Mycorrhiza Type 7; Dominik (1959), genus Af or genus Ga; Fontana (1962), Forma micorrizica n.1 or Forma micorrizica n.4; Haug and Oberwinkler (1987): Type 9: prosenchymatous type or type 15: puzzle type with cystidia; Ingleby *et al.* (1990), ITE. 1 or ITE. 4; Melin (1927), Mykorrhiza A or Mykorrhiza C; Voiry (1981), Mycorhize a feutrage + Hetre or Mycorhize a cordons epais + Hetre.

When such binary names for ectomycorrhiza are used it is necessary to avoid using identical names for different species. To help researchers an annual report will be published in the journal *Mycorrhiza*, compiling names already in use (Agerer, 1991b).

The binary names have already shown their scientific importance. Haug (1989) refers to *Piceirhiza gelatinosa*, indicating that this species of ectomycorrhiza was infected by another fungus.

E. *Colour atlas of ectomycorrhizae*

The number of species of ectomycorrhiza likely to occur throughout the world is very great. Estimates based upon assumed or actually proven ectomycorrhizal fungi result in more than 2000 species. Not included in this number are ectomycorrhiza of one fungal species on different trees that might influence their morphology as well as some anatomical features (see above). On the other hand several species are concentrated on only some tree genera, either because of compatibility, ecology or distribution. Thus such a high number of different ectomycorrhiza is not to be expected for each tree genus.

The *Colour Atlas of Ectomycorrhizae* is intended to be an international manual for identification of ectomycorrhiza. Thus—based on the tree genera—an ever-increasing number of ectomycorrhiza will be described and provided with keys for identification. A representative number of ectomycorrhiza will also be depicted in half-tone plates as well as in natural colour photographs. It is expected to include plates of ectomycorrhiza of all important woodland, forest and nursery situations throughout the world. Therefore all scientists who characterize ectomycorrhiza are invited to contribute to the *Colour Atlas of Ectomycorrhizae*. The following requirements should be satisfied (for further information already published material should be consulted):

- Comprehensive descriptions must already have been published elsewhere.
- Two half-tone plates of high quality, showing anatomical features in plan views of mantles, rhizomorphs and/or emanating hyphae, as well as characteristics of sections, must be provided.
- Four colour slides of ectomycorrhiza in natural colours, showing the most important features with respect to ramification, colour and surface, preferably at four different magnifications, must be provided.
- Legends must be provided for the pictures.
- If possible some additional information should be provided on the ecology of the ectomycorrhiza and on the fruit bodies.

Each author is responsible for his own plates. The publisher and the editor reserve the right to accept or reject plates for publication.

Acknowledgement

This chapter is dedicated to the late Professor Jack L. Harley, FRS.

References

Agerer, R. (1980). *Mycologia* **72,** 908–915.
Agerer, R. (1986a). *Mycotaxon* **26,** 473–492.
Agerer, R. (1986b). *Mycotaxon* **27,** 1–59.
Agerer, R. (1987a). In *Proceedings of the 7th North American Conference on Mycorrhizae* (D. M. Sylvia, L. L. Hung and J. H. Graham, eds), p. 78. Gainesville, FL.

Agerer, R. (1987b). *Mycologia* **79,** 524–539.
Agerer, R. (1987c). *Nova Hedwigia* **44,** 69–89.
Agerer, R. (1987d). *Mycotaxon* **28,** 327–360.
Agerer, R. (ed.) (1987–1990). *Colour Atlas of Ectomycorrhizae*, 1st–4th edn. Einhorn-Verlag, Schwäbisch Gmünd.
Agerer, R. (1988a). *Can. J. Bot.* **66,** 2068–2078.
Agerer, R. (1988b). *Nova Hedwigia* **47,** 311–334.
Agerer, R. (1990a). *Nova Hedwigia* **50,** 1–63.
Agerer, R. (1990b). *Crypt. Bot.* (in press).
Agerer, R. (1991a). *Nova Hedwigia* (in press).
Agerer, R. (1991b). *Mycorrhiza* **1** (submitted).
Agerer, R., Brand, F. and Gronbach, E. (1986). *AFZ* **20,** 497–503; 509.
Agerer, R. and Gronbach, E. (1988). In *Colour Atlas of Ectomycorrhizae* (R. Agerer, ed.), Plate 12. Einhorn-Verlag, Schwäbisch Gmünd.
Agerer, R. and Weiss, M. (1989). *Mycologia* **81,** 444–453.
Agerer, R., Treu, R. and Brand, F. (1990). *Proceedings of the 10th CEM*, Tallinn, 22 pp. (in press).
Alexander, I. J. and Fairley, R. J. (1983). *Plant Soil* **71,** 49–54.
Alexander, I. J., Taylor, A. F. S. and Ryan, E. A. (1990). IMC4, Abstract 105/2.
Ammirati, J. (1979). *Mycologia* **71,** 437–441.
Baral, H. -O. (1991). *Mycotaxon* (submitted).
Berg, B. (1990). Charakterisierung und Vergleich von Ektomykorrhizen gekalkter Fichtenbestände. Dissertation, University of Munich.
Berndt, R., Kottke, I. and Oberwinkler, F. (1990). *New Phytol.* **115,** 471–482.
Blaschke, H. (1987). *Z. Mykol.* **53,** 283–288.
Bohus, G. (1973). *Ann. Hist. Nat. Mus. Nation. Hung.* **65,** 63–81.
Bommer, C. (1896). *Mem. Acad. Roy. Sci. Belg.* **54,** 1–116.
Bonfante-Fasolo, P. (1973). *Myc. Appl.* **49,** 161–167.
Brand, F. (1989). *Nova Hedwigia* **48,** 469–483.
Brand, F. (1991). *Bibl. Mycol.* (in press).
Brand, F. and Agerer, R. (1986). *Z. Mykol.* **52,** 287–320.
Brand, F. and Agerer, R. (1988). *Sydowia Ann. Mycol.* **40,** (1987), 1–37.
Cairney, J. W. G., Jennings, D. H. and Agerer, R. (1991). *Crypt. Bot.* (submitted).
Chilvers, G. A. (1968). *Austral. J. Bot.* **16,** 49–70.
Chu-Chou, M. and Grace, L. J. (1983). *Austral. For. Res.* **13,** 121–132.
Chu-Chou, M. and Grace, L. J. (1985). *N. Z. J. Bot.* **23,** 417–424.
Clemençon, H. (1978). *Persoonia* **10,** 83–96.
Cunningham, G. (1972). *Mycologia* **64,** 906–911.
Danielson, R. M. (1982). *Can. J. Bot.* **60,** 7–18.
Danielson, R. M. (1984a). *Can. J. Bot.* **62,** 932–939.
Danielson, R. M. (1984b). *Mycologia* **76,** 454–461.
Dominik, T. (1959). *Mycopathol. Mycol. Appl.* **11,** 359–367.
Duddridge, J. A., Malibari, A. and Read, D. J. (1980). *Nature (Lond.)* **287,** 834–836.
Egli, S. (1981). *Schweiz. Z. Forstwes.* **132,** 345–353.
Erb, B. and Matheis, W. (1983). *Pilzmikroskopie.* Kosmos, Stuttgart.
Fassi, B. and De Vecchi, E. (1962). *Allionia* **8,** 133–151.
Fleming, L. V. (1983). *Plant Soil* **71,** 263–267.

Fontana, A. (1962). *Allionia* **8,** 67–85.
Frank, A. B. (1885). *Ber. Dtsch. Bot. Ges.* **3,** 128–145.
Froidevaux, L. (1973). *Can. J. For. Res.* **31,** 601–603.
Gerlach, D. (1969). *Botanische Mikrotechnik. Eine EinfÜhrung*. Thieme, Stuttgart.
Gibelli, G. (1883). *Mem. Reale Acad. Sci. Inst. Bologna* **4,** 287–314.
Giovanenetti, G. and Fontana, A. (1981). *Allionia* **24,** 13–17.
Godbout, C. and Fortin, J. A. (1983). *New Phytol.* **94,** 249–262.
Godbout, C. and Fortin, J. A. (1985). *Can. J. Bot.* **63,** 252–262.
Gronbach, E. (1988). *Bibl. Mycol.* **125,** 1–216.
Gronbach, E. and Agerer, R. (1986). *Forstwiss. Cbl.* **105,** 230–233.
Haug, I. (1989). *New Phytol.* **111,** 203–207.
Haug, I. and Oberwinkler, F. (1987). *Trees* **1,** 172–188.
Haug, I., Kottke, I. and Oberwinkler, F. (1986). *Z. Mykol.* **52,** 373–392.
Holmgren, P. K., Keuken, W. and Schofield, K. (1981). *Regn. Veget.* **106,** 1–452.
Ingleby, K., Mason, P. A., Last, F. T. and Fleming, L. V. (1990). *Identification of Ectomycorrhizas*. HMSO, London.
Jülich, W. (1985). *Int. J. Mycol. Lichenol.* **21,** 123–140.
Kammerbauer, H., Agerer, R. and Sandermann, H. (1989). *Trees* **31,** 78–84.
Kottke, I. and Oberwinkler, F. (1986). *Trees* **1,** 1–24.
Kreger-Van Rij, N. J. W. and Veenhuis, M. (1971). *J. Gen. Microbiol.* **68,** 87–95.
Malajczuk, N. and Cromack, K. (1982). *New Phytol.* **92,** 527–531.
Massicotte, H. B., Ackerley, C. A. and Peterson, R. L. (1985). *Proc. Microsc. Soc. Can.* **12,** 68–69.
Massicotte, H. B., Peterson, R. L., Ackerley, C. A. and Piché, Y. (1986). *Can. J. Bot.* **64,** 177–192.
Massicotte, H. B., Melville, L. H. and Peterson, R. L. (1987). *Scann. Microsc.* **1,** 1439–1454.
Masui, K. (1926a). *Mem. Coll. Sci. Kyoto Imp. Univ. Ser. B* **2,** 161–187.
Masui, K. (1926b). *Mem. Coll. Sci. Kyoto Imp. Univ. Ser. B* **2,** 190–209.
Meixner, A. (1975). *Chemische Farbreaktionen von Pilzen*. J. Cramer, p. 66.
Melin, E. (1927). *Medd. Skogsf. Anst. Stockh.* **23,** 433–494.
Miller, O. K. (1983). *Can. J. Bot.* **61,** 909–916.
Miller, S. L. and Miller, O. K. (1984). *Can. J. Bot.* **62,** 2363–2369.
Moser, M. (1978). In *Kleine Kryptogamenflora* (H. Gams, ed. IIb/2. Fischer, Stuttgart and New York.
Müller, W. and Agerer, R. (1990a). *Nova Hedwigia* **51,** 381–410.
Müller, W. and Agerer, R. (1990b). *Crypta. Bot.* **2,** 64–68.
Ogawa, M. (1981). *Proc. IUFRO* **17,** 89–95.
Peyronel, B. (1934). *Nuov. Giorn. Bot. Ital. n. S.* **41,** 744–746.
Peyronel, B. (1963). In: *Mykorrhiza*, International Symposium, Weimar, 1960, pp. 16–25. Fischer, Jena.
Read, D. A. (1984). In *The Ecology and Physiology of the Fungal Mycelium* (D. H. Jennings and A. D. Rayner, eds), pp. 215–240. Cambridge University Press, Cambridge.
Read, D. J., Francis, R. and Finlay, R. D. (1985). In *Ecological Interactions in Soil*, Special Publication of the British Ecological Society, Vol. 4, pp. 193–217.

Schenck, N. (ed.) (1982). *Methods and Principles of Mycorrhizal Research*. American Phytopathological Society, St Paul, MN.
Schenck, N. C. and Perez, Y. (1987). *Manual for the Identification of VA Mycorrhizal Fungi*. INVAM, Gainesville, FL.
Schramm, J. R. (1966). *Trans. Am. Phil. Soc.* **56,** 1–190.
Singer, R. (1986). *Agaricales in Modern Taxonomy*. Koeltz, Koenigstein.
Stenström, E. (1990). Ecology of mycorrhizal *Pinus silvestris* seedlings—aspects of colonization and growth. Dissertation, Uppsala University.
Strullu, D. G. (1985). *Encycl. Plant Anat.* **13,** 1–198.
Taylor, A. F. S. and Alexander, I. J. (1989). *Mycol. Res.* **92,** 103–107.
Thoen, D. (1977). *Bull. Br. Mycol. Soc.* **11,** 39–43.
Trappe, J. (1965). *For. Sci.* **11,** 27–32.
Trappe, J. (1967). *Proc. Int. Union For. Res.* **24,** 46–59.
Trappe, J. (1977). *Ann. Rev. Phytopathol.* **15,** 203–222.
Trappe, J. and Schenck, N. C. (1982). In *Methods and Principles of Mycorrhizal Research* (N. Schenck, ed.), pp. 1–9. American Phytopathological Society, St Paul, MN.
Treu, R. (1990a). *Bibl. Mycol.* **134,** 1–196.
Treu, R. (1990b). In *Colour Atlas of Ectomycorrhizae* (R. Agerer, ed.), Plate 46. Einhorn-Verlag, Schwäbisch Gmund.
Treu, R. (1990c). In *Colour Atlas of Ectomycorrhizae* (R. Agerer, ed.), Plate 47. Einhorn-Verlag, Schwäbisch Gmünd.
Tyler, G. (1985). *For. Ecol. Management* **10,** 13–29.
Uhl, M. (1988a). Identifizierung und Charakterisierung von Ektomykorrhizen an *Pinus silvestris* und von Ektomykorrhizen aus der Gattung Tricholoma. Dissertation.
Uhl, M. (1988b). *Persoonia* **13,** 449–458.
Uhl, M. and Agerer, R. (1987). *Nova Hedwigia* **45,** 509–527.
Voiry, H. (1981). *Eur. J. For. Pathol.* **11,** 284–299.
Walker, C. (1985). *Trans. Br. Mycol. Soc.* **84,** 353–355.
Weiss, M. (1990). *Hedwigia* **50,** 361–393.
Weiss, M. and Agerer, R. (1988). *Eur. J. For. Pathol.* **18,** 26–43.
Wilcox, H. (1982). In *Methods and Principles of Mycorrhizal Research* (N. Schenck, ed.), pp. 103–113. American Phytopathological Society, St Paul, MN.
Zak, B. (1969). *Can. J. Bot.* **47,** 1833–1840.
Zak, B. (1971). *Can. J. Bot.* **49,** 1079–1084.
Zak, B. (1973). *Ectomycorrhizae, Their Ecology and Physiology* (G. Marks and T. T. Kozlowski, eds), pp. 43–78. Academic Press, New York and London.

3

Techniques in Synthesizing Ectomycorrhiza

R. L. PETERSON and P. CHAKRAVARTY

Department of Botany, University of Guelph, Guelph, Ontario N1G 2W1, Canada

I. Introduction

In order to exploit ectomycorrhiza for increased yield of commercially important tree species, more research is required on the interaction between the symbionts and the screening of potential host–fungus combinations. This can only be accomplished by being able to synthesize ectomycorrhiza between known symbionts under controlled conditions. Depending on the questions to be asked, either sterile or non-sterile methods can be used. An understanding of the structure and functioning

METHODS IN MICROBIOLOGY
VOLUME 23 ISBN 0-12-521523-1

Copyright © 1991 by Academic Press Limited
All rights of reproduction in any form reserved

of ectomycorrhiza in natural ecosystems is the ultimate goal of research on ectomycorrhiza and therefore methods of synthesis approximating conditions in nature are desirable. However, for some questions including early events in colonization, elicitation of biochemical responses by ectomycorrhizal fungi, and ultimately the genetic control of mycorrhiza formation, it is necessary to use sterile, and therefore somewhat artificial, conditions. Although there have been previous discussions on techniques of synthesizing ectomycorrhiza (Molina and Palmer, 1982; Fortin *et al.*, 1983; Piché and Peterson, 1988), this chapter attempts to bring together all of the major methods developed with illustrations of the apparatus used. In addition, each method will be discussed in terms of advantages and disadvantages for particular studies.

II. Sterile techniques

A. Erlenmeyer flasks

Some of the first attempts at synthesizing ectomycorrhiza under sterile conditions were made by Melin (1921, 1922) using Erlenmeyer flasks with sand as a substrate for addition of nutrient solution and for root growth. Subsequently, Hacskaylo (1953) used vermiculite while Marx and Zak (1965) used finely ground peat moss mixed with vermiculite in a ratio of 1:15 as the substrate. The substrates provided better aeration for the root system and the addition of peat lowered the pH. Chakravarty and Unestam (1987a,b) achieved excellent ectomycorrhiza formation with *Pinus sylvestris* using "Leca" brick pellets as the substrate. Recently, Chakravarty *et al.* (1990) have used Promix Bx, a commercial mix of vermiculite–peat moss–perlite (Les Tourbières Premier Ltd, Riviere-du-Loup, Quebec, Canada) as the substrate.

In setting up the system, any size of flask can be used and the amount of substrate and nutrient solution added can be adjusted accordingly. In earlier experiments (Marx and Zak, 1965), 2-litre flasks containing a ratio of 840 ml vermiculite: 60 ml ground peat moss moistened with 550 ml nutrient solution were used. In later experiments (Chakravarty *et al.*, 1990), 250 ml flasks containing 100 ml Promix Bx moistened with 90 ml nutrient solution were used for 8-week experiments using *Pinus resinosa* Ait. seedlings.

Flasks of any size containing substrate and nutrient solution are plugged either with foam or absorbent cotton and capped with a small beaker or aluminium foil (see Fig. 1) before autoclaving at 120 °C, 1 atm, for at least 15 min.

After the substrate has cooled, one or more aseptic seeds or seedlings are introduced into each flask under sterile conditions. Fungal inoculum can be added as plugs of actively growing mycelium or as a slurry of fungal hyphae at this time or after seedling establishment, depending on the objective of the experiment.

1. Advantages

This method is very easy to set up so that a large number of replicates can be used. Sterility is easily maintained during all phases of establishing this culture system and, once set up, maintenance requires little effort. Root development and ectomycorrhiza formation are excellent in a substrate such as vermiculite–peat moss or Promix Bx that allows good aeration and a buffered acidic environment conducive to growth of conifer seedlings. This system has been used very effectively in studying interactions between ectomycorrhizal fungi and root pathogens.

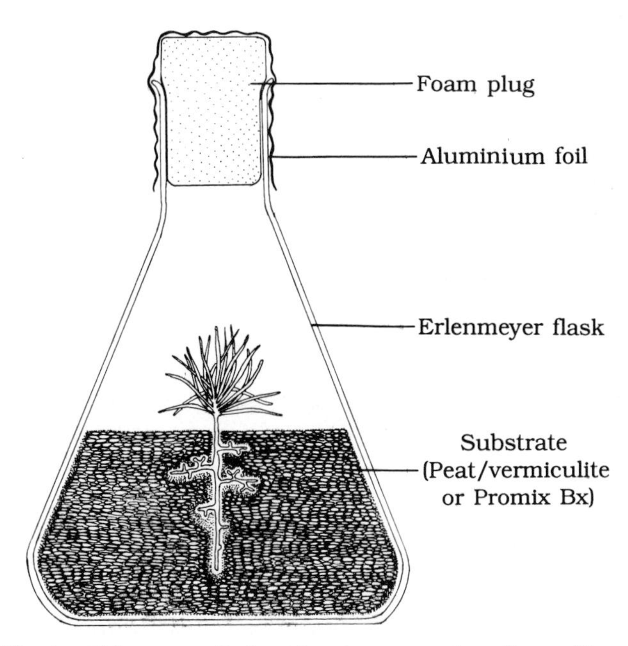

Fig. 1. Erlenmeyer flask with substrate and sterile seedling.

2. Disadvantages

Probably the biggest disadvantage of this system is the artificial environment imposed on the shoot system because, in a closed system, CO_2 might be limiting and ethylene might be expected to accumulate. Similarly, root exudates would accumulate in the substrate. This system is also not suitable for developmental studies of ectomycorrhiza and the extramatrical mycelium, because the root system is difficult to monitor without sacrificing the seedling. The flasks also occupy a considerable surface area of a growth chamber.

B. Mason jars

Trappe (1967), realizing the potential problems in growing conifer seedlings in enclosed chambers, devised a method to keep the root system sterile yet allowing the shoot system to grow into the environment. The assembly (Fig. 2) consists of a wide-mouth quart mason jar (canning jar) filled with perlite sieved to include particles 1–8 mm diameter. Nutrient solution is added to the perlite to moisten it and to leave about 5 cm depth of excess solution at the bottom. A steel vent tube (2 cm diameter, 3 cm long) is inserted into the perlite so that the top of the tube is flush with the lip of the jar. Perlite is removed from the vent tube before the jar top is covered with lightweight aluminium foil. A hole is poked through the foil at the site of the vent tube and the tube sealed to the foil by rubber cement (sterile lanolin could be used after autoclaving the assembly). The foil is folded over the outside of the jar's lip and held in place by a mason jar ring of appropriate size screwed down tightly over it. The vent tube is plugged with absorbent cotton, a glass Petri plate is placed over the top of the jar, and the whole assembly is autoclaved.

After cooling, the Petri plate is removed and a small hole poked aseptically into the foil and into the perlite near the rim of the jar. A sterile seedling radicle is placed in the hole and the hypocotyl is sealed into the chamber. Although Trappe (1967) used rubber cement to seal the seedling, some necrosis occurred. Sterile lanolin should probably be used. Jars are wrapped in foil and placed in growth chambers for seedling growth.

A later modification of the system involved lining the jar with paper towelling to improve capillary movement of nutrient solution.

Seedlings are inoculated with test fungi by inserting the needle of a syringe containing a slurry of fungal mycelium through the foil and injecting the slurry into the perlite adjacent to the root system.

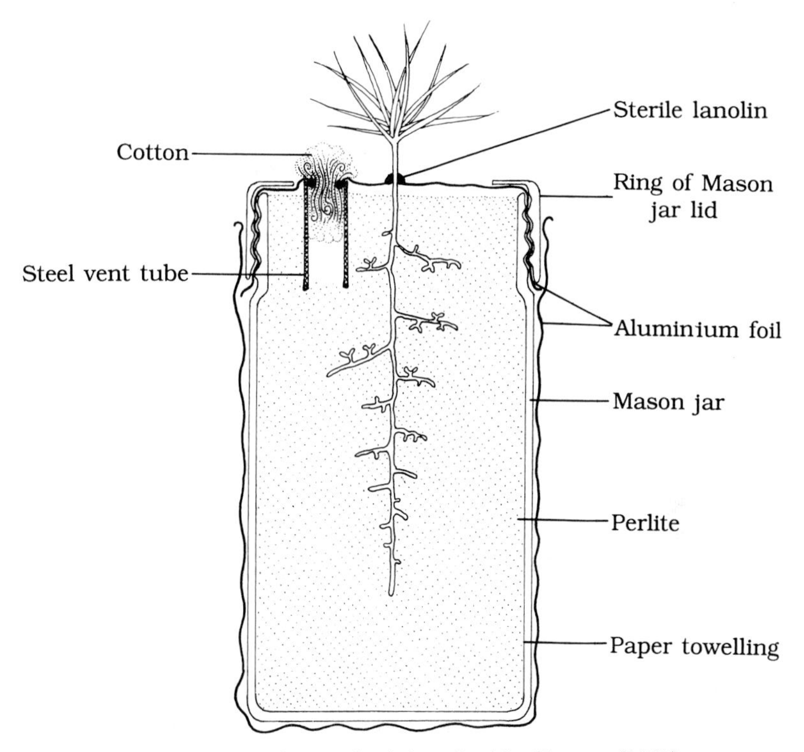

Fig. 2. Mason jar method described by Trappe (1967).

1. *Advantages*

The main advantages of this system are that the shoot system grows under ambient conditions, roots growing against the glass can be observed conveniently, and the fungus can be distributed to a large portion of the root system.

2. *Disadvantages*

These assemblies take some effort to set up, particularly when adding the vent system. The perlite substrate has poor capillarity and some fungal species do not recover very well after fragmentation. A simpler system involves using a mason jar filled with either peat moss–vermiculite or Promix Bx moistened with nutrients, leaving out the vent tube, and using the lid assembly of the jar with a hole punched in it (see Fig. 3). The top of the assembly can be covered with aluminium foil during

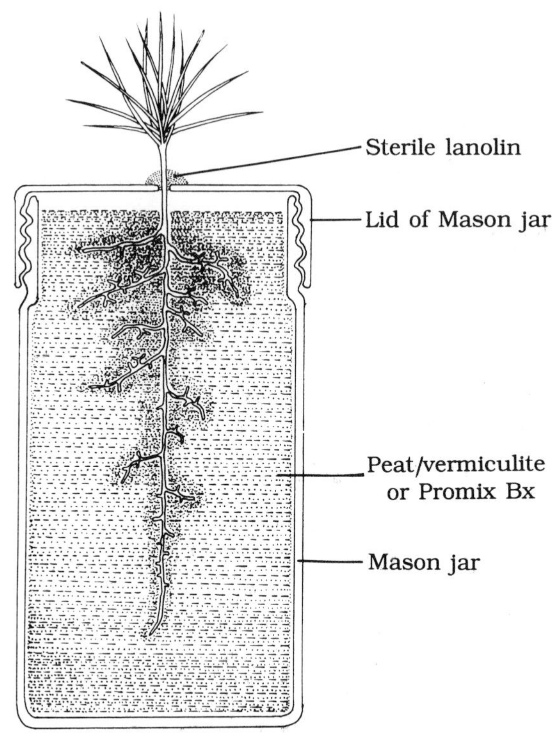

Fig. 3. Simpler mason jar method.

autoclaving which can be removed prior to inserting the seedling into the hole and sealing it in with autoclaved lanolin.

C. Leonard jars

Mullette (1976) adapted a method first used by Vincent (1970) in studying nodulation to synthesize ectomycorrhiza between *Eucalyptus gummifera* (Gaertn. & Hochr.) and *Pisolithus tinctorius* (Pers.) Coker & Couch under different phosphorus concentrations. The apparatus, illustrated in Fig. 4, consists of a pyrex jar containing nutrient solution into which an inverted bottle with its base removed and its neck stoppered with absorbent cotton is inserted. The inverted bottle contains crushed quartz (quartz sand) or soil overlain by coarse quartz. The entire assembly can be covered with aluminium foil and autoclaved. After cooling, the aluminium foil is removed from the top, a sterile seedling is placed in the quartz, and fungal inoculum placed in the quartz. Mullette

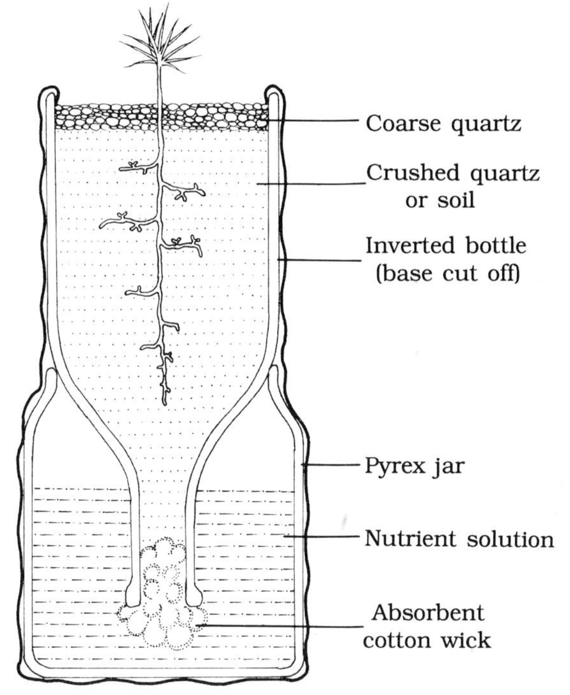

Fig. 4. Leonard jar method. Redrawn from Mullette (1976).

(1976) used crushed fruit bodies of *Pisolithus tinctorius* as inoculum. The cotton acts as a wick to provide nutrients to the developing root system.

1. Advantages

Mullette (1976) claims that it is easy to retain sterility during long-term experiments and since a quantity (e.g. 1 litre) of nutrient solution can be used there is very little maintenance of the system. Another advantage is that the shoot system is exposed to ambient conditions.

2. Disadvantages

The main disadvantages of the system are the glassware required, the time involved to set up the system and the space required to maintain a suitable number of replicates for experiments. Sterility may be a problem during long-term experiments.

D. Test tubes

Pachlewska (1968) and Pachlewski and Pachlewska (1974) used test tubes containing water agar plus thiamine but without minerals to synthesize ectomycorrhiza between *Pinus sylvestris* and fungal species. This method was modified by Mason (1975, 1980) by adding minerals and glucose in addition to thiamine to the agar medium. Molina (1979, 1981) used 300 × 38 mm glass test tubes containing vermiculite and peat moistened with nutrient solution (see Fig. 5) to test a wide range of ectomycorrhizal fungi on *Alnus* species. He reported excellent seedling growth under these conditions. Promix Bx could be used instead of vermiculite–peat. Sohn (1981) used a mixture of quartz sand and ion-exchange resins in a similar setup.

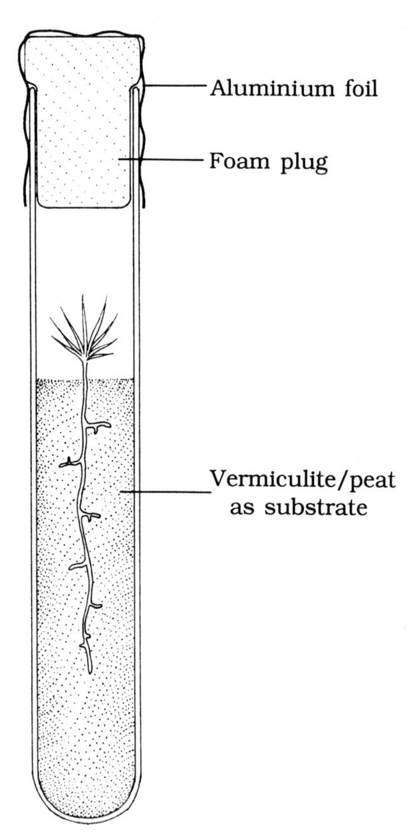

Fig. 5. Test tube system using vermiculite–peat as substrate. After Molina (1979).

Sylvia and Sinclair (1983a) designed a wick-culture system to test the effects of *Laccaria laccata* (Scop.:Fr.) Berk & Br. on *Pseudotsuga menziesii* (Mirb.) Franco (Douglas fir) seedling growth. This system was used subsequently (Sylvia and Sinclair, 1983b) to study the interaction between *L. laccata* and root pathogens, again with Douglas fir. The wick-culture system (Fig. 6) is constructed by lining a 200 × 32 mm test tube with a layer of cellophane, Whatman No. 3 chromatography paper and polypropylene in that order. The chromatography paper acts as a wick to keep the root system moist while the cellophane and polypropylene allow the root system to be removed from the tube without damage. A stainless steel rod is placed between the cellophane and chromatography paper for subsequent insertion of the seedling radicle

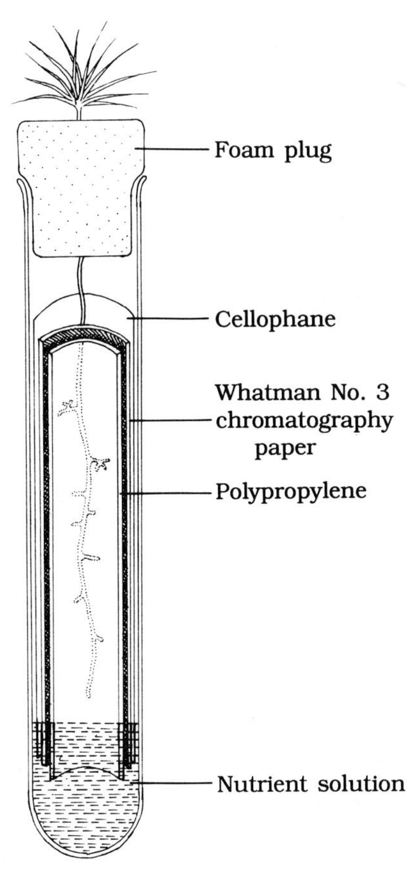

Fig. 6. Test tube system described by Sylvia and Sinclair (1983a).

and for adding the slurry of fungal hyphae. Nutrient solution (40 ml) is added, the tube plugged with a foam plug and the assembly autoclaved. A sterile seedling is added between the foam plug and the wall of the test tube so that the shoot system is exserted into the environment. The shoot system is enclosed within a plastic bag for 2 weeks to prevent desiccation.

Yang and Wilcox (1984) used an apparatus (Fig. 7) combining features of the simple test tube system used by Molina (1979, 1981) with some aspects of the system designed by Sylvia and Sinclair (1983a). These authors lined test tubes with chromatography paper and then

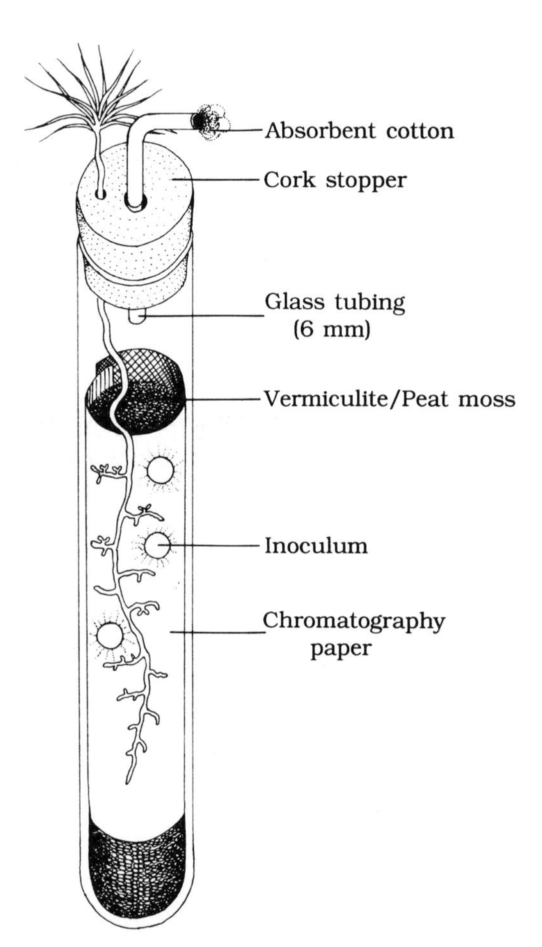

Fig. 7. Test tube system combining features shown in Figs 5 and 6. After Yang and Wilcox (1984).

filled the tube with a vermiculite–peat mixture moistened with sterile water. Tubes were plugged with a cork stopper in which a bent glass tube plugged with cotton was inserted for aeration. Another hole bored into the cork near the rim of the tube was plugged with cotton during autoclaving. After autoclaving, the cotton was removed from the latter hole and a sterile seedling was inserted so that the radicle was between the test tube wall and the chromatography paper. Nutrient solution and 0.5% glucose was added at this time, as were two agar plugs of mycelium, one at the base of the seedling and the other in the vermiculite–peat mixture. Although not mentioned in the description, the photograph of the system indicates that parafilm was wrapped around the base of the cork.

Duchesne *et al.* (1988a) devised a simple test tube system to study the short-term effects of the ectomycorrhiza fungus, *Paxillus involutus* Fr,. on the root-rot organism *Fusarium oxysporum* in combination with *Pinus resinosa* Ait. seedlings. The apparatus (Fig. 8) consists of 1.5 × 18 cm test tubes lined either completely or partially with Whatman No. 2 chromatography paper, a foam plug and either a plastic cap or aluminium foil. After autoclaving, 5 ml of sterile nutrient solution is added to each tube and a seedling placed between the filter paper and test tube wall. This apparatus was very useful for studying the effect of root exudates on anti-fungal compound secretion by *P. involutus* (Duchesne *et al.*, 1988b) and for structural studies (Farquhar and Peterson, 1989).

1. Advantages

The test tube systems in which the shoot system is allowed to grow into the environment have the advantage that shoots are not stressed by high levels of accumulated ethylene or low levels of CO_2. The test tubes can be maintained at constant temperature by immersing them in a water bath. All the test tube systems have the advantage that a large number of samples can be run for any experiment because of the small amount of space required and the root systems can be monitored as they grow against the test tube wall. The systems employing chromatography paper wicks are excellent for studies on root exudates and interactions between ectomycorrhizal fungi and pathogenic organisms.

2. Disadvantages

The test tube systems in which the shoot is enclosed suffer, like all similar apparatus with enclosed shoots, in that CO_2 may be limiting to

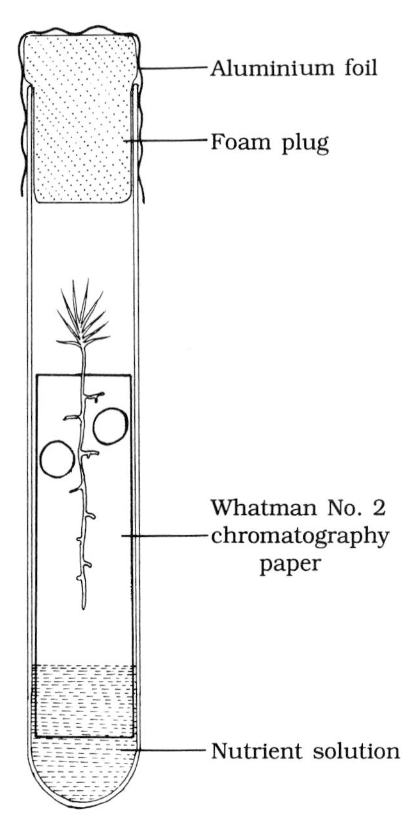

Fig. 8. Simple test tube system using a paper wick. Redrawn from Duchesne *et al*. (1988a).

growth and ethylene might accumulate. All of the systems have the potential for the accumulation of phytotoxic substances in the nutrient medium.

E. Petri plates

1. Simple system

Petri plates of various sizes, configurations and modifications have been used extensively in the synthesis of ectomycorrhiza. A simple but effective method used by Duddridge (1986) and later by Stenström and Unestam (1987) involves filling the plates with either a mixture of peat–vermiculite (1:4) (Duddridge, 1986) or "Leca" brick pellet-peat (Stenström and Unestam, 1987) moistened with a suitable nutrient solution and inserting a sterile seedling into a cut in the plate so that the shoot remains in the environment (Fig. 9). The shoot is sealed in with sterile lanolin and the plates wrapped with PVC tape (Duddridge, 1986) or any suitable alternative. This method was very useful for studying the effects of flooding on ectomycorrhiza formation since roots in the bottom of an upright plate can be flooded while the roots towards the upper part of the plate can be kept under moist but not flooded conditions and therefore act as controls (Stenström and Unestam, 1987).

2. *Divided Petri plates*

Brownlee *et al.* (1983) and Boyd *et al.* (1986) used divided Petri plates to study the effect of mycelium strands on leaf water potential and photosynthesis. The divided Petri plate system (Fig. 10) contained moistened peat, usually on both sides of the plate, into which the root system of the seedling is allowed to grow. Plates with roots in the upper half but only mycelium of the inoculated fungus in the lower half were used to determine the effect of severing the mycelial strands on water potential (Brownlee *et al.*, 1983; Boyd *et al.*, 1986) and photosynthesis in the shoot (Boyd *et al.*, 1986). This system is very adaptable in that various substrates can be used on either side of the plate, depending on the objectives of the experiment.

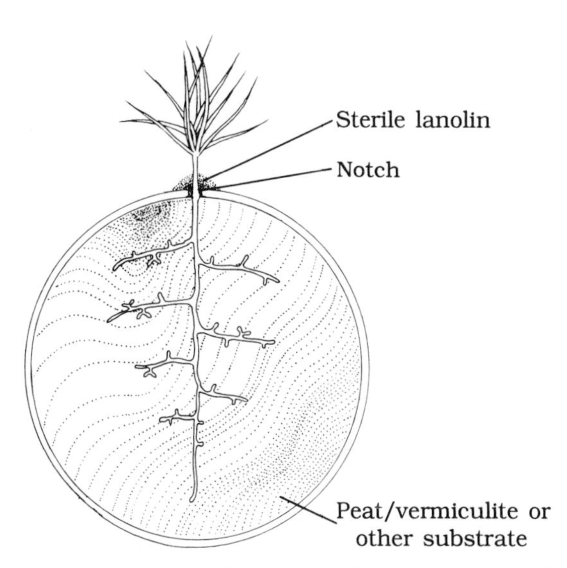

Fig. 9. Petri plate method to maintain a sterile root system. After Duddridge (1986).

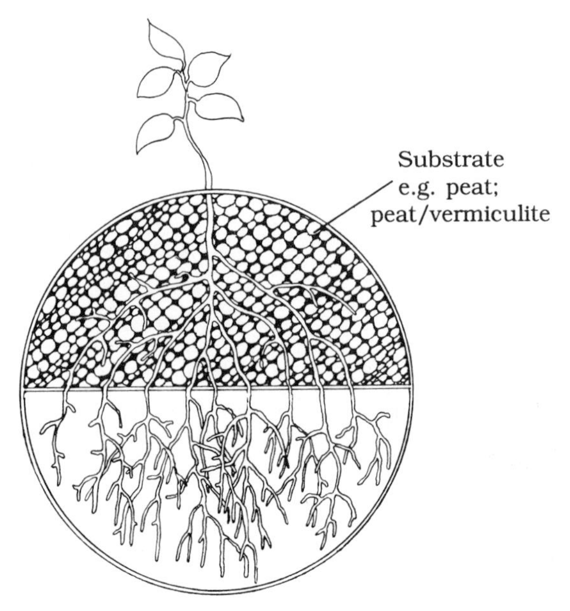

Fig. 10. Divided Petri plate method. This method can be used to produce either mycelium only in the lower half or roots and mycelium. After Brownlee *et al*. (1983).

3. *Sandwich technique*

Chilvers *et al*. (1986) developed a paper sandwich method (Fig. 11) for synthesizing ectomycorrhiza of *Eucalyptus*. Large (150 mm diameter) Petri plates containing a 5 mm layer of mineral agar had sterile Whatman No. 1 filter papers laid over the agar surface. Aseptic seedlings were placed on this filter paper towards one edge of the plate. The radicle was then covered with another sterile piece of filter paper cut across the top to leave the shoot exposed. The plates were then sealed and placed in the growth chamber to allow root development. In preparation of inoculation of seedlings, fungal inoculum was grown on autoclaved squares (60 × 60 mm) of stiff absorbent paper ("beer-mat board") laid over a nutrient medium. At the time of seedling inoculation, the filter paper covering the seedling was removed and the cardboard squares were placed, inoculum side down, over the root system. Plates were then returned to the growth chamber.

A slight modification of this technique has been used by Boudarga *et al*. (1990) for the dual colonization of *Eucalyptus* roots by ectomycorrhizal and vesicular-arbuscular mycorrhizal fungi.

4. Nylon mesh method

Wong and Fortin (1989) have modified the Petri plate method by using large round plastic plates (150 × 15 mm) filled with 80 ml sugar-free nutrient medium. A piece (9.5 cm^2) of sterilized Swiss Nitex nylon membrane (150 μm mesh; obtained from B & SM Thompson & Co. Ltd., Ville Mont-Royal, Quebec, Canada) is placed on the surface of the agar and another of the same size on top of the root system of a seedling placed on the bottom membrane but with the shoot protruding into the environment through a notch in the plate (Fig. 12). A piece of sterilized Whatman No. 1 filter paper is placed over the upper nylon membrane and four (1 × 4 cm) sterilized cotton rolls are placed in the dish opposite the shoot to absorb excess moisture. Plates are then sealed with parafilm, placed upright (standing) in a closed acrylic box with two small holes cut into each long side to allow some air exchange but to reduce transpiration, and the box placed in a growth chamber. After 4 weeks, Petri plates are opened under aseptic conditions, the filter paper removed, and a plug of fungal mycelium placed on the upper nylon membrane adjacent to lateral roots.

5. Seedlings in plates of fungal mycelium

Malajczuk *et al.* (1990) have used a very simple system for rapid colonization of *Eucalyptus* seedlings. Small sterile seedlings are placed directly on mycelial mats in culture plates and within 3–6 h of contact, hyphae colonize the root surface. Within 3 days the mycelium envelopes seedling roots.

(a) Advantages. Petri plates in general are advantageous in that they are inexpensive, easy to prepare and take up little room. It is easy to modify them so that the shoot system can be grown under ambient conditions. Root systems can be monitored easily for ectomycorrhiza development and samples can be obtained at different stages for structural and developmental studies.

(b) Disadvantages. The sandwich technique and the nylon mesh method require a considerable amount of manipulation and therefore the chances of contamination are high. The latter method, however, provides easy access to the ectomycorrhiza for external observations and for structural studies.

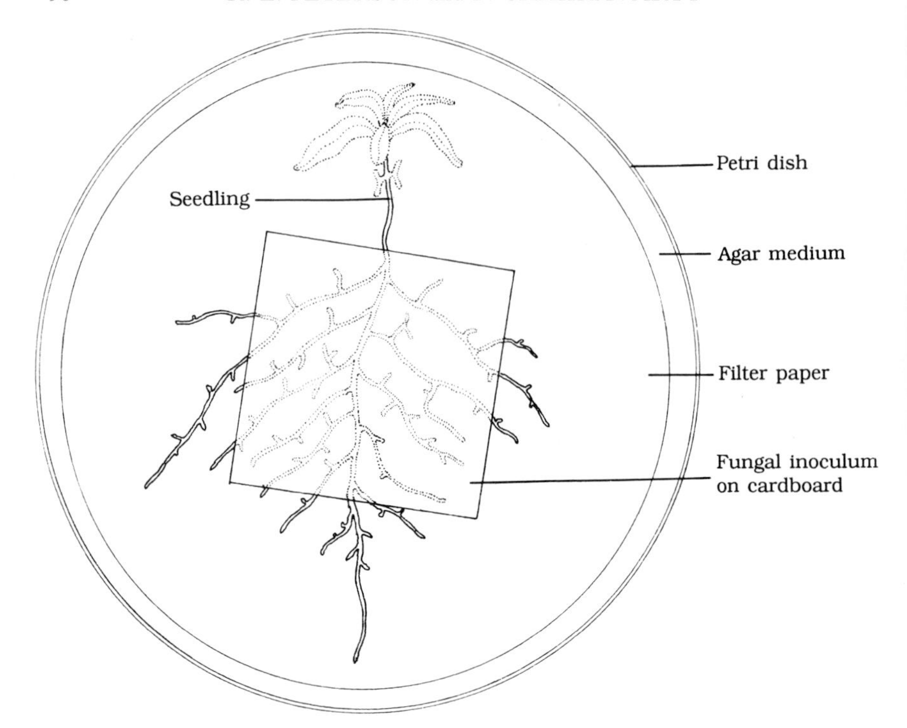

Fig. 11. Paper sandwich method described by Chilvers *et al*. (1986).

F. Petri plates—organ explants

Raggio and Raggio (1956) introduced a method to treat the base and apex of excised roots simultaneously with media of different composition. This method has been used frequently for studies of nodulation and factors controlling root development. Fortin (1966) and Fortin and Piché (1979) modified this method slightly by retaining a portion of the hypocotyl of *Pinus sylvestris* L. or *Pinus strobus* L., respectively, on the excised root. The cut hypocotyl is placed in a small vial containing an organic medium including sucrose to mimic shoot-derived nutrients. The root portion, with the apical meristem removed to induce lateral roots, is placed on sterilized vermiculite moistened with an inorganic nutrient medium (Fig. 13). Fungal inoculum in the form of a mycelium slurry is then added to the vermiculite.

Bailey and Peterson (1988) modified the system by using divided Petri plates with notches cut into the partition so that the base of an excised root, in this case *Eucalyptus pilularis*, is exposed to nutrient medium

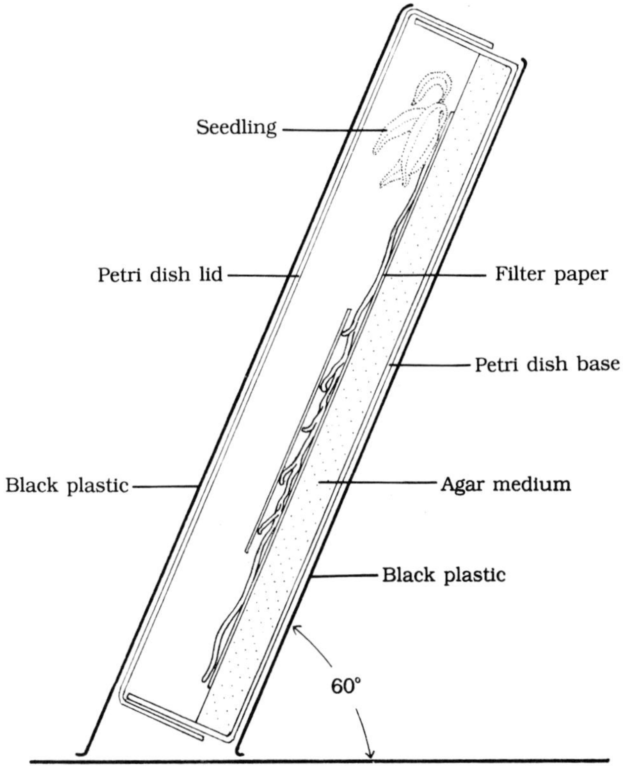

Fig. 11.

while the apex is grown on a filter paper placed over the nutrient medium (Fig. 14). A plug of fungal mycelium (*Pisolithus tinctorius* (Pers.) Coker & Couch) is added onto the filter paper. Excellent mantle and Hartig net formation occurred along a considerable length of the excised root.

Louis and Scott (1987) induced adventitious roots on cultured excised cotyledon pieces of *Shorea roxburghii* G. Don, a member of the Dipterocarpaceae, and used these to synthesize ectomycorrhiza with the fungus, *Rhodophyllus* sp. The technique involves suspending excised roots over U-shaped glass rods placed on the surface of agar medium so that only the apical portion of the root is in contact with agar medium (Fig. 15). A plug of mycelium is placed in the vicinity of the root apices, the plates sealed and wrapped in aluminium foil.

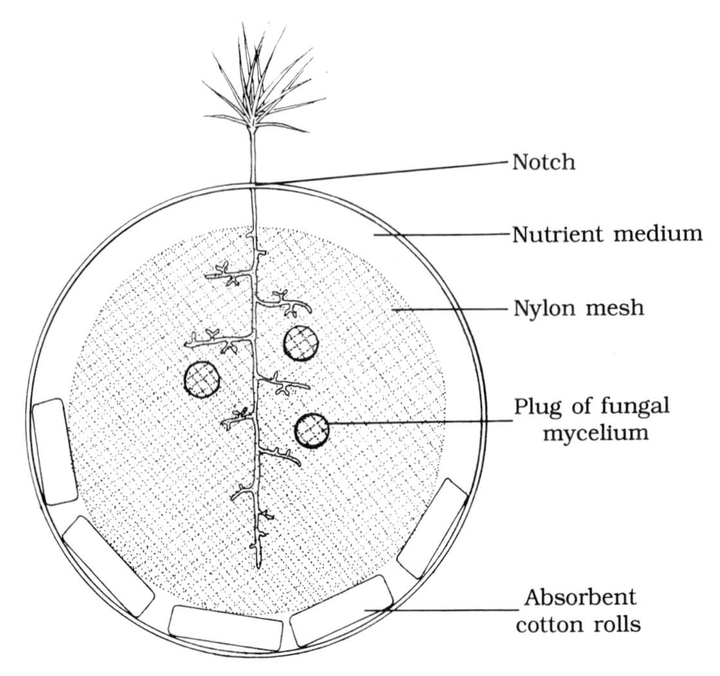

Fig. 12. Petri plate method using large (150 × 15 mm) plates and nylon mesh. After Wong and Fortin (1989).

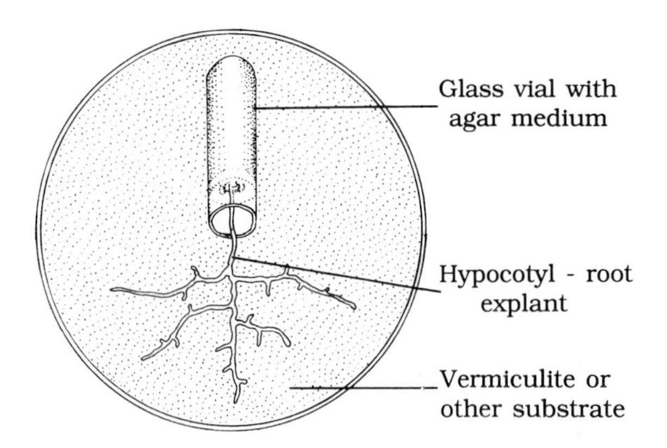

Fig. 13. Method for growing root or hypocotyl–root explants, originally described by Raggio and Raggio (1956).

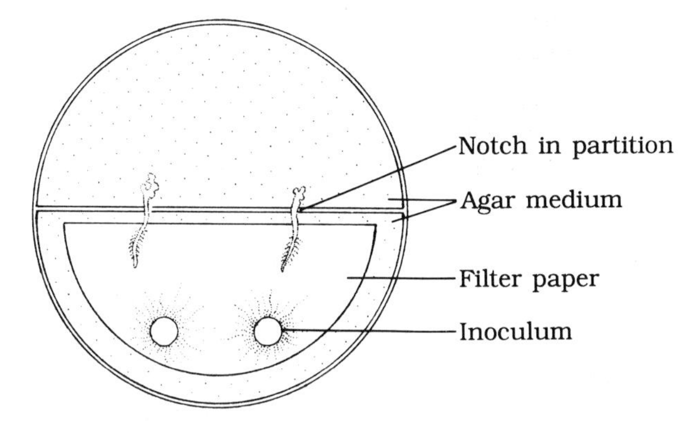

Fig. 14. Divided Petri plate method for colonization of root explants. After Bailey and Peterson (1988).

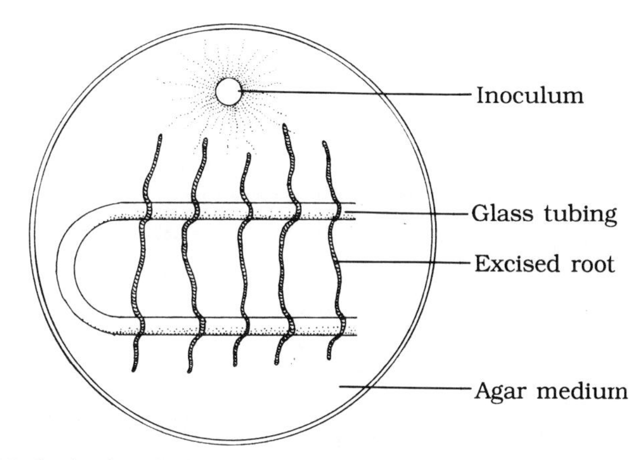

Fig. 15. Method of colonizing excised roots described by Louis and Scott (1987).

1. Advantages

Root or root–hypocotyl explants are useful to study early biochemical and structural changes in roots colonized by ectomycorrhizal fungi since any interference from the shoot is avoided. The systems used are axenic so changes in surface exudates of both symbionts can be studied. If adventitious roots are used, clonal material is available.

2. *Disadvantages*

Unfortunately excised roots of woody species, except *Pinus radiata*, are difficult to grow *in vitro*. Until this problem can be overcome, only a limited number of species can be cultured.

III. Non-sterile systems

A. Growth pouches

The details of the use of growth pouches for the synthesis of ectomycorrhiza were first published by Fortin, Piché and Lalonde (1980) and discussed subsequently by Fortin, Piché and Godbout (1983) and Piché and Peterson (1988). The method, first described by Porter *et al*. (1966),

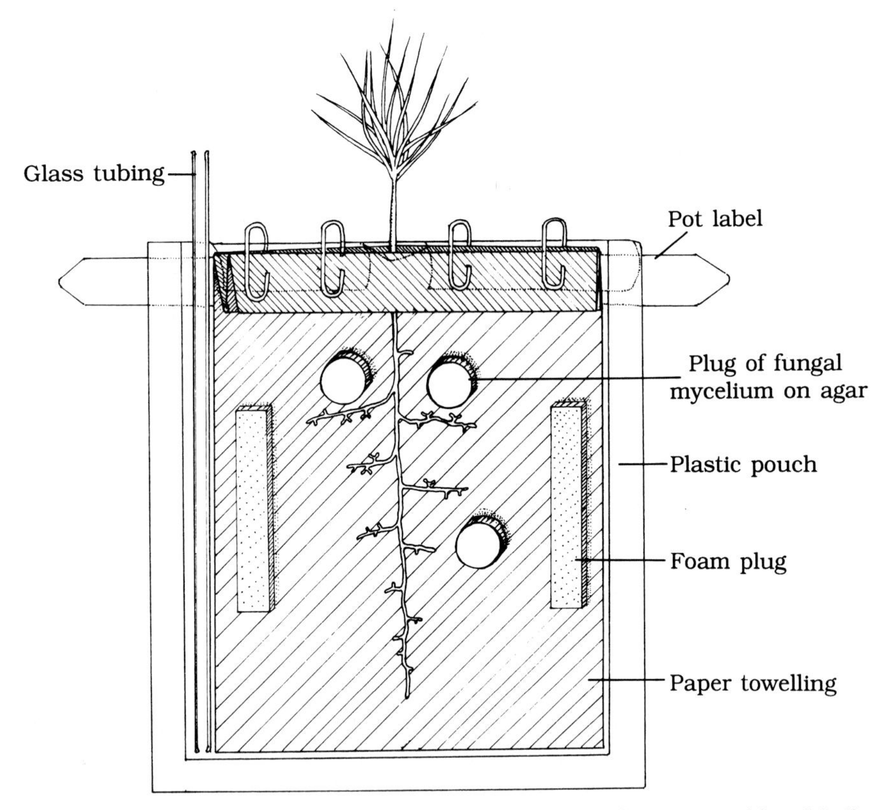

Fig. 16. Plastic growth pouch technique for synthesis of ectomycorrhiza. Modified from Fortin *et al*. (1980).

had been used previously in studying nodulation in legumes (Weavers and Frederick, 1972) and in actinorrhizal plants (Lalonde, 1979). The initial design by Fortin *et al.* (1980) has been modified in various ways over the years so that the apparatus currently used routinely by us is illustrated in Fig. 16. The pouches consist of a transparent plastic bag enclosing an absorbent paper wick (available from Northrup King Co., PO Box 959, Minneapolis, MN 55440, USA). These can be modified by placing a glass tube between the plastic sheets through which nutrients can be added by a hypodermic needle attached to a syringe, adding foam plugs between the plastic sheets to provide better aeration for the root system, and by stapling two pot labels to the top of the pouch so that the pouches can either be supported on wires or placed in boxes of appropriate size. If very delicate seedlings are to be placed in the pouches, a V-shaped notch can be cut into the plastic at the top so that the seedling will not be damaged. The paper wick is then moistened with distilled water and groups of pouches are wrapped in aluminium foil before autoclaving on a "wet cycle" setting.

For the synthesis of ectomycorrhiza, seedlings grown under sterile conditions are placed between the plastic and the paper wick. Seedlings should have a well-established root system. Pouches are then placed in a growth room for approximately one or more weeks before inoculation with the appropriate fungal species. Figure 17 outlines the procedure for preparing the fungal inoculum and inoculating the seedling root system. The pouches should be monitored daily to ensure that the paper wick remains moist and to remove pouches that may show contamination. Nutrients can be added weekly once the root system is established.

1. Advantages

Numerous seedlings can be grown in a very small space so that large trials can be run for growth experiments or for developmental studies. This system is ideal for developmental studies since the root system can be viewed under a stereo-binocular microscope without disturbing the roots or the fungus and roots showing developmental stages of particular interest can be removed by making a small slit in the pouch and patching it subsequently with sticky tape. Since soil or other substrate is not involved, the roots are clean and therefore surface features of the mantle can be studied and if embedded in resin for transmission electron microscopy, sectioning is not usually a problem. Since the extramatrical mycelium develops on the surface of the paper wick, the development of structures such as rhizomorphs and sclerotia (see Grenville *et al.*, 1985a,b) can be studied and radioactively labelled compounds can be

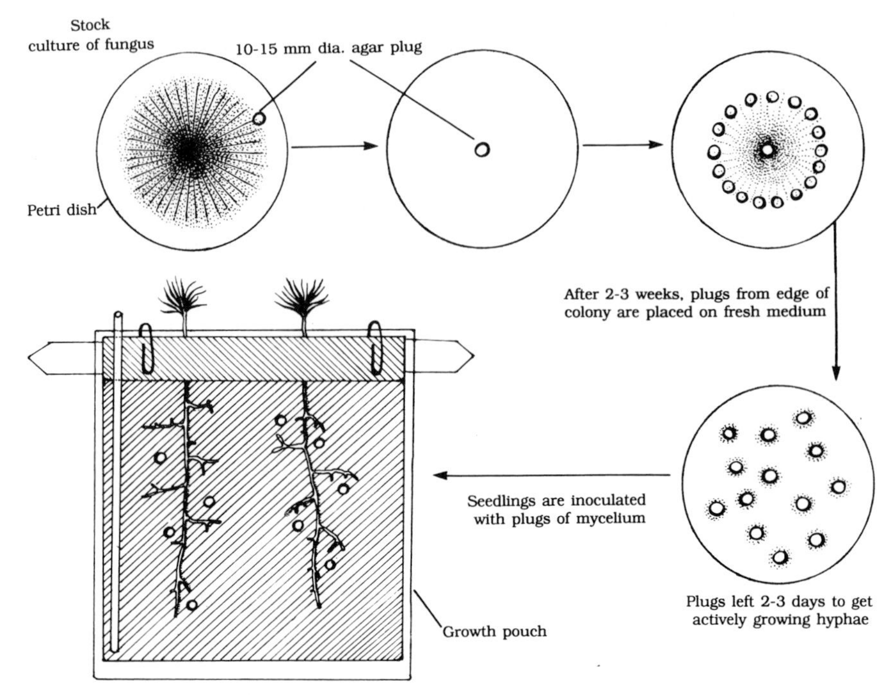

Fig. 17. Procedure used for preparing fungal inoculum and colonizing roots in growth pouches.

added to the nutrient solution for uptake and translocation studies (see Cairney *et al*., 1989).

2. *Disadvantages*

The main disadvantage of the growth pouch technique is that it is difficult to maintain sterility for the duration of an experiment. If one is careful, however, this is not a major problem but for certain studies, as for example the formation of surface exudates, the presence of even a small population of bacteria may compromise the results.

B. Minirhizotrons

Reidaker (1974) described plexiglas minirhizotrons for studies of roots and the rhizosphere; these were used by Kropp *et al*. (1987) to study mycorrhiza formation between *Pinus banksiana* and both monokaryotic and dikaryotic cultures of *Laccaria bicolor*.

Skinner and Bowen (1974a,b) constructed wooden boxes with removable perspex (polymethylmethacrylate) sides to study the growth of and translocation phenomena in mycelial strands of *Rhizopogon luteolus* associated with *Pinus radiata*. Excellent development of mycelial strands was achieved and the effect of soil type on mycelial strand development was assessed.

Finlay and Read (1986) describe in detail the construction of similar but simpler chambers (Fig. 18), which basically are as follows: the

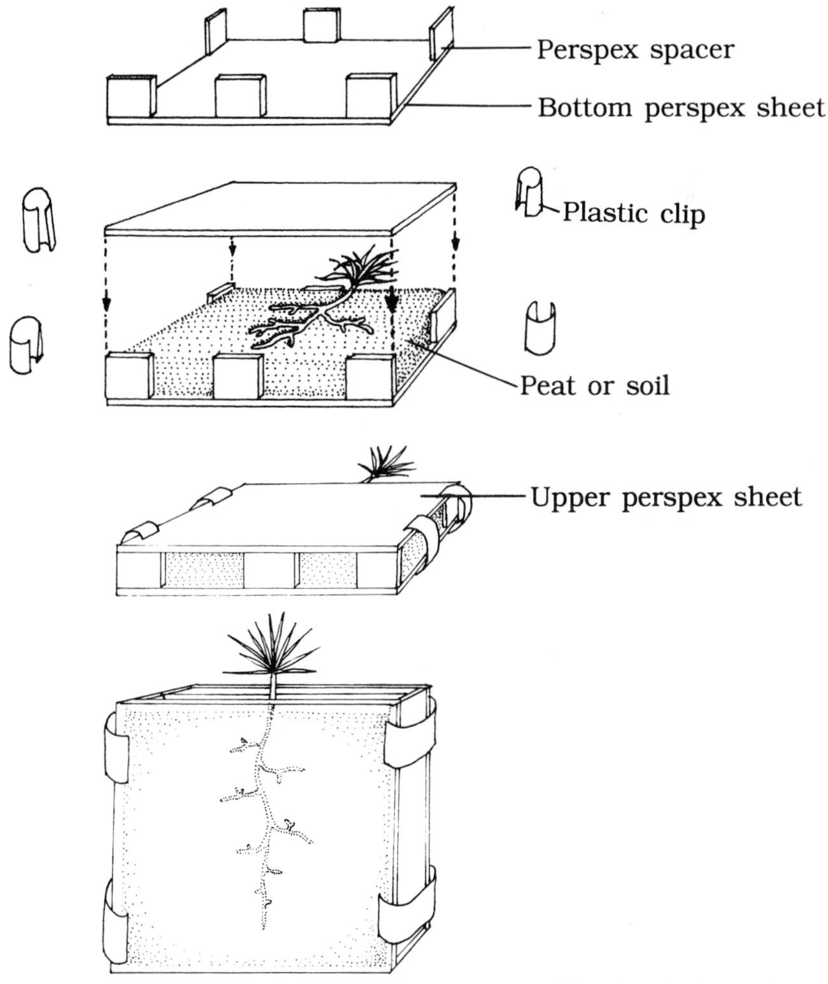

Fig. 18. Perspex chamber for synthesizing ectomycorrhizas in soil. Draw from description given in Finlay and Read (1986).

chambers are constructed of plates of transparent perspex (polymethylmethacrylate), or Darvic (ICI) of 1.5 mm thickness of various sizes, e.g. 12 cm × 12 cm, 20 cm × 20 cm or 40 cm × 40 cm. Small perspex spacers are glued to one of the plates so that the second plate rests on these and not on the peat or soil layer, which is between 3 and 6 mm deep. It is imperative that a gap is present to allow space for mycelial growth. After moistening the peat with nutrient medium, the peat is flattened to a level to allow for a small space between the peat and the upper perspex sheet. The lid and base of each chamber are held together using 3 cm long clips cut from plastic document binding spines. Chambers are then wrapped in aluminium foil and kept in an upright position inside plastic plant propagators which in turn are transferred to a growth room. Additional nutrient medium is added as required. Usually an infected plant is introduced centrally in each chamber to allow development of a mycelial system and subsequent study of mycelial colonization, nutrient translocation, infection processes, etc.

1. Advantages

These systems are *not* used as a routine method for batch synthesis of mycorrhizal plants, but rather to examine infection processes or nutrient translocation when uninfected seedlings are introduced and become infected by mycelium growing from a previously established mycorrhizal plant. Since these chambers favour the development of extensive mycelium fans on the surface of the peat next to the perspex, experiments involving the effects of severing the extramatrical mycelium can be designed. Also two or more plants can be grown in one chamber so that transport of labelled compounds supplied to one plant can be traced through the interconnecting mycelium network to adjacent plants. This is made possible by opening the chamber, covering the peat surface with a thin acrylic film (Melinex, ICI) and freezing the entire chamber prior to incubation at below zero temperature in contact with X-ray film. The autoradiograph, then records the position of the labelled compounds at the time the experiment was terminated by freezing the chamber.

Infection processes and compatibility studies are possible without an exogenous carbon source since infection is from a growing mycelium in contact with an autotrophic seedling.

C. Plastic pots

Although plastic pots with a variety of substrates are routinely used for synthesis of ectomycorrhiza when sterility is not of concern, Mullette

(1976) devised an apparatus that is sterile when first set up, but after some time becomes non-sterile. Figure 19 illustrates this apparatus. It consists of a 12 cm diameter plastic pot filled with quartz sand or soil set in a saucer of nutrients which are absorbed by cotton wicks plugging the holes in the bottom of the pot. After a sterile seedling is inserted into the substrate, a piece of foam is placed over the substrate surface to reduce moisture loss and algal growth.

1. Advantages

This apparatus is easy to assemble and allows the shoot to grow into the environment.

2. Disadvantages

It is impossible to keep the system sterile and the root system is not visible unless the apparatus is disassembled.

D. Perforated soil system

Van den Tweel and Schalk (1981) described an apparatus in which the growth and architecture of root systems can be examined in a soil system without excavation of the root system. Bosch (1984) extended

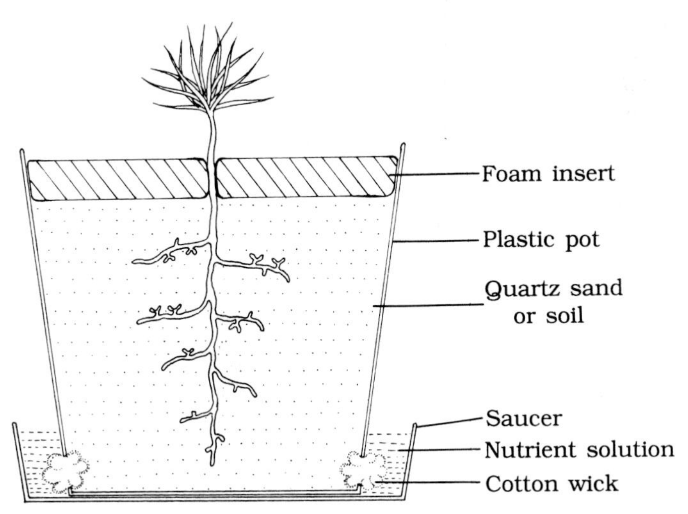

Fig.19. Plastic pot system described by Mullette (1976) for ectomycorrhiza synthesis.

these initial studies in a competition study between *Chamaenerion angustifolia* and *Quercus robur*. Limonard and Smits (1984) have used this apparatus (illustrated in Fig. 20) to study mycorrhiza development. The apparatus consists of boxes through which holes less than 15 mm diameter are drilled though the front and back plates. After the box is filled with soil of reasonable cohesive properties, a small tube is pushed through the holes, and then removed, to provide open channels along which roots can develop. The boxes are kept moist and in the dark for best root development. Subsequent to planting a seedling in the box, roots can be examined *in situ* by the use of an intrascope directly or interfaced with a TV monitor. Small samples can be taken from the root system by long biopsy forceps. The root boxes and ancillary equipment are available from Mechalectron International B.V., PO Box 2800, BE Eonda, The Netherlands, or can be constructed.

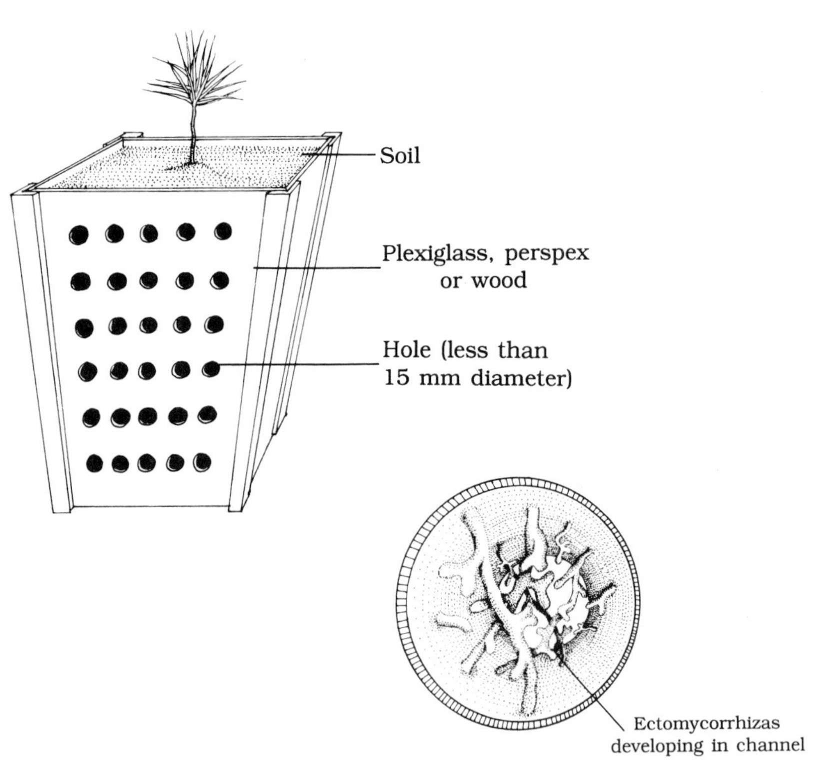

Fig. 20. Perforated soil system for growing root systems under natural conditions. Drawing is based on that found in Limonard and Smits (1984).

1. *Advantages*

Root systems can be examined *in situ* under conditions simulating those found in the field and portions of the root system can be removed for further study without excavating the entire root system. Undisturbed profiles of soil from the field can be placed in the boxes so that effects of various soil parameters on root growth can be examined.

2. *Disadvantages*

The main disadvantages are the cost if units are purchased and the time involved to construct the units.

E. Rootrainers

Egli and Kälin (1984) used Rootrainers to study the competition between introduced and indigenous species of ectomycorrhizal fungi on *Picea abies* and Farquhar (1990) found them to be very convenient in studying the interaction between the ectomycorrhizal fungus, *Paxillus involutus*, and the root-rot organism, *Fusarium oxysporum*, on *Pinus resinosa* Ait. These units are manufactured by Spencer Lemaire Industries Ltd, 11413-120 St, Edmonton, Alberta, Canada T5G 2Y3 and can also be purchased in the USA at 160 South Dellrosa Avenue, Wichita, Kansas 67218.

The apparatus consists of polypropylene fold-up liners ("books") of various sizes that are held upright by polypropylene trays, or various frames (see Fig. 21). The liners are filled with a substrate (soil, Promix Bx, etc.) and a single seed or seedling placed in each unit. After root

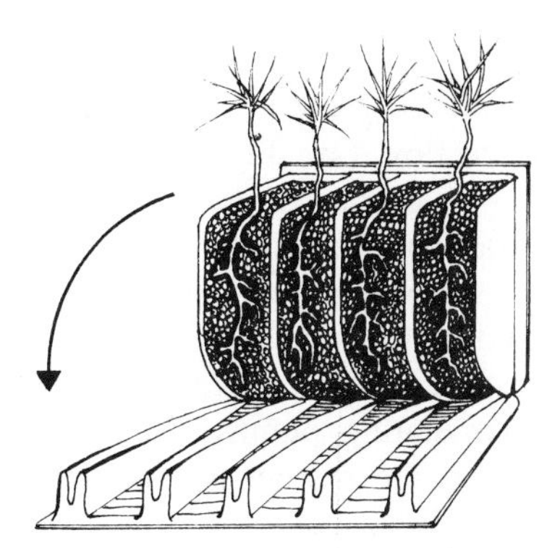

Fig. 21. Rootrainers manufactured by Spencer-Lemaire.

establishment, the unit can be opened without disturbing the root system or the extramatrical mycelium of ectomycorrhiza and then can be closed to allow further development of the root system.

1. Advantages

These units are inexpensive, easy to set up, and large numbers of seedlings can be grown in limited space. Since the liners can be opened without disturbing the root system they are convenient in studying the effects of various parameters on ectomycorrhiza development.

2. Disadvantages

These units cannot be used if sterility is essential.

IV. Hydroponic systems

One of the concerns when conducting physiological experiments using any of the closed systems described is that it is impossible to flush the system periodically to prevent the accumulation of high levels of particular nutrients and metabolites that might be detrimental to seedling growth or the establishment of ectomycorrhiza. Also, none of the systems described regulate the input of nutrients in terms of seedling demands. Ingestad and Lund (1979) used a rather complex aeroponics system to maintain plants in a state of balanced exponential growth or steady state (i.e. internal nutrient concentration, relative growth rate and shoot/root ratio remain constant) by supplying the nutrients at an exponential rate. Ingestad *et al.* (1986) synthesized ectomycorrhiza between *Pinus sylvestris* L. and *Suillus bovinus* (L. ex. Fr.) O. Kuntze using the system. Kähr and Arveby (1986) changed the system to synthesize ectomycorrhiza by allowing the roots to grow on the surface of inclined plastic sheets over which nutrients flow (see Fig. 22), since ectomycorrhiza did not form normal mantles or extramatrical mycelium in the aeroponics system described by Ingestad and Lund (1979). In this modification, the seedling is inserted into a styrofoam plug so that the root system grows on a plastic sheet bathed in nutrient solution. The nutrient solution is recirculated through the pipe system attached to the lid of the apparatus and connected to recirculating pumps. Recently Nylund and Wallander (1989) further modified the system by allowing the roots to grow into a solid substrate, "Leca" brick pellets, placed in a plastic tray with a combined inlet–outlet in the bottom. The tray is

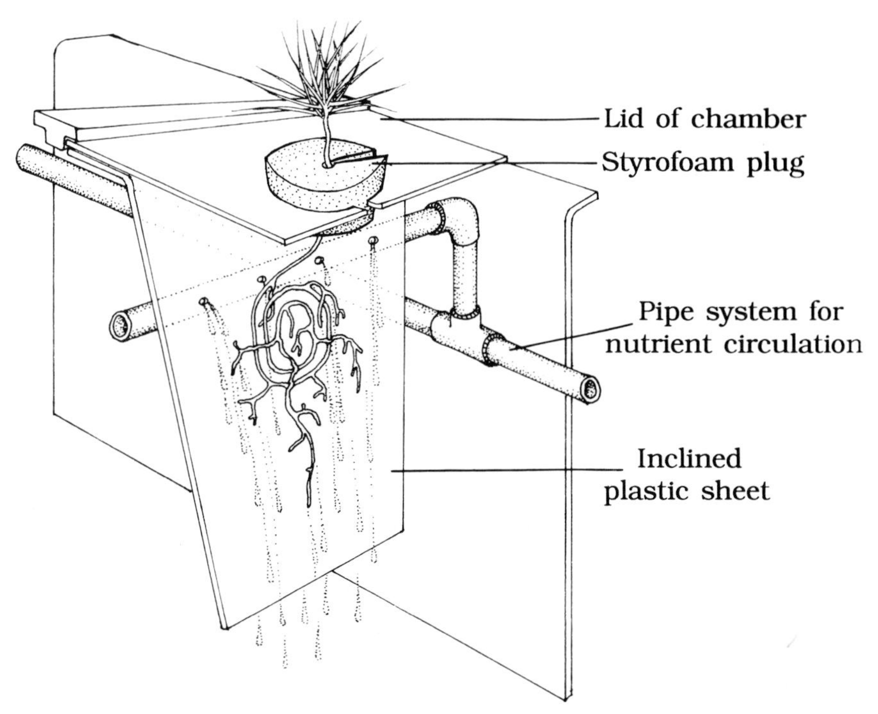

Fig. 22. Hydroponic system for synthesis of ectomycorrhiza under steady state conditions. Modified from a drawing in Kähr and Arveby (1986).

connected via centrifugal aquarium pumps to reservoir bottles containing nutrient solution. The relay-operated pumps flood the substrate for 2 min every hour. The authors report excellent ectomycorrhiza development with this apparatus.

1. Advantages

The main advantage of the systems based on Ingestad's pioneering system is the control of nutrient input into the system to provide a steady-state nutritional status. Meaningful data can thus be obtained in measuring the carbon cost of ectomycorrhiza development (see Rousseau and Reid, 1989, for an excellent discussion of this topic) and the effects of nutrient levels on this balance and other physiological aspects of seedling growth. A second advantage is that potentially phytotoxic components are not allowed to accumulate. Root systems are easily observed and changes in root architecture and mass can be determined without having to remove substrate.

2. Disadvantages

One disadvantage is the amount of work required to set up the system initially and to maintain it in a sterile condition. Mycelial development may not be natural in these systems because of the waterlogging and therefore caution should be used in interpreting results related to "carbon cost".

V. Conclusions

It is evident from the number of methods that have been developed for the colonization of roots with ectomycorrhizal fungi that researchers have either developed new methods or modified previous methods to answer particular questions related to this important symbiotic association. Many of the methods described are of use for physiological studies, while others are best utilized for structural studies. With the increased interest in molecular biology it has become imperative to be able to synthesize ectomycorrhiza between particular symbionts under very strictly defined aseptic conditions. While one may argue that these conditions are hardly relevant to natural ecosystems, this approach is necessary if the contributions of the genome of each symbiont to the final structure, the ectomycorrhiza, are to be realized. Within the constraints of a sterile system, one can mimic to some extent, nutritional and abiotic conditions, but the influence of other rhizosphere microorganisms is not taken into consideration.

The detailed information being published on the morphology of ectomycorrhizal associations from natural ecosystems (e.g. Agerer, 1987) should provide a background against which to measure the similarity of synthesized ectomycorrhiza to those occurring under natural conditions.

Acknowledgements

We thank Lewis Melville for the exceptional care taken in producing the illustrations, Melanie Chapple for editorial assistance, Dr Yves Piché and Dr Roger Finlay for comments on the manuscript, and Carol Schlaht for her skilled typing. Costs involved were covered, in part, from the Natural Sciences and Engineering Research Council of Canada.

References

Agerer, R. (1987). *Colour Atlas of Ectomycorrhizae*. Einhorn-Verlag Eduard Dietenberger, Schwabisch, Emund, West Germany.
Bailey, S. R. and Peterson, R. L. (1988). *Can. J. Bot.* **66**, 1237–1239.
Bosch, A. L. (1984). *Acta Oecologica Oecol. Plant.* **5**, 61–74.
Boudarga, K., Lapeyrie, F. and Dexheimer, J. (1990). *New Phytol.* **114**, 73–76.
Boyd, R., Furbank, R. T. and Read, D. J. (1986). In *Physiological and Genetical Aspects of Mycorrhizae, Proceedings of the 1st European Symposium on Mycorrhizae* (V. Gianinazzi-Pearson and S. Gianinazzi, eds). pp. 689–693. INRA, Paris.
Brownlee, C., Duddridge, J. A., Malibari, A. and Read, D. J. (1983). *Plant Soil* **71**, 433–443.
Cairney, J. W. E., Ashford, A. E. and Allaway, W. E. (1989) *New Phytol.* **112**, 495–500.
Chakravarty, P. and Unestam, T. (1987a). *Phytopathol. Z.* **120**, 104–120.
Chakravarty, P. and Unestam, T. (1987b). *Phytopathol. Z.* **118**, 335–340.
Chakravarty, P., Peterson, R. L. and Ellis, B. E. (1990). *Can. J. For. Res.* **20**, 1283–1288.
Chilvers, G. A., Douglass, P. A. and Lapeyrie, F. F. (1986). *New Phytol.* **103**, 397–402.
Duschesne, L. C., Peterson, R. L. and Ellis, B. E. (1988a). *Can. J. Bot.* **66**, 558–562.
Duchesne, L. C., Peterson, R. L. and Ellis, B. E. (1988b). *New Phytol.* **108**, 470–476.
Duddridge, J. A. (1986). *New Phytol.* **103**, 457–464.
Egli, S. and Kälin, I. (1984). In *Proceedings of the 6th North American Conference on Mycorrhizae* (R. Molina, ed.), p. 226. Forest Research Laboratory, Oregon State University, Corvallis, OR.
Farquhar, M. L. (1990). *Interactions between* Fusarium oxysporum, Paxillus involutus *and* Pinus resinosa. Ph.D. Thesis, University of Guelph, Ontario, Canada.
Farquhar, M. L. and Peterson, R. L. (1989). *Can. J. Plant Pathol.* **11**, 221–228.
Finlay, R. D. and Read, D. J. (1986). *New Phytol.* **103**, 143–156.
Fortin, J. A. (1966). *Can. J. Bot.* **44**, 1087–1092.
Fortin, J. A. and Piché, Y. (1979). *New Phytol.* **83**, 109–119.
Fortin, J. A., Piché, Y. and Lalonde, M. (1980). *Can. J. Bot.* **58**, 361–365.
Fortin, J. A., Piché, Y. and Godbout, C. (1983). *Plant Soil* **71**, 275–284.
Grenville, D. J., Peterson, R. L. and Piché, Y. (1985a). *Can. J. Bot.* **63**, 1402–1411.
Grenville, D. J., Peterson, R. L. and Piché, Y. (1985b). *Can. J. Bot.* **63**, 1412–1417.
Hacskaylo, E. (1953). *Mycologia* **45**, 971–975.
Ingestad, T. and Lund, A. B. (1979). *Physiol. Plant.* **45**, 137–148.
Ingestad, T., Arveby, A. S. and Kähr, M. (1986). *Physiol Plant.* **68**, 575–582.
Kähr, M. and Arveby, A. (1986). *Physiol. Plant.* **67**, 333–339.
Kropp, B. R., McAffee, B. J. and Fortin, J. A. (1987). *Can. J. Bot.* **65**, 500–504.
Lalonde, M. (1979). In *Recent Advances in Biological Nitrogen Fixation* (N. S. Subba Rao, ed.). Oxford and IBH Publication, New Delhi.

Limonard, T. and Smits, W. (1984). In *Proceedings of the 6th North American Conference on Mycorrhizae* (R. Molina, ed.), p. 254. Forest Research Laboratory, Oregon State University, Corvallis, OR.
Louis, I. and Scott, E. (1987). *Trans. Br. Mycol. Soc.* **88**, 565–568.
Malajczuk, N., Lapeyrie, F. and Garbaye, J. (1990). *New Phytol.* **114**, 627–631.
Marx, D. H. and Zak, B. (1965). *For. Sci.* **11**, 66–75.
Mason, P. A. (1975). In *The Development and Function of Roots* (J. G. Torrey and D. T. Clarkson, eds), pp. 567–574 Academic Press, New York.
Mason, P. A. (1980). In *Tissue Culture Methods* (D. S. Ingram and J. P. Helgeson, eds), pp. 173–178. Blackwell Scientific, Oxford.
Melin, E. (1921). *Karst. Svensk. Bot. Tidskr.* **15**, 192–203.
Melin, E. (1922). *Svensk. Bot. Tidskr.* **16**, 161–196.
Molina, R. (1979). *Can. J. Bot.* **57**, 1223–1228.
Molina, R. (1981). *Can. J. Bot.* **59**, 325–334.
Molina, R. and Palmer. J. G. (1982). In *Methods and Principles of Mycorrhizal Research* (N. C. Scenck, ed.), pp. 115–129. American Phytopathology Society, St Paul, MN.
Mullette, K. J. (1976). *Austral. J. Bot.* **24**, 193–200.
Nylund, J. E. and Wallander, H. (1989). *New Phytol.* **112**, 389–398.
Pachlewska, J. (1968). *Pr. Inst. Badaw Lesn.* **345**, 3–76.
Pachlewski, R. and Pachlewska, J. (1974). *Studies on Symbiotic Properties of Mycorrhizal Fungi of Pine* (Pinus sylvestris L.) *with the Aid of the Method of Mycorrhizal Synthesis in Pure Cultures on Agar.* Forest Research Institute Warsaw, Poland. 228 pp.
Piché, Y. and Peterson, R. L. (1988). In *Forest and Crop Biotechnology. Progress and Prospects* (F. A. Valentine, ed.), pp. 298–313. Springer-Verlag, New York.
Porter, F. E., Nelson, I. S. and Wold, E. K. (1966). *Crops Soils* **18**, 10–11.
Raggio, M. and Raggio, N. (1956). *Physiol. Plant.* **9**, 466–469.
Reidaker, A. (1974). *Ann. Sci. For.* **31**, 129–134.
Rousseau, J. V. D. and Reid, C. P. P. (1989). In *Application of Continuous and Steady-state Methods to Root Biology* (J. G. Torrey and L. J. Winship, eds), pp. 183–186. Kluwer Academic Publishers, Dordrecht.
Skinner, M. F. and Bowen, G. D. (1974a). *Soil Biol. Biochem.* **6**, 53–56.
Skinner, M. F. and Bowen, G. D. (1974b). *Soil Biol. Biochem.* **6**, 57–61.
Sohn, R. F. (1981). *Can. J. Bot.* **59**, 2129–2134.
Stenström, E. and Unestam, T. (1987). *Proceedings of the 7th North American Conference on Mycorrhizae* (D. M. Sylvia, L. L. Hung and J. H. Graham, eds), p. 104. University of Florida, Gainesville, Florida.
Sylvia, D. M. and Sinclair, W. A. (1983a). *Phytopathology* **73**, 384–389.
Sylvia, D. M. and Sinclair, W. A. (1983b). *Phytopathology* **73**, 390–397.
Trappe, J. M. (1967). *For. Sci.* **13**, 121–130.
Van den Tweel, P. A. and Schalk, B. (1981). *Plant Soil* **59**, 163–165.
Vincent, J. M. (1970). *A Manual for the Practical Study of Root-nodule Bacteria.* Blackwell Scientific, Oxford.
Weavers, R. W. and Frederick, L. R. (1972). *Plant Soil* **36**, 219–222.
Wong, K. K. Y. and Fortin, J. A. (1989). *Can. J. Bot.* **67**, 1713–1716.
Yang, C. S. and Wilcox, H. E. (1984). *Can. J. Bot.* **62**, 251–254.

4
Histochemistry of Ectomycorrhiza

R. L. PETERSON
Department of Botany, University of Guelph, Guelph, Ontario N1G 2W1, Canada

I. Introduction

Although the majority of papers published on ectomycorrhiza concern physiological or ecological topics, it is often imperative that basic information pertaining to the interaction between the symbionts at the structural level be included so that one knows the stage in the development of the symbiosis. Many researchers, however, avoid considering the structural characteristics of systems they work with because of the perceived difficulty in handling tissues for microscopic examination. It is certainly true that in order to study some aspects of the development and structure of ectomycorrhiza skilled microscopists and often sophisticated equipment are required. There are available, however, a range of techniques to study structural and histochemical features of ectomycorrhiza.

In this chapter, some rather simple methods developed to study basic features of the structure of ectomycorrhiza will be discussed first. Some

METHODS IN MICROBIOLOGY
VOLUME 23 ISBN 0-12-521523-1

Copyright © 1991 by Academic Press Limited
All rights of reproduction in any form reserved

more complex methods used to study particular aspects of the cytochemistry of ectomycorrhiza will then be discussed and the results obtained from these approaches will be considered.

II. Hand-sectioning and staining

Frequently, although it is important to assess the mycorrhizal status of roots of seedlings used in experiments for the presence of mantle and Hartig net, the sample size precludes embedding materials prior to sectioning. A considerable amount of information can be obtained, however, using simple techniques. Very small roots can be hand-sectioned and stained using modifications of methods published by Frolich (1984) for sectioning and Brundrett *et al.* (1988) for staining. The procedure is outlined in Fig. 1. Roots to be placed between sheets of parafilm on a dental wax base are kept moist with water. Thin sections cut with a sharp two-sided razor blade are picked up with fine forceps and placed in a small staining basket constructed either by cutting the lid and end off a Beem™ capsule or cutting polyethylene tubing into appropriate lengths and gluing nylon screen (50 μm mesh) to one end with fast-drying epoxy. This basket with sections can be transferred from solution to solution, blotting the basket on absorbent tissue during the transfer. Brundrett *et al.* (1988) constructed a multiple chamber system by fusing several polyethylene "rings" onto small rings placed on the underside of a larger sheet of nylon screen (see their Fig. 1) for the handling of many samples at the same time.

For determination of the presence of a mantle and Hartig net, sections can be stained for approximately 1 min in an aqueous solution (0.05%) of toluidine blue O, rinsed in water, removed from the basket with a toothpick and mounted in water under a cover-glass on a slide. An alternate and effective method of demonstrating the mantle and Hartig net involves clearing of root hand-sections in 10% KOH for 6–12 h at 90 °C, rinsing, staining in 0.01% chlorazol black E and then mounting them in glycerol under a cover glass for observation (Brundrett *et al.*, 1990). Observation with Nomarski interference contrast optics gives particularly clear images of the Hartig net (Fig. 2). Wilcox and Marsh (1964) used a combination of chlorazol black E and Pianase III-B to demonstrate fungal hyphae in sections of mycorrhizal *Pinus* roots embedded in paraffin. If it is important to demonstrate features of the root such as Casparian bands in the endodermis and exodermis and lignified tissues, a very effective method has been published by Brundrett *et al.* (1988). Freehand sections are placed into a basket and stained

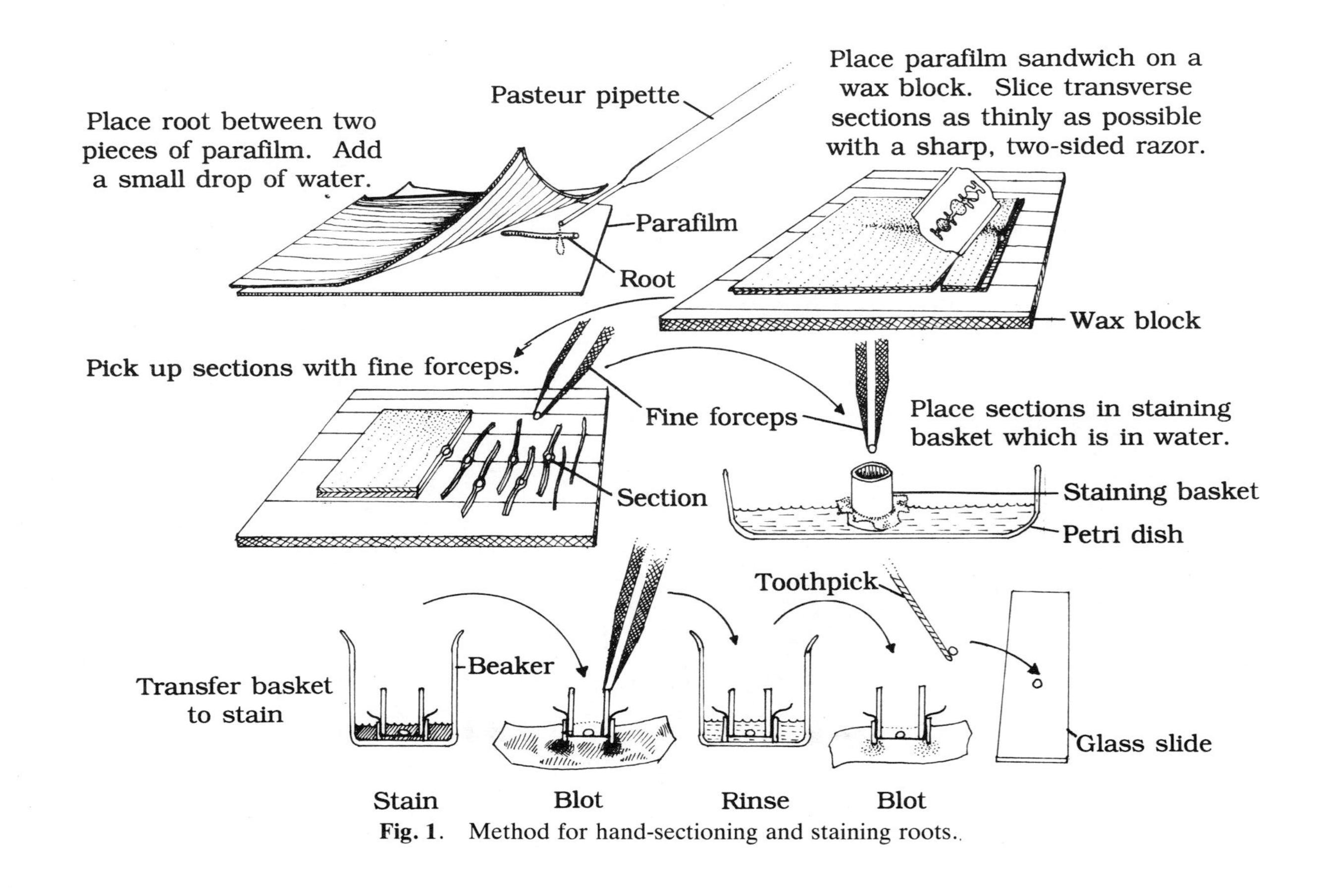

Fig. 1. Method for hand-sectioning and staining roots.

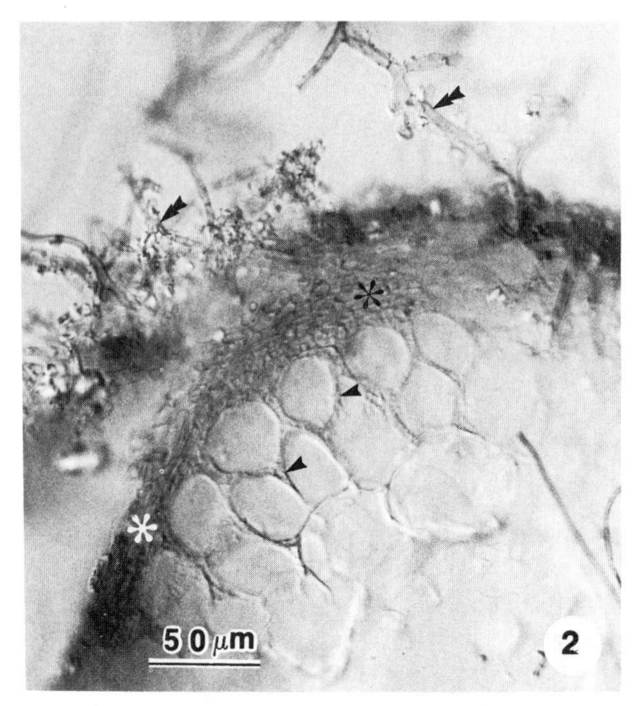

Fig. 2. Transverse hand section of field-collected *Fagus grandifolia* Ehrh. ectomycorrhiza, cleared in KOH and stained with chlorazol black E. A thick mantle (*), Hartig net hyphae (arrowheads) and extraradical hyphae (double arrowheads) are evident. Photograph courtesy of Martin Damus.

in 0.1% (w/v) berberine hemisulphate (Sigma, CI No. 75160) in distilled water for 1 h, rinsed in several changes of distilled water, blotting each time, stained in 0.5% (w/v) aniline blue WS (Polysciences, CI No. 42755) in distilled water for 30 min and then rinsed as above. Finally, sections are transferred into 0.1% (w/v) $FeCl_3$ in 50% glycerine (prepared by first dissolving $FeCl_3$ in distilled water (0.1%), filtering, and adding 50% glycerine to this solution). Sections are mounted in this solution and observed with ultraviolet illumination on an epifluorescence microscope (Fig. 6). Any number of histochemical procedures, many considered below for embedded tissues, can be used on freehand sections.

III. Embedding roots for light microscopy

Excellent results of embedding ectomycorrhiza for light and electron microscopy can be obtained by using LR White, a resin introduced by the London Resin Co. Ltd, Basingstoke, Hampshire, England. This resin works extremely well with ectomycorrhiza because of its very low viscosity, thus achieving good infiltration of both symbionts. In addition, because the cured resin is hydrophilic, most histochemical procedures can be used. A protocol that has been used routinely in my laboratory for light microscopy is as follows:

1. Ectomycorrhiza are excised and fixed in 2.5% glutaraldehyde in N-2 Hydroxyethyl piperazine-N′-2-Ethane Sulfonic Acid (HEPES) buffer (0.10 M, pH 6.8) for 3 h at room temperature (Massicotte *et al.*, 1985). A phosphate buffer system can be substituted for HEPES. Following several rinses in the same buffer, tissue is dehydrated in a graded ethanol series (20, 40, 60, 80, 100%) for 30 min each and in a further 100% ethanol overnight.
2. Tissue is infiltrated with LR White gradually over one day in the following sequence: 100% ethanol–LR White (1:3); 100% ethanol–LR White (1:2); 100% ethanol–LR White (1:1); LR White. Tissue should be left in a second change of LR White overnight, all at room temperature.
3. Tissue is embedded in one of three ways: in gelatin capsules or aluminium weighing dishes, (see Figs 3 and 4) and the resin polymerized at 60 °C ± 2 °C for 20–24 h, or in open aluminium

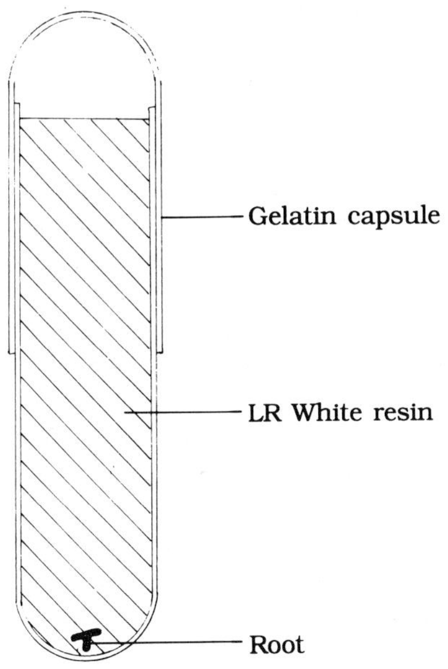

Fig. 3. Embedding ectomycorrhiza in LR White resin in a gelatin capsule.

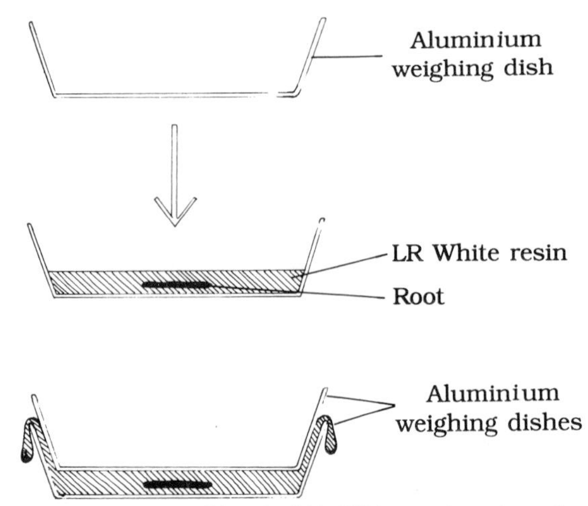

Fig. 4. Embedding ectomycorrhiza in LR White resin using aluminium weighing dishes.

weighing dishes and the resin polymerized for 1–2 h in a nitrogen atmosphere under ultraviolet light (see Fig. 5). The latter polymerizing procedure is advantageous in that it can be modified for polymerization under cold temperatures, thus preserving enzyme activity and antigenicity in tissues.

4. Tissue embedded in gelatin capsules can be mounted directly in the chuck of a microtome after the gelatin capsule is removed, or, if the tissue is not oriented properly, it can be cut out of the block and handled in the same way as the tissue embedded in weighing dishes. For the latter, tissue is cut out of the block with a fine coping saw and mounted with 5-min epoxy on a resin stub prepared by polymerizing left-over resin in polyethylene disposable pipettes and cutting these into appropriate lengths.
5. Sections are cut with glass knives at a thickness of 0.5–2.0 μm into water, picked up with a toothpick or loop and placed in a drop of water on a microscope slide, heated on a warming plate until the water evaporates and the sections adhere firmly to the slide, and then stained. Two staining schedules have produced excellent results with most ectomycorrhiza using bright-field microscopy. The simplest protocol is to stain sections in a 0.05% (w/v) solution of toluidine blue O either in distilled water or 1% (w/v) sodium tetraborate in distilled water for approximately 1 min and then rinsing in H_2O. An alternative method that gives striking differentiation between fungal hyphae and root tissue is as follows: sections

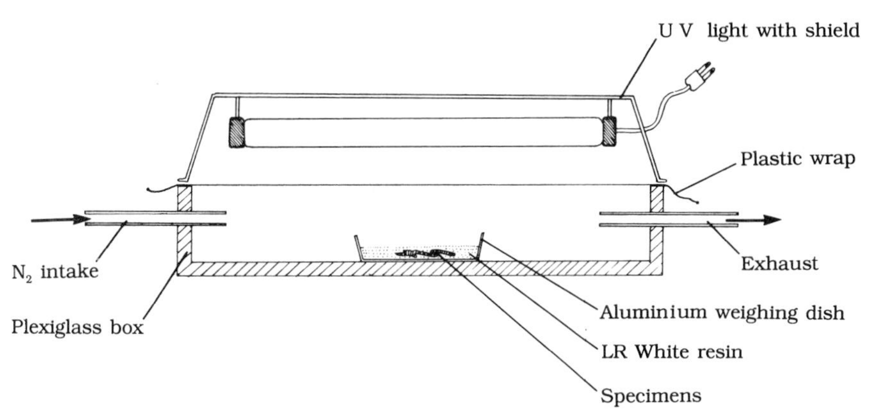

Fig. 5. Plexiglass chamber for polymerizing LR White or glycolmethacrylate resin in an atmosphere of nitrogen.

are first stained in a mixture of 1% (w/v) methylene blue/1% (w/v) azure B (1:1) in 1% (w/v) sodium tetraborate in distilled water, rinsed in distilled water, counterstained in a 0.5% (w/v) solution of basic fuchsin in distilled water, and rinsed in distilled water. After slides are air-dried, permanent mounts are made by placing a drop of immersion oil on the sections, applying a cover-glass, inverting the slide on an absorbent surface (Kimwipes, Kleenex) and squeezing the excess oil from beneath the cover glass before ringing the cover-glass with nail varnish. Figure 7 shows an ectomycorrhiza stained with toluidine blue O whereas Figs 8 and 9 are of ectomycorrhiza stained with the combination stain.

One of the most useful techiques for demonstrating Hartig net and mantle hyphae in LR White-embedded ectomycorrhiza is a simple fluorescence technique modified from Culling (1974). Sections are prepared as above and heated to ensure that they adhere to the slide. Slides are then passed through a series of solutions in coplin jars in the following sequence:

1. Filtered saturated solution of 2,4-dinitrophenyl-hydrazine (DNPH) in 15% acetic acid for 30 min. This step is to block the aldehyde groups introduced during fixation or those inherent in the tissue (O'Brien and McCully, 1981).
2. Rinse in running tap water for 10 min.
3. 1% (w/v) periodic acid in distilled water for 10 min.
4. Rinse in running tap water for 5 min.
5. Stain in 1% (w/v) acriflavine–HCl (BDH No. 27043) for 20 min.

6. Rinse in acid-alcohol (a few drops of 1 N HCl in 95% ethanol) for 5 min.
7. Rinse in a second acid-alcohol for 10 min.
8. Transfer slides to absolute ethanol and rinse briefly.
9. Rinse again in absolute ethanol and then allow slides to air-dry.
10. Mount in low fluorescence immersion oil, ring cover-glass with nail varnish and view with blue light on an epifluorescence microscope.

Figures 10 and 11 illustrate results with this technique.

IV. Cytochemistry

A. Polysaccharides

1. Storage polysaccharides

The two storage polysaccharides demonstrated histochemically in ectomycorrhiza are starch and glycogen, the former in root cells, the latter in fungal hyphae. Both can be visualized in sections embedded in either LR White resin or glycol methacrylate (another hydrophilic resin) by either the well-known PAS (periodic acid Schiff's) or IKI (iodine–potassium iodide) procedures. Details of these are given in O'Brien and McCully (1981). Ling-Lee *et al.* (1977) used both methods in studying polysaccharide distribution in *Eucalyptus* ectomycorrhiza. They showed that the use of a 5% α-amylase solution in acetate buffer at pH 5.0 for 4 h at 37 °C or ptyalin found in saliva at 37 °C for 4 h, removed both starch and glycogen from tissue sections, providing good controls for these procedures. In *E. fastigata*, the colonization of roots by an ectomycorrhizal fungus resulted in the suppression of starch accumulation in the root cap and meristematic regions of the root. Inner mantle hyphae and Hartig net hyphae contained glycogen.

Piché *et al.* (1981), using the PAS reaction to demonstrate starch distribution in *Pinus strobus* L. roots, showed that inoculated short roots and short roots colonized by *Pisolithus tinctorius* lacked starch in cortical cells but had some starch grains in stellar parenchyma. Uncolonized long roots, however, had a considerable quantity of starch within cortical, stellar parenchyma and root cap cells. Inner mantle and Hartig net hyphae contained PAS-positive material, presumably glycogen. These authors suggest that the lack of starch accumulation in cortical cells of short roots might be a prerequisite for ectomycorrhiza formation.

The acriflavine–HCl fluorescence procedure described earlier is also excellent for demonstrating starch and glycogen since even the smallest starch grains or glycogen granules are visible (see Figs 10 and 11). The proper controls must be included.

2. *Secreted polysaccharides*

Since ectomycorrhizal fungi contact the root surface and presumably receive some signal from this, some effort has gone into studying secretions by both symbionts. Ling-Lee *et al.* (1977a) compared ectomycorrhizal and uncolonized *Eucalyptus fastigata* roots for surface secretions using several histochemical procedures on tissues embedded in glycol methacrylate (GMA). Based on the staining reactions with toluidine blue O, alcian blue 8GX, the PAS reaction, and Mayer's tannic acid–ferric chloride stain, the authors identified a mucilage on the surface of *E. fastigata* roots that contains uronic acid residues. The acid groups attracting dye molecules are carboxyl rather than sulphate groups, based on the reactions of the mucilage to alcian blue and toluidine blue at low pH. Roots colonized by ectomycorrhizal fungal hyphae had less mucilage than uncolonized roots, as determined from tissue conventionally fixed and embedded.

Piché *et al.* (1983a) studied surface secretions in non-mycorrhizal and mycorrhizal (fungal symbiont *Pisolithus tinctorius*) short roots of *Pinus strobus* L. using a histochemical procedure for polysaccharides combined with scanning electron microscopy. This procedure, the thiocarbohydrazide–silver proteinate method (Thiery, 1967), uses silver as an electron-opaque marker in determining the sites of polysaccharides with 1,2-glycol groups. The distribution of silver, and therefore polysaccharides with 1,2-glycol groups, can be determined by using a back-scatter electron detector and the identification of the marker as silver can be determined by energy dispersive spectroscopy. Results from these techniques show that both uncolonized and colonized root apices have surface polysaccharides but that with colonization the pattern is more heterogeneous. Sloughing cap cells showed particularly intensive binding of silver, consistent with the known secretory function of peripheral root cap cells.

An analysis of thin sections of *Pinus strobus* L. short roots colonized by *Pisolithus tinctorius* following treatment of sections with the silver proteinate method (Piché *et al.*, 1983b) showed that silver proteinate-positive material bridges the hyphae and the root surface. These hyphae appear to have modified cell walls and these, along with lomasomes, react strongly for polysaccharides with 1,2-glycol groups.

B. Phenolics

In almost all ectomycorrhiza studied histologically, some cells at the interface between root and mantle accumulate substances that appear electron-dense in the electron microscope and stain with various staining reagents, including toluidine blue O, for light microscopy. These cells form part of the "tannin layer" described by Marks and Foster (1973) but the nature of the substances within peripheral root cells of ectomycorrhiza has rarely been characterized.

One of the most thorough investigations in this area (Ling-Lee *et al.*, 1977b) documented the presence of phenolic compounds in various cell types of mycorrhizal and non-mycorrhizal *Eucalyptus fastigata* Deane and Maiden roots. Eight histochemical procedures were used on sections of root tissue embedded in GMA. Although it is necessary to use a battery of histochemical procedures to localize phenolics in plant tissues because of the diversity in chemical structure, probably the two most useful and simplest procedures to recommend on hand-sectioned fresh material or material embedded in LR White resin or GMA are the following:

(1) Toluidine blue O prepared at a concentration of 0.05% (w/v) in 0.1 M acetate buffer at pH 4.4 (Feder and O'Brien, 1968) reacts with a wide range of phenolic compounds producing blue to bluish green colouration which persists at low pH, i.e. 1.0 (O'Brien *et al.*, 1964).

(2) Ferric chloride prepared as a 10% (w/v) aqueous solution reacts with various phenolics to produce a bluish coloration.

Many phenolic compounds show autofluorescence when viewed with ultraviolet or blue light, particularly in hand-sectioned fresh material. Some of this autofluorescence is quenched subsequent to embedding in LR White or GMA resin and therefore this techique is less useful on embedded tissues.

Techniques to identify phenolics at the ultrastructural level are general and rely primarily on the interaction between these compounds and OsO_4 during fixation to produce electron opaque deposits.

The distribution of phenolics in ectomycorrhiza and the relationship between phenolic deposition and compatible/incompatible reactions have received some attention in the literature. Piché *et al.* (1981) studied phenolic substances, identified by staining with toluidine blue O, in *Pinus strobus* L. roots and found no difference between uncolonized short roots and short roots colonized by *Pisolithus tinctorius*. Cortical

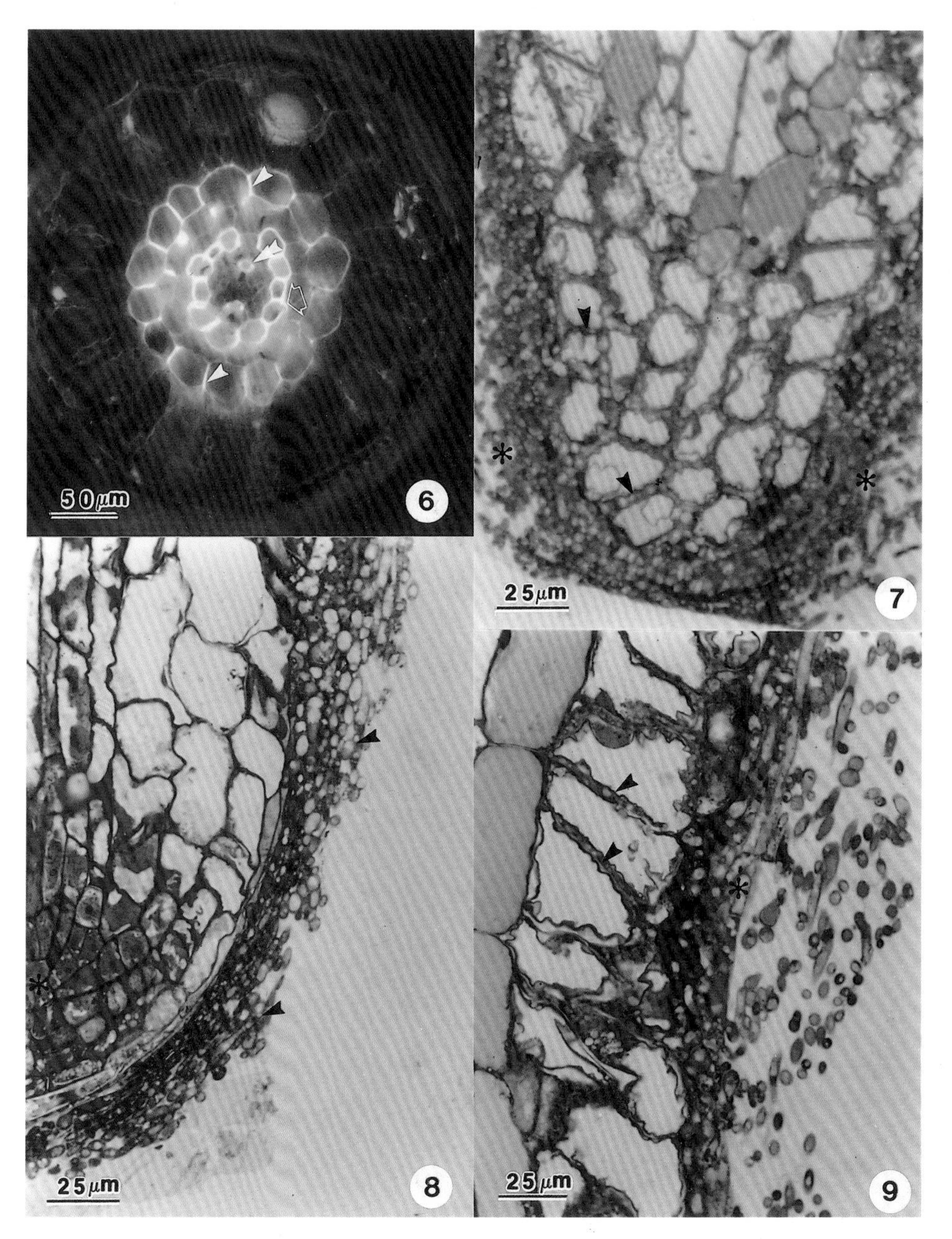

Fig. 6. Hand-section of *Betula alleghaniensis–Wilcoxina mikolae* var. *mikolae* ectomycorrhiza stained with berberine hemisulphate and aniline blue and viewed with ultraviolet light. A suberized exodermis with Casparian strips (arrowheads), a suberized endodermis (arrows), and tracheids (double arrowhead) are evident.
Photograph courtesy of P. F. Scales.

Figs 7–9. Ectomycorrhiza of *Betula alleghaniensis–Paxillus involutus* embedded in LR White resin.

Fig. 7. Glancing section of epidermis showing mantle (*) and Hartig net (arrowheads). Stained with toluidine blue O.

Fig. 8. Portion of root apex showing apical meristem (*) and mantle (arrowheads). Section stained with methylene blue/azure B followed by basic fuchsin.

Fig. 9. Paraepidermal Hartig net (arrowheads) and mantle (*). Section stained with methylene blue/azure B followed by basic fuchsin.

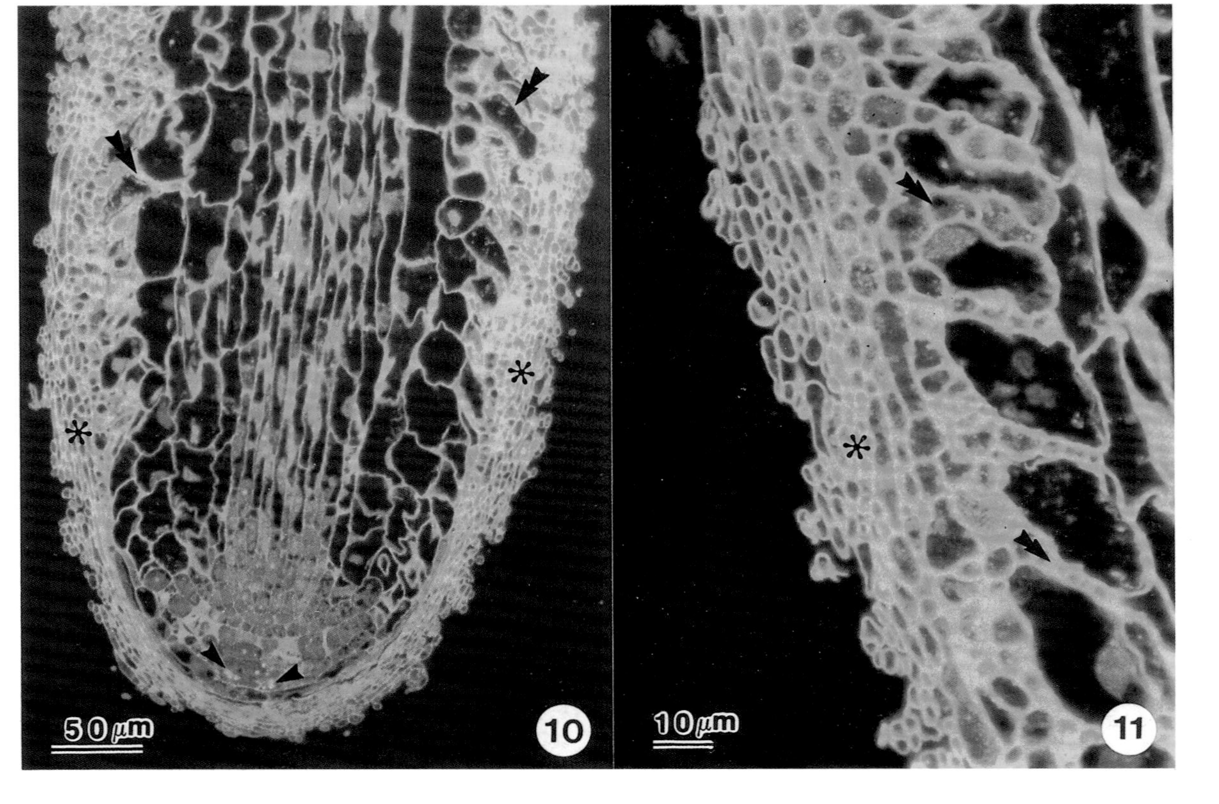

Figs 10 and 11. Ectomycorrhiza of *Eucalyptus pilularis–Hydnangium carneum* embedded in LR White resin and stained with acriflavine–HCl.

Fig. 10. Longitudinal section of root apex showing a thick fungal mantle (∗), some Hartig net hyphae (double arrowheads) and small starch grains in the root cap (arrowheads).

Fig. 11. Thick mantle (∗) and Hartig net hyphae (double arrowheads). Hyphae in both regions have glycogen granules (arrowheads).

cells of long roots, however, had much less phenols than either category of short root.

Malajczuk *et al.* (1982) showed that in pure culture synthesis experiments, most *Eucalyptus* species reacted with *Rhizopogon* and *Suillus* spp. by a darkening of root epidermal and cortical cells, presumably as a result of phenol deposition, but ectomycorrhiza did not form. In a subsequent paper, Malajczuk *et al.* (1984) extended these observations by electron microscopy and showed that *Eucalyptus* species colonized by compatible ectomycorrhizal fungi reacted by depositing "tannins" in root cortical cell vacuoles adjacent to Hartig net hyphae, whereas interaction with incompatible fungi led to "tannin" deposition within epidermal, cortical, endodermal and stelar parenchyma cells.

Recently, Horan *et al.* (1988) showed that an early response of *Eucalyptus globulus* subsp. *bicostata* (Maiden, Blakely and J. Sims) Kirkpatr. roots to either *Pisolithus tinctorius* Coker & Couch or *Paxillus involutus* (Batsch. ex Fr.) Fr. was the deposition of phenolic substances in root cap cells.

The interhyphal matrix material in the outer mantle of *Pisonia grandis* ectomycorrhiza was determined to be of a phenolic nature based on toluidine blue O staining of GMA-embedded tissues (Ashford *et al.*, 1988). These authors also showed, using the fluorescent tracer Cellufluor, that this portion of the mantle was impermeable to this apoplastic dye, presumably due to the deposits of phenolics.

C. Polyphosphate

It has been demonstrated repeatedly that ectomycorrhizal fungi, like mycorrhizal fungi in general, are able to absorb phosphorus and concentrate it as polyphosphate granules, particularly in the mantle. The first demonstration of polyphosphate in ectomycorrhiza by histochemical procedures was in *Eucalyptus fastigata* Deane and Maiden (Ashford *et al.*, 1975). Sections of ectomycorrhiza embedded in GMA and stained with toluidine blue O either at pH 4.4 or pH 1.0 showed red-stained granules in mantle and Hartig net hyphae. Treatment of sections with cold trichloroacetic acid removed these granules and confirmed that they were polyphosphate.

Sections treated with lead nitrate followed by ammonium sulphide showed granules of similar distribution to those stained red with toluidine blue O. These simple histochemical procedures can be used, therefore, to localize polyphosphate in ectomycorrhiza at the light microscope level. Confirmation that granules localized histochemically at the light microscope level do indeed contain phosphate and associated cations can be made with energy-dispersive X-ray microanalysis (EDX) on the same material (Ashford *et al.*, 1986). In conventionally fixed material, the cations usually associated with phosphorus in polyphosphate granules are Ca^{2+} and sometimes Mg^{2+} (Ashford *et al.*, 1986; Moore *et al.*, 1989; Grellier *et al.*, 1989).

Recently, Orlovich *et al.* (1989) demonstrated that in freeze-substituted embedded hyphae of the ectomycorrhizal fungus, *Pisolithus tinctorius* (Pers.) Coker & Couch, K^+ was the usual cation associated with phosphorus in polyphosphate granules. The implication of this finding is that K^+ is replaced by Ca^{2+} during tissue preparation if water is not excluded, and that the repeated reports of Ca^{2+} being the predominant cation associated with polyphosphate are probably describing artefacts.

Grellier *et al.* (1989), using phosphorus-31 nuclear magnetic resonance (^{31}P-NMR), showed that the predominant form of phosphate in cultures of *Paxillus involutus* (Batsch) Fr. mycelium is orthophosphate with a minor component of polyphosphate, whereas in hyphae associated with roots of *Betula pendula* Roth there is a dramatic increase in polyphosphate.

D. Enzyme localization

The importance of ectomycorrhizal fungi in phosphate absorption and storage has prompted several studies related to the presence and localization of phosphatases in isolated mycorrhizal fungi and within ectomycorrhiza.

The research on isolated ectomycorrhizal fungi involves biochemical techniques for extraction and identification of phosphatases (see Ho, 1989), whereas histochemical procedures involving lead precipitation on sections prepared for transmission electron microscopy have been used in some studies of ectomycorrhiza. The discussion on phosphatases will be limited to those involving histochemical procedures. Dexheimer *et al.* (1986) showed that acid phosphatases were localized along the plasma membranes of Hartig net fungal hyphae and adjacent cortical cells in two host–fungus combinations as long as the hyphae were not senescent. Mantle hyphae also showed this distribution unless they were undergoing autolysis, in which case, like old Hartig net hyphae, acid phosphatase was present throughout the fungal cytoplasm. Some activity of acid phosphatase was also present in the matrix material of the mantle and along the interface between the mantle and the rooting medium.

Lei and Dexheimer (1988), in a study of the localization of ATPase in *Pinus sylvestris–Laccaria laccata* ectomycorrhiza, showed that this enzyme was located along the plasma membrane of cortical cells, external hyphae, Hartig net hyphae and mantle hyphae. Degenerating cortical cells did not have ATPase activity and neither did the plasma membrane of Hartig net hyphae adjacent to senescing cortical cells.

V. Conclusions

Although this is not an exhaustive treatment of histochemical methods to study ectomycorrhiza, enough information is presented to enable a novice to obtain basic structural information using the simple method of staining hand-sectioned root material. In recent excellent publications, Brundrett and Kendrick (1988) and Brundrett *et al.* (1990) have shown the usefulness of simple techniques in studying the structure of field-collected ectomycorrhiza. Considerably more information on storage compounds, phenol distribution and plant cell wall characteristics could be obtained from hand-sectioned material.

Sections of ectomycorrhiza embedded in LR White resin are excellent for histochemical procedures, some of which have been illustrated in this chapter. All of the histochemical procedures for plant material embedded in glycol methacrylate discussed by O'Brien and McCully (1981) can be applied to LR White-embedded ectomycorrhiza. Although we have used LR White-embedded tissues for transmission electron microscopy (Massicotte *et al.*, 1987) the results were not entirely satisfactory.

Future work on the localization of enzymes, storage proteins, glycoproteins and other compounds in ectomycorrhiza will undoubtedly involve immunocytochemistry at both the light and electron microscope levels.

Acknowledgements

I thank Lewis Melville and Melanie Chapple for help with the illustrations, and Carol Schlaht for the care in typing the manuscript. Financial support from the Natural Sciences and Engineering Research Council of Canada is acknowledged.

References

Ashford, A. E., Ling-Lee, M. and Chilvers, G. A. (1975). *New Phytol.* **74**, 447–453.

Ashford, A. E., Peterson, R. L., Dwarte, D. and Chilvers, G. A. (1986). *Can. J. Bot.* **64**, 677–687.
Ashford, A. E., Peterson, C. A., Carpenter, J. L., Cairney, J. W. G. and Allaway, W. E. (1988). *Protoplasma* **147**, 149–161.
Brundrett, M. C. and Kendrick, B. (1988). *Can. J. Bot.* **66**, 1153–1173.
Brundrett, M. C., Enstone, D. W. and Peterson, C. A. (1988). *Protoplasma* **146**, 133–142.
Brundrett, M., Murase, G. and Kendrick, B. (1990). *Can. J. Bot.* **68**, 551–578.
Culling, D. F. A. (1974). *Modern Microscopy: Elementary Theory and Practice*. Butterworths, London.
Dexheimer, J., Aubert-Dufresne, M-P., Gerard, J., LeTacon, F. and Mousain, D. (1986). *Bull. Soc. Bot. Fr.* **133**, 343–352.
Feder, N. and O'Brien, T. P. (1968). *Am. J. Bot.* **55**, 123–142.
Frolich, M. W. (1984). *Stain Technol.* **59**, 61–62.
Grellier, B., Strullu, D. G., Martin, F. and Renaudin, S. (1989). *New Phytol.* **112**, 49–54.
Ho, I. (1989). *Can. J. Bot.* **67**, 750–753.
Horan, D. P., Chilvers, G. A. and Lapeyrie, F. F. (1988). *New Phytol.* **109**, 451–458.
Lei, J and Dexheimer, J. (1988). *New Phytol.* **108**, 329–334.
Ling-Lee, M., Ashford, A. E. and Chilvers, G. A. (1977a). *New Phytol.* **78**, 329–335.
Ling-Lee, M., Chilvers, G. A. and Ashford, A. E. (1977b). *New Phytol.* **78**, 313–328.
Malajczuk, N., Molina, R. and Trappe, J. M. (1982). *New Phytol.* **91**, 467–482.
Malajczuk, N., Molina, R. and Trappe, J. M. (1984). *New Phytol.* **96**, 43–53.
Marks, G. C. and Foster, R. C. (1973). In *Ectomycorrhizae* (G. C. Marks and T. T. Kozlowski, eds), pp. 1–41. Academic Press, New York.
Massicotte, H. B., Ackerley, C. A. and Peterson, R. L. (1985). *Proc. Microsc. Soc. Can.* **12**, 68–69.
Massicotte, H. B., Peterson, R. L., Ackerley, C. A. and Ashford, A. E. (1987). *Can. J. Bot.* **65**, 1940–1947.
Moore, A. E. P., Massicotte, H. B. and Peterson, R. L. (1989). *New Phytol.* **112**, 193–204.
O'Brien, T. P., Feder, N. and McCully, M. E. (1964). *Protoplasma* **59**, 368–373.
O'Brien, T. P. and McCully, M. E. (1981). *The Study of Plant Structure: Principles and Selected Methods*. Termarcarphi Pty Ltd, Victoria.
Orlovich, D. A., Ashford, A. E., Cox, G. C. and Moore, A. E. P. (1989). In *Endocytobiology IV* (P. Nardon, V. Gianinazzi-Pearson, A. M. Grenier, L. Margulis and D. C. Smith, eds), pp. 139–143. INRA, Paris.
Piché, Y., Fortin, J. A. and Lafontaine, J. E. (1981). *New Phytol.* **88**, 695–703.
Piché, Y., Peterson, R. L. and Ackerley, C. A. (1983a). *Scan. Electron Microsc.* **3**, 1467–1474.
Piché, Y., Peterson, R. L., Howarth, M. J. and Fortin, J. A. (1983b). *Can. J. Bot.* **61** 1185–1193.
Thiery, J. P. (1967). *J. Microsc.* **6**, 987–1018.
Wilcox, H. E. and Marsh, L. C. (1964). *Stain Technol.* **39**, 81–86.

5

Nuclear Magnetic Resonance Studies of Ectomycorrhizal Fungi

F. M. MARTIN

Laboratoire de Microbiologie Forestière, Centre de Recherches Forestière de Nancy, Institut National de la Recherche Agronomique, 54280 Champenoux, France

I. Introduction

In recent years, *in vivo* nuclear magnetic resonance spectroscopy (NMR) has emerged as a non-invasive tool for studying the biochemistry, structure and environment of intact biological entities. It is now widely used to study cellular physiology and energetics in micro-organisms (Norton, 1983; Martin, 1985a; Campbell-Burk and Shulman, 1987; Lundberg *et al.*, 1990), plants (Loughman and Ratcliffe, 1984;

METHODS IN MICROBIOLOGY
VOLUME 23 ISBN 0-12-521523-1

Copyright © 1991 by Academic Press Limited
All rights of reproduction in any form reserved

Ratcliffe, 1987; Lundberg *et al.*, 1990), cell suspensions (Balaban, 1984), intact animals and humans (Avison *et al.*, 1986). This chapter will focus on the use of multinuclear NMR as an analytical metabolic tool to study cellular compounds and pathways in mycorrhizal tissues and discusses the physiological significance of such studies. Representative examples from the work of our own laboratory are used for illustrative purposes where appropriate.

II. Nuclear magnetic resonance spectroscopy

Nuclear magnetic resonance (NMR) spectroscopy is a spectroscopic technique that exploits the magnetic properties of the atomic nucleus. The physics and biophysics of NMR have been elucidated for the physiologist by several authors (Gadian and Radda, 1981). Briefly, NMR is a form of spectroscopy based on the properties of nuclides that possess a non-zero spin and associated magnetic moment so that a resonance occurs at a characteristic frequency. The nuclides that are detectable in an NMR experiment and are of most interest to the physiologist are ^{1}H, ^{13}C, ^{23}Na, ^{31}P, ^{15}N, ^{14}N and ^{39}K (Table I). This resonance can be detected by means of an appropriate experimental arrangement including: a static magnetic field B_0 necessary for polarizing the nuclear spins, an oscillating magnetic field at the characteristic frequency to induce the resonance, and a receiving coil for detecting the NMR signal. For magnetic field strengths that are commonly used (1–14 T), ν_0 is in the radiofrequency range of 1 to 600 MHz. Molecules

TABLE I
Biochemically interesting nuclei and their NMR properties

Nucleus	Spin quantum number	Natural abundance	Relative sensitivity at constant field[a]
^{1}H	1/2	99.98	100
^{2}D	1	0.0156	1.5×10^{-4}
^{13}C	1/2	1.1	1.6×10^{-2}
^{15}N	1/2	0.36	3.7×10^{-4}
^{19}F	1/2	100	83
^{23}Na	3/2	100	9.3
^{31}P	1/2	100	6.6

[a] Relative to an equal number of protons, multiplied by the percentage natural abundance.

have many magnetic nuclei precessing at slightly different frequencies representative of their different chemical environments, and the resulting time-dependent NMR signal is rather complex. However, a simple Fourier transformation, followed by computer processing of the data, provides an NMR spectrum.

III. A typical labelling experiment for NMR analysis

A typical labelling experiment where ectomycorrhizal mycelia are provided with $[^{15}N]NH_4^+$ and the amino acid biosynthesis analysed by NMR is usually performed as follows (Fig. 1):

1. Cultures of an ectomycorrhizal fungus are grown in 1-litre flasks containing 500 ml of nutrient medium (e.g. Pachlewski's medium) with aeration provided by gyratory shaking. Cultures are maintained at 25 °C for 16 days or until growth of the culture is logarithmic as measured by the fungal biomass.
2. The mycelia are collected by filtration and 10 g of mycelia are transferred to 100 ml of N-free medium supplemented with 2.5 mM $[^{15}N]$ammonium sulphate (50% ^{15}N, CEA, France). The culture is shaken for specified periods (usually 2, 4, 8, 16, 24 and 48 h) with aeration in 250-ml flasks at room temperature.
3. For NMR measurements, 1 g of mycelia is collected by filtration and transferred to a 10 mm o.d. NMR sample tube and gently packed, so that mycelia occupy the bottom 3 cm of the tube. The sample tube is then transferred into the probe head of the spectrometer and maintained at 0 °C.
4. The ^{15}N-NMR spectrum of the fungal sample is obtained with a Bruker AM400 spectrometer operating in the Fourier transform mode at 40.5 MHz with proton broadband decoupling. The operating conditions are: 17 μs pulse width with a 2 s delay time and 8 k data storage array. Chemical shifts are reported in parts per million downfield of a solution made up to have 1 M $(NH_4)_2SO_4$ in D_2O.
5. After spectra have been recorded, mycelia are collected and soluble compounds extracted using a methanol–water (70:30) solution according to Martin and Canet (1986).
6. The fungal extracts are analysed by ^{15}N-NMR as described above except that the accumulation time is usually longer (up to 16 h).
7. Assignments of ^{15}N resonances are made by comparison with known spectra of amino acids. The assignments are additionally

supported by proton-decoupled ^{15}N spectra of the samples and by degradation of putative amino acids of the fungal extracts by enzymes (glutaminase, glutamate dehydrogenase, aminotransferases).

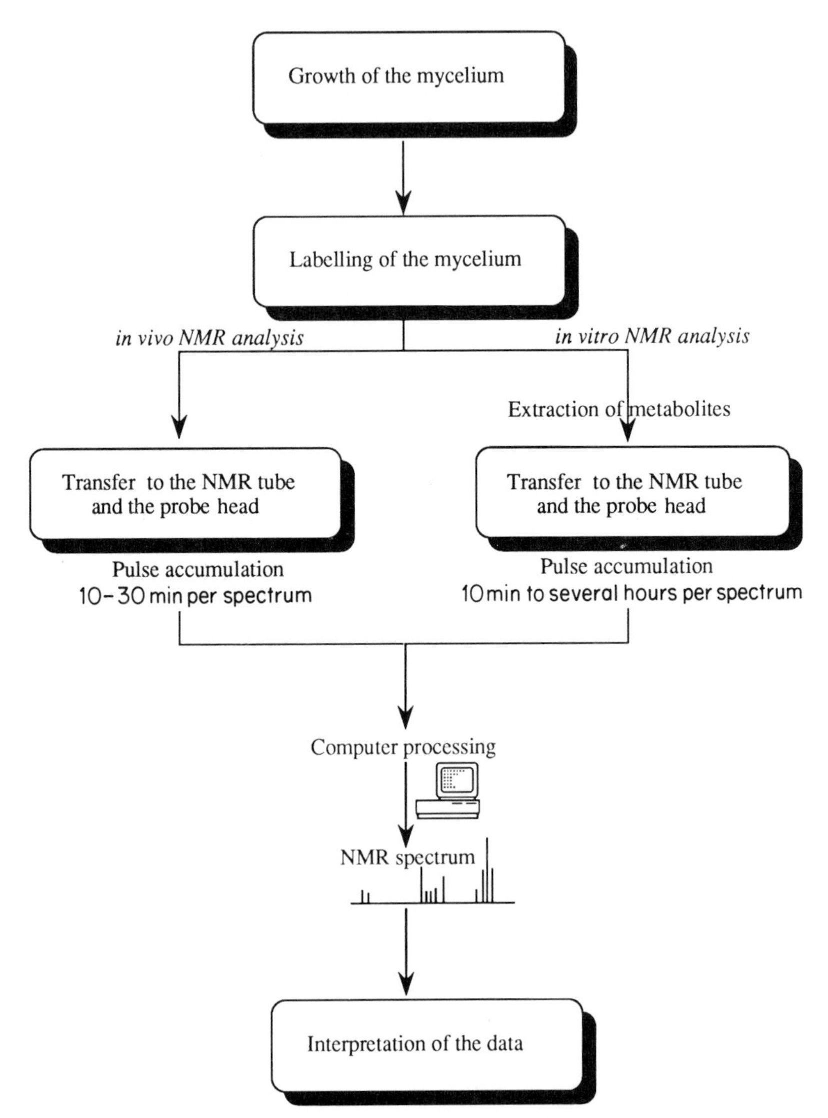

Fig. 1. Flow diagram illustrating the main steps involved from producing the material through to analysis of the data.

IV. Interpretation of NMR spectra

NMR resonances (or peaks) are characterized by their frequency, intensity, and lineshape. The *frequency* of a resonance peak depends on the splitting of the nuclear energy levels and this splitting depends on: (1) the nucleus; (2) the strength of the applied magnetic field; and (3) the local chemical environment of the nucleus. This local chemical environment is affected not only by the applied magnetic field, but also by magnetic effects associated with the electron density and other magnetic nuclei. Similar nuclei in different chemical environments (e.g. ^{13}C in CH_2OH [C1/C6] and CHOH [C2/C5]; [C3/C4] groups of mannitol; Fig. 2) are magnetically different and absorb energy at different frequencies. Line positions on NMR spectra are by convention specified relative to a reference standard. The chemical shift, δ, is defined as:

$$\delta = \frac{\nu_s - \nu_r}{\nu_r \times 10^6}$$

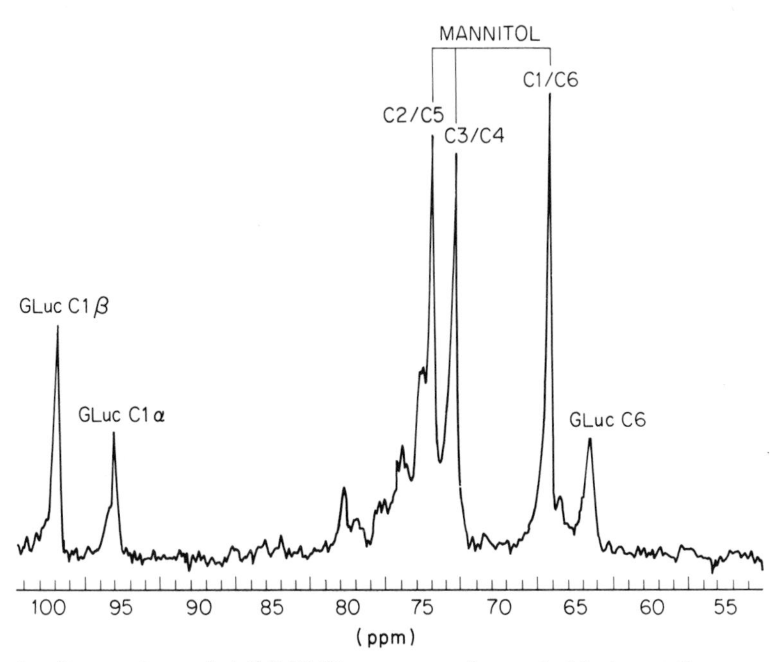

Fig. 2. Proton-decoupled ^{13}C-NMR spectrum of mannitol in intact *Cenococcum geophilum*, illustrating the resolution that can be obtained from intact mycelium. Natural abundance spectrum at 50.4 MHz (1000 scans) from 1.0 g wet weight.

where ν_s and ν_r are the absolute resonance frequencies of the line of interest and of the reference, respectively. A reference may be internal, i.e. in the same compartment as the sample, or external, i.e. separated from the sample by a diffusion barrier. Tetramethylsilane has been used as an external reference compound for *in vivo* ^{13}C-NMR while phosphoric acid (85%) is useful for ^{31}P-NMR studies.

The *intensity* of the resonance line relies on the concentration of the nuclei, the numbers of equivalent nuclei contributing to the resonance line and relaxation times of the observed nuclei. The overriding problem with most NMR techniques is insensitivity. The most favourable nucleus for quantitative analysis is ^{1}H, but even with a 600 MHz spectrometer the lower limit for detection is 0.1 mM. With regard to the signal-to-noise ratio, various factors have to be considered:

- the nucleus under study and its relative sensitivity (Table I);
- the magnetic field strength of the spectrometer;
- the isotopic enrichment where possible (^{13}C, ^{15}N, . . .);
- the intracellular level of investigated metabolites;
- the sample volume;
- splitting and width of the resonance peaks;
- the relaxation times;
- the total accumulation time.

The signal-to-noise ratio of a spectrum can be increased by accumulating a series of spectra obtained by repeatedly exciting and recording the transitions. Increasing the number of scans improves dramatically the signal-to-noise ratio and the insensitivity of the technique may be relatively insignificant, but if the goal is to detect intracellular changes, then the sensitivity problem becomes important since it is the rate of acquisition of signal-to-noise which determines the maximum rate of metabolic change that can be detected.

The intensity of an NMR resonance represents the nuclear magnetization which was colinear to the static magnetic field B_0. When disturbed from its equilibrium position, it will recover according to a generally exponential process whose time constant defines the longitudinal relaxation time, T_1. An infinite time would be necessary for a complete recovery. However, after $5T_1$, more than 99% of the magnetization has recovered. If a transition is excited again before it recovers, then it becomes saturated and contributes less intensity to the spectrum. Since different metabolites can have different T_1 relaxation times (e.g., 0.01 s for P in inorganic polyphosphates and 20 s for ^{15}N in amino acids), the intensities of the different lines can be reduced to a different extent in the same pulse sequence. Saturation can be avoided if the interval between pulses is long (approximately $5T_1$). In order to maximize the spectral signal-to-noise ratio it is common practice to repeat pulses much more frequently than $5T_1$; saturation must therefore be quantitatively accounted for, requiring an estimation of the relaxation time.

The *lineshape* of a resonance may take various forms, but in the simplest case it is a Lorentzian with a characteristic linewidth at half height. The linewidth is related to the rate at which the magnetic order decreases in the plane perpendicular to the external field after a transition has been excited. The narrowest lines are found for small molecules in non-viscous solvents, but as the mobility and the transverse (spin-spin) relaxation time T_2 of the nucleus decrease, the linewidth increases. More complicated lineshapes than the simple Lorentzian are frequently observed and examples include: (1) the resonances that show multiplet structure arising from through-bond interactions between magnetic nuclei (see Fig. 9), and (2) the non-Lorentzian lines obtained from motionally restricted nuclei in solids and membranes.

NMR spectra can be readily interpreted once NMR peaks have been identified. Valid assignment of resonances in a spectrum is an essential, but often problematical, first step in any *in vivo* NMR study. The strategy for the assignment of peaks is as follows. The resonances are tentatively assigned or narrowed down to a few specific cellular compounds by analysing chemical shifts, guided by a general knowledge of the identity and approximate amounts of metabolites previously measured in the fungus by traditional methods (e.g. high performance liquid chromatography). Peaks are then identified in the extracts by addition of a small amount of the purified compound, or addition of an enzyme which degrades the suspected metabolite, or a combination of these.

V. Making an NMR measurement and keeping cells viable

A. Choosing a sample

In commercial high-resolution NMR spectrometers, the sample tube is placed at the bottom of the superconducting magnet. The receiver coil in the probehead is generally 2–3 cm long and is designed to accommodate an NMR tube 5–25 mm in diameter. The geometry of the spectrometer therefore imposes a limitation on the size and type of the sample. A whole mycorrhizal plant would be difficult to fit in a conventional spectrometer. Cell suspensions and excised roots are the most convenient samples for high-resolution investigations. So far, high-resolution NMR has been performed on free-living mycelium of ectomycorrhizal fungi (Martin and Canet, 1986; Martin *et al.*, 1983, 1984a,b, 1985a,b, 1988), excised mycorrhizal roots (Loughman and Ratcliffe, 1984; Grellier *et al.*, 1989) and spores of an endomycorrhizal fungus (Bécard *et al.*, 1991).

In contrast, the greater flexibility of the usual probehead configuration in an electromagnet and magnetic resonance imager (Bottomley *et al.*, 1986; Connelly *et al.*, 1987; Lohman and Ratcliffe, 1988), together with the higher sensitivity of ^{1}H-NMR and the high water content of most tissues, makes it possible to record low-resolution ^{1}H-NMR spectra of intact mycorrhizal plants (Mc Fall *et al.*, 1990).

B. Keeping the cell viable

As stressed above, NMR is an insensitive analytical technique. While this is not too serious a limitation when large amounts of material are available as in studies of purified metabolites, it is one of the main limitations of the technique *in vivo*. In order to detect physiological levels of many metabolites, it is necessary to perform NMR investigations on dense suspensions of cells, generally 10–50% wet weight/volume. In a wide-bore NMR instrument, large diameter (20 mm) sample tubes can be accommodated; thus large sample volumes can be observed, but dense suspensions of cells are nearly always needed. It is not easy to maintain cells in a well-oxygenated physiological state at these high concentrations, but the goal is now routinely met. Isolated plant cells and fungal mycelia have been studied by NMR using free suspensions of cells in non-perfused chambers, or cultured cells in a perfused chamber. Clearly, the perfusion method is the preferred approach, as it allows a constant supply of oxygen and substrates as well as the monitoring of the preparation through determinations of oxygen consumption.

The first studies of ectomycorrhizal mycelium in a NMR spectrometer were the ^{31}P-NMR experiments performed on the ascomycete *Cenococcum geophilum* by Martin *et al.* (1983). In these studies, the mycelium was packed in a 10 mm diameter NMR tube and simply placed in the magnet at 0 °C in the same way as any other chemical sample. The pool of phosphorylated metabolites remained stable under these conditions, due to their low metabolic rate at 0 °C. Many NMR studies on yeast cell suspensions (Campbell-Burk and Shulman, 1987) and ectomycorrhizal fungi (Martin, unpubl. data) have been performed using a double bubbler apparatus. One set of bubblers of about 100 μm diameter was situated at the bottom of the cell suspension to help oxygenate the cells and prevent them from settling. A set of larger diameter (0.5 mm) upper bubblers positioned 1–2 cm above the detection coil provided most of the oxygen without producing undue heterogeneity in the detection region. This apparatus was fixed in the NMR tube using a teflon holder. Active cells could thus be maintained in dense populations for NMR studies after the metabolic rate had been reduced by lowering the temperature. In studies of *C. geophilum* carried out using bubblers, assimilation of glucose proceeded for several hours. However, accumulation of metabolic end-products, such as lactate, indicated that mycelia experienced anaerobic conditions.

A perfused cell preparation is therefore preferred for the study of dense plant cell and micro-organism populations (Karczmar *et al.*, 1983; Reid *et al.*, 1985; Ratcliffe, 1987; Roby *et al.*, 1988; Bécard *et al.*, 1991). The oxygenated buffer containing substrates is supplied to the bottom of the NMR tube. The level of buffer is maintained by removal of the effluent buffer containing metabolic wastes to an external reservoir with the aid of a peristaltic pump. The oxygen content and pH in the influent and effluent buffer can be measured with an electrode. The rate of substrate consumption and production of extracellular metabolites can be estimated by sampling the external reservoir as a function of time, either by NMR or chemical assay. Viability of tissues and cells within the magnet has usually been judged by the quality of the NMR spectrum obtained. This is especially true with ^{31}P-NMR, as ATP is sensitive to metabolic perturbations.

Other parameters to consider when choosing a tissue sample, and interpreting the spectra, include the cell water content, the probable levels of NMR-observable metabolites, the relative volume of the cytoplasm and vacuoles, the broadening of paramagnetic ions and the heterogeneity of the tissue.

A limited number of NMR experiments have been performed on mycorrhizal tissues (Table II). The examples I have chosen will illustrate specific applications of NMR as some of the more significant findings and should not be considered an exhaustive review of this rapidly developing field. The references provided will allow the reader to investigate areas of interest.

TABLE II
Examples of NMR-observable cellular compounds in mycorrhizal systems

NMR-observable compounds	Nuclides	References[a]
Phosphorylated compounds		
Inorganic P	^{31}P, natural abundance	1, 2
Polyphosphates	^{31}P, natural abundance	1, 2
Fatty acids		
saturated	^{13}C, natural abundance	3, 6
unsaturated	^{13}C, natural abundance	3, 6
Carbohydrates		
Glycogen	^{13}C, natural abundance, labelling	3, 6
Mannitol	^{13}C, natural abundance, labelling	3
Trehalose	^{13}C, natural abundance, labelling	3, 6
Glucose	^{13}C, natural abundance, labelling	3, 6
Amino acids		
Glutamate	^{13}C-labelling, ^{15}N-labelling	4, 5
Glutamine	^{13}C-labelling, ^{15}N-labelling	4, 5
Alanine	^{13}C-labelling, ^{15}N-labelling	4, 5
γ-Aminobutyrate	^{13}C-labelling, ^{15}N-labelling	4, 5
Arginine	^{13}C-labelling, ^{15}N-labelling	
NH_4^+	^{15}N-labelling	

[a] References: 1, Martin *et al.* (1983, 1985a); 2, Loughman and Ratcliffe (1984); 3, Martin *et al.* (1984a, b); 4, Martin and Canet (1986, 1988); 5, Martin (1985); 6, Bécard *et al.* (1991).

VI. ^{31}P-NMR studies of phosphate metabolism

Phosphorus-31 has been the most extensively studied nuclide in intact cells by NMR (for reviews see Loughman and Ratcliffe, 1984; Ratcliffe, 1987; Campbell-Burk and Shulman, 1987). These studies have generally sought information about the energetic status and the cytoplasmic pH of the cell or tissue, the former from the levels of intracellular high- and low-energy phosphates (primarily ATP and Pi), and the latter from the chemical shift of Pi and some other endogenous or added titratable phosphorus-containing compounds. In yeasts (Gillies *et al.*, 1981), mycorrhizal fungi (Martin *et al.*, 1983), and algae (Kuesel *et al.*, 1989) ^{31}P-NMR has also been used to study polyphosphate (PolyP) metabolism.

A. Compartmentation and cellular environment

The ^{31}P-NMR spectra of fungal and plant tissues generally contain a number of identifiable resonances: three arising from the α, β, and γ phosphates of ATP (and other NTPs); two Pi resonances arising from the cytosolic and vacuolar phosphates with a pH-dependent chemical shift; and sugar phosphoester peaks. In addition to Pi, the spectra of ectomycorrhizal fungi show resonances from the terminal and penultimate phosphates of PolyP (Fig. 3).

One of the earliest and still most widespread uses of ^{31}P-NMR has been the measurement of intracellular pH (Roberts *et al.*, 1980). Several methods are available to measure the intracellular pH in living cells—micro-pH electrodes, fluorescent dyes, the distribution of weak acids or bases and NMR are the most commonly used techniques. The first three

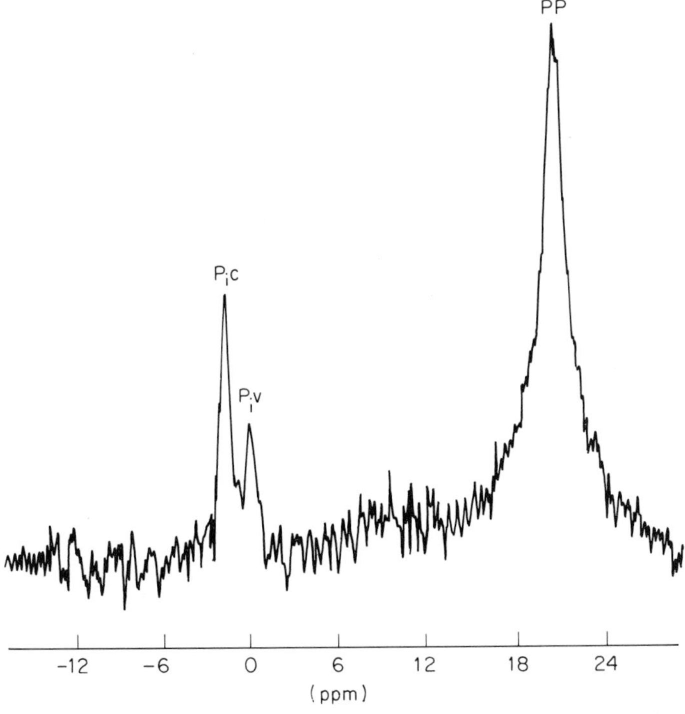

Fig. 3. Representative ^{31}P-NMR spectrum at 162 MHz of *C. geophilum* mycelium (approx. 0.5 g fresh weight). 4000 scans were accumulated with a recycle time of 2 s and a 90 ° pulse. The peak assignments are denoted: Pi_c, cytoplasmic inorganic phosphate; Pi_v, vacuolar inorganic phosphate, PP, inner phosphates of polyphosphates.

methods are all invasive and have the potential of bringing about undesired alterations inside the cell. NMR is the only method for determining the intracellular pH which can be completely non-invasive. For this purpose, any NMR-detectable nucleus which is in close proximity to a protonation state in an intracellular weak acid can generally be utilized as an NMR probe. Inorganic phosphate is the most popular probe to determine intracellular pH since it is well resolved, easily identified, and usually occurring in large concentrations. At neutral pH, inorganic phosphate exists mainly in the $H_2PO_4^-$ form (at 0.58 ppm) and the divalent form HPO_4^{2-} (at 3.14 ppm).

If the pH difference between any of the cell compartments (cytosol, mitochondria, vacuoles, etc.) is larger than 0.3 pH units, Pi resonances will be resolved. For example, there are two Pi peaks in the *in vivo* spectrum of *C. geophilum* (Fig. 3), whereas only one resonance is observed in the spectrum of the fungal extract (not shown). The lower-field Pi resonance was assigned to Pi originating from the cytosolic pH. Pi in organelles such as mitochondria is usually neglected since it appears to be present in an NMR-invisible form. Roberts and Jardetzky (1981) have shown that the pK_a of Pi is sensitive to the ionic strength of the solution. However, Gillies *et al.* (1981) have pointed out that this dependence, within the physiological range, is very small, corresponding to less than 0.1 pH unit. Several pH-dependent phosphomonoesters can function as alternative pH probes including cellular metabolites (fructose-1-phosphate, glucose-6-phosphate, polyphosphates, etc.) and chemicals (methylenediphosphonic acid, etc.). Tissue pH has now been measured in many non-mycorrhizal roots. The cytoplasmic pH of corn and pea root tips was 7.0–7.4 while the vacuolar pH was 5.5–6.0 (Loughman and Ratcliffe, 1984). In ectomycorrhizal fungi the cytoplasmic pH was 6.0–6.5 whereas the vacuolar pH was acidic (5.0) (F. Martin, unpubl. data). The importance of maintaining the tissue in a well aerated medium in the NMR tube has been stressed (Loughman and Ratcliffe, 1984).

The large size of the vacuole in mature roots and fungal cells prevents the analysis of cytoplasmic metabolites. In spectra of ectomycorrhiza and ectomycorrhizal mycelium (Fig. 3), Pi and PolyP resonances were the only significant resonances (Martin *et al.*, 1983, 1985b; Loughman and Ratcliffe, 1984; Grellier *et al.*, 1989). Monitoring of cytoplasmic events by NMR therefore requires samples containing an important cytosolic compartment. It is for this reason that many of the high-resolution NMR investigations on roots have been carried out using root tips (see Loughman and Ratcliffe, 1984).

Excess intracellular phosphate is stored as PolyP in most fungi, including those that are ectomycorrhizal (Martin *et al.*, 1983, 1985b). This large amount of PolyP gives rise to prominent resonances in the ^{31}P-NMR spectrum of these micro-organisms and NMR is thus an excellent approach for investigating PolyP metabolism and PolyP physicochemical properties in mycorrhizal tissue. Most PolyP (80%) occurs in mycorrhizal fungi as oligophosphates with an average chain length of 10 phosphate residues (Martin *et al.*, 1983). ^{31}P-NMR has been used to estimate the quantities of PolyP in the ectomycorrhizal fungi *C. geophilum*, *Hebeloma crustuliniforme* (Martin *et al.*, 1983, 1985b), *Paxillus involutus* (Grellier *et al.*, 1989) and *Pisolithus tinctorius* (Tillard *et al.*, 1990). The *in vivo* physicochemical state of PolyP was determined by Martin *et al.* (1985b). Comparison of measured nuclear magnetic relaxation parameters, such as resonance linewidths and spin-lattice relaxation times, T_1, of PolyP *in vivo*, with those observed in model solutions, suggested that PolyP exist in mycorrhizal fungi as aggregates with reduced correlation time.

VII. ^{13}C-NMR studies of carbon metabolism

Carbon-14 has been used to enable measurements of flux through metabolic pathways for decades. Significant information is provided by purification and chemical degradation of metabolic intermediates and measurement of the specific activity of each carbon of a metabolite. Due to a practical limit on the number of measurements that can be made, the metabolic analysis provided by most ^{14}C investigations is limited. NMR offers several advantages for metabolic studies, including identification and quantitation of multiple metabolites (e.g. amino acids, carbohydrates, organic acids, etc.) from a single spectrum (Fig. 4). Carbon-13 has a natural abundance of only 1.1%; therefore most ^{13}C-NMR studies require the preparation to be enriched with ^{13}C, or are restriced to investigations of highly concentrated metabolites such as fatty acids, glycogen and storage carbohydrates (Fig. 4). For many experiments this is an advantage, as specifically labelled compounds can be added to cell preparations and their metabolism characterized and quantified by following the movement of ^{13}C into other compounds. The same information can thus be obtained from a single ^{13}C-NMR spectrum as is obtained from radioisotopic tracer studies but with less effort.

A further advantage of NMR is that it is non-invasive, and therefore serial measurements of intact mycelium and roots are possible. Due to

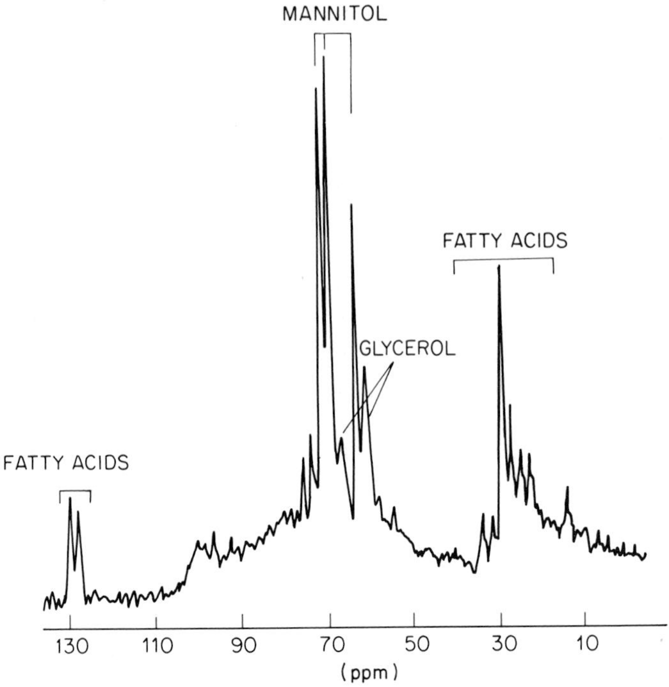

Fig. 4. Natural abundance ^{13}C-NMR spectra at 50.4 MHz of intact mycelium of *Cenococcum geophilum*, illustrating the occurrence of various storage carbon compounds (fatty acids, mannitol, glycerol, glycogen) in the mycelium. As a result of its restricted molecular freedom, glycogen gave rise to the broad resonance underneath mannitol and glycerol. The spectrum represents the time average of 1000 scans of 30° flip angle pulses with a 0.77 s recycle time.

the large range of chemical shifts of ^{13}C, ^{13}C-NMR has been shown to be capable of resolving and monitoring most metabolic intermediates of an added ^{13}C-labelled substrate, within the mM range of ^{13}C. These intermediates include most amino acids, carbohydrates, organic acids and lipids. An advantage over ^{14}C techniques is that the positions of the label within the metabolite can be easily determined, thereby allowing a wider variety of metabolic applications. Several reviews detailing the application of ^{13}C-NMR to studies of the metabolism of micro-organisms, animal and plant cells have appeared (Martin, 1985a; Campbell-Burk and Shulman, 1987; Lundberg *et al.*, 1990; Jeffrey *et al.*, 1991). The present discussion of the application of ^{13}C-NMR to mycorrhizal metabolism is focused on the use of ^{13}C substrates in the elucidation of metabolic fluxes.

A. Examination of metabolite fluxes and interconversion

Numerous experiments have been conducted employing ^{13}C-NMR to elucidate carbohydrate and amino acid metabolism in ectomycorrhizal fungi. This section describes the type of NMR spectra obtained from ^{13}C-fed mycelia and how this information can be used to resolve some metabolic questions. When cultures of ectomycorrhizal fungi were fed [^{13}C]glucose, ^{13}C-NMR spectra of sufficient signal-to-noise ratio were obtained for changes in levels of certain metabolites to be determined with a time resolution of 20 min using a 200 MHz spectrometer (Martin *et al.*, 1985a, 1988). These studies allowed the characterization of biochemical pathways for assimilation of glucose and the identification of the compounds accumulated during glucose and acetate assimilation.

B. Carbohydrate biosynthesis

Mannitol is the main soluble carbon storage compound in several fungi, including those that are ectomycorrhizal (Martin *et al.*, 1987), and it has been suggested that it plays an important role in the carbon transfer between the mycorrhizal symbionts. However, its metabolism is poorly known. A problem that is suitable for NMR analysis is the determination of its pathways of synthesis and degradation. When *C. geophilum*, *Sphaerosporella brunnea* and *Piloderma croceum* were fed with [1-^{13}C]glucose, this polyol contained the highest proportion of carbon from assimilated glucose (Martin *et al.*, 1985a, 1988) (Fig. 5), indicating that it is an important component of carbohydrate conversion and biosynthesis. More than 70% of the ^{13}C label accumulated in mannitol. Mannitol has been proposed to be metabolized in a cyclic pathway in certain Fungi Imperfecti (Hult and Gatenbeck, 1978) and in ectomycorrhizal fungi (Ramstedt *et al.*, 1987). The enzymes mannitol 1-P dehydrogenase (NADH-dependent), mannitol 1-P phosphatase, and mannitol dehydrogenase (NADPH-dependent) are thought to be involved, although this is an area of much controversy. This problem is of interest since the occurrence of the cycle may influence the fate of reducing power (NADPH) through transhydrogenation processes. Measurements of enzyme activity have produced conflicting results. As discussed above the use of ^{13}C-NMR analysis in the determination of ^{13}C enrichment overcomes these problems by its ability to measure the fate of added label in various pathways. Due to the symmetry of the mannitol molecule, operation of the mannitol cycle leads to randomization of the ^{13}C in fructose 6-P and glucose 6-P (Fig. 6). In these conditions, both C1 and C6 of these carbohydrates become ^{13}C enriched in ^{13}C-labelled

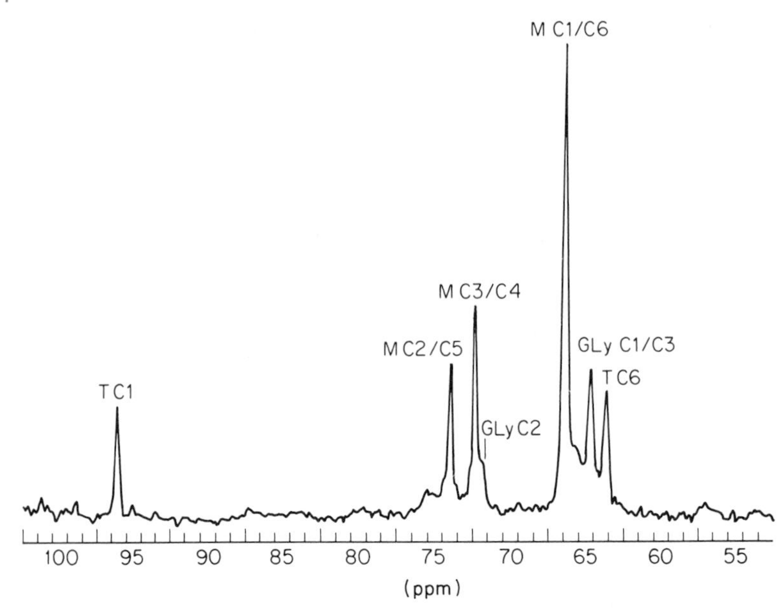

Fig. 5. Carbohydrate biosynthesis in the ectomycorrhizal *Cenococcum geophilum* as studied by ^{13}C-NMR. ^{13}C-NMR (50.4 MHz) spectrum recorded from 1.0 g wet weight mycelium of *C. geophilum* after a 6 h incubation with [1-^{13}C]glucose-containing medium. Note the labelling of C1/C6 of mannitol (M) and of C1 and C6 of trehalose (T). Carbon-13 in C2/C5 and C3/C4 of mannitol, and all carbons of glycerol corresponded to natural abundance ^{13}C.

C. geophilum (Fig. 5) (for a discussion of this mechanism, see Holligan and Jennings, 1972). In *S. brunnea*, glucose and trehalose are also enriched at the C1 and C6 carbons. The occurrence of mannitol cycle enzymes (Ramstedt *et al.*, 1987) and the observed intramolecular ^{13}C labelling pattern of glucose and trehalose suggest that the mannitol cycle is operative in *S. brunnea*, *C. geophilum* and *P. croceum*. We have therefore proposed that the mannitol cycle plays a major role in the regulation of glucose metabolism and carbon storage in several ectomycorrhizal fungi.

C. The tricarboxylic acid cycle, amino acid biosynthesis and anaplerotic pathways

Carbon flux through the tricarboxylic acid (TCA) cycle and the occurrence of anaplerotic pathways in ectomycorrhizal fungi are another area of much interest. Since glutamate and glutamine have a strong signal and their labelling patterns reflect the isotopic distribution of α-ketoglu-

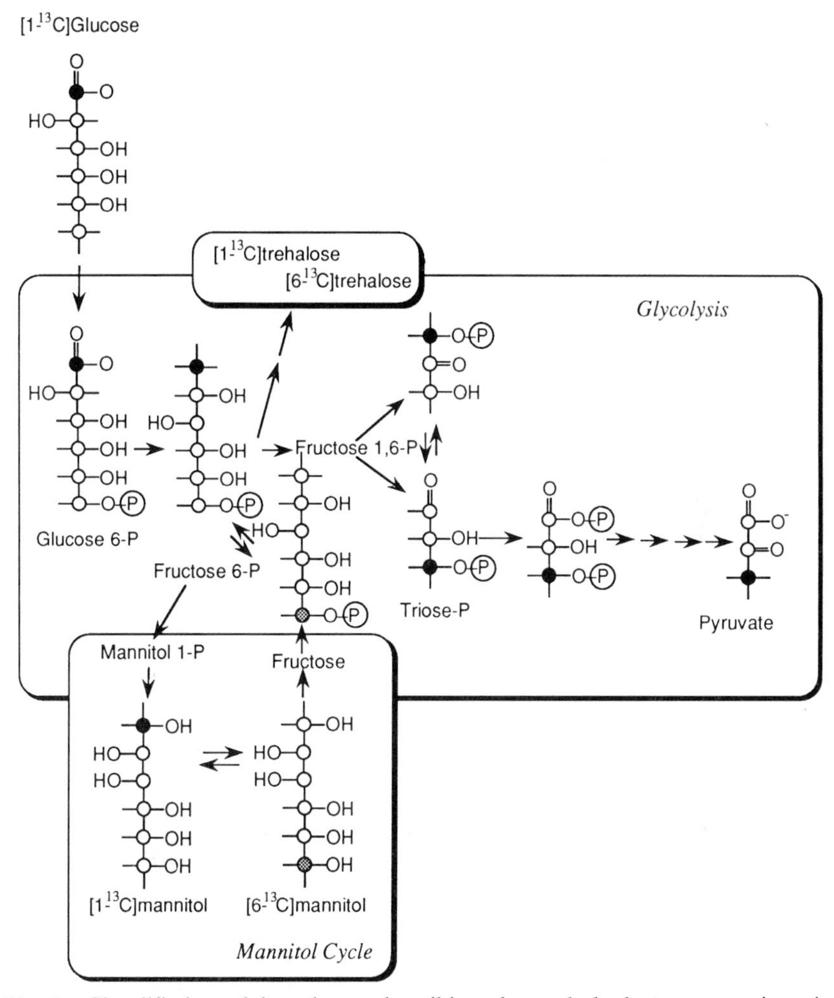

Fig. 6. Simplified model pathway describing the carbohydrate conversions in ectomycorrhizal fungi. The original label at C1 of glucose is incorporated in mannitol C1 through [1-^{13}C]fructose-6-P. The label from [1-^{13}C]fructose is randomized by the mannitol cycle (due to the symmetry of the polyol) giving rise to [1-^{13}C]- or [6-^{13}C]mannitol, so that the label is found with equal probability at either the C1 and C6 of fructose-6-P and then C1 and C6 of glucose and trehalose (after Martin *et al.*, 1985a, 1988).

tarate, they could be used to track the label through Krebs cycle intermediates (Den Hollander and Shulman, 1983; Dickinson *et al.*, 1983; Jeffrey *et al.*, 1991). The metabolism of [1-^{13}C]glucose by ectomycorrhizal fungi resulted in the formation of labelled alanine, glutamine, glutamate and arginine (Martin and Canet, 1986; Martin *et al.*, 1988). Intramolecular ^{13}C labelling patterns of glutamate and glutamine (Fig. 7) were similar and are in agreement with the operation of the Krebs cycle (Martin and Canet, 1986; Martin *et al*, 1988). Since no spin–spin coupling was observed between adjacent ^{13}C atoms, it was concluded that the majority of the α-ketoglutarate was used directly for glutamate biosynthesis rather than for TCA cycle activity. α-Ketoglutarate, used to synthesize glutamate and glutamine, arises thus from sequential action of citrate synthase, aconitase, and isocitrate dehydrogenase. Glutamate dehydrogenase and/or aminotransferases use [2-^{13}C]-, [3-^{13}C]- or [4-^{13}C]-α-ketoglutarate as a substrate. Figure 8 shows putative ^{13}C intramolecular labelling of glutamate. The intramolecular labelling of glutamate in Fig. 8a is obtained when [3-^{13}C]pyruvate, from [1-^{13}C]glucose, enters the TCA cycle via the pyruvate dehydrogenase. Because acetyl-CoA derived from [1-^{13}C]glucose will be labelled at the C3, only glutamate labelled at the C4 position will be formed. On the other hand, [3-^{13}C]pyruvate entering the TCA cycle via the pyruvate carboxylase (or any carboxylating enzymes) will give rise to glutamate labelled

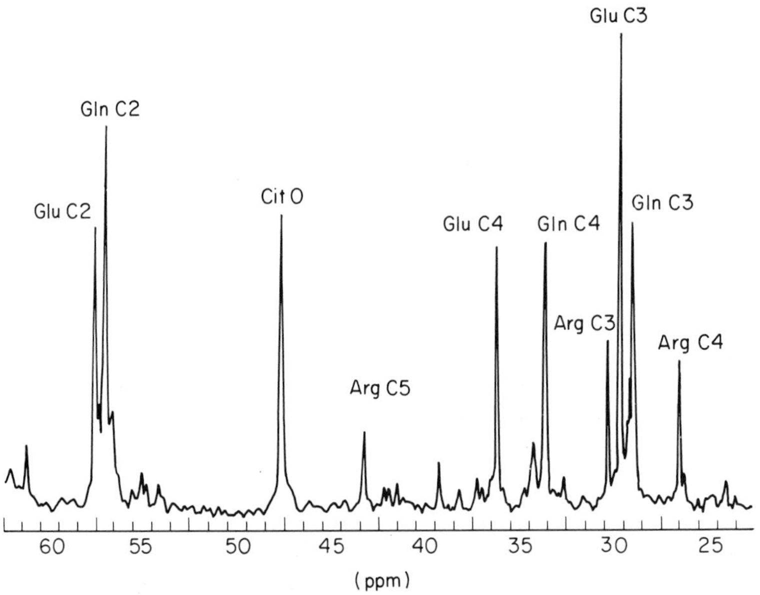

Fig. 7. Amino acid biosynthesis in ectomycorrhizal fungi studied by ^{13}C-NMR (50.4 MHz) spectrum recorded from 1.0 g wet weight mycelium of *Cenococcum geophilum* after a 6 h incubation with [1-^{13}C]glucose-containing medium. Note the labelling of C2, C3 and C4 of glutamine (Gln), glutamate (Glu) and arginine (Arg) and of C2,4 of citrate (CitO).

at the C2 and C3 positions (Fig. 8b). The ratio [^{13}C4/(^{13}C3 + ^{13}C2)] of glutamate and glutamine from [1-^{13}C]glucose-fed *C. geophilum* was approximately 1.0 during the time course experiment (Fig. 7). This is consistent with equivalent contributions of pyruvate carboxylase (or phosphoenolpyruvate carboxykinase) and pyruvate dehydrogenase to the production of Krebs cycle intermediates. The high flux of carbon through the carboxylation step indicates that CO_2 fixation is an important component of carbon metabolism in *C. geophilum* (Martin and Canet, 1986) and *S. brunnea* (Martin *et al.*, 1988). It is likely that this anaplerotic role is particularly significant under conditions of amino acid accumulation since it will enable the replenishment of intermediates of the Krebs cycle that are drawn off for biosynthesis during active growth. [^{13}C]Acetate utilization by *C. geophilum* gives rise to complex ^{13}C-NMR spectra (Fig. 9) as a result of multicycling of the TCA and glyoxylate intermediates (Boudot *et al.*, 1988). Methods have been developed to measure these isotopomers and to use them to obtain important metabolic information (Boudot *et al.*, 1988; Jeffrey *et al.*, 1990).

In vivo ^{13}C-NMR studies of ectomycorrhizal fungi, used in conjunction with other biochemical data, have provided knowledge that has quantitatively and qualitatively brought us beyond the previous level of understanding in the field.

VIII. ^{15}N-NMR studies of nitrogen metabolism

From the viewpoint of spectroscopic sensitivity, ^{15}N-NMR appears an unpromising means of monitoring ^{15}N enrichment of labelled metabolites. The inherent signal strength of a ^{15}N nucleus is one order of magnitude lower than ^{13}C and ^{15}N is only 0.37% abundant (Table I). On the other hand, the chemical shift dispersion of ^{15}N nuclei is so large that each nucleus at a distinct chemical site gives rise to a separate ^{15}N resonance whose intensity is proportional to the concentration at the site. In addition, favourable nuclear Overhauser effect and relaxation times result in ^{15}N spectra having high signal intensity, and an increasing variety of specifically, selectively, and uniformly labelled ^{15}N compounds has become commercially available. Finally, an important consideration is that the low NMR sensitivity of the ^{15}N nucleus has recently been overcome by using spectrometers with a large field magnet (e.g. 40.5 MHz for ^{15}N).

The diverse cellular mechanisms studied by this approach have included glutamine, alanine and glycine metabolism (Legerton *et al.*,

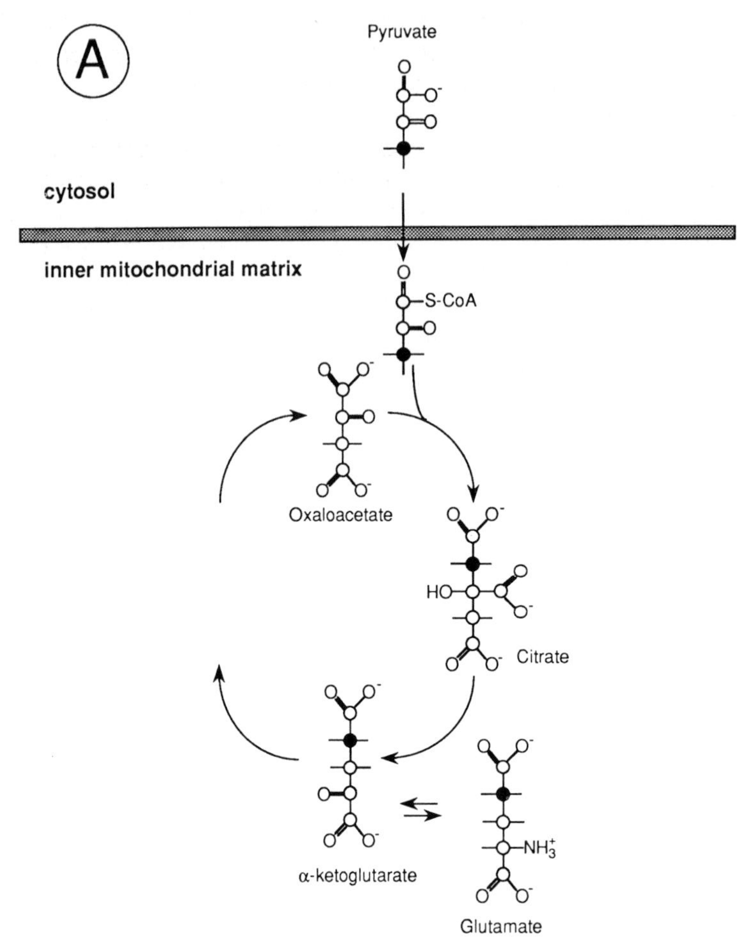

Fig. 8. Flow diagram showing ^{13}C isotope distribution during one turn of the TCA cycle with [1-^{13}C]glucose as the supplied substrate. This simplified model describes the flow of label from [3-^{13}C]pyruvate to glutamate. (A) The carbon originating from [3-^{13}C]pyruvate through the pyruvate dehydrogenase (acetyl-CoA pool) gives rise to [4-^{13}C]glutamate.

1981; Kanamori *et al.*, 1982a) and probing of the intracellular environment (pH, viscosity, molecular interactions) (Legerton *et al.*, 1983; Kanamori *et al.*, 1982b). The potential value of ^{15}N-NMR for observing relative rates of biosynthesis of amino acids in intact ectomycorrhizal fungi is evident. Determination of the *in vivo* rates, combined with knowledge from *in vitro* studies of the enzymes involved, allows evaluation of the relative importance of substrate availability and

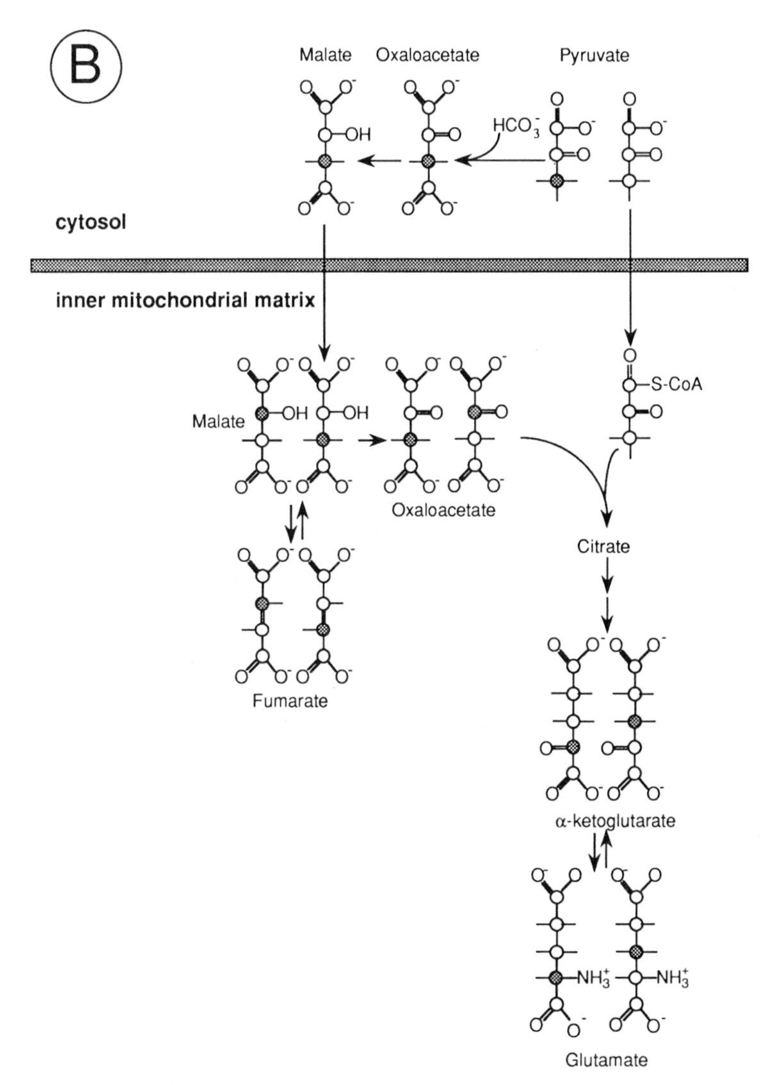

Fig. 8. (B) On the other hand the carbon originating from [3-^{13}C]pyruvate through the pyruvate carboxylase (oxaloacetate pool) is randomized by malate dehydrogenase and fumarase exchange, so that the label is found with equal probability at either the methyl or the carbonyl carbon (due to the symmetry of fumarate). This anaplerotic pathway gives rise to [2-^{13}C]- and [3-^{13}C]glutamate.

repression and induction of enzymes regulating the rates of a given pathway under physiological conditions.

The feasibility of observing amino acids in intracellular compartments

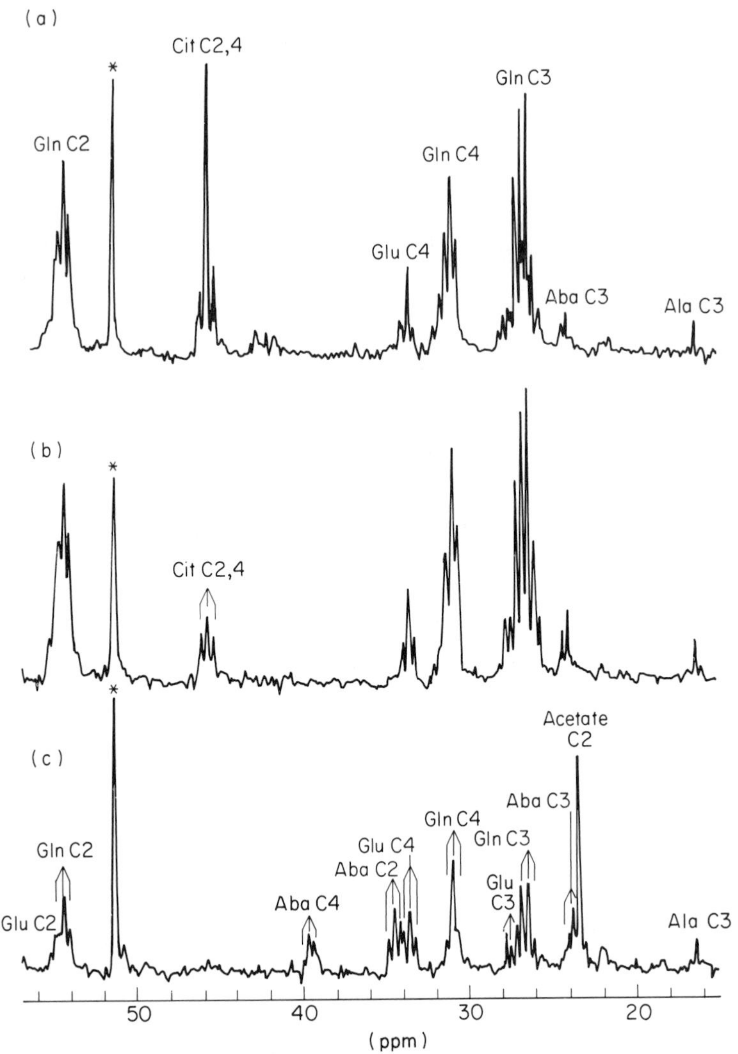

Fig. 9. ^{13}C-NMR (50.4 MHz) spectrum of an aqueous mycelial extract taken (a) 5 h, (b) 24 h and (c) 48 h after addition of 50 mM [2-^{13}C]acetate to a suspension of *Cenococcum geophilum*. The region between 15 and 60 ppm, which contains most of the amino acid and organic peaks, is shown. The spectrum is the Fourier transform of the accumulation of 1000 scans of 45° free induction decays, with a 2 s recovery time between pulses. Mycelial extracts were dissolved in 50 mM EDTA solution (pH 7.0) containing 25% D_2O. Identified are signals from acetate, alanine (Ala), 4-aminobutyrate (Aba), glutamine (Gln), glutamate (Glu) and citrate (Cit). Note the presence of triplet structures arising from ^{13}C–^{13}C spin–spin couplings between neighbouring carbons. These arise from the multicycling of the TCA cycle intermediates.

of intact mycelial suspensions of ectomycorrhizal fungi by ^{15}N-NMR has recently been demonstrated. To investigate the types of ^{15}N-labelled metabolites in *Cenococcum geophilum* (Martin, 1985b) and *Laccaria laccata* (unpublished data) that can be observed *in vivo* by ^{15}N-NMR, spectra were taken of intact mycelium grown in NH_4Cl-containing medium. The mycelium was grown in minimal medium containing 5 mM $^{15}NH_4Cl$ as the sole source of nitrogen, and a mycelial suspension (1.0 g fresh weight) was prepared for NMR measurement (Fig. 1). The resonances were assigned by comparison of the chemical shifts with those of the pure compounds and by acquisition of proton-coupled ^{15}N-NMR spectra. It was found that the resonances arose from the glutamine Nγ and the α-amino nitrogen of alanine and glutamate. The resonance at 0 ppm is assigned to intracellular ammonium N. The course of the biosynthesis could be inferred from the increase in the resonance intensities of γ-^{15}N of glutamine, α-^{15}N alanine, and α-^{15}N of glutamate, which is resolvable from α-^{15}N of glutamine when plotted on an expanded scale (Fig. 10).

In biosynthetic pathways of nitrogen metabolism in ectomycorrhizal fungi glutamate and glutamine play crucial roles in assimilating NH_4^+ and act as precursors for many important nitrogenous cellular components (Martin, 1985b). A large pool of glutamine is maintained in *C. geophilum* (Martin, 1985b; Martin *et al.*, 1988). The size of this pool is determined by a steady state between synthesis and utilization. Each pool turns over depending on its relative rates of synthesis and degradation. The turnover rate can be estimated by the method of isotopic dilution where the ^{15}N-labelled pool is subjected to a chase of unlabelled precursor, and decrease of ^{15}N label from the pool is monitored. This type of study is quite easy to perform by ^{15}N-NMR. Figure 11b shows the results of such an isotope dilution experiment in which the precursor $^{14}NH_4Cl$ is allowed to chase ^{15}N-labelled metabolites from the amino acid pools observed in Fig. 11a. The ^{15}Nγ-glutamine exhibits a very rapid turnover rate. From comparison of the ^{15}Nγ-glutamine resonance intensities in Figs. 11a and 11b it could be deduced that the half-life of glutamine in *L. laccata* growing in NH_4Cl must be less than 3 h.

For each amino acid, the intensity of its ^{15}N resonance is proportional to its intracellular concentration because all of the spectra were taken under identical operating conditions. To estimate the relative intracellular concentrations of different ^{15}N-labelled amino acids from their resonance intensities in the proton-decoupled spectra, it is necessary to take into account the respective nuclear Overhauser effect and spin lattice relaxation times to the resonance intensities. Determination of

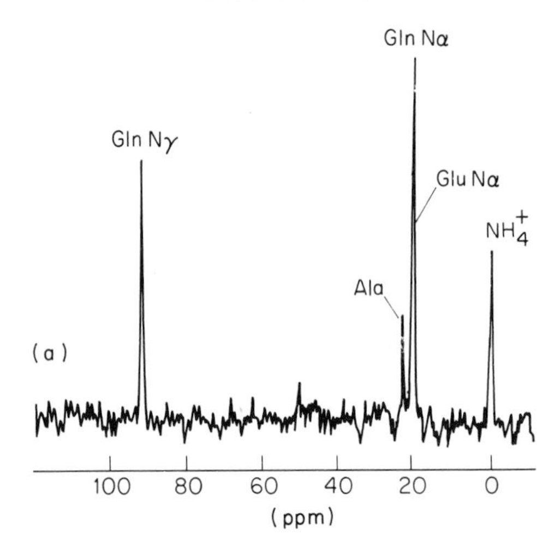

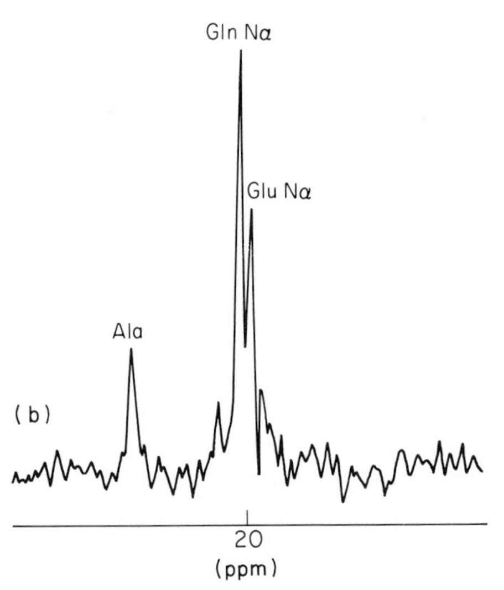

Fig. 10. Nitrogen metabolism in mycelium from the ectomycorrhiza ascomycete *Cenococcum geophilum*. (a) ^{15}N-NMR spectrum (40.5 MHz) obtained for 1.0 g wet weight mycelium after a 24 h incubation with [^{15}N]NH_4^+-containing acetate medium. Note the large magnitudes of the glutamine Nα and Nγ resonances. In acetate-grown *C. geophilum*, these peaks exhibited similar intensity, suggesting an equivalent contribution of glutamine synthetase and glutamate dehydrogenase to NH_4^+ assimilation. (b) Expanded ^{15}N-NMR spectrum of the Nα region showing the Nα of glutamine and glutamate.

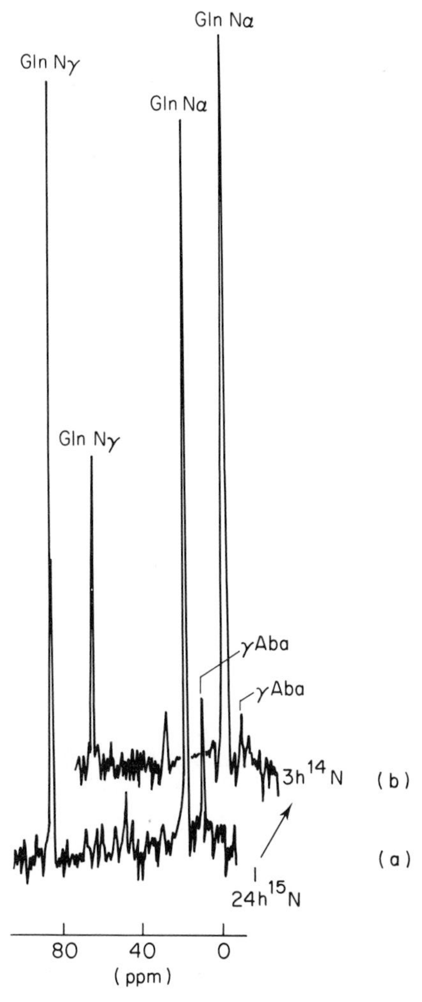

Fig. 11. Glutamine turnover in mycelium from the ectomycorrhizal basidiomycete *Laccaria laccata* S238. (a) ^{15}N-NMR spectrum (40.5 MHz) obtained for 1.0 g wet weight mycelium after a 24 h incubation with $[^{15}N]NH_4^+$. Note the large magnitude of the glutamine Nα and Nγ resonance. The mycelium was then incubated in a medium containing $[^{14}N]NH_4^+$ for 3 h (b).

these parameters for each organism under study is a major precaution to be taken in the use of *in vivo* ^{15}N-NMR (Martin and Ben Driss, 1986).

NMR spectroscopy is a useful technique for studying the dynamic properties of cellular compounds in intact organisms. ^{15}N-NMR has been used to probe the intracellular environments (microviscosity, pH, intermolecular interactions) of amino acids in *Neurospora crassa* through measurements of the spin-lattice relaxation time, T_1, resonance linewidths, and nuclear Overhauser enhancements of their ^{15}N nuclei (Kanamori *et al.*, 1982b; Legerton *et al.*, 1983). In *in vivo* spectra of the ectomycorrhizal fungus *C. geophilum* resonances representing arginine were not found (Martin, 1985b), although large arginine peaks appeared in cell extracts. This appears to be the result of restricted rotational freedom of arginine because of its complexation with polyphosphates within vacuoles *in vivo*.

Acknowledgements

The author would like to thank Professor Daniel Canet (University of Nancy) for numerous discussions concerning many of the topics considered here. This chapter is dedicated to Dr Denis Boudot.

References

Avison, M. J., Hetherington, H. P. and Shulman, R. G. (1986). *Ann. Rev. Biophys. Biophys. Chem.* **15,** 377–402.
Balaban, R. S. (1984). *Am. J. Physiol.* **246,** C10–C19.
Bécard, G., Doner, L. W., Rolin, D. B., Douds, D. D. and Pfeffer, P. E. (1991). *New Phytol.* (in press).
Bottomley, P. A., Rogers, H. H. and Foster, T. H. (1986). *Proc. Natl. Acad. Sci. USA* **83,** 87–89.
Boudot, D., Canet, D., Brondeau, J. and Martin, F. (1988). *FEBS Lett.* **231,** 11–15.
Campbell-Burk, S. L. and Shulman, R. G. (1987). *Ann. Rev. Microbiol.* **41,** 595–616.
Connelly, A., Lohman, J. A. B., Loughman, B. C., Quiquampoix, H. and Ratcliffe, R. G. (1987). *J. Exp. Bot.* **38,** 1713–1723.
Den Hollander, J. A. and Shulman, R. G. (1983). *Tetrahedron* **39,** 3529–3538.
Dickinson, J. R., Dawes, I. W., Boyd, A. S. F. and Baxter, R. L. (1983). *Proc. Natl. Acad. Sci. USA* **80,** 5847–5851.
Gadian, D. G. and Radda, G. K. (1981). *Ann. Rev. Biochem.* **50,** 69–83.
Gillies, R. J., Ugurbil, K., Den Hollander, J. A. and Shulman, R. G. (1981). *Proc. Natl. Acad. Sci. USA* **78,** 2125–2129.
Grellier, B., Strullu, D. G., Martin, F. and Renaudin S. (1989). *New Phytol.* **112,** 49–54.
Holligan, P. M. and Jennings, D. H. (1972). *Phytochemistry* **11,** 3447–3451.
Hult, K. A. and Gatenbeck, S. (1978). *Eur. J. Biochem.* **88,** 607–612.
Jeffrey, F. M. H., Rajagopal, A., Malloy, C. R. and Sherry, A. D. (1991). *Trends Biochem. Sci.* **16,** 5–10.

Kanamori, K., Legerton, T. L., Weiss, R. L. and Roberts, J. D. (1982a). *J. Biol. Chem.* **257,** 14168–14172.

Kanamori, K., Legerton, T. L., Weiss, R. L. and Roberts, J. D. (1982b). *Biochemistry* **21,** 4916–4920.

Karczmar, G. S., Koretsky, A. P., Bissell, M. J., Klein, M. P. and Weiner, M. W. (1983). *J. Magn. Res.* **53,** 123–128.

Kuesel, A. C., Sianoudis, J., Leibfritz, D., Grimme, L. H. and Mayer, A. (1989). *Arch. Microbiol.* **152,** 167–171.

Legerton, T. L., Kanamori, K., Weiss, R. L. and Roberts, J. D. (1981). *Proc. Natl. Acad. Sci. USA* **78,** 1495–1498.

Legerton, T. L., Kanamori, K., Weiss, R. L. and Roberts, J. D. (1983). *Biochemistry* **22,** 899–903.

Lohman, J. A. B. and Ratcliffe, R. G. (1988). *Experientia* **44,** 666–672.

Loughman, B. C. and Ratcliffe, R. G. (1984). In *Advances in Plant Nutrition* (P. B. Tinker and A. Läuchli, eds), Vol. 1, pp. 241–285. Praeger Scientific, New York.

Lundberg, P., Harmsen, E., Ho, C. and Vogel, H. J. (1990). *Anal. Biochem.* **191,** 193–222.

MacFall, J. S., Johnson, G. A. and Kramer, P. J. (1990). *Proc. Natl. Acad. Sci. USA* **87,** 1203–1207.

Martin, F. (1985a). *Physiol. Vég.* **23,** 463–490.

Martin, F. (1985b). *FEBS Lett.* **182,** 350–354.

Martin, F. (1991).

Martin, F. and Ben Driss, M. (1986). In *Fundamental, Ecological and Agricultural Aspects of Nitrogen Metabolism in Higher Plants* (H. Lambers, J. J. Neeteson and I. Stulen, eds), pp. 191–195. Martinus Nijhoff Publishers, Dordrecht.

Martin, F. and Canet, D. (1986). *Physiol. Vég.* **24,** 209–218.

Martin, F., Canet, D., Rolin, D., Marchal, J. P. and Larher, F. (1983). *Plant Soil* **71,** 469–476.

Martin, F., Canet, D. and Marchal, J. P. (1984a). *Physiol. Vég.* **22,** 733–743.

Martin, F., Canet, D., Marchal, J. P. and Brondeau, J. (1984b). *Plant Physiol.* **75,** 151–153.

Martin, F., Canet, D. and Marchal, J. P. (1985a). *Plant Physiol.* **77,** 499–502.

Martin, F., Marchal, J. P., Timinska, A. and Canet, D. (1985b). *New Phytol.* **101,** 275–290.

Martin, F., Ramstedt, M. and Söderhäll, K. (1987). *Biochimie* **69,** 569–581.

Martin, F., Ramstedt, M., Söderhäll, K. and Canet, D. (1988). *Plant Physiol.* **86,** 935–940.

Nicolay, K., Scheffers, W. A., Bruinenberg, P. M. and Kaptein R. (1983). *Arch. Microbiol.* **134,** 270–275.

Norton, R. S. (1983). *Bull. Magn. Reson.* **3,** 29–48.

Pfeffer, P. E., Tu, S. I., Gerasimowicz, W. V. and Cavanaugh, J. R. (1986). *Plant Physiol.* **80,** 77–84.

Pfeffer, P. E., Tu, S. I., Gerasimowicz, W. V. and Boswell, R. T. (1987). *Planta* **172,** 200–208.

Ramstedt, M., Niehaus, W. G. and Söderhäll, K. (1986). *Exp. Mycol.* **10,** 9–18.

Ramstedt, M., Jirjis, R. and Söderhäll, K. (1987). *New Phytol.* **105,** 281–287.

Ratcliffe, R. G. (1987). *Meth. Enzymol.* **148,** 683–700.

Reid, R. J., Loughman, B. C. and Ratcliffe, R. G. (1985). *J. Exp. Bot.* **36,** 889–897.

Roberts, J. K. M. and Jardetski, O. (1981). *Biochim. Biophys. Acta.* **639,** 53–76.

Roberts, J. K. M., Ray, P. M., Wade-Jardetski, N. and Jardetski, O. (1980). *Nature* **283,** 870–872.

Roby, C., Bligny, R., Douce, R., Tu, S.-I. and Pfeffer, P. E. (1988) *Biochem. J.* **252,** 401–408.

Scott, A. I. and Baxter, R. L. (1981). *Ann. Rev. Biophys. Bioeng.* **10,** 151–174.

Tillard, P., Bousquet, N., Mousain, D., Martin, F. and Salsac, L. (1990). *Agric. Ecosys. Environ.* **28,** 525–528.

6
Carbon Metabolism in Mycorrhiza

IVER JAKOBSEN
Plant Biology Section, Risø National Laboratory, DK-4000 Roskilde, Denmark

I. Introduction

In most types of mycorrhiza the host plant represents the major carbon source for the fungus and it has not yet been possible to grow the fungi forming arbuscular mycorrhiza in pure culture. Orchid mycorrhiza represent an exception where carbon is usually transferred in the opposite direction from the fungus to the host plant (Harley and Smith, 1983). The orchid mycorrhiza are not considered further in this context but some of the methods discussed in this chapter would be generally applicable to them. Beneficial effects of mycorrhiza on nutrient uptake, stress tolerance or general fitness of plants can be explained and maximized only if we understand in detail the processes involved in the acquisition and utilization of carbon by the mycobiont. We need to

METHODS IN MICROBIOLOGY
VOLUME 23 ISBN 0-12-521523-1
Copyright © 1991 by Academic Press Limited
All rights of reproduction in any form reserved

quantify the carbon flow to the fungus, to obtain qualitative information on the carbon compounds involved, and to study the fate of this carbon. Furthermore, we need more information on the importance of the fungal carbon drain on the carbon physiology of the host. The subject has been comprehensively reviewed (Lewis, 1975; Harley and Smith, 1983; Cooper, 1984; Lewis, 1986; Harris and Paul, 1987; Martin *et al.*, 1987; Rousseau and Reid, 1989; Finlay and Söderström, 1991; Schwab *et al.*, 1991). The intention of the present chapter is to provide a framework for some practical approaches to the study of carbon metabolism in mycorrhiza; readers should consult the references to obtain more detailed information on materials and methods.

II. Carbon nutrition of mycorrhizal fungi

A. External organic carbon sources

The ability of a fungus to utilize different carbon compounds may be tested by growing the fungus on agar or in liquid medium. So far only fungi forming ericoid mycorrhiza and ectomycorrhiza can be continuously grown in pure culture. Examples of substrates and methods used are given by Pearson and Read (1973, 1975) for ericoid mycorrhizal fungi and by Molina and Palmer (1982), Marx and Kenney (1982), Taber and Taber (1987), and Hutchinson (1990) for ectomycorrhizal fungi. Degradation of ^{14}C-labelled lignin by mycorrhizal fungi in pure culture was studied by Trojanowski *et al.* (1984) and Haselwandter *et al.* (1990). The latter authors measured both fungal dry matter yields and the proportion of total ^{14}C which could be collected as respired $^{14}CO_2$. They showed that lignin degradation by ectomycorrhizal fungi was small compared with the ericoid mycorrhizal fungi.

The very limited *in vitro* growth of hyphae from germinated spores of arbuscular mycorrhizal fungi may be somewhat stimulated by addition of nutrients to the medium. Although some early work indicated that sugars are inhibitory or non-stimulatory (see Hepper, 1987), it now appears that sucrose and glucose can be stimulatory at low concentrations (Siqueira *et al.*, 1982; Carr *et al.*, 1985; Siqueira and Hubbell, 1985). The optimum levels for stimulation of around $2\ g\,kg^{-1}$ fresh wt are similar to the concentrations of ethanol-soluble carbohydrates recorded in mycorrhizal leek roots (Amijee *et al.*, 1990).

The saprotrophic abilities of external hyphae of mycorrhizal fungi growing in association with a plant may be investigated by measuring the uptake of radiolabelled compounds placed in a special hyphal compartment containing no roots (see Section III.C.2). Hirrel and Gerdemann (1979) and Nelson and Khan (1990) used such an approach to show that ^{14}C from labelled glucose or a labelled insecticide, respectively, applied to the hyphal compartment, was transported to the host plant via hyphae of arbuscular fungi. However, as their systems were not sterile, it cannot be ascertained whether the incorporation of ^{14}C into the host plant indicates a saprophytic ability of the fungus or had occurred via hyphal uptake of $^{14}CO_2$ resulting from the activity of mineralizing micro-organisms. Recent work reveals that hyphae of arbuscular fungi may have the capacity to perform dark CO_2 fixation (see Section II.C). Ectomycorrhizal and especially ericoid mycorrhizal fungi have a well developed ability to utilize different complex organic compounds (Bajwa and Read, 1985; Abuzinadah and Read, 1986). Heterotrophic fungal carbon assimilation may thus account for 10 and 50% of the carbon gained by ectomycorrhizal and ericoid mycorrhizal hosts, respectively (S. Davies and D. Read, pers. commun.).

B. Carbon transfer from the host

The utilization of host carbon by the mycobiont can be demonstrated by exposing the shoots of the host plant to an atmosphere containing $^{14}CO_2$. Radioactivity is subsequently detected in the fungal tissue. This was first done for ectomycorrhiza by Melin and Nilsson (1957), for arbuscular mycorrhiza by Ho and Trappe (1973) and Bevege *et al.* (1975), and for ericoid mycorrhiza by Stribley and Read (1974). None of these studies provided information on the quantities of carbon transferred (see Section III). The dependency of the mycobiont on host photosynthate may be indirectly demonstrated by studying effects of treatments which reduce the production of photosynthates. Reduced root colonization by arbuscular fungi has thus been observed after defoliation (Daft and El-Giahmi, 1978; Same *et al.*, 1983), lowered light intensity (Hayman, 1974; Johnson, 1976; Diederichs, 1982, 1983a; Bethlenfalvay and Pacovsky, 1983; Tester *et al.*, 1986), shortened photoperiod (Ferguson and Menge, 1982; Johnson *et al.*, 1982; Diederichs, 1983b), girdling (Davis and Fucik, 1986) and ozone stress (McCool and Menge, 1983). The content of carbohydrates and amino acids in root extracts and root exudates might accordingly be expected to be related to the extent of root colonization. Available data are contradictory, however: While Jasper *et al.* (1979), Same *et al.* (1983)

and Thompson *et al.* (1986) all found a correlation between the content of ethanol-extractable soluble carbohydrates in the roots and the extent of root colonization, Ocampo and Azcon (1985), and Amijee *et al.* (1990) found no such relationship. Root colonization has also been found to be correlated with the amount and composition of carbon compounds in root exudates (Ratnayake *et al.*, 1978; Graham *et al.*, 1981; Schwab *et al.*, 1983). In future studies attention should be paid to possible ontogenetic effects on these relationships, their variation with host plant species, and to standardization of methods for extraction and analysis of compounds.

Photosynthate is transported to the root mainly as sucrose but the form and mode of the carbon transfer from the plant to the fungus are as yet unknown. Elevated levels of invertase activity have been recorded in roots of ectomycorrhiza (Lewis and Harley, 1965) and arbuscular mycorrhiza (Dehne, 1986; Snellgrove *et al.*, 1987), while the possible role of sucrose synthase has not been investigated. Gel electrophoresis showed that the increased invertase activity in mycorrhizal roots was due to a stimulation of the activity of host enzymes (Snellgrove *et al.*, 1987), but otherwise it is not known whether the sucrose-hydrolysing enzymes are of plant or of fungal origin.

Autoradiography was used by Cox *et al.* (1975) to show that host-derived ^{14}C was concentrated in cortex cells containing arbuscules, suggesting that the main site for carbon exchange is the host–fungus interface formed by the modified host plasmalemma, an interfacial apoplast and the fungal membrane. It is, as yet, a matter of speculation as to how the transfer takes place, but both passive and active transport may be involved, and the processes may be linked to the oppositely directed transfer of inorganic nutrients from the fungus to the host (Harley and Smith, 1983). However, the transfer mechanisms between the symbiotic partners are undoubtedly far more complex than a simple exchange of organic carbon for inorganic nutrients (see Smith and Smith, 1990).

When seeking useful methods for use in investigations of mycorrhizal transfer processes, it is worthwhile considering methods used for the study of other biotrophic systems. Recent work on transport across the membrane system of peribacteroid units of legume nodules (Day *et al.*, 1989; McRae *et al.*, 1989), *Frankia* vesicles from *Alnus* nodules (Vikman and Huss-Danell, 1991) and haustoria of powdery mildew (I. Donaldson, pers. commun.) are of relevance. For a detailed discussion of problems and methods related to the interface transfer of carbon, readers should consult other chapters in this series (Methods in Microbiology, Vol. 24 Chapters 7, 8 and 10).

C. Carbon metabolism in the mycobiont

1. Identification of carbon storage compounds

The ability of mycorrhizal fungi to convert readily host-derived carbohydrates into forms specific for the fungi may be of importance to the carbon transfer from the host by maintaining a concentration gradient across the interface. The polyols glycerol, mannitol, and arabitol, the disaccharide trehalose and glycogen have been identified in ectomycorrhiza by chromatographic methods (Lewis and Harley, 1965; Bevege *et al.*, 1975; Söderström *et al.*, 1988). The presence of these fungal carbohydrates has also been studied *in vivo* and in mycelial extracts of free-living ectomycorrhizal fungi by means of nuclear magnetic resonance (NMR) spectroscopy. Carbohydrates and fatty acids were thus identified using the natural abundance of ^{13}C (Martin *et al.*, 1984ab) and the synthesis of fungal carbohydrates was studied using ^{13}C-labelled sucrose (Martin *et al.*, 1985). Early attempts to detect these compounds in arbuscular mycorrhizal fungi failed (Hayman, 1974; Bevege *et al.*, 1975). However, trehalose, along with mannitol/sorbitol and inositol, was reported by Cooper (1984) to be present in external mycelium stripped from root surfaces and recently trehalose has been detected in both infected roots and sporocarps by means of gas chromatography (A. Schubert *et al.*, pers. commun.) and in roots and spores of arbuscular mycorrhizas by means of ^{13}C-NMR and high performance liquid chromatography (HPLC) (G. Bécard *et al.*, pers. commun.). Earlier attempts to detect trehalose probably failed due to a low ratio of fungal to root biomass, which would have resulted in the masking of fungal products by those of the roots. Furthermore, there are problems in distinguishing trehalose from inositols on paper chromatograms (Lewis, 1975). The enzymes of trehalose metabolism trehalose-6-P synthase and neutral trehalose were detected in roots of arbuscular mycorrhizas (A. Schubert *et al.*, pers. commun.).

It is important to realize that storage as fungal carbohydrates or lipids is likely to occur only if the rate of photosynthesis is high. Carbon-14 from labelled CO_2 will only be detected in the storage compounds if photosynthesis proceeds at a high rate during and after labelling. Although no storage can be detected even in this case, the mycobiont may still be able to maintain a certain sink strength by incorporating host-derived carbon into organic acids, amino acids, proteins and cell wall material (Bevege *et al.*, 1975).

The major storage compounds in arbuscular mycorrhiza are lipids, which are present in large amounts in hyphae, spores and vesicles. Cox *et al.* (1975) detected the presence of lipids by light microscopy, transmission electron microscopy and autoradiography, Pacovsky and Fuller (1988) used gas chromatography, while Cooper and Lösel (1978), Nagy *et al.* (1980), Nordby *et al.* (1981), Beilby (1980, 1983), and Beilby and Kidby (1980b) used thin layer chromatography in different combinations with gas chromatography, liquid scintillation counting and mass spectrometry. Some fatty acids in the 16:1, 18:3, 20:3 and 20:4 classes seem to be specific to arbuscular mycorrhizal fungi (Beilby, 1980; Beilby and Kidby, 1980b; Nordby *et al.*, 1981; Pacovsky, 1988, 1989; Pacovsky and Fuller, 1988) and might be used to localize and quantify arbuscular mycorrhizal fungi in roots and soil (see Section III.C.1).

2. *Hexose catabolism*

The catabolism of hexoses (or other compounds transferred) in the mycobiont is important for the production of energy and reducing power, as well as for providing carbon skeletons for use in anabolism. Only limited information is available on these important catabolic pathways, and much more research is needed. Most work has been done on ectomycorrhizal fungi and enzymological studies indicate the presence of both the Embden–Meyerhof and the pentose-phosphate pathways (Martin *et al.*, 1987). By feeding specifically labelled ^{13}C-glucose to ectomycorrhizal fungi and monitoring its fate using NMR spectroscopy, Martin *et al.* (1987) have further verified the presence of the Embden–Meyerhof pathway in ectomycorrhizal systems. Cytochemical studies of the germ tubes from spores of *Glomus mosseae* (Nicol and Gerd.). Gerdemann and Trappe revealed the presence of glyceraldehyde-3-phosphate dehydrogenase, succinate dehydrogenase and glucose-6-phosphate dehydrogenase which were considered as "indicator" enzymes to imply the presence of the Embden–Meyerhof pathway, the Krebs cycle and the pentose-phosphate pathway, respectively (MacDonald and Lewis, 1978). However, confirmation that these pathways are operative would require that the whole sequence of enzymes is shown to be present in the hyphae in the correct relative amounts. The pentose-phosphate pathway in ectomycorrhiza is perhaps a major source of NADPH for nitrate assimilation and lipid synthesis (Martin *et al.*, 1987). There is an urgent need to study in detail the pathways of hexose catabolism, especially in arbuscular mycorrhiza. The necessary enzyme assays may readily be performed not only on isolated spores, but also on mycelium washed from hyphal compartments of intact mycorrhizal systems.

Anaplerotic pathways for the replenishment of tricarboxylic acid cycle intermediates are important as these intermediates are extensively used in the synthesis of amino acids. The process of dark CO_2 fixation requires energy, but it may have important ecological consequences by increasing the chance of a germinating hypha encountering a host root (France and Reid, 1983). Pyruvate carboxylase, phosphoenolpyruvate carboxykinase, and phosphoenolpyruvate carboxylase may all catalyse the dark fixation of CO_2, but only the two former have been detected in filamentous fungi (Casselton, 1976). Labelling experiments with ^{13}C and ^{14}C have indicated their presence in ectomycorrhizal (Martin *et al.*, 1987) as well as in arbuscular mycorrhizal fungi (Bécard and Piché, 1989).

3. *Respiration of the mycobiont*

Some theoretical considerations on growth and respiration in arbuscular fungi are given by Harris and Paul (1987). Respiration can be separated into growth and maintenance components. The latter may be fairly high due to a high turnover rate of arbuscules and high energy costs for active transport and translocation by cytoplasmic streaming. Consequently methods for the measurement of respiration of the mycobiont are important for estimating the carbon flow between host and fungus.

Root respiration may occur via the "normal" energy-yielding cytochrome pathway or via a non-phosphorylating pathway (Lambers, 1985). The latter, which is resistant to metabolic inhibitors such as cyanide, has been shown to exist in both ectomycorrhizal and endomycorrhizal root systems (Antibus *et al.*, 1980), and has been suggested to represent an energy overflow system which enables the removal of an excess of carbohydrates (Lambers, 1982). This non-phosphorylating pathway has been demonstrated in mitochondrial preparations from beech mycorrhiza (Coleman and Harley, 1976). Its presence in arbuscular fungi has not been directly demonstrated; however, the non-phosphorylating pathway represented a similar proportion of total root respiration in mycorrhizal and non-mycorrhizal plants of *Plantago major* L. (Baas *et al.*, 1989). Total respiration was measured as O_2 uptake by excised roots inserted in an O_2-saturated solution, and was found to be 75% higher in mycorrhizal plants than in P-fed non-mycorrhizal plants. Several other workers have compared respiration of mycorrhizal and non-mycorrhizal roots (see Harley and Smith, 1983). Methods used to study respiration include the use of O_2 electrodes (Baas *et al.*, 1989), measurement of CO_2 production by infrared gas analysis (Silsbury *et al.*, 1983; Söderström and Read, 1987; Jakobsen and Rosendahl, 1990a), and measurement of ^{14}C content in CO_2 traps connected to the root containers of

$^{14}CO_2$ labelled plants (see Section III). In general the fungal contribution to respiration is estimated as the difference between respiration by mycorrhizal and non-mycorrhizal roots and is expressed on a weight basis. Such comparisons may be facilitated by using matched non-mycorrhizal plants, i.e. plants which have received additional nutrients to achieve a growth pattern similar to that of mycorrhizal plants (Pacovsky and Fuller, 1986; Pacovsky *et al.*, 1986). However, it is possible that the presence of the mycobiont in the root cells also affects respiration of the roots *per se*. Consequently there is a need for exclusive measurements of fungal respiration.

Martin *et al.* (1987) refer to some unpublished results which indicate that ^{14}C-labelled glucose fed to ectomycorrhizal roots of *Picea abies* (L.) Karst. was respired at a higher rate by the external hyphae than by the mantle, and Söderström and Read (1987) report that 30% of the total respiration in an ectomycorrhizal system was likely to be attributable to the external hyphae. So far nobody has been able to measure exclusively the respiration of mycorrhizal hyphae. Such measurements require that hyphae are allowed to grow into a root-free compartment which can be sealed gas tight—with the hyphal connections intact—during the period of measurement. Söderström and Read (1987) used the lower half of a vertically placed Petri dish as the hyphal compartment. Roots were confined to the upper half while hyphae were allowed to grow into the lower compartment. For measuring CO_2 exchange by infrared gas analysis anhydrous lanolin was placed as a seal between the lid and the perspex barrier. The periphery of the dish was sealed by Terostat (Teroson GmbH, Heidelberg, Germany), a very plastic and non-drying material which is widely used as a sealing agent in gas-flow studies. Unfortunately Söderström and Read were not able to express respiration on a dry weight basis as hyphae could not be separated from the peat medium sufficiently to enable gravimetric analysis. A more advanced system, although similar in principle, was developed by Rygiewicz *et al.* (1988) and modified by Andersen and Rygiewicz (1991). Here a narrow passage connects the root compartment and the hyphal compartments. This passage should restrict root growth but allow for the growth of external hyphae. Before the measurement of respiration in the various compartments it should be possible to seal the passage with anhydrous lanolin without damaging the crossing hyphae. No results on hyphal respiration rates are yet available, but a continuing effort along similar lines seems worthwhile. Microelectrodes initially developed to measure O_2 profiles in sediments (Revsbech and Jørgensen, 1986) may also be used to study profiles of oxygen depletion at root surfaces (Zahka, 1989); the electrodes could be useful in studying respiration profiles around hyphal strands of ectomycorrhizal fungi, and if sensitivity can be somewhat increased, respiration profiles around single hyphae and germinating spores may even be achieved.

The overall growth yield of fungi ranges from 0.2 to 0.4 $g C g^{-1} C$ (Perlman, 1965). It has been estimated as 0.18–0.21 $g C g^{-1} C$ in arbuscular fungi (Kucey and Paul, 1982a; Harris and Paul, 1987). Such values may be used to obtain an indirect estimate of fungal respiration based on the incorporation of ^{14}C by a hypha of known biomass (Jakobsen and Rosendahl, 1990b).

III. The carbon balance of mycorrhiza

Due to the carbon drain by the mycobiont, the pattern of carbon partitioning differs between mycorrhizal and non-mycorrhizal plants. It is of considerable interest to study and quantify the carbon flow to the fungus in relation to the carbon balance of the host and in relation to the influence of mycorrhiza on nutrient acquisition. Sometimes mycorrhiza result in growth depressions; such events may be predicted only on the basis of a thorough understanding of the carbon balance of the symbiotic system.

Comparisons of carbon balance in mycorrhizal and non-mycorrhizal plants are subject to methodological problems, particularly involving the choice of control treatment. Obviously it is not appropriate to compare a large mycorrhizal plant with a small nutrient-deficient control plant. Some kind of matching must be attempted. Control plants may be supplied with additional nutrients to mimic growth of the mycorrhizal plants (Pacovsky *et al.*, 1986). The necessary quantities to be added are determined from growth response curves obtained in earlier experiments involving addition of increasing amounts of the nutrient in question to both mycorrhizal and non-mycorrhizal plants (Abbott and Robson, 1984). However, mycorrhizal plants often contain considerably more phosphorus than non-mycorrhizal plants of similar size (Stribley *et al.*, 1980) and host physiology may be affected in several other ways (see Smith and Gianinazzi-Pearson, 1988). In some cases it may therefore be better to compare plants with the same nutrient status but of different size (Baas and Lambers, 1988). The physiological differences between mycorrhizal and non-mycorrhizal plants are so pronounced that it may possibly be more rewarding to study the mycorrhizal plants independently of any non-mycorrhizal control. Changes in carbon balance could thus be monitored after exposing a non-mycorrhizal plant to a heavy colonization pressure or after stressing or eliminating the mycobiont of an already established mycorrhiza. Benomyl and bavistan, systemic fungicides with carbendazim as the active ingredient, have been used with some success for decreasing the activity of the mycobiont without apparently affecting the plant directly (Hale and Sanders, 1982; Carr and Hinkley, 1985; Kough *et al.*, 1987; Dodd and Jeffries, 1989).

An alternative approach was taken by Koch and Johnson (1984) and Douds *et al.* (1988) to eliminate the problems of differences in shoot phosphorus status or other physiological parameters in studies of mycorrhizal effects on carbon allocation patterns in different citrus species. Carbon-14 was fed to shoots of plants with their roots split into a mycorrhizal and non-mycorrhizal half, and more carbon was allocated to mycorrhizal than to non-mycorrhizal root halves, both when expressed as total and specific (dpm g^{-1} root wt) radioactivity. This directly showed that mycorrhizal development increases the sink strength of roots.

A. Photosynthesis

The increased below-ground carbon allocation in mycorrhizal plants is in most cases accompanied by an increased rate of net photosynthesis. This may be measured by analysis of CO_2 exchange rate of whole shoots or single leaves or by their rate of $^{14}CO_2$ uptake. Alternatively the net assimilation rate of CO_2 (NAR) may be calculated from the relative growth rate (RGR) and the leaf area ratio (LAR) as NAR = RGR / LAR (Baas *et al.*, 1989). In relation to mycorrhizal effects on photosynthesis, there are two major problems which need further clarification: it is so far unclear which of several possible mechanisms is responsible for the photosynthetic responses observed, and it is unknown why photosynthesis cannot always compensate for the carbon drain by the mycobiont to prevent mycorrhiza-induced growth depressions. The possible mechanisms involved in mediating photosynthetic responses will be identified in this section. Some guidelines on experimental approaches to these problems are also provided.

1. Possible mechanisms

A primary limitation on the rate of photosynthesis is known to be the accumulation of its end-products in the cytoplasm of leaf cells (Herold, 1980). The drain of carbohydrates by the mycobiont may offset end-product limitation and so facilitate an increased photosynthetic rate. The greater phosphorus concentration in mycorrhizal than in non-mycorrhizal plants of similar size (Stribley *et al.*, 1980) may also affect photosynthesis because the process is known to be influenced by the phosphate level of the chloroplast (Sivak and Walker, 1986). Furthermore, since formation of mycorrhiza often leads to increases in the leaf area ratio (Harris *et al.*, 1985; Baas *et al.*, 1989) and to leaf hydration (Snellgrove *et al.*, 1982; Son and Smith, 1988; Smith and Gianinazzi-Pearson, 1990), there may still be a response when photosynthesis is expressed on the basis of leaf dry weight, even though no such response is evident when photosynthesis is expressed on a leaf area basis. Probably the effect of mycorrhiza on leaf morphology is partly caused by enhanced phosphorus nutrition. An additional supply of phosphorus to non-mycorrhizal plants has been shown to increase the rate of leaf expansion of *Glycine max* L. (Fredeen *et al.*, 1989) and the specific leaf area of *G. max* L. (Pacovsky *et al.*, 1986), as well as the leaf area ratio of *Plantago major* L. (Baas *et al.*, 1989). Finally, ectomycorrhizal fungi produce hormones (Ek *et al.*, 1983) and hormone levels in roots or shoots are influenced both by ectomycorrhiza (Mitchell *et al.*, 1986) and by arbuscular mycorrhiza (Allen *et al.*, 1980, 1982; Baas and Kuiper, 1989). Such hormonal effects may be involved in photosynthetic responses to mycorrhiza as hormones are potential messengers between sinks and sources (Herold, 1980) [see Nylund (1988) and Nylund and Wallander (1989) for a detailed discussion of the hormone theory].

2. *Identification of mechanisms*

Further investigations of the nature of mycorrhizal effects on photosynthesis may benefit from the following experimental approaches. Studies on the role of phosphorus should include several levels of phosphorus supply (Johnson, 1984; Fredeen and Terry, 1988), especially to non-mycorrhizal treatments (Rousseau and Reid, 1990), and the highest level must exceed the level necessary for maximum plant growth. Additionally, it is essential to run preliminary experiments to facilitate the selection of matched control plants; the matching should preferably be made on the basis of similar leaf phosphorus concentrations. The most appropriate control of nutrient supply and plant growth is probably achieved by means of a nutrient flow system, where nutrients are continuously supplied at the same rate as they are removed by the roots (Ingestad and Lund, 1986). The nutrient supply can be adjusted to allow for a constant and maximum relative growth rate without inhibiting mycorrhiza formation (Kähr and Arveby, 1986). Nylund and Wallander (1989) used the system for studying the carbon balance of ectomycorrhizal *Pinus sylvestris* L. grown in expanded clay pellets and the system should also be suitable for studies of arbuscular mycorrhiza, which have been established in nutrient flow culture (Höweler *et al.*, 1982) and in aeroponic culture (Sylvia and Hubbell, 1986). In general, photosynthetic responses to colonization by arbuscular mycorrhizal fungi are not apparant when mycorrhizal and non-mycorrhizal plants are carefully matched with respect to plant size and shoot-phosphorus content (Graham and Syvertsen, 1985; Fredeen and Terry, 1988; Douds *et al.*, 1988; Nemec and Vu, 1990; Syvertsen and Graham, 1990). However, there is a need for including the ontogenetic factor in future studies as mycorrhizal effects on host photosynthesis might vary according to the age of the symbiotic association.

Manipulation of the intensity of mycorrhizal colonization is suitable to provide information on sink effects not related to improved phosphorus nutrition. Rousseau and Reid (1990) supplied different concentrations of inoculum of *Pisolithus tinctorius* (Pers.) Coker and Couch to seedlings of *Pinus taeda* L. and rates of photosynthesis were increased considerably more by the highest infection level than could be ascribed to improved phosphorus nutrition. Different degrees of mycorrhization due to the use of different mycobionts (Nylund and Wallander, 1989; Dosskey *et al.*, 1990, 1991) have also been used to indicate pure sink effects of the mycobionts. Finally, information on the mycobiotic sink size may be obtained from measurements of photosynthesis in response to excision of major parts of the external mycelium.

Conclusions from the experiments discussed above can be correctly drawn and results be properly compared only if growth conditions for the host plants were optimal. While most experiments are done at ambient CO_2 concentrations, the photosynthetic photon flux densities employed vary considerably between experiments. Studies of photosynthetic response mechanisms require that the rate of photosynthesis is not already limited by suboptimal light conditions. While light saturation of photosynthesis may require 1000–1500 $\mu mol\, m^{-2}\, s^{-1}$ more realistic experimental light levels ranging from 600–800 $\mu mol\, m^{-2}\, s^{-1}$ may still be appropriate.

3. *Growth depressions*

Plant growth responses to mycorrhiza disappear when the external nutrient supply or the root density is so high that the plant gains no benefit from the mycobiont. This situation is not common in natural systems but may be present in intensive agricultural systems or pot experiments. Provided there is full compensation for the carbon drain by the mycobiont in the form of an increased photosynthetic rate, plant growth should not be affected by the carbon drain. However, there are several reports on growth depressions in both arbuscular mycorrhiza

(Cooper, 1975; Hall *et al.*, 1977; Buwalda and Goh, 1982; Koide, 1985; Baas and Lambers, 1988; Son and Smith, 1988; Smith and Gianinazzi-Pearson, 1990) and ectomycorrhizas (Molina and Chamard, 1982; Ingestad *et al.*, 1986; Nylund and Wallander, 1989; Rousseau and Reid, 1990). In addition to effects from the physical environment and ontogenetic effects, further attention should be paid to genetically determined differences among mycobionts and host plants. It is known that the extent of growth depression varies with fungal species in ectomycorrhizal species (Nylund and Wallander, 1989: Dosskey *et al.*, 1990, 1991). Furthermore, citrus genotypes producing the largest growth responses to arbuscular mycorrhizas under phosphorus-limiting conditions also exhibited the greatest growth depressions due to mycorrhiza when grown in high phosphorus soils (D.M. Eissenstat and J.H. Graham, pers. commun.). Mycorrhizal and non-mycorrhizal seedlings of *Pinus taeda* L. with a relatively high shoot/root ratio (2.8 on average) had the same relative growth rate at conditions where phosphorus was not limiting plant growth (Rousseau and Reid, 1991). By means of a simple model based on Ledig *et al.* (1976) it was shown that increased below-ground carbon allocation due to mycorrhiza would result in non-detectable effects on relative growth rate if the shoot/root ratio was about 3:1. In contrast, the mycobiont would be expected to consume a larger proportion of the photosynthate if the shoot/root ratio was 1:1 and the mycobiont constituted a similar proportion of root dry matter as in the first case. Under those conditions mycorrhiza-induced increases in photosynthesis may not be sufficient to compensate for the increased drain (J.V.D. Rousseau and C.P.P. Reid, pers commun.). This hypothesis finds some support in the observation that the mycorrhizal dependency was highest in those maize (Hall, 1978) and wheat (Azcon and Ocampo, 1981) cultivars which had the highest shoot/root ratios. Consequently the hypothesis could be further tested by using the natural variation in shoot/root ratios among plant species and cultivars.

B. Shoot/root distribution of carbon

The application of labelled carbon isotopes to shoots followed by analysis of the distribution of the isotope in the plant, fungus and soil has proved to be a superior method for the analysis of the effects of mycorrhiza formation on host carbon balance. Such studies can elucidate mycorrhizal effects on changes in net transfer of photosynthates from shoot to root, i.e. the total additional below-ground costs due to the presence of the mycobiont. Part of these changes may be due to altered partitioning patterns within the plant tissues. Methods for estimating the net transfer of carbon from the host to the fungus are described in Section III.C.

1. Supply of labelled carbohydrates

Carbon-14 labelled substrates may be fed directly to the plants by foliar application (L'Annunziata, 1979). This approach was used by Schumacher and Smucker (1985) to study effects of localized anoxia on carbon partitioning in *Phaseolus vulgaris* L. They applied [U-^{14}C] sucrose directly to abraded areas of the leaves and the uptake of ^{14}C was quantified by removal of the source leaf and determination of the radiolabel in and on that leaf. Obviously these foliar applications of labelled compounds have a limited potential for the study of carbon balance as the pattern of translocation of photosynthates will be different from the normal phloem loading–unloading pattern. The methods may be of some use for studying the below-ground carbon balance but they find their main use when specifically labelled carbohydrates are used for analysis of metabolic pathways.

2. Labelling of shoots with $^{14}CO_2$

Pulse-labelling with $^{14}CO_2$ has been widely used in mycorrhizal research (Bevege *et al.*, 1975; Pang and Paul, 1980; Kucey and Paul, 1982a; Snellgrove *et al.*., 1982; McCool and Menge, 1983; Reid *et al.*, 1983; Koch and Johnson, 1984; Harris *et al.*, 1985; Finlay and Read, 1986; Douds *et al.*, 1988; Smith and Paul, 1988; Cairney *et al.*, 1989; Miller *et al.*, 1989; Nylund and Wallander, 1989; Jakobsen and Rosendahl, 1990b). The general approach employed in all of these studies involves pulse-labelling for relatively short periods (minutes–hours) of shoots placed in a sealed container, a subsequent chase period for equilibration of assimilated ^{14}C in the symbiotic system, and finally a quantification of radiolabel in the different components after harvest. A portable field labelling apparatus for *in situ* labelling of small trees is described by Smith and Paul (1988). Otherwise labellings have been performed under controlled environmental conditions and the system used by Jakobsen and Rosendahl (1990b) is described below as an example (Fig. 1).

Shoots of ten plants are contained by a common perspex canopy in order to avoid working with several flow lines containing radioactive air. The dark respiration of individual shoots cannot be directly measured with this system, but has to be estimated from the assumption that the contribution of each shoot to the total dark respiration is directly related

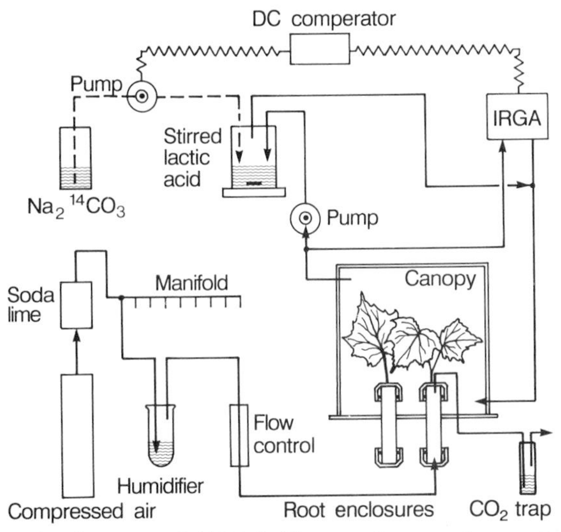

Fig. 1. Schematic diagram of ^{14}C-labelling equipment for studying effects of mycorrhiza on carbon allocation patterns. —, - - - - and ⌇ represent gas flow, liquid flow and electric circuit, respectively.

to the shoot dry weight. Carbon-14 labelled CO_2 is passed from a CO_2 generating unit through the shoot canopy in a closed gas circuit. The CO_2 concentration is continuously monitored by infrared gas analysis (IRGA). A dc comparator is connected to an analogue output of the IRGA and to the peristaltic pump of the CO_2 generator. The circuit diagram of the dc comparator is given in Fig. 2. When CO_2 levels are decreased due to photosynthesis, causing the mV output from the IRGA to drop below a pre-fixed level, the dc comparator activates the pump; this system maintains the CO_2 concentration in the canopy at 390 ppm +/− 5%. Alternatively, the CO_2 concentration may be controlled by a Geiger–Müller tube inserted in the gas circuit and connected to the solenoid valve of a burette containing the $^{14}CO_3$ solution (Harris *et al.*, 1985). The maintenance of both a constant CO_2 concentration and specific radioactivity during the labelling period facilitates the correct interpretation of results. A separate gas-flow system passes CO_2-free air through a manifold connected to hermetically sealed root containers, and the $^{14}CO_2$ released by the individual root systems is collected in 30% ethanolamine in methanol. The glass beads in the central glass tube of the traps provide a large surface area for the absorption of CO_2 in the alkaline solution, and makes the trap 100% effective (Fig. 3).

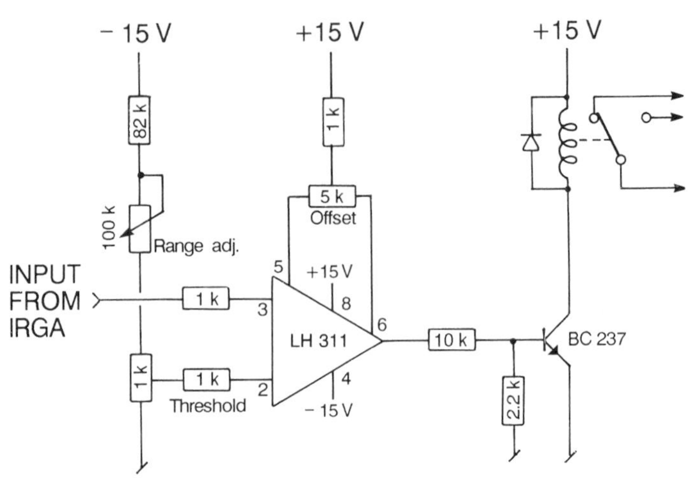

Fig. 2. Circuit diagram of the dc comparator for controlling the CO_2 concentration in the shoot canopy. When the mV input from the IRGA drops below a pre-set threshhold voltage, the limit comparator (LH311) activates the relay and consequently the CO_3 pump.

The labelling period in the example described here was 16 h (one photoperiod), but shorter periods have been used by others. These short-term labellings give information on the movement and losses of recently fixed carbon and are useful for studying both fluxes through different pools in the plant and losses of exudates; however, they give no information on the degradation of whole cells or dying parts of the roots (Whipps, 1990). A uniform labelling of all plant carbon can be obtained only if plants are maintained from seedlings up to maturity in an atmosphere with a constant specific activity of $^{14}CO_2$. Methodological details of such more advanced systems are given by Whipps and Lynch (1983) and Merckx *et al.* (1986). Pulse-labelling is adequate in most mycorrhizal studies where it is not intended to quantify long-term rhizodeposition. However, it is important to extend the chase period until the rate of $^{14}CO_2$ respired from the root system has approached a constant low level. This will ensure that only minor amounts of readily available ^{14}C are still circulating in the plant. Consequently, the $^{14}CO_2$ output from the roots must be monitored by frequent ^{14}C measurements in the CO_2 traps (Snellgrove *et al.*, 1982; Harris *et al.*, 1985; Jakobsen and Rosendahl, 1990b). Carbon-14 in plant or soil components is measured by liquid scintillation counting after wet or dry combustion of homogenized samples (L'Annunziata, 1979). The preparation of samples for scintillation counting is greatly facilitated by the use of a sample oxidizer (e.g. Packard Model 307), where the sample is flame oxidized

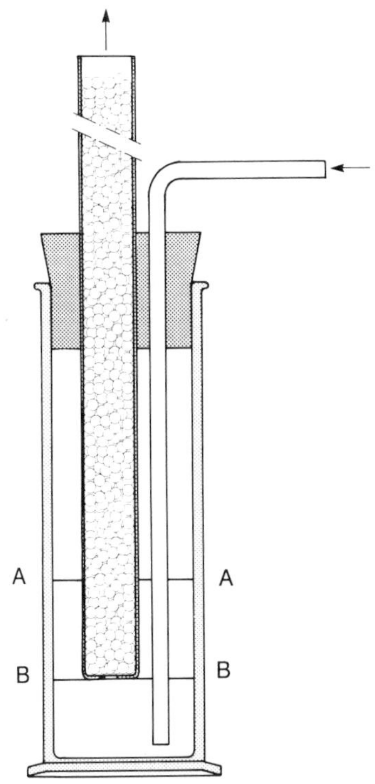

Fig. 3. Carbon dioxide absorber developed by H. Sørensen, Risø, Denmark. Direction of gas flow is indicated by arrows. AA indicates the solution level when at rest, BB when gas is passing through.

and the $^{14}CO_2$ automatically transferred with scintillant to a scintillation vial, ready for counting.

3. *Labelling with* $^{11}CO_2$

The short-lived ^{11}C isotope ($t_{1/2} = 20.3$ min) yields γ rays and can therefore be used with relative ease for measurements *in vivo*; in contrast the use of ^{14}C usually requires destructive techniques. A system using ^{11}C for studying uptake and allocation of carbon in plants is described in detail by Magnuson *et al.* (1982) and Fares *et al.* (1988). Several pairs of scintillation detectors placed at appropriate positions in the shoot–root system make it possible to carry out real-time measurements of allocation of very small amounts of photosynthate. This system has been successfully used for studying effects of arbuscular mycorrhiza on carbon partitioning patterns in *Panicum coloratum* L. (Wang *et al.*, 1989).

C. Carbon consumption by the mycobiont

The isotope labelling methods described above also apply for the measurement of host–fungus carbon flow; however, as the mycobiont is an integral component of the root system it is impossible to distinguish ^{14}C in root matter from ^{14}C in intraradical fungal material; likewise the respiratory contribution of the individual symbiotic partners cannot be readily distinguished. The carbon flow to the fungus therefore has to be estimated on the basis of obtainable measurements in combination with some assumptions. Parameters which can be directly measured include intra- and extraradical fungal biomass, carbon incorporation by extraradical hyphae and loss of organic carbon from the roots. The respiration of the mycobiont and the carbon incorporation by its intraradical phase can be indirectly estimated by comparison of mycorrhizal and matched non-mycorrhizal plants, with the limitation that mycorrhiza may well affect the respiration of root cells. Methods for measurement of total respiration and possible methods for the direct measurement of hyphal respiration are discussed in Section II.C.3.

1. *Fungal biomass*

It is possible to isolate external mycelium of arbuscular mycorrhizal fungi from soil by hand-picking with forceps (Sanders *et al.*, 1977) or by wet-sieving (Jakobsen and Rosendahl, 1990b), but it is difficult to get rid of all fragments of soil organic matter. The peat growth substrates used in ectomycorrhizal research amplify this problem. Consequently gravimetric methods are of limited use for quantifying the biomass of internal fungal components of the roots. The biomass of hyphae is related to their biovolume, which may be calculated from microscope measurements of length and diameter of the hyphae. Kucey and Paul (1982b) measured the length of intraradical hyphae after high-speed blending of colonized roots. Methods for the measurement of external hyphae are discussed elsewhere (*Methods in Microbiology*, Vol. 24, Chapter 3). It is desirable to group the hyphae in classes according to their diameter as the use of an overall mean diameter may greatly underestimate the true biovolume (Bååth and Söderström, 1979; Schnürer *et al.*, 1985). Biovolume may be converted to biomass on the basis of density and dry matter content. Conversion factors (g dry wt cm^{-3}) may be obtained from pure cultures of fungi (Van Veen and Paul, 1979; Bakken and Olsen, 1983) or from hyphae picked from their natural substrates (Lodge, 1987). The conversion factors for hyphae from leaf litter were 0.19–0.23 (Lodge, 1987) while an average of 0.23 was obtained for 10 different fungi (Bakken and Olsen, 1983).

Fungal biomass may also be determined from chemical measurement of compounds which are specific to fungal tissue (Whipps *et al.*, 1982). Chitin, a polymer of *N*-acetylglucosamine and the major wall component in all mycorrhizal fungi, may be quantified colorimetrically after alkaline (Hepper, 1977; Bethlenfalvay *et al.*, 1981) or acid (Vignon *et al.*, 1986) hydrolysis. The analytical procedure involving acid hydrolysis is faster than the alkaline hydrolysis, but has the disadvantage that it is not strictly specific as aldehydes derived from plant material are also sensitive to the colour reaction (Vignon *et al.*, 1986). Glucosamine contents in roots may be converted to fungal biomass using the specific glucosamine content of clean external mycelium. This conversion factor ranged from 21–40 μg glucosamine mg^{-1} fungal dry weight in five fungi forming arbuscular mycorrhiza (Hepper, 1977; Bethlenfalvay *et al.*, 1982), indicating that conversion factors should be determined in each experimental situation. The use of conversion factors obtained from pure cultures of fungi forming ericoid and ectomycorrhiza should be avoided until it has been shown that the conversion factors are unaffected by changed environmental conditions. The use of these conversion factors further necessitates the assumption that the specific glucosamine content is similar for intra- and extraradical hyphae. The chitin method is suited for determination of biomass of mycorrhizal fungi in pathogen-free roots, while its use on external hyphae is hampered by the native amino sugar component (Parsons, 1981) and chitin-containing non-mycorrhizal organisms in soil. Pacovsky and Bethlenfalvay (1982) used a rather complicated wet-sieving procedure to get rid of the amino sugar component and obtained reasonably large differences between chitin content in soil from mycorrhizal and non-mycorrhizal plants. In general, the method is suitable only for soils low in organic matter and a considerable degree of variability may be introduced due to the many steps in the procedure (Bethlenfalvay and Ames, 1987).

The chitin assay measures total chitin in both viable and dead hyphae. Methods for measuring the biomass of viable hyphae are important in relation to functional aspects of the mycorrhiza. Living biomass may be determined microscopically from total biovolume and the proportion of biovolume showing metabolic activity after staining with fluorescein diacetate (Schubert *et al.*, 1987) or iodotetrazolium (Sylvia, 1988). Recently, image-analysis techniques have been applied to quantify the activity of arbuscules measured as the intensity of staining for succinate dehydrogenase with nitroblue tetrazolium (S.E. Smith, pers. commun.).

The availability of biochemical assays for detecting compounds which are specific to living mycorrhizal hyphae would facilitate a more exact measurement of viable biomass of mycorrhizal fungi. Ergosterol is the dominant sterol in most fungi (Weete, 1974) and is becoming increasingly popular as an index of fungal biomass in soil (Grant and West, 1986) and in plant material (Newell *et al.*, 1988). Salamanowicz and Nylund (1988) measured the ergosterol content of hyphae of ectomycorrhizal fungi by HPLC and found only small differences due to fungal species and time; values ranged from 2.88–4.03 mg ergosterol g^{-1} dry matter. Ergosterol analysis may be performed on samples as small as 2 mg fresh wt and was therefore superior to the chitin method for quantifying fungal growth during early stages of ectomycorrhiza formation (Martin *et al.*, 1990). Ergosterol analysis is not adequate for measuring biomass of external hyphae of ectomycorrhizal fungi in unsterile substrates as the results would be influenced by the ergosterol content of the background fungi. Ergosterol has also been identified in arbuscular mycorrhizal fungi but in general concentrations seem to be very low (Beilby, 1980; Beilby and Kidby, 1980a; Nagy *et al.*, 1980; Nordby *et al.*, 1981). The arbuscular mycorrhizal fungi contain large quantities of a number of fatty acids which are not found in plants (see Section II.C.1). When quantification is achieved these fatty acids, especially the 16:1(11*c*) and 18:3(6*c*,9*c*,12*c*) may turn out to be useful indicators of the living biomass of arbuscular mycorrhizal fungi. The composition of fatty acids differs between fungal taxonomic groups (Weete, 1974). In Ascomycetes and Basidiomycetes the 18:3 component has a (9*c*,12*c*,15*c*) conformation in contrast to the (6*c*,9*c*,12*c*) found in Phycomycetes. Furthermore, the fatty acids of Zygomycetes differ in their degree of unsaturation from that of Chytridomycetes and Oomycetes, as the two latter groups have a higher potential for long-chain polyunsaturated ($>$ 18:3) fatty-acid synthesis than the Zygomycetes. There is a need for more detailed comparisons of the fatty acids in arbuscular mycorrhizal fungi and in saprotrophs isolated from soil, in order to assess the potential value of the fatty acids as markers for external mycorrhizal hyphae.

2. *Carbon incorporation by external hyphae*

The carbon flow to the external hyphae is of importance not only for the study of the carbon balance of the symbiosis but also to the distribution and cycling of carbon in the soil ecosystem. This subject has been reviewed recently by Finlay and Söderström (1991). The use of ^{14}C labelling has made it possible to quantify the amount of carbon which is incorporated by the hyphae.

Methods used for the sampling of hyphae for scintillation measurements are not always well described. Bevege *et al.* (1975) quantified the ^{14}C activity in the total amount of external hyphae from pulse-labelled *Trifolium subterraneum* L. and Kucey and Paul (1982a) picked hyphae for ^{14}C determination from soil cores taken from pots with *Vicea faba* L. but no details on sampling techniques were provided. Visible hyphal strands were picked with forceps from a sandy soil of a 27 cm × 27 cm × 2 cm Plexiglass chamber with *Pinus ponderosa* Laws. and specific activity determined (Norton *et al.*, 1990). In a carbon translocation study with mycorrhizal *Pinus sylvestris* L. grown in peat in a similar Plexiglass chamber the peat with hyphae was dissected in squares of equal size and the ^{14}C activity determined for each square (S. Erland, pers. commun.). A quantitative sampling of external hyphae is facilitated from growth systems including a special hyphal compartment, where roots are prevented from crossing the barrier between the hyphal compartment and the root compartment. The hyphal compartments of the systems developed by Rygiewicz *et al.* (1988) (see Section II.C.3) contained glass fibre filter paper moistened with nutrient solution. At harvest hyphal dry weights were determined gravimetrically and their ^{14}C content measured (Miller *et al.*, 1989; Andersen and Rygiewicz, 1991). This system is designed also to allow for measuring $^{14}CO_2$ development in the hyphal compartments and consequently the carbon flow to the hyphae and their growth efficiency may be determined.

The following method is suitable for the quantification of carbon incorporation by external hyphae of arbuscular mycorrhiza (Jakobsen and Rosendahl, 1990b). A soil container is divided into two compartments by a nylon or stainless steel mesh which prevents root growth but allows hyphae to penetrate. A 25–35 μm mesh will eliminate root penetration, even of fine-rooted grasses. The compartmentation may take different forms depending on the container used. A simple solution is to grow the roots in a mesh bag inserted in a pot or PVC tube filled with soil (Fig. 4). The PVC tubes may be hermetically sealed for $^{14}CO_2$ labelling very easily by using standard PVC closing sockets (Fig. 4) in combination with O rings and Terostat. At harvest the external mycelium may be quantitatively extracted by repeated wet-sieving and decanting from either the whole soil volume or from a subsample of the soil. Hyphae are collected on a fine sieve (40–50 μm mesh). At least 5–10 resuspensions of the soil in water followed by decanting are needed. Care must be taken to collect hyphal aggregates trapped in the

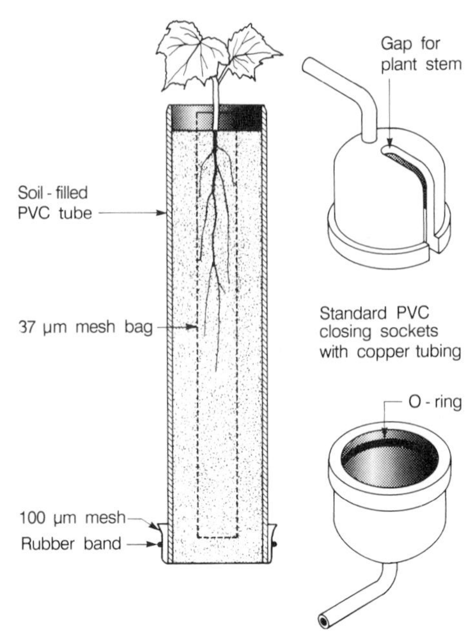

Fig. 4. An example of a growth system suitable for measuring incorporation of ^{14}C by the external hyphae of arbuscular mycorrhiza. Hyphae can grow out into the hyphal compartment surrounding the mesh bag containing the roots. The two closing sockets are mounted before labelling.

sand precipitate in the suspension beaker; a magnifying glass may be useful in this respect. The method is probably not adequate for soils high in organic matter. The hyphae on the sieve are washed and analysed and their ^{14}C content expressed on the basis of dry weight obtained gravimetrically or by conversion from biovolume (Bakken and Olsen, 1983). The radioactivity in the total amount of external hyphae may be calculated if the hyphal density is assumed to be similar inside and outside the root compartment. This will depend on the size of the hyphal compartment and on the mycorrhizal fungus used, but in any case the hyphal density is never smaller in the root compartment than in the hyphal compartment (I. Jakobsen, unpubl. res.). Another problem is that it is not yet known whether similar amounts of hyphae inside and outside the root bag are likely to incorporate equal amounts of ^{14}C. Adaptation of the system with a hermetically sealed hyphal compartment (Andersen and Rygiewicz, 1991) to arbuscular mycorrhiza would allow for respiration measurements and the direct measurement of carbon flow to the hyphae.

3. *Extra-mycorrhizal loss of organic carbon from roots*

In addition to respired CO_2 and carbon contained in external mycorrhizal hyphae, roots also lose considerable amounts of carbon as exudates (carbohydrates, organic acids, amino acids), secretions (polymeric carbohydrates) and lysates (sloughed cells or part of cells) (see Whipps, 1990). This non-hyphal flow of organic carbon may be influenced by mycorrhiza (Graham *et al.*, 1981) and should be quantified in studies of carbon balance in mycorrhiza. However, exudates are rapidly metabolized by micro-organisms in the growth medium (Minchin and McNaughton, 1984) and meaningful data can therefore be obtained only from aseptically grown plants (Laheurte and Berthelin, 1986; Lipton *et al.*, 1987). This problem was addressed in studies of arbuscular mycorrhiza by removing roots from soil, washing and treating roots with antibiotics and subsequently collecting exudates in $CaCl_2$ for a number of hours (Graham *et al.*, 1981; McCool and Menge, 1983; Schwab *et al.*, 1983). However, it is likely that the mechanical disturbance of roots and the removal of micro-organisms would have affected exudation patterns (Whipps, 1990). Quantification of carbon loss from undisturbed roots of soil-grown plants is best performed by means of $^{14}CO_2$ pulse-labelling of shoots. Carbon-14 in soil samples including external mycelium of arbuscular mycorrhiza was measured by Snellgrove *et al.* (1982), Harris *et al.* (1985) and Douds *et al.* (1988). Before analysis of ^{14}C, the external hyphae may be quantitatively removed (see Section III.C.2). After filtration of the washing water to remove all clay, silt and sand fractions, soluble ^{14}C compounds are measured in subsamples from the filtrate while insoluble ^{14}C is measured in subsamples from the material retained by the filter (Whipps and Lynch, 1983).

D. Interplant transfer of carbon

Due to a relatively weak host specificity, mycorrhizal fungi are able to form hyphal links between individuals of the same or of different plant species. It has been shown that ^{14}C fed to one shoot may appear in the shoot of a receiver plant connected by a common mycelium (Read *et al.*, 1985). There is a need to establish whether a net transfer of carbon actually occurs in these cases (Newman, 1988). It should be possible to study this by running two parallel labelling experiments. In one the "donor" is labelled with $^{14}CO_2$ and the "receiver" is analysed for the presence of ^{14}C. In the other $^{14}CO_2$ is applied to the shoot of the "receiver" and the fate of the assimilated ^{14}C is monitored. Comparing the interplant transfer of ^{14}C in the two cases should enable an assessment of whether net transfer has occurred. Even if net transfer actually does not happen, hyphal interconnections may still be important by providing carbon for the function of the hyphal network supporting the "receiver" and thereby ensuring the ability of this mycelium to take up inorganic nutrients (Finlay and Read, 1986). This may be tested by studying the nutrient uptake capacity of a "receiver" plant with and without the presence of hyphal interconnections. The stable carbon isotope ^{13}C has been used to study the carbon source of mycorrhizal fungi in mixed stands of C3 and C4 plants which differ in their discrimination against ^{13}C (Rundel *et al.*, 1989). Measurement of $^{13}C/^{12}C$ ratios in plant components and in spores collected from the rhizosphere of the C4 plants *Atriplex gardneri* (Moq.) D.Dietr. and *Salsola kali* L. and the C3 plants *Agropyrum dasystachum* (Hook.) Scribn. and *A. smithii* Rydb. showed that mycorrhiza of the supposedly non-mycorrhizal *Chenopodiaceae* supported carbon for spore formation in their rhizosphere (Allen and Allen, 1990).

IV. Carbon costs in relation to nutrient uptake and growth response

The carbon costs involved in nutrient absorption by a unit length of mycorrhizal hypha would generally be expected to be lower than that of a corresponding length of root, which is much thicker, and the amount of phosphorus uptake per unit of carbon allocated below-ground would consequently be expected to be higher in mycorrhizal than in non-mycorrhizal plants. On the other hand ^{14}C studies have indicated that the maintenance costs of the mycobiont may be higher than expected from its biomass (Harris *et al.*, 1985), perhaps due to costs imposed by translocation of phosphorus through hyphae and its transfer to the host.

A theoretical framework for studying the carbon efficiency in mycorrhiza was given by Koide and Elliot (1989). They defined the efficiency of below-ground carbon utilization, $\Delta C^{w}/\Delta C^{b}$, where ΔC^{w} is the total carbon accumulation in the whole plant per unit time while ΔC^{b} is the total below-ground allocation of carbon per unit time. This ratio, however, is the product of the efficiency of phosphorus utilization ($\Delta C^{w}/\Delta P^{w}$) and the efficiency of phosphorus acquisition ($\Delta P^{w}/\Delta C^{b}$), where ΔP^{w} represents the total phosphorus uptake per unit time. Arbuscular mycorrhiza had no influence on the efficiency of phosphorus acquisition by the roots of citrange seedlings (Douds *et al.*, 1988). Below-ground carbon allocation was measured by ^{14}C pulse-labelling and this was compared to the total plant phosphorus content at harvest of the labelled plants. The results would be easier to interpret if carbon partitioning and phosphorus uptake had been measured during the same time interval. This was attempted by Jones *et al.* (1991) in a similar study with ectomycorrhizal and non-mycorrhizal *Salix viminalis* L. The amount of phosphorus taken up by the plants over defined time intervals was related to the below-ground carbon allocation during the same intervals. The carbon allocation was estimated from $^{14}CO_2$ pulse-labellings performed once during each time interval. The efficiency of phosphorus acquisition was found to be highest in mycorrhizal plants during the initial 50-day growth period, while it was highest in non-mycorrhizal plants during the subsequent 48-day period. In order to obtain realistic results in these studies of efficiency it is important that root and fungal development occur at levels which are comparable to the field situation. The effects of mycorrhiza on phosphorus uptake are known to decrease with increasing root densities (Bååth and Hayman, 1984). Optimal conditions for root growth, combined with a limited soil volume often produce root densities in pots which are an order of magnitude larger than in the field, while hyphal lengths are likely to be proportionally reduced.

In a study of "mycorrhizal efficiency", below-ground carbon allocation measured by ^{14}C techniques was similar in leek plants inoculated with three different arbuscular fungi (Menge *et al.*, 1985). In this case the production of external hyphae and the mycorrhizal effects on phosphorus uptake were also unaffected by the species of mycobiont. However, the total amount of hyphae per unit colonized root length and the spatial distribution of the hyphae may vary considerably with species of arbuscular fungi (I. Jakobsen, unpubl. res.); these fungi also differed in their capacity for uptake of phosphorus per unit length of hyphae. Consequently, there is scope for further comparisons of carbon allocation to the external hyphae of mycorrhiza when different species of mycobionts are involved. An hermetically sealed hyphal compartment containing ^{32}P-labelled soil, in combination with ^{14}C pulse-labelling, would facilitate simultaneous measurement of carbon flow to and phosphorus uptake by external hyphae of different fungi.

V. Conclusions

Methods for studying the carbon metabolism of mycorrhizal fungi in pure culture and of entire mycorrhiza are discussed. It is shown that some problems may be investigated by means of methods already developed. Among these are enzymatic studies of hexose catabolism in the arbuscular fungi including the possibly important role of dark CO_2 fixation providing anaplerotic pathways to Krebs cycle intermediates, as well as ^{14}C studies of below-ground carbon allocation, carbon incorporation by the external hyphae and interplant carbon flow mediated by connecting hyphal networks. There is an obvious need for comparative work using well-defined species of both plants and fungi. Other important areas are methodologically less well developed. These include studies of mechanisms involved in the host–fungus carbon transfer and direct measurements of host–fungus carbon flow. Approaches already developed for the study of other symbiotic associations may be an important source of new experimental initiatives in mycorrhizal research.

References

Abbott, L. K. and Robson, A. D. (1984). In *VA Mycorrhiza* (C. Ll. Powell and D. J. Bagyaraj, eds), pp. 113–130. CRC Press, Boca Raton, FL.

Abuzinadah, R. A. and Read, D. J. (1986). *New Phytol* **103**, 481–493.

Allen, M. F. and Allen, E. B. (1990). *Ecology* **71**, 2019–2021.

Allen, M. F., Moore, T. S. Jr. and Christensen, M. (1980). *Can. J. Bot.* **58** 371–374.

Allen, M. F., Moore, T. S. Jr. and Christensen, M. (1982). *Can. J. Bot.* **60**, 468–471.

Amijee, F., Stribley, D. P. and Tinker, P. B. (1990). *Plant Soil* **124**, 195–198.

Andersen, C. P. and Rygiewicz, P. T. (1991). *Environ. Pollution*. (in press).

Antibus, R. K., Trappe, J. M. and Linkins, A. E. (1980). *Can. J. Bot.* **58**, 14–20.

Azcón, R. and Ocampo, J. A. (1981). *New Phytol.* **87**, 677–685.

Baas, R. and Kuiper, D. (1989). *Physiol. Plant.* **76**, 211–215.

Baas, R. and Lambers, H. (1988). *Physiol. Plant.* **74**, 701–707.

Baas, R., van der Werf and Lambers, H. (1989). *Plant Physiol.* **91**, 227–232.

Bååth, E. and Hayman, D. S. (1984). *Plant Soil* **77**, 373–376.

Bååth, E. and Söderström, B. (1979). *Rev. d'Ecol. Biol. Sol.* **16**, 477–489.

Bajwa, R. and Read, D. J. (1985). *New Phytol.* **101**, 459–467.

Bakken, L. R. and Olsen, R. A. (1983). *Appl. Environ. Microbiol.* **45**, 1188–1195.

Bécard, G and Piché, Y. (1989). *Appl. Environ. Microbiol.* **55**, 2320–2325.

Beilby, J. P. (1980). *Lipids* **15**, 949–952.

Beilby, J. P. (1983). *Can. J. Microbiol.* **29**, 596–601.

Beilby, J. P. and Kidby, D. K. (1980a). *Lipids* **15**, 375–378.

Beilby, J. P. and Kidby, D. K. (1980b). *J. Lipid Res.* **21**, 739–750.

Bethlenfalvay, G. J. and Ames, R. N. (1987). *Soil Sci. Soc. Am. J.* **51**, 834–837.

Bethlenfalvay, G. J. and Pacovsky, R. S. (1983). *Plant Physiol.* **73**, 969–972.

Bethlenfalvay, G. J., Pacovsky, R. S. and Brown, M. S. (1981). *Soil Sci. Soc. Am. J.* **45**, 871–875.

Bethlenfalvay, G. J., Brown, M. S. and Pacovsky, R. S. (1982). *New Phytol.* **90**, 537–543.
Bevege, D. I., Bowen, G. D. and Skinner, M. F. (1975). In *Endomycorrhizas* (F. E. Sanders, B. Mosse and P. B. Tinker, eds), pp. 149–174. Academic Press, London.
Buwalda, J. G. and Goh, K. M. (1982). *Soil Biol. Biochem.* **14**, 103–106.
Cairney, J. W. G. , Ashford, A. E. and Allaway, W. G. (1989). *New Phytol.* **112**, 495–500.
Carr, G. R. and Hinkley, M. A. (1985). *Soil Biol. Biochem.* **17**, 313–316.
Carr, G. R., Hinkley, M. A., Le Tacon, F., Hepper, C. M., Jones, M. G. K. and Thomas, E. (1985). *New Phytol.* **101**, 417–426.
Casselton, P. J. (1976). In *The Filamentous Fungi* (J. E. Smith and D. R. Berry, eds), Vol. 2, pp. 121–136. Edward Arnold, London.
Coleman, J. O. D. and Harley, J. L. (1976). *New Phytol.* **10**, 317–330.
Cooper, K. M. (1975). In *Endomycorrhizas* (F. E. Sanders, B. Mosse and P. B. Tinker, eds), pp. 391–408. Academic Press, London.
Cooper, K. M. (1984). In *VA Mycorrhiza* (C. Ll. Powell and D. J. Bagyaraj, eds), pp. 155–186. CRC Press, Boca Raton, FL.
Cooper, K. M. and Lösel, D. M. (1978). *New Phytol.* **80**, 143–151.
Cox, G., Sanders, F. E., Tinker, P. B. and Wild, J. A. (1975). In *Endomycorrhizas* (F. E. Sanders, B. Mosse and P. B. Tinker, eds), pp. 298–312. Academic Press, London.
Daft, M. J. and El-Giahmi, A. A. (1978). *New Phytol.* **80**, 365–372.
Davis, R. M. and Fucik, J. E. (1986). *Hort. Sci.* **21**, 302–304.
Day, D. A., Price, G. D. and Udvardi, M. K. (1989). *Austral. J. Plant. Phys.* **16**, 69–84.
Dehne, H. W. (1986). In *Physiological and Genetical Aspects of Mycorrhizae* (V. Gianinazzi-Pearson and S. Gianinazzi, eds), pp. 431–435. INRA Press, Paris.
Diederichs, C. (1982). *Angew. Bot.* **56**, 325–333.
Diederichs, C. (1983a). *Angew. Bot.* **57**, 45–53.
Diederichs, C. (1983b). *Angew. Bot.* **57**, 55–67.
Dodd, J. C. and Jeffries, P. (1989). *Biol. Fertil. Soils* **7**, 120–128.
Dosskey, M. G., Linderman, R. G. and Boersma, L. (1990). *New Phytol.* **115**, 269–274.
Dosskey, M. G., Boersma, L. and Linderman, R. G. (1991). *New Phytol.* **117**, 327–334.
Douds, D. D. Jr., Johnson, C. R. and Koch, K. E. (1988). *Plant Physiol.* **86**, 491–496.
Ek, M., Ljungquist, P. O. and Stenström, E. (1983). *New Phytol.* **94**, 401–407.
Fares, Y., Goeschl, J. D., Magnuson, C. E., Scheld, H. W. and Strain, B. R. (1988). *J. Radioanal. Nucl. Chem.* **124**, 103–122.
Ferguson, J. J. and Menge, J. A. (1982). *New Phytol.* **92**, 183–191.
Finlay, R. D. and Read, D. J. (1986). *New Phytol.* **103**, 143–156.
Finlay, R. and Söderström, B. (1991). In *Mycorrhizal Functioning* (M. Allen, ed.). Chapman and Hall, London (in press).
France, R. C. and Reid, C. P. P. (1983). *Can J. Bot.* **61**, 964–984.
Fredeen, A. L. and Terry, N. (1988). *Can. J. Bot.* **66**, 2311–2316.
Fredeen, A. L., Rao, I. M. and Terry, N. (1989). *Plant Physiol.* **89**, 225–230.
Graham, J. H. and Syvertsen, J. P. (1985). *New Phytol.* **101**, 667–676.

Graham, J. H., Leonard, R. T. and Menge, J. A. (1981). *Plant Physiol.* **68**, 548–552.
Grant, W. D. and West, A. W. (1986). *J. Microbial Meth.* **6**, 47–53.
Hale, K. A. and Sanders, F. E. (1982). *J. Plant. Nutr.* **5**, 1355–1367.
Hall, I. R. (1978). *N. Z. J. Agric. Res.* **21**, 517–519.
Hall, I. R., Scott, R. S. and Johnstone, P. D. (1977). *N. Z. J. Agric. Res.* **20**, 349–355.
Harley, J. L. and Smith, S. E. (1983). *Mycorrhizal Symbiosis*. Academic Press, London.
Harris, D. and Paul, E. A. (1987). In *Ecophysiology of VA Mycorrhizal Plants* (G. R. Safir, ed.), pp. 93–106. CRC Press, Boca Raton, FL.
Harris, D., Pacovsky, R. S. and Paul, E. A. (1985). *New Phytol.* **101**, 427–440.
Haselwandter, K., Bobleter, O. and Read, D. J. (1990). *Arch. Microbiol.* **153**, 352–354.
Hayman, D. S. (1974). *New Phytol.* **73**, 71–80.
Hepper, C. M. (1977). *Soil Biol. Biochem.* **9**, 15–18.
Hepper, C. M. (1987). In *Proceedings of the 7th North American Conference on Mycorrhizae* (D. M. Sylvia, L. L. Hung and J. A. Graham, eds), pp. 172–174. IFAS, Gainesville, FL.
Herold, A. (1980). *New Phytol.* **86**, 131–144.
Hirrel, M. C. and Gerdemann, J. W. (1979). *New Phytol.* **83**, 731–738.
Ho, I. and Trappe, J. M. (1973). *Nature* **244**, 30–31.
Höweler, R. H., Asher, C. J. and Edwards, D. G. (1982). *New Phytol.* **90**, 229–238.
Hutchinson, L. J. (1990). *Can. J. Bot.* **68**, 1522–1530.
Ingestad, T. and Lund, A. -B. (1986). *Scan. J. For. Res.* **1**, 439–453.
Ingestad, T., Arveby, A. S. and Kähr, M. (1986). *Physiol. Plant.* **68**, 575–582.
Jakobsen, I. and Rosendahl, L. (1990a). *Agric. Ecosyst. Environ.* **29**, 205–209.
Jakobsen, I. and Rosendahl, L. (1990b). *New Phytol.* **115**, 77–83.
Jasper, D. A., Robson, A. D. and Abbott, L. K. (1979). *Soil Biol. Biochem.* **11**, 501–505.
Johnson, C. R. (1984). *Plant Soil* **80**, 35–42.
Johnson, C. R., Menge, J. A., Schwab, S. and Ting, I. P. (1982). *New Phytol.* **90**, 665–669.
Johnson, P. N. (1976). *N. Z. J. Bot.* **14**, 333–340.
Jones, M. D., Durall, D. M. and Tinker, P. B. (1991). *New Phytol.* (in press).
Koch, K. E. and Johnson, C. R. (1984). *Plant Physiol.* **75**, 26–30.
Koide, R. (1985). *New Phytol.* **99**, 449–462.
Koide, R. and Elliot, G. (1989). *Functional Ecol.* **3**, 252–255.
Kough, J. L., Gianinazzi-Pearson, V. and Gianinazzi, S. (1987). *New Phytol.* **106**, 707–715.
Kucey, R. M. N. and Paul, E. A. (1982a). *Soil Biol. Biochem.* **14**, 407–412.
Kucey, R. M. N. and Paul, E. A. (1982b). *Soil Biol. Biochem.* **14**, 413–414.
Kähr, M. and Arveby, A. (1986). *Physiol. Plant.* **67**, 333–339.
Laheurte, F. and Berthelin, J. (1986). In *Physiological and Genetical Aspects of Mycorrhizae* (V. Gianinazzi-Pearson and S. Gianinazzi, eds), pp. 339–344. INRA Press, Paris.
Lambers, H. (1982). *Physiol. Plant.* **55**, 478–485.
Lambers, H. (1985). In *Encyclopedia of Plant Physiology, New Series* (R. Douce and D. A. Day, eds), Vol. 18, pp. 418–473. Springer-Verlag, Berlin.

L'Annunziata, M. F. (1979). *Radiotracers in Agricultural Chemistry*. Academic Press, London.
Ledig, F.T., Drew, A. P. and Clark, J. G. (1976). *Ann. Bot.* **40**, 289–300.
Lewis, D. H. (1975). In *Endomycorrhizas* (F. E. Sanders, B. Mosse and P. B. Tinker, eds), pp. 119–148. Academic Press, London.
Lewis, D. H. (1986). In *Physiological and Genetical Aspects of Mycorrhizae* (V. Gianinazzi-Pearson and S. Gianinazzi, eds), pp. 85–100. INRA Press, Paris.
Lewis, D. H. and Harley, J. L. (1965). *New Phytol.* **64**, 256–269.
Lipton, D. S., Blanchar, R. W. and Blevins, D. G. (1987). *Plant Physiol.* **85**, 315–317.
Lodge, D. J. (1987). *Soil Biol. Biochem.* **19**, 727–733.
MacDonald, R. M. and Lewis, M. (1978). *New Phytol.* **80**, 135–141.
Magnuson, C. E., Fares, Y., Goeschl, J. D., Nelson, C. E., Strain, B. R., Jaeger, C. H. and Bilpuch, E. C. (1982). *Rad. Environ. Biophys.* **21**, 51–65.
Martin, F., Canet, D. and Marchal, J. P. (1984). *Physiol. Vég.* **22**, 733–743.
Martin, F., Canet, D. and Marchal, J. P. (1985). *Plant Physiol.* **77**, 499–502.
Martin, F., Ramstedt, M. and Söderhäll, K. (1987). *Biochimie* **69**, 569–581.
Martin, F., Delaruelle, C. and Hilbert, J. -L. (1990). *Mycol. Res.* **94**, 1059–1064.
Martin, F., Canet, D., Marchal, J. P. and Brondeau, J. (1984). *Plant Physiol.* **75**, 151–153.
Marx, D. H. and Kenney, D. S. (1982). In *Methods and Principles of Mycorrhizal Research* (N. C. Schenck, ed.), pp. 131–146. The American Phytopathological Society, St. Paul, MN.
McCool, P. M. and Menge, J. A. (1983). *New Phytol.* **94**, 241–247.
McRae, D. G., Miller, R. W., Berndt, W. B. and Jay, K. (1989). *Mol. Plant–Microbe Interactions* **2**, 273–278.
Melin, E. and Nilsson, H. (1957). *Svensk Bot. Tidsskr.* **51**, 166–186.
Menge, J. A., Tinker, P. B., Stribley, D. and Snellgrove, R. (1985). In *Proceedings of the 6th North American Conference on Mycorrhizae* (R. Molina, ed.), p. 394. Forest Research Laboratory, Oregon State University, Corvallis, OR.
Merckx, R., van Ginkel, J. H., Sinnaeve, J. and Cremers, A. (1986). *Plant Soil* **96**, 85–93.
Miller, S. L., Durall, D. M. and Rygiewicz, P. T. (1989). *Tree Physiol.* **5**, 239–249.
Minchin, P. E. H. and McNaughton, G. S. (1984). *J. Exp. Bot.* **35**, 74–82.
Mitchell, R. J., Garrett, H. E., Cox, G. S. and Atalay, A. (1986). *Tree Physiol.* **1**, 1–8.
Molina, R. and Chamard, J. (1982). *Can J. For. Res.* **13**, 89–95.
Molina, R. and Palmer, J. G. (1987). In *Methods and Principles of Mycorrhizal Research* (N. C. Schenck, ed.), pp. 115–130. The American Phytopathological Society, St. Paul, MN.
Nagy, S., Nordby, H. E. and Nemec, S. (1980). *New Phytol.* **85**, 377–384.
Nelson, S. D. and Khan, S. U. (1990). *J. Agric. Food Chem.* **38**, 894–898.
Nemec, S. and Vu, J. C. V. (1990). *Plant Soil* **128**, 257–263.
Newell, S. Y., Arsuffi, T. L. and Fallon, R. D. (1988). *Appl. Environ. Microbiol.* **54**, 1876–1879.
Newman, E. I. (1988). *Adv. Ecol. Res.* **18**, 243–270.
Nordby, H. E. Nemec, S. and Nagy, S. (1981). *J. Agric. Food Chem.* **29**,

396–401.
Norton, J. M., Smith, J. L. and Firestone, M. K. (1990). *Soil Biol. Biochem.* **22**, 449–455.
Nylund, J. -E. (1988). *Scan, J. For. Res.* **3**, 465–470.
Nylund, J. -E. and Wallander, H. (1989). *New Phytol.* **112**, 389–398.
Ocampo, J. A. and Azcon, R. (1985). *Plant Soil* **68**, 548–552.
Pacovsky, R. S. (1988). *Plant Soil* **110**, 283–287.
Pacovsky, R. S. (1989). *Soil Biol. Biochem.* **21**, 953–960.
Pacovsky, R. S. and Bethlenfalvay, G. J. (1982). *Plant Soil* **68**, 143–147.
Pacovsky, R. S. and Fuller, G. (1986). *Plant Soil* **95**, 361–377.
Pacovsky, R. S. and Fuller, G. (1988). *Physiol. Plant* **72**, 733–746.
Pacovsky, R. S., Bethlenfalvay, G. J. and Paul, E. A. (1986). *Crop Sci.* **26**, 151–156.
Pang, P. C. and Paul, E. A. (1980). *Can. J. Soil Sci.* **60**, 241–250.
Parsons, J. W. (1981). In *Soil Biochemistry* (E. A. Paul and N. J. Ladd, eds), Vol. 5, pp. 197–227. Marcel Dekker, New York.
Pearson, V. and Read, D. J. (1973). *New Phytol.* **72**, 371–379.
Pearson, V. and Read, D. J. (1975). *Trans. Br. Mycol. Soc.* **64**, 1–7.
Perlman, D. (1965). In *The Fungi—An Advanced Treatise* (G. C. Ainsworth and A. S. Sussmann, eds), Vol. 1, pp. 479–489. Academic Press, New York.
Ratnayake, M., Leonard, R. T. and Menge, J. A. (1978). *New Phytol.* **81**, 543–552.
Read, D. J., Francis, R. and Finlay, R. D. (1985). In *Ecological Interactions in Soil* (A. H. Fitter, ed.), pp. 193–217. Blackwell, Oxford.
Reid, C. P. P., Kidd, F. A. and Ekwebelam, S. A. (1983). *Plant Soil* **71**, 415–432.
Revsbech, N. P. and Jørgensen, B. B. (1986). In *Advances in Microbial Ecology* (K. C. Marshall, ed.), Vol. 9, pp. 293–352. Plenum Press, New York.
Rousseau, J. V. D. and Reid, C. P. P. (1989). In *Applications of Continuous and Steady-State Methods to Root Biology* (J. G. Torrey and L. J. Winship, eds), pp. 183–196. Kluwer Academic Publishers, Dordrecht.
Rousseau, J. V. D. and Reid, C. P. P. (1990). *Forest Sci.* **36**, 101–112.
Rousseau, J. V. D. and Reid, C. P. P. (1991). *New Phytol.* **117**, 319–326.
Rundel, P. W., Ehleringer, J. R. and Nagy, K. A. (1989). *Stable Isotopes in Ecological Research.* Springer-Verlag, New York.
Rygiewicz, P. T., Miller, S. L. and Durall, D. M. (1988). *Plant Soil* **109**, 281–284.
Salmanovicz, B. and Nylund, J. -E. (1988). *Eur. J. For. Pathol.* **18**, 291–298.
Same, B. I., Robson, A. D. and Abbott, L. K. (1983). *Soil Biol. Biochem.* **15**, 593–597.
Sanders, F. E., Tinker, P. B., Black, R. L. B. and Palmerley, S. (1977). *New Phytol.* **78**, 257–268.
Schnürer, J., Clarholm, M. and Rosswall, T. (1985). *Soil Biol. Biochem.* **17**, 611–618.
Schubert, A., Marzachi, C., Mazzitelli, M., Cravero, M. C. and Bonfante-Fasolo, P. (1987). *New Phytol.* **107**, 183–190.
Schumacher, T. E. and Smucker, A. J. M. (1985). *Plant Physiol.* **78**, 359–364.
Schwab, S. M., Menge, J. A. and Leonard, R. T. (1983). *Plant Physiol.* **73**, 761–765.

Schwab, S. M., Menge, J. A. and Tinker, P. B. (1991). *New Phytol.* **117**, 387–398.
Silsbury, J. H., Smith, S. E. and Oliver, A. J. (1983). *New Phytol.* **93**, 555–566.
Siqueira, J. O. and Hubbell, D. H. (1985). In *Proceedings of the 6th North American Conference on Mycorrhizae*, (R. Molina, ed), p. 368. Forest Research Laboratory, Oregon State University, Corvallis, OR.
Siqueira, J. O., Hubbell, D. H. and Schenck, N. C. (1982). *Mycologia* **74**, 952–959.
Sivak, M. N. and Walker, D. A. (1986). *New Phytol.* **102**, 499–512.
Smith, J. L. and Paul, E. A. (1988). *Plant Soil* **106**, 221–229.
Smith, S. E. and Gianinazzi-Pearson, V. (1988). *Ann. Rev. Plant. Physiol. Plant Mol. Biol.* **39**, 221–244.
Smith, S. E. and Gianinazzi-Pearson, V. (1990). *Austral. J. Plant Physiol.* **17**, 177–188.
Smith, S. E. and Smith, F. A. (1990). *New Phytol.* **114**, 1–38.
Snellgrove, R. C., Splittstoesser, W. E., Stribley, D. P. and Tinker, P. B. (1982). *New Phytol.* **92**, 75–87.
Snellgrove, R. C., Stribley, D. P. and Hepper, C. M. (1987). *Rothamsted Rep. 1986* (1), p. 142.
Son, C. L. and Smith, S. E. (1988). *New Phytol.* **108**, 305–314.
Stribley, D. P. and Read, D. J. (1974). *New Phytol.* **73**, 731–741.
Stribley, D. P., Tinker, P. B. and Rayner, J. H. (1980). *New Phytol.* **86**, 261–266.
Sylvia, D. M. (1988). *Soil Biol. Biochem.* **20**, 39–43.
Sylvia, D. M. and Hubbell, D. H. (1986). *Symbiosis* **1**, 259–267.
Syvertsen, J. P. and Graham, J. H. (1990). *Plant Physiol.* **94**, 1424–1428.
Söderström, B. and Read, D. J. (1987). *Soil Biol. Biochem* **19**, 231–236.
Söderström, B., Finlay, R. D. and Read, D. J. (1988). *New Phytol.* **109**, 163–166.
Taber, W. A. and Taber, R. A. (1987). *Trans. Br. Mycol. Soc.* **89**, 13–26.
Tester, M., Smith, S. E., Smith, F. A. and Walker, N. A. (1986). *New Phytol.* **103**, 375–390.
Thompson, B. D., Robson, A. D. and Abbott, L. K. (1986). *New Phytol.* **103**, 751–765.
Trojanowski, J., Haider, K. and Hüttermann, A. (1984). *Arch. Microbiol.* **139**, 202–206.
Van Veen, J. A. and Paul, E. A. (1979). *Appl. Environ. Microbiol.* **37**, 686–692.
Vignon, C., Plassard, C., Mousain, D. and Salsac, L. (1986). *Physiol. Vég.* **24**, 201–207.
Vikman, P. -Å. and Huss-Danell, K. (1991). *J. Exp. Bot.* **42**, 221–228.
Wang, G. M., Coleman, D. C., Freckman, D. W., Dyer, M. I., McNaughton, S. J., Acra, M. A. and Goeschl, J. D. (1989). *New Phytol.* **112**, 489–493.
Weete, J. D. (1974). *Fungal Lipid Biochemistry. Distribution and Metabolism.* Plenum Press, New York.
Whipps, J. M. (1990). In *The Rhizosphere* (J. M. Lynch, ed.), pp. 59–98. John Wiley and Sons, Chichester.
Whipps, J. M. and Lynch, J. M. (1983). *New Phytol.* **95**, 605–623.
Whipps, J. M., Haselwandter, K., McGee, E. E. M. and Lewis, D. H. (1982).

Trans. Br. Mycol. Soc. **79**, 385–400.
Zahka, G. (1989). *The establishment of obligate plant parasites in axenic culture with* Agrobacterium rhizogenes *induced roots.* PhD Thesis, University of Copenhagen.

7

Enzymology of Nitrogen Assimilation in Mycorrhiza

IFTIKHAR AHMAD and JOHAN A. HELLEBUST

Centre for Plant Biotechnology, Department of Botany, University of Toronto, Toronto, Ontario M5S 3B2, Canada

I. Introduction

It has long been recognized that the widespread occurrence of mycorrhiza, in both natural and managed fields is a symbiotic relationship based on the exchange of nutrients between fungal mycelia and host roots. However, the conventional view of mycorrhizal associations as an extension of the root system to increase the absorption of free nutrients in the soil appears to be an oversimplification. Read *et al.* (1989) have recently pointed out that most of the biomass of the fungus in some forms of mycorrhizal associations is located within or close to the root; a formation not extremely efficient in increasing the absorption area. It is becoming evident that the fungal symbiont may be playing an important

METHODS IN MICROBIOLOGY
VOLUME 23 ISBN 0-12-521523-1

Copyright © 1991 by Academic Press Limited
All rights of reproduction in any form reserved

role in mobilizing resources that are not directly accessible to the host root.

Ectomycorrhizal fungi have long been known to be efficient utilizers of amino acids (Melin and Nilsson, 1953; Carrodus, 1966; Abuzinadah and Read, 1988), and a number of recent studies have demonstrated their release of proteinases (protein degrading enzymes) to free amino acid from the soil litter (El-Badaoui and Botton, 1989; Leake and Read, 1989). These features would render mycorrhizal fungi in direct competition with their saprotrophic counterparts and it has been speculated that the supply of carbohydrates by the host plant gives an advantage to the mycorrhizal fungi in this competition (Abuzinadah *et al.*, 1986).

This raises many important questions about the relationship between carbon balance and nitrogen metabolism in mycorrhizal systems. First of all, there is a need for the characterization of various fungal symbionts for their ability to utilize different nitrogen sources. Such data may be obtained for intact mycorrhiza and where possible, also for pure cultures of the fungal symbiont. The utilization of nitrogen sources by the fungal symbiont should be characterized in terms of uptake, assimilation, and the release of assimilated or waste products to the external medium.

The question arises as to the regulation of enzymes of nitrogen metabolism in the fungal symbiont. The pathway of ammonium assimilation in higher plants is well studied, and in the root tissue of mycorrhiza the incorporation of ammonium into amino acids is expected to be carried out primarily via the combined action of glutamine synthetase (GS) and glutamate synthase (GOGAT) (Miflin and Lea, 1980; Oak and Hirel, 1985). The operation of a GS-GOGAT cycle in mycorrhizal fungi is still under scrutiny however, and so far no clear evidence has been presented for the presence of GOGAT in these micro-organisms (see Ahmad *et al.*, 1990). There is evidence that the incorporation of ammonium in mycorrhizal fungi is at least in part carried out via the alternative pathway involving the aminating activity of glutamate dehydrogenase (GDH) (Martin *et al.*, 1986; Dell *et al.*, 1989). Deaminating activities of GDH enzymes, on the other hand, are known to play a major catabolic role generating 2-oxoglutarate from glutamate for carbohydrate metabolism. Thus GDH enzymes may play a key role in nitrogen metabolism of the fungal symbiont and may also act at branch points between nitrogen and carbon metabolism. Coupled with these pathways, the presence of highly active aminotransferases in mycorrhizal fungi (Khalid *et al.*, 1988; Ahmad *et al.*, 1990) is indicative of an efficient metabolic machinery present in these symbionts for the utilization of diverse nitrogen sources. Figure 1 presents a summary diagram of major pathways of nitrogen metabolism in mycorrhizal fungi. Fungi

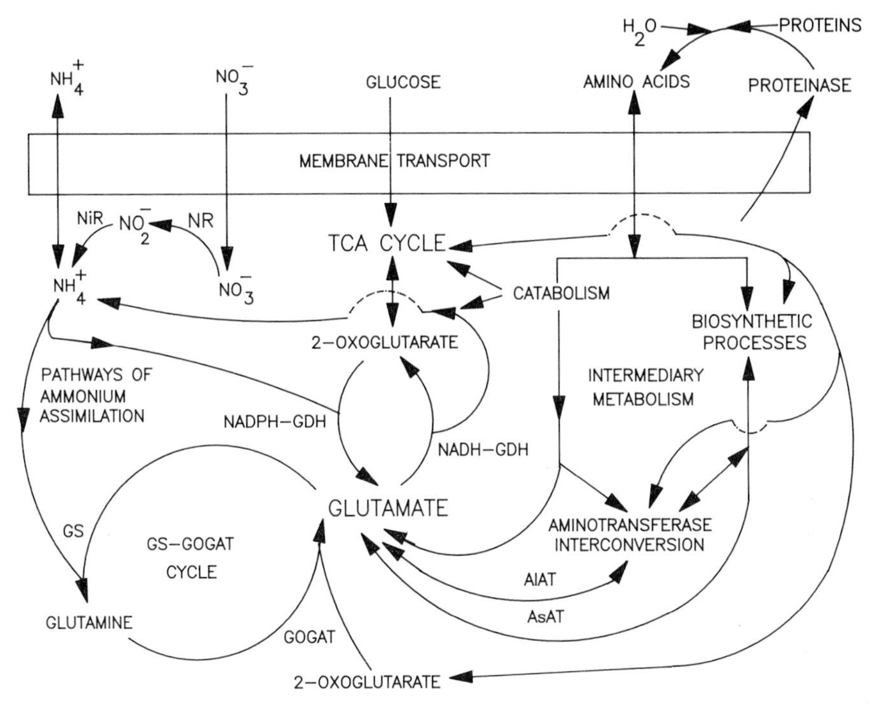

Fig. 1. Summary diagram of major pathways of nitrogen metabolism. The scheme presented here emphasises the role of both the GS-GOGAT cycle and the NADPH-GDH pathway in inorganic nitrogen assimilation and that of the NADH-GDH in amino acid catabolism. The TCA cycle is shown to be linked to both ammonium assimilation and intermediary nitrogen metabolism. The scheme also signifies the role of membrane transport in the exchange of small molecular weight metabolites and the excretion of proteinase activities. The hydrolysis of soil proteins by excreted proteinases contributes to the extracellular amino acid pool.

are able to adapt to changing nutritional conditions by invoking diverse anabolic and catabolic pathways, which, because of subcellular localization of enzyme activities, are compartmentalized in cell organelles. So far little attention has been paid to the subcellular organization of metabolic pathways in mycorrhizal fungi. There is also a need for a better understanding of the synthesis and the regulatory properties of proteolytic enzymes released by fungal symbionts. All of these studies are essential for elucidating nitrogen metabolism of symbiotic fungi and would lead to a better understanding of the basis of the plant–fungus mutualism in mycorrhizal associations.

II. Pure culture manipulation of mycorrhizal fungi for studies of nitrogen metabolism

A. Fungal cultures

Many ectomycorrhizal fungi grow readily in pure cultures (Molina and Palmer, 1982). Over the years a number of artificial media have been formulated for propagating these symbiotic fungi under defined conditions (Melin and Rama Das, 1954; Palmer and Hacskaylo, 1970; Martin *et al.*, 1983, 1988). Surprisingly, all of these so-called defined media omit at least one major nutrient, calcium, and even when supplemented with trace elements as suggested by Molina and Palmer (1982), do not contain a full complement of micronutrients and vitamins. The use of these incomplete nutrient media can hamper the study of nitrogen metabolism seriously since enzymes such as glutamate dehydrogenase and nitrate reductase are known to be affected adversely by the lack of calcium and molybdenum, respectively. We have recently formulated a buffered medium (Ahmad *et al.*, 1990) that contains all major mineral nutrients and is enriched with micronutrients and vitamins. The medium can be prepared from the following solutions.

Macronutrients and buffer stock solution. 65 mM KCl + 20 mM $MgSO_4$ + 3 mM $CaCl_2$ + 15 mM NaCl + 20 mM 2-(*N*-morpholino) ethanesulphonic acid (MES) adjusted to pH 5.5 with NaOH.

Micronutrient stock solution. 2 mM NaH_2PO_4 + 1 mM boric acid + 18 μM $MnCl_2$ + 16 μM $ZnSO_4$ + 0.6 μM Na_2MoO_4 + 1 μM $CoCl_2$ + 0.8 μM $CuSO_4$ + 230 μM sodium iron salt of EDTA + 200 μg litre^{-1} thiamine hydrochloride + 10 μg litre^{-1} biotin + 10 μg litre^{-1} vitamin B_{12}.

Carbon stock solution. 500 mM glucose.

Inorganic nitrogen stock solution. 50 mM $NaNO_3$ (nitrate media) or 50 mM NH_4Cl (ammonium media).

The stock solutions are kept under refrigeration. The medium is prepared by adding 100 ml each of the four stock solutions to a final volume of 1 litre. In experiments where amino acids are used as nitrogen sources, the inorganic nitrogen stock solution is substituted by solid additions of amino acids. Glucose addition can also be substituted or omitted according to experimental conditions. Media are made ready for inoculation by autoclaving 100 ml fractions in 500 ml wide-mouth conical flasks except for solutions of heat-labile nitrogen sources such as glutamine, which are filter-sterilized separately.

B. Nitrogen utilization

Most ectomycorrhizal fungi are able to utilize both inorganic and organic nitrogen sources. Growth rates and final biomass production give an overall estimate of the fungal efficiency in assimilating a given nitrogen source. The disappearance of nutrients from the growth media as a function of growth rate (μ) has been used for estimating the long-term uptake of compounds in microbial cultures. The most conventional way of measuring nutrient uptake by micro-organisms is by determining the radioactivity of filtered cells after a period of incubation in the presence of a radiolabelled nutrient source (Hellebust and Lin, 1978). The uptake of ^{14}C- and ^{3}H-labelled compounds can be quantified conveniently by scintillation counting. In laboratories with facilities for heavy isotope analysis, the incorporation of ^{15}N-labelled compounds allows a direct and more specific measure of nitrogen utilization (Rhodes *et al.*, 1981; Genetet *et al.*, 1984).

III. Isolation and characterization of enzyme activities

A. Isolation

Most nitrogen assimilatory enzymes are soluble proteins and thus are readily extracted by any of the standard tissue homogenization techniques. Grinding with acid-washed sand has been successfully used for extracting various mycorrhizal fungal enzymes (Khalid *et al.*, 1988; Ahmad *et al.*, 1990). Mycorrhiza and non-mycorrhizal roots may be ground in the presence of polyvinylpyrrolidone (Dell *et al.*, 1989) to protect enzymes against phenolic compounds released during tissue maceration. The macerate is usually clarified by centrifugation. These crude enzyme preparations can be purified partially either by ammonium sulphate fractionation and/or by one of the standard chromatographic techniques (Martin *et al.*, 1983; Dell *et al.*, 1989; Ahmad *et al.*, 1990). Glutamine synthetase, glutamate dehydrogenase and aminotransferases show good resolution on anion-exchange columns, and chromatographic separations have been improved considerably by the introduction of fast protein liquid chromatography. The stability of enzyme activities during purification and the level of purification required for assay procedures are the two important criteria for selecting appropriate extraction and purification protocols. These issues are addressed in the following discussion of the characterization of various enzyme activities.

B. Enzyme characterization and assay procedures

1. Nitrate reductase

Nitrate reductase (NR) is a labile enzyme and for this reason a number of protectants are added to the extraction media (Wray and Fido, 1990). These protectants usually include nitrate, which is an NR substrate; FAD, which is a prosthetic group present in the protease-sensitive region of NR; EDTA, which chelates toxic metals released during cell breakage; and sulphydryl compounds such as dithiothreitol and mercaptoethanol, which prevent the oxidation of essential SH-groups of NR. Protease inhibitors such as phenylmethylsulphonyl fluoride are usually added to inhibit proteolysis during extraction. Exogenous proteins such as bovine serum albumin and casein are routinely used to enhance the stability of NR. A detailed study of *in vitro* measurement of NR in the basidiomycete *Hebeloma cylindrosporum* has been carried out by Plassard *et al.* (1984a).

Nitrate reductase can be assayed directly in crude cell-free preparations. The overall physiological reaction of NR is usually determined by the reduction of nitrate in the presence of NAD(P)H followed by the colorimetric measurement of the nitrite produced:

$$\text{Nitrate} + \text{NAD(P)H} \xrightarrow{\text{NR}} \text{Nitrite} + \text{NAD(P)}$$

In higher plants, nitrate reductase is NADH-specific, whereas in fungi the enzyme shows a preference for NADPH (Beevers and Hageman, 1980). In *Hebeloma cylindrosporum* the activity of NR is shown to be strictly dependent on NADPH as the electron donor (Plassard *et al.*, 1984a).

The following assay procedure is a modification of Scholl *et al.* (1974). The reaction mixture contains 25 mM potassium phosphate (pH 75), 10 mM potassium nitrate, 0.5 mM NADH or NADPH and up to 0.25 ml enzyme in a final volume of 0.5 ml. The reaction is carried out for 20–30 min at 25 °C and stopped by adding 0.25 ml of a 200 mM zinc acetate solution. The precipitated material is cleared by centrifugation. The residual pyridine nucleotide in the reaction mixture is oxidized by adding 0.25 ml of a freshly prepared 50 μM phenazine methosulphate solution and leaving the mixture at room temperature for about 20 min. The colour development is initiated by adding 1 ml each of the diazo-coupling reagents 1% sulphanilamide solution in 3 N HCl and 0.02%

N-(1-naphthyl)-ethylenediamine dihydrochloride solution in 0.1 N HCl. After 20 min the absorbance of the pink diazo dye is read at 540 nm and the amount of nitrite is determine from a standard curve for 10–100 nmol nitrite.

The K_m for nitrate and NADPH of the fungal nitrate reductase range from 60–200 μM and 9–60 μM, respectively (Beevers and Hageman, 1980). The K_m values for nitrate and NADH are similarly low for the plant enzyme. Nitrate reductase is a substrate-induced enzyme and is usually found only in the presence of nitrate. In fungi, the presence of ammonium in the culture medium completely abolishes the nitrate-mediated induction of nitrate reductase (Lewis and Fincham, 1970). A similar inhibition of nitrate reductase induction has been shown in plant roots (Smith and Thompson, 1971; Radin, 1975).

2. *Nitrite reductase*

One of the major differences between the plant and fungal nitrite reductase (NiR) is in their specificity for electron donors. The plant enzyme accepts electrons from reduced ferredoxin whereas the fungal enzyme utilizes pyridine nucleotides as the electron donor and generally shows a preference for NADPH. The reaction catalysed by NADPH-NR is considered to involve a sequential transfer of electrons:

$$\text{NADPH} \rightarrow \text{FAD} \rightarrow \text{siroheme} \rightarrow \text{nitrite}$$

NADPH-NR extracted from *Hebeloma cylindrosporum* shows a rapid loss of activity but can be reactivated by the addition of sodium dithionite and methylviologen as the electron donor (Plassard *et al.*, 1984b). These authors have suggested that the loss of NADPH-NR activity in the fungal extract was associated with the step involved in the transfer of electrons from NADPH to FAD, and the reduction of siroheme by methylviologen bypassed the need for electrons from NADPH. The overall reaction of NiR can be summarized as:

$$\text{Nitrite} + \text{reduced ferredoxin or NAD(P)H} \xrightarrow{\text{NiR}} NH_4^+ + \text{oxidized ferredoxin or NAD(P)}$$

The activity of nitrite reductase is usually measured by determining the disappearance of nitrite in the assay system. For pyridine nucleotide-dependent assay, the following modification of Garret (1972) may be used. The reaction mixture contains 50 mM potassium phosphate (pH 7.5), 1 mM potassium nitrite, 0.05 mM FAD, 0.5 mM NADPH or NADH and enzyme in a final volume of 0.5 ml. After 15 or 20 min incubation at 25 °C, the reaction is stopped and the amount of nitrite in the sample is determined according to the procedure described for nitrate reductase assay. For assaying ferredoxin-dependent nitrite reductase activity, the above reaction mixture is modified to substitute NAD(P)H by 0.5 mM ferredoxin and 12 mM sodium dithionite. The reaction is started by the addition of dithionite and stopped by vigorous shaking. Residual nitrite is measured as above.

As with nitrate reductase, nitrite reductase is a high affinity enzyme and shows K_m values for nitrite in the micromolar range. The fungal and plant NiR differ considerably in their molecular size. In higher plants NiR has a molecular mass of about 63 kDa (Wray and Fido, 1990) which is four times lower than the molecular weight of 290 kDa reported for NiR isolated from *Neurospora crassa* (Nason *et al.*, 1954). The molecular mass of the purified NiR is usually determined by gel chromatography.

3. *Glutamate dehydrogenase*

The *in vitro* activity of glutamate dehydrogenase (GDH) enzymes can be assayed both in the aminating (assimilatory) and deaminating (catabolic) directions.

$$NH_4^+ + \text{2-oxoglutarate} + \text{NAD(P)H} \overset{\text{GDH}}{\rightleftharpoons} \text{Glutamate} + \text{NAD(P)}$$

The root tissue of higher plants usually contains a mitochondrial NADH-linked GDH, which functions primarily in the deaminating direction. Many fungi possess both the NADH-linked catabolic enzyme and an NADPH-linked assimilatory GDH which can be separated in a one step anion-exchange chromatographic procedure (Fig. 2).

The usual assay system of GDH activities is based on spectrophotometric monitoring of the oxidation of NAD(P)H, which gives a stoichiometric measure of the amount of glutamate produced. The following is a modification of Ahmad and Hellebust (1984). The reaction mixture of 1 ml contains (final concentration) 50 mM potassium phosphate (pH 7.5), 100 mM ammonium chloride, 20 mM 2-oxoglutarate, 0.2 mM NADH or NADPH and 0.2 ml enzyme. The oxidation of NAD(P)H is monitored at 340 nm using an extinction coefficient (ε) of 6.2. The assay must be corrected for non-specific oxidase activities present in the enzyme extract. These contaminating activities can be determined by omitting 2-oxoglutarate from the reaction mixture. Membrane-bound oxidases can be removed either by 20% ammonium sulphate precipitation followed by centrifugation at 10 000 *g* for 10 min (Ahmad and Hellebust, 1986) or by high speed (30 000 *g*) centrifugation for 30 min (Ahmad *et al.*, 1990).

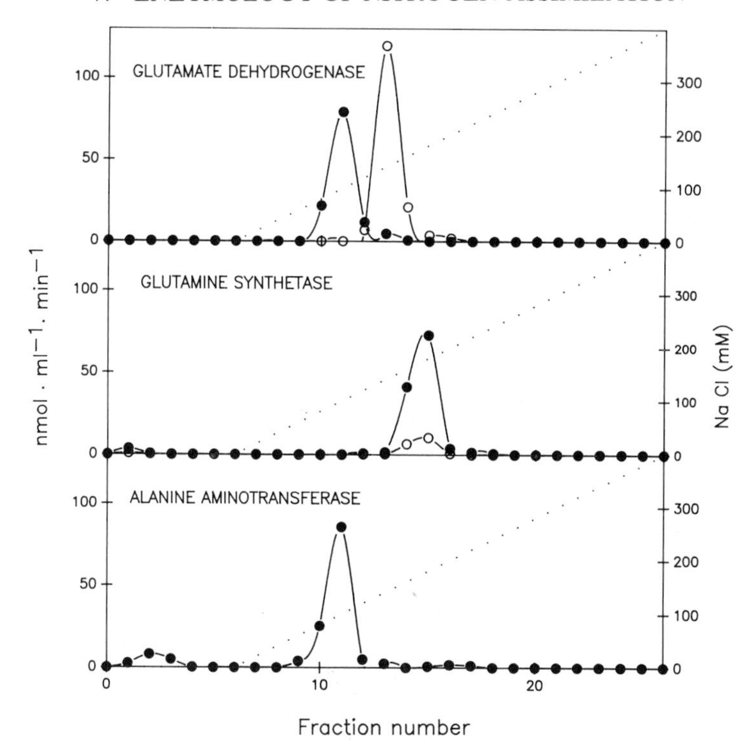

Fig. 2. Elution profiles for the fractionation by anion-exchange chromatography of glutamate dehydrogenase (NADH-GDH, closed symbols; NADPH-GDH open symbols); glutamine synthetase (transferase activity, closed symbols; synthetase activity, open symbols) and alanine aminotransferase. A 4 ml extract prepared by grinding 1 ml packed volume of *Laccaria bicolor* was clarified by centrifugation (30 000 *g*, 30 min) and a 2 ml fraction loaded onto a Mono-Q column attached to a fast protein liquid chromatography assembly. The composition of the elution media was as described by Ahmad and Hellebust (1987).

The pH optima of the amination reactions of both NADH-GDH and NADPH-GDH from various sources fall in the pH range 7–8, whereas the pH optima of the deaminating reactions are usually 1.0–1.5 pH units higher. These differences in pH optima appear to have broad implications for the functional relations of these enzymes. NADH-GDH from various sources have a high K_m for ammonium (10–80 mM). The K_m for ammonium of NADPH-GDH from higher plants and many fungi is also high. However, NADPH-GDH from some ectomycorrhizal fungi show biphasic kinetics with different K_m values for ammonium (Martin *et al.*, 1983; Ahmad *et al.*, 1990). The lower K_m value for these NADPH-GDH enzymes is in the range of 2–5 mM ammonium. The

NADH-linked GDH of higher plants is a metalloprotein and requires added calcium to prevent inactivation by EDTA or other chelating agents. No calcium requirement has been reported for the fungal GDH enzymes.

4. *Glutamine synthetase*

Two separate forms of glutamine synthetase (GS) exist in higher plants; one located in the cell cytosol and the other in plastids. GS activity in plant roots is shown to be predominantly cytosolic (Emes and Fowler 1979; Suzuki *et al.*, 1981). The anion-exchange study of GS in the ectomycorrhizal fungus, *Laccaria bicolor*, shows the presence of a single molecular form (Fig. 2) with an elution profile on a sodium chloride gradient similar to that observed for cytosolic enzymes from many plants and algal sources (McNally *et al.*, 1983; Casselton *et al.*, 1986). The plastid enzyme requires the presence of sulphydryl reagents to prevent inactivation during isolation. However, these reagents are shown to suppress the activity of cytosolic GS from some sources (Wallsgrove *et al.*, 1983; Ahmad and Hellebust, 1987). It may therefore be necessary to isolate the two GS forms by separate extraction procedures with appropriate sulphydryl composition of the isolation media. The two GS isoforms are also reported to differ in their thermal stability with the cytosolic form being considerably less labile (Mann *et al.*, 1979; Ahmad *et al.*, 1982).

GS catalyses the formation of glutamine from ammonium and glutamate at the expense of ATP hydrolysis; the overall reaction can be summarized as:

$$\text{Glutamate} + NH_4^+ + \text{ATP} \xrightarrow{\text{GS}} \text{Glutamine} + \text{ADP} + \text{Pi} + H_2O$$

This reaction can be determined by a coupled spectrophotometric assay procedure where the ADP produced in GS reaction is linked to the conversion of phosphoenolpyruvate by pyruvate kinase to pyruvate, which in turn is linked to the oxidation of NADH by lactate dehydrogenase. The following is a modification of Stewart and Rhodes (1977a). The reaction mixture of 1 ml is prepared with 50 mM sodium phosphate buffer (pH 7.5) containing 100 mM sodium glutamate, 10 mM ATP, 10 mM ammonium chloride, 20 mM magnesium sulphate, 1 mM phosphoenolpyruvate, 0.2 mM NADH, 5 units pyruvate kinase (PK), 2 units lactate dehydrogenase (LDH) and 200 μl enzyme extract. The oxidation of NADH monitored at 340 nm gives a stoichiometric estimate of the amount of glutamine produced by the GS reaction.

A colorimetric procedure has been developed by substituting hydroxylamine (NH_2OH) for ammonium and determining the amount of

γ-glutamylhydroxamate produced spectrophotometrically following the development of a brown colour with ferric chloride.

$$\text{Glutamate} + NH_2OH + ATP \xrightarrow{GS} \gamma\text{-Glutamylhydroxamate} + ADP + Pi + H_2O$$

This alternative assay gives a measure of GS reaction in the biosynthetic direction and is usually termed the synthetase assay. The V_{max} and K_m values determined by this method are usually similar to those obtained by a coupled assay system. The reaction is started in a 1 ml mixture prepared with 100 mM Tris-HCl buffer (pH 7.6) and containing 80 mM glutamate, 20 mM hydroxylamine, 20 mM ATP, 50 mM magnesium sulphate and 200 μl enzyme extract (Ahmad and Hellebust, 1984). After a desired period of incubation (20–30 min) at 25 °C, the reaction is stopped by the addition of 1 ml of acidified ferric chloride solution (26 g of ferric chloride and 40 g of TCA in one litre of 1 N HCl). It is essential to run zero time controls by stopping the reaction immediately after the addition of enzyme extract to the reaction mixture. Where necessary, the precipitated material is removed by sedimentation in a bench-top centrifuge before reading the aborbance at 540 nm. A standard plot using a commercial glutamylhydroxamate preparation is established as a reference for GS activity.

A third assay has been developed on the basis of the ability of GS to catalyze the γ-glutamyl transfer reaction that results in the formation of γ-glutamylhydroxamate from glutamine and hydroxylamine:

$$\text{Glutamine} + NH_2OH \xrightarrow{GS} \gamma\text{-Glutamylhydroxamate} + NH_4^+$$

This reaction—termed transferase assay—gives rates several times higher than those obtained by the above biosynthetic reactions. After establishing the transferase:synthetase ratio of a given GS enzyme, this method can be used as a sensitive indicator of its activities at different stages of purification protocols. The transferase assay is based on the colorimetric determination of glutamylhydroxamate produced. The reaction is started in a 1 ml mixture with 100 mM Tris-acetate buffer (pH 6.4) containing 100 mM glutamine, 30 mM hydroxylamine, 30 mM sodium arsenate, 1.5 mM $MnCl_2$, 0.2 mM ADP and 200 μl enzyme (Ahmad and Hellebust, 1984). The reaction is stopped after 30 min by the addition of acidified ferric chloride and GS activity is quantified as described for the synthetase assay.

The kinetics of GS have been extensively studied. Various purified preparations have been shown to have high affinity for ammonium, but a considerably lower affinity for glutamate (Stewart *et al.*, 1980). The ATP-dependent reaction of glutamine synthetase has a specific requirement for magnesium. The enzyme shows complex kinetics with respect to concentrations of ATP and magnesium and pH of the reaction system (Stewart *et al.*, 1980).

5. *Glutamate synthase*

Glutamine produced via the glutamine synthetase pathway is utilized by glutamate synthase (GOGAT), which catalyses the transfer of amido nitrogen to 2-oxoglutarate, resulting in the formation of two molecules of glutamate. The enzyme from green plants is specific for ferredoxin or ferredoxin-like proteins (Miflin and Lea, 1980; Suzuki *et al.*, 1985) as the electron donor

$$\text{Glutamine} + \text{2-oxoglutarate} + \text{reduced ferredoxin} \xrightarrow{\text{GOGAT}} \text{2 Glutamate} + \text{oxidized ferredoxin}$$

whereas GOGAT from various non-green micro-organisms is found to be NAD(P)H-specific (Stewart *et al.*, 1980):

$$\text{Glutamine} + \text{2-oxoglutarate} + \text{NAD(P)H} \xrightarrow{\text{GOGAT}} \text{2 Glutamate} + \text{NAD(P)}$$

The net result of the combined action of GS and GOGAT is the synthesis of glutamate from ammonium and 2-oxoglutarate; this combined action is frequently referred to as the GS-GOGAT cycle or simply as the glutamate synthase pathway.

GOGAT can be measured directly in crude extracts. The enzyme is usually extracted in 50 mM phosphate buffer (pH 7.5) containing protectants such as sulphydryl reagents, mercaptoethanol and DTT, substrate 2-oxoglutarate and protease inhibitor PMSF (Marquez *et al.*, 1988; Wallsgrove *et al.*, 1977). In some cases it is necessary to solubilize membrane-associated ferredoxin-dependent GOGAT from plant tissue by the addition of 0.05% to 0.5% Triton X-100 to the extraction media. GOGAT activity is usually measured by the quantitative measurement of glutamate produced. For assaying *in vitro* ferredoxin-dependent GOGAT activity the electron donor can be replaced by reduced methyl viologen without significant change in activity (Marquez *et al.*, 1988). The enzyme extract is pre-incubated for 20 min in a 0.5 ml mixture containing 100 mM phosphate buffer, pH 7.5, 10 mM glutamine, 10 mM 2-oxoglutarate and 15 mM methyl viologen. The reaction is started by adding 100 μl of freshly prepared dithionite reductant mixture (235 mg sodium dithionite, 250 mg sodium bicarbonate in 5 ml water). In NAD(P)H-dependent GOGAT assays, methyl viologen in the reaction mixture and its reduction by dithionite mixture are omitted, and the reaction is started by the addition of 0.5 mM NADH or NADPH. After

20 min at 25 °C, the reaction is stopped either by boiling or by the addition of 1 ml ethanol. Glutamate produced can be separated from glutamine and quantified by one of the chromatographic procedures employing paper chromatography (Wallsgrove *et al.*, 1977), thin layer chromatography (see Lea *et al.*, 1990), high-performance liquid chromatography (Martin *et al.*, 1982), anion-exchange column chromatography (Hecht *et al.*, 1988) or paper electrophoresis (Chen and Cullimore, 1988). Ninhydrin-based colorimetric determination of glutamate is routinely applied in these studies. The use of radiolabelled ^{14}C glutamine in the reaction mixture allows a more sensitive determination of glutamate by scintillation counting (Wallsgrove *et al.*, 1982). Pyridine nucleotide-dependent GOGAT activity can be measured spectrophotometrically by following the oxidation of NAD(P)H provided the enzyme preparation does not contain non-specific NAD(P)H oxidase activities. The composition of the reaction mixture for spectrophotometric measurements is the same as that described above for NAD(P)H GOGAT activity except for the concentration of NAD(P)H, which is lowered to 0.25 mM to give an initial absorbance reading of less than 2.

6. *Aminotransferases*

Glutamate synthesized either via the GDH pathway or the GS-GOGAT cycle is the primary amino donor for the synthesis of most other amino acids. Several enzymes catalysing the transfer of amino groups from glutamate to different keto acids have been identified in plants and fungi. The most active of these aminotransferases are aspartate aminotransferase (AsAT) and alanine aminotransferase (AlAT). The reactions of both aminotransferases are reversible:

$$\text{Glutamate} + \text{oxaloacetate} \overset{\text{AsAT}}{\rightleftharpoons} \text{Aspartate} + \text{2-oxoglutarate}$$

$$\text{Glutamate} + \text{pyruvate} \overset{\text{AlAT}}{\rightleftharpoons} \text{Alanine} + \text{2-oxoglutarate}$$

These enzymes are routinely measured spectrophotometrically by coupling the production of keto acids to the oxidation of NAD(P)H catalyzed by an auxiliary dehydrogenase enzyme. A typical reaction mixture for AsAT assay contains in a final volume of 1 ml, 50 mM sodium phosphate buffer (pH 7.5), 100 mM sodium aspartate, 10 mM 2-oxoglutarate, 0.25 mM NADH, 2 units malate dehydrogenase and purified enzyme extract (Ahmad and Hellebust, 1989).

The reaction mixture for AlAT assay is similar in its composition to the reaction mixture for AsAT except for aspartate and for malate dehydrogenase, which are replaced by 100 mM alanine and 2 units of lactate dehydrogenase, respectively. When the coupled reaction proceeds linearly, the oxidation of NADH gives a stoichiometric measure of aminotransferase activity. The spectrophotometric procedure for measuring aminotransferase activities using glutamate as the amino donor has been described elsewhere (Ahmad and Hellebust, 1989). Enzyme preparations clarified by high speed centrifugation (30 000 *g*, 30 min) and/or ammonium sulphate precipitation are usually adequate for the coupled enzyme assays. The reactions catalysed by aminotransferases require pyridoxal 5′-phosphate as a coenzyme, but in plants and fungi it usually remains tightly bound to the enzyme during different stages of purification and therefore its addition to the reaction mixture is generally not needed. Several isoforms of both alanine and aspartate aminotransferases have been identified from various plant and microbial sources. Figure 2 shows the elution profile by anion-exchange chromatography of alanine aminotransferase from the ectomycorrhizal fungus, *Laccaria bicolor*. The fungal extract is fractionated into two peaks of AlAT activity; eluting first a cationic minor isoform followed by a major anionic isoform. Cationic aminotransferases are considered mitochondrial in origin whereas anionic isoforms are thought to be cytosolic enzymes (Givan, 1980).

IV. Pathways of ammonium assimilation

Prior to the demonstration that GOGAT occurs in higher plants (Lea and Miflin, 1974), GDH was considered to be the primary enzyme of ammonium assimilation in both plants and micro-organisms. There is now considerable evidence that in vascular plants ammonium assimilation occurs almost exclusively via the GS-GOGAT cycle (see Miflin and Lea, 1980 for review). Incubating plants with methionine sulphoximine—a specific inhibitor of GS—almost invariably results in a complete inhibition of nitrogen assimilation in both root and shoot tissue. Feeding plants with ^{15}N-labelled nitrate or ammonium and following the enrichment of amino acid pools by gas chromatography/mass spectrometry (GC/MS) or by nuclear magnetic resonance spectroscopy shows a rapid labelling of amido nitrogen of the glutamine pool that is indicative of the entry of inorganic nitrogen to organic molecules primarily via GS activity. These results exclude GDH from playing a role in ammonium assimilation by higher plants.

The relative importance of the GDH pathway and GS-GOGAT cycle for ammonium assimilation in mycorrhizal fungi is, however, not clear. It may be noted that a large group of fungi lack GOGAT activities. Furthermore, ^{15}N labelling studies have produced convincing evidence for the assimilation of ammonium via the GDH pathway in the fungus

Candida utilis (Folkes and Sims, 1974). There are, however, conflicting reports about the presence of GOGAT in mycorrhizal fungi. Recently, Vezina *et al.* (1989) reported an active NADH-dependent GOGAT in *Laccaria bicolor*. However, the levels of GOGAT activity measured by Vezina *et al.* (1989) appear extremely low compared with the activity of GS and other enzymes of nitrogen assimilation present in the same fungal extract. Using a combination of ^{15}N labelling of *Cenococcum geophilum* and the inhibition of its GS activity by methionine sulphoximine, the work of Martin *et al.* (1988) has excluded a role of GOGAT in the synthesis of glutamate by this ectomycorrhizal ascomycete. The incorporation of nitrogen into amino acids in this fungus probably occurs via both the reductive amination reaction of GDH producing glutamate and the amination of glutamate by GS producing glutamine. A concurrent role of GS and NADPH-linked GDH pathways was also evident in our study of *L. bicolor* (Ahmad *et al.*, 1990) which showed that the activity of these enzymes reached maximum levels in rapidly growing mycelia and declined rapidly during the onset of the stationary growth phase. A highly active NADPH-GDH has also been reported in *Hebeloma* spp. (Dell *et al.*, 1989). These basidiomycetes and possibly other mycorrhizal fungi with similar enzymic characteristics provide useful material for elucidating the relative importance of the two potential pathways in the assimilation of nitrogen by fungal symbionts. It will be rewarding to adopt the isotopic procedures described by Martin and his co-workers (Martin, 1985; Martin *et al.*, 1986) for *C. geophilum* in monitoring ^{15}N labelling patterns of other fungal symbionts. The use of the GS inhibitor, methionine sulphoximine, has afforded information regarding the contribution of the GS pathway in the nitrogen assimilation of higher plants and green algae (Fentem *et al.*, 1983; Ahmad and Hellebust, 1985, 1986), and one expects that it can be applied equally successfully in mycorrhizal studies. Selection of GS and GDH fungal mutants has surprisingly not been undertaken so far in mycorrhizal research. Intraspecific variability of the NADPH-GDH from *Hebeloma* has been studied by Wagner *et al.* (1989).

V. Induction of anabolic and catabolic pathways

For the purpose of this chapter, anabolic pathways of nitrogen metabolism refer to activities associated with primary nitrogen assimilation leading to the biosynthesis of glutamate and the conversion of glutamate to other primary amino acids. Catabolic pathways refer to activities associated with the breakdown of amino acids for supplying carbon skeletons of amino acids to support energy production and growth. In this sense, the trophic condition of the organisms becomes the distinguishing mark for the expressed pathway. Fungi are usually able to utilize a variety of nitrogen sources including nitrate, ammonium, amino acids and proteins. In the fungal literature these nitrogen sources are often categorized as either inducers, repressors or neutrals according to their effects on the expression of a given pathway. The pathway leading to the utilization of a given nitrogen source may be either substrate inducible, product repressible or both. For example, in the ascomycetes, *Emericella* (*Aspergillus*) *nidulans* and *Neurospora crassa*, NR is induced by its substrate, nitrate, but is repressed by the product of nitrate reduction, ammonium, and also by the product of ammonium assimilation, glutamine (Tomsett, 1989). It may be noted, however, that the basidiomycete *Hebeloma cylindrosporum* has been reported to show similar levels of NR activity when incubated in the presence of either ammonium or nitrate as the sole nitrogen source (Scheromm *et al.*, 1990c). Nitrogen sources such as urea and uric acid are considered neutral for NR as they neither induce nor repress NR activity. Substrate inducibility in *E. nidulans* and *N. crassa* is shown to be pathway specific. Their product-repression mechanisms, however, have been found to influence several pathways simultaneously. Thus a number of amino acid-catabolizing enzymes are completely repressed when these fungi are supplied either with ammonium or some primary amino acids such as glutamate and glutamine as nitrogen sources (Marzluf, 1981; Jennings, 1989).

The non-filamentous ascomycete *Saccharomyces cerevisiae* has also been found to possess a similar control mechanism (Cooper, 1982). This control—termed nitrogen catabolite control—facilitates a preferred assimilation of ammonium, glutamate and glutamine by maintaining high levels of anabolic activities of GS and NADPH-GDH, and only after a complete utilization of these preferred nitrogen sources permits the synthesis and/or activation of enzymes and transport systems necessary for catabolizing other nitrogen sources. These observations, based on combined biochemical and genetic analysis, allow a clear understanding of how the various processes of anabolism and catabolism of nitrogen are regulated in these non-symbiotic ascomycetes.

Little is known, however, about the regulation of nitrogen metabolism in mycorrhizal fungi. Ectomycorrhizal fungi are generally considered to be poor utilizers of nitrate nitrogen. Contrary to such assumptions, the growth of the *Hebeloma cylindrosporum* thalli is shown to be much faster in the presence of nitrate than in the presence of ammonium (Scheromm *et al.*, 1990b). A recent study showed that the ectomycorrhizal basidiomycete, *Laccaria bicolor* is able to grow on both ammonium and nitrate as well as on several organic nitrogen sources (Ahmad *et al.*, 1990). Of a large number of amino acids tested, glutamate was the most efficient nitrogen source for this fungus, and during exponential growth the utilization of glutamate was accompanied by high mycelial levels of GS, NADPH-GDH and AsAT activities. On the other hand, mycelia growing under strict catabolic conditions, where media contained either arginine or alanine as sole carbon and nitrogen sources, contained high levels of NADH-GDH and AlAT activities and very low levels of GS, NADPH-GDH and AsAT activities. These results indicate that the regulation of nitrogen metabolism in the fungal symbiont is highly integrated and may be subject to elaborate transcriptional control, as has been observed in the well-studied non-symbiotic ascomycetes. Furthermore, the work of Wagner *et al.* (1988, 1989) suggests significant intraspecific genetic variations in the levels of nitrogen-assimilating enzymes in fungal symbionts. There appears to be an urgent need for detailed biochemical and genetic analysis of the regulatory features of nitrogen metabolism in mycorrhizal fungi.

Our knowledge of the regulatory features of nitrogen metabolism in plant roots is very limited at present. In vascular green plants, GS-GOGAT constitutes an integral part of nitrogen anabolism in both roots and leaves (Miflin and Lea, 1980; Stewart *et al.*, 1980), and is usually present in excess to ensure rapid ammonium assimilation (Blackwell *et al.*, 1987). GS is, however, strongly inhibited by various amino acids, notably alanine and glycine. Glutamine, the product of GS reaction, has been identified as the key regulator of GS levels in plants (Rhodes *et al.*, 1976; Stewart and Rhodes, 1977b). Glutamine has also been implicated in the control of NR levels in plants. However, Oak and Hirel (1985) have argued that the substrate nitrate is the main regulator of both the synthesis of NR protein and its activity. Plants growing under autotrophic conditions are not dependent on exogenous carbon supplies. However, they usually possess a significant capacity for the degradation of a number of amino acids to TCA cycle intermediates. The activity of the mitochondrial NADH-GDH in both leaves and roots is considered to be catabolic, and the enzyme is usually more active in the root tissue than in the leaves (Lee and Stewart, 1978).

Some detailed studies of the regulatory diversity of nitrogen metabolism in the fungal symbiont and the plant root are clearly needed. However, any extrapolation of the regulatory mechanisms seen in pure cultures of the fungal symbiont and the host plant to the mycorrhizal tissue must be undertaken with caution and, wherever possible, attempts should be made to include the analysis of the mycorrhizal tissue in these studies. As the work on several mycorrhizal species (Martin, 1985; Genetet *et al.*, 1984; Martin *et al.*, 1986, 1988; Dell *et al.*, 1989) suggests, overarching the endogenous control of nitrogen metabolism in the fungal symbiont there may be a control by the host root. The biochemical and genetic basis of such regulation of nitrogen metabolism by plant roots in symbiotic associations is not clear. It may be noted that a regulatory control of nitrogen metabolism exerted by plant roots is also found in legume–*Rhizobium* symbioses, where a number of nodule-specific proteins (nodulins) are encoded by specific plant genes (Verma *et al.*, 1983). Interestingly, among these proteins is a GS isoform designated as GS_N (Lara *et al.*, 1983) that is synthesized by the plant root after nodulation where it functions together with the root-specific GS isoform (GS_R) to ensure rapid ammonium assimilation during nitrogen fixation. Free living *Rhizobium* contain two GS isoforms; both are suppressed after nodulation (Ludwig, 1980). It must be added that any study of the fungus–plant interaction in mycorrhizal formations should consider the long-term nature of such symbiosis. This is an important consideration for monitoring the beneficial effects of the mycorrhizal association as short-term studies often fail to show any nutritional or growth difference between mycorrhizal and non-mycorrhizal plants (see Scheromm *et al.*, 1990a).

VI. Extracellular enzymes

A special case of nitrogen catabolism in mycorrhiza is the utilization of soil proteins by some fungal symbionts. A number of ericoid and ectomycorrhizal fungi are able to grow on proteins as sole nitrogen sources (Read *et al.*, 1989). These fungi—some of which have been termed "protein fungi" because of their preference for proteinaceous nitrogen sources (Abuzinadah and Read, 1986)—excrete acid proteinase enzymes when incubated in the presence of pure proteins such as bovine serum albumin (Leake and Read, 1989), or protein extracted from soil litter (El-Badaoui and Botton, 1989). Amino acids released by proteinase activity are taken up by the fungi and assimilated.

For the induction of proteinase secretion the fungus is grown for several days in pure liquid cultures containing 0.5–1 $g\,litre^{-1}$ protein. The appearence of proteinase activity can be monitored by determining the concentration of residual proteins in the culture filtrates at regular intervals. Protein can be quantified by the procedure described by Bradford (1976) or Lowry *et al.* (1951). A procedure allowing direct measurements of proteinase activities in the culture filtrate has been described by Twining (1984). This assay is based on the spectrofluorometric measurement of fluorescein isothiocyanate (FITC)-labelled proteins. A modification of this procedure for mycorrhizal proteinase activities has been described by Leake and Read (1989). The kinetics of extracellular acid proteinases from mycorrhizal fungi are similar to those of acid and alkaline proteinases from other microbial sources (Read *et al.*, 1989). The work of El-Badaoui and Botton (1989) suggests that ectomycorrhizal fungi are capable of releasing alkaline proteinases. These activities are highly stimulated when the fungus is incubated with proteins extracted from forest litter, and suppressed by low concentration of ammonium.

VII. Discussion

The complexity of regulation of nitrogen assimilation in fungi has been clearly recognized in extensive studies of the three non-symbiotic fungi, *Emericella* (*Aspergillus*) *nidulans*, *Neurospora crassa* and *Saccharomyces cerevisiae*. These ascomycetes, in spite of sharing common pathways of nitrogen metabolism and displaying an overriding catabolite control, exhibit distinct patterns of regulation of nitrogen metabolism in response to changes in trophic conditions (Cooper, 1982; Tomsett, 1989). It has been realized that one of the key elements of the regulatory complexity in these fungi is the separation of certain enzymes in different cell or hyphal compartments. This allows simultaneous activation of several anabolic and catabolic processes, enabling the fungus to explore diverse nitrogen and carbon sources. In comparison with these studies of non-symbiotic fungi, work on the regulation of nitrogen metabolism in symbiotic fungi is still in its infancy.

Recent developments in mycorrhizal research include the establishment of appropriate culturing and inoculating techniques, isotopic studies of inorganic and organic nitrogen utilization, and preliminary characterization of certain enzymes active in primary nitrogen assimilation. That litter proteins may play a major role in the nitrogen nutrition of mycorrhiza is being increasingly recognized lately. There is, therefore, a need to appreciate the importance of the regulatory diversity and compartmentation of various pathways of nitrogen metabolism in mycorrhizal fungi. Efforts are needed for the selection of appropriate mutants to evaluate the relative importance of various pathways in the nitrogen nutrition of mycorrhizal fungi. The roles of NADPH-GDH and GS in ammonium assimilation, for example, can be compared by using mutants specifically lacking one or other of these enzymes. Furthermore, many novel techniques to study the genetic control of nitrogen metabolism in plant–microbe association have been developed recently in the study of nodulated roots. These genetic approaches can be expected to provide great assistance in elucidating the precise biochemical mechanisms involved in nitrogen assimilation by mycorrhiza.

References

Abuzinadah, R. A. and Read, D. J. (1986). *New Phytol.* **103**, 507–514.

Abuzinadah, R. A. and Read, D. J. (1988). *Trans. Br. Mycol. Soc.* **91**, 437–479.

Abuzinadah, R. A., Finlay, R. D. and Read, D. J. (1986). *New Phytol.* **103**, 495–506.

Ahmad, I. and Hellebust, J. A. (1984). *Plant Physiol.* **76**, 658–663.

Ahmad, I. and Hellebust, J. A. (1985). *Marine Biol.* **86**, 85–91.

Ahmad, I. and Hellebust, J. A. (1986). *New Phytol.* **103**, 57–68.

Ahmad, I. and Hellebust, J. A. (1987). *Plant Physiol.* **83**, 259–261.

Ahmad, I. and Hellebust, J. A. (1989). *Anal. Biochem.* **180**, 99–104.

Ahmad, I., Larher, F., Mann, A. F., S. M. McNally, S. F. and Stewart, G. R. (1982). *New Phytol.* **91**, 585–595.

Ahmad, I., Carleton, T. J., Malloch, D. W. and Hellebust, J. A. (1990). *New Phytol.* **116**, 431–440.

Beevers, L. and Hageman, R. H. (1980). In *The Biochemistry of Plants* (B. J. Miflin, ed.), Vol. 5, pp. 115–168. Academic Press, New York.

Blackwell, R. D., Murray, A. J. S. and Lea, P. J. (1987). *J. Exp. Bot.* **38**; 1799–1809.

Bradford, M. M. (1976). *Anal. Biochem.* **72**, 248–254.

Carrodus, B. B. (1966). *New Phytol.* **65**, 358–371.

Casselton, P. J., Chandler, G., Shah, N., Stewart, G. R. and Sumar, N. (1986). *New Phytol.* **102**, 261–270.

Chen, F.-L. and Cullimore, J. V. (1988). *Plant Physiol* **88**, 1411–1417.

Cooper, T. J. (1982). In *The Molecular Biology of the Yeast* Saccharomyces: *Metabolism and Gene Expression* (J. N. Strachem, E. W. Jones and J. R. Broach, eds,), pp. 399–461. Cold Spring Harbor Laboratory, New York.

Dell, B., Botton, B., Martin, F. and Le Tacon, F. (1989). *New Phytol.* **111**, 683–692.

El-Badaoui, K. and Botton, B. (1989). *Ann. Sci. For.* **46** (Suppl.), 728s–730s.

Emes, M. J. and Fowler, M. J. (1979). *Planta* **144**, 249–253.

Fentem, P. A., Lea, P. J. and Stewart, G. R. (1983). *Plant Physiol.* **71**, 502–506.

Folkes, B. S. and Sims, A. P. (1974). *J. Gen. Microbiol.* **80**, 159–171.

Genetet, I., Martin, F. and G. R. Stewart, G. R. (1984). *Plant Physiol.* **76**, 395–399.

Garret, R. H. (1972). *Biochim. Biophys. Acta* **264**, 481–489.

Givan, C. G. (1980). In *The Biochemistry of Plants* (B. J. Miflin, ed,), Vol. 5, pp. 329–357. Academic Press, New York.

Hecht, U., Oelmuller, R., Schmidt, S. and Mohr, H. (1988). *Planta* **175**, 130–138.

Hellebust, J. A. and Lin, Y. (1978). In *Handbook of Phycological Methods: Physiological and Biochemical Methods* (J. A. Hellebust and J. S. Craigie, eds), pp. 379–388. Cambridge University Press, Cambridge.

Jennings, D. H. (1989). In *Nitrogen, Phosphorous and Sulphut Utilization by Fungi* (L. Boddy, R. Merchant and D. J. Read), pp. 1–31. Cambridge University Press, Cambridge.

Khalid, A., Boukroute, A., Botton, B. and Martin, F. (1988). *Plant Physiol. Biochem* **26**, 17–28.

Lara, M., Cullimore, J. V., Lea, P. J., Miflin, B. J., Johnston, A. W. B. and Lamb, J. W. (1983). *Planta* **157**, 254–258.

Lea, P. J. and Miflin, B. J. (1974). *Nature* **251**, 614–616.

Lea, P. J., Blackwell, R. D., Chen, F.-L. and Hecht, U. (1990). In *Methods in Plant Biochemistry* (P. J. Lea, ed.), Vol. 3, pp. 257–276. Academic Press, London.

Leake, J. R. and Read, D. J. (1989). *New Phytol.* **113**, 535–544.

Lee, J. A. and Stewart, G. R. (1978). *Adv. Bot. Res.* **6**, 1–43.

Lewis, C. M. and Fincham, J. R. S. (1970). *J. Bacteriol.* **103**, 55–61.

Lowry, O. H., Rosenburg, N. J., Farr, A. L. and Randall, R. J. (1951). *J. Biol. Chem.* **193**, 256–275.

Ludwig, R. A. (1980). *J. Bacteriol.* **141**, 1209–1216.

Mann, A. F., Fentem, P. A. and Stewart, G. R. (1979). *Biochem. Biophys. Res. Commun.* **88**, 515–521.

Marquez, A. J., Avila, C., Forde, B. G. and Wallsgrove, R. M. (1988). *Plant Physiol. Biochem.* **26**, 645–651.

Martin, F. (1985). *FEBS Lett.* **182**, 350–354.

Martin, F., Suzuki, A. and Hirel, B. (1982). *Anal. Biochem.* **125**, 24–29.

Martin, F., Msatef, Y. and Botton, B. (1983). *New Phytol.* **93**, 415–422.

Martin, F., Stewart, G. R., Genetet, I. and Le Tacon, E. (1986). *New Phytol.* **102**, 85–94.

Martin, F., Stewart, G. R., Genetet, I. and Mourot, B. (1988). *New Phytol.* **110**, 541–550.

Marzluf, G. A. (1981). *Microbiol. Rev.* **45**, 437–461.

McNally, S. F., Hirel, B., Gadal, P., Mann, A. F. and Stewart, G. R. (1983). *Plant Physiol.* **72**, 22–25.

Melin, E. and Nilsson, H. (1953). *Nature* **171**, 134.

Melin, E. and Rama Das, V. S. (1954). *Physiol. Plant* **7**, 851–858.

Miflin, B. J. and Lea, P. J. (1980). In *The Biochemistry of Plants* (B. J. Miflin, ed.), Vol. 5, pp. 169–202. Academic Press, New York.

Molina, R. and Palmer, J. G. (1982). In *Methods and Principles of Mycorrhizal Research* (N. C. Schenck, ed.), pp. 115–129. The American Phytopathological Society, St Paul, MN.

Nason, A., Lee, K. Y. and Averbach, B. C. (1954). *Biochim. Biophys. Acta* **15**, 159–161.

Oak, A. and Hirel, B. (1985). *Ann. Rev. Plant Physiol.* **36**, 354–365.

Palmer, J. and Hacskaylo, E. (1970). *Physiol. Plant.* **23**, 1187–1197.

Plassard, C., Mousain, D. and Salsac, L. (1984a). *Physiol. Veg.* **22**, 67–74.

Plassard, C. Mousain, D. and Salsac, L. (1984b). *Physiol. Veg.* **22**, 147–154.
Radin, J. W. (1975). *Plant Physiol.* **55**, 178–182.
Read, D. J., Leake, J. R. and Langdale, A. R. (1989). In *Nitrogen, Phosphorus and Sulphur Utilization by Fungi* (L. Boddy, R. Merchant and D. J. Read, eds), pp. 181–204. Cambridge University Press, Cambridge.
Rhodes, D., Rendon, G. A. and Stewart, G. R. (1976). *Planta* **129**, 203–210.
Rhodes, D. Meyer, A. C. and Jamieson, G. (1981). *Plant Physiol.* **68**, 1197–1205.
Scholl, R. L., Harper, J. E. and Hegeman, R. H. (1974). *Plant Physiol.* **53**, 825–828.
Scheromm, P., Plassard, C. and Salsac, L. (1990a). *New Phytol.* **114**, 93–98.
Scheromm, P., Plassard, C. and Salsac, L. (1990b). *New Phytol.* **114**, 227–234.
Scheromm P., Plassard, P. and Salsac, L. (1990c). *New Phytol.* **114**, 441–447.
Smith, F. W. and Thompson, J. F. (1971). *Plant Physiol.* **48**, 219–223.
Stewart, G. R. and Rhodes, D. (1977a). *New Phytol.* **79**, 257–268.
Stewart, G. R. and Rhodes, D. (1977b). In *Regulation of Enzyme Synthesis and Activity in Higher Plants* (H. Smith, ed.), pp. 1–19. Academic Press, New York.
Stewart, G. R., Mann, A. F. and Fentem, P. A. (1980). In *The Biochemistry of Plants* (B. J. Miflin, ed.), Vol. 5, pp. 271–327. Academic Press, New York.
Suzuki, A., Gadal, P. and Oaks, A. (1981). *Planta* **151**, 547–461.
Suzuki, A., Oaks, A., Jacquot, J. P., Vidal, J. and Gadal, P. (1985). *Plant Physiol.* **78**, 347–378.
Tomsett, A. B. (1989). In *Nitrogen, Phosphorus and Sulphur Utilization by Fungi* (L. Boddy, R. Merchant and D. J. Read, eds), pp. 33–57. Cambridge University Press, Cambridge.
Twining, S. S. (1984). *Anal. Biochem.* **143**, 30–34.
Verma, D. P. S., Fuller, F., Lee, J., Kunstner, P., Brisson, N. and Nguyen, T. (1983). In *Structure and Function of Plant Genomes* (O. Ciferri and L. Dure III, eds), pp. 269–283. Plenum Press, New York.
Vezina, L.-P., Margolis, H. A., McAfee, B. J. and Delany, S. (1989). *Physiol. Plant* **75**, 55–62.
Wagner, F., Gay, G. and Debaud, J. C. (1988). *Appl. Microbiol. Biotechnol.* **28**, 566–571.
Wagner, F., Gay, G. and Debaud, J. C. (1989). *New Phytol.* **113**, 259–264.
Wallsgrove, R. M., Harel, E., Lea, P. J. and Miflin, B. J. (1977). *J. Exp. Bot.* **28**, 588–596.
Wallsgrove, R. M., Lea, P. J. and Miflin, B. J. (1982). *Planta* **154**, 473–476.
Wallsgrove, R. M., Keys, A. F., Lea, P. J. and Miflin, B. J. (1983). *Plant Cell Environ.* **6**, 301–309.
Wray, J. L. and Fido, R. J. (1990). In *Methods in Plant Biochemistry* (P. J. Lea, ed.), Vol. 3, pp. 241–256. Academic Press, New York.

8

Techniques for the Study of Nitrogen Metabolism in Ectomycorrhiza

BERNARD BOTTON and MICHEL CHALOT

Université de Nancy I, Laboratoire de Physiologie Végétale et Forestière, BP 239, 54506 Vandoeuvre les Nancy Cedex, France

METHODS IN MICROBIOLOGY
VOLUME 23 ISBN 0-12-521523-1

Copyright © 1991 by Academic Press Limited
All rights of reproduction in any form reserved

I. Introduction

Progress in science is made by observation and experiment. This is particularly evident in the biochemistry of mycorrhiza, where the remarkable advances of the last decade have been made possible largely by the introduction of new methods and techniques.

We shall attempt to evaluate some of the methods used to investigate the main metabolic pathways involved in nitrogen metabolism in plants and fungi. Details of a few of these procedures are given elsewhere in this volume and many of them make use of principles that have also been fruitfully applied to other organisms.

In addition to more generally used techniques such as those for the estimations of the different forms of nitrogen, we have tried to cover a large range of enzymological techniques such as purification, quantification, localization and the use of enzyme inhibitors which appear to have considerable future potential in the study of nitrogen metabolism in mycorrhiza.

Much research has been devoted to electrophoretic and immunological procedures, which are very sensitive methods for elucidating metabolic pathways and are used after the protein purification steps which are the key to many areas of biochemical research on mycorrhiza.

However, with the recent advances in gene cloning and expression it is not only protein researchers who need to be able to use these techniques, but also biochemists and molecular biologists, working on a wide range of problems.

II. Extraction and recovery of the nitrogen fractions

A. Total nitrogen

The extraction and recovery of total nitrogen is based on the universal Kjeldahl method (Paech and Tracey, 1956) in which organic matter is oxidized by sulphuric acid and nitrogen converted quantitatively to ammonia according to the following reaction:

$$\text{Organic matter} + H_2SO_4 \rightarrow H_2O + CO_2 + \tfrac{1}{2}O_2 + NH_4HSO_4$$

This method has undergone many modifications which have increased both speed and accuracy and can be run as a two-step procedure.

1. Principle

The first step consists of the digestion of the nitrogen-containing material according to the Kjeldahl digestion method using concentrated sulphuric acid as the oxidizing agent and selenium reagent as the catalyst. All the nitrogen is then converted into ammonium sulphate. Several authors recommend running a pre-digestion with salicylic acid or reduced iron if the material is suspected to contain nitrate (Guiraud and Fardeau, 1977).

The second step requires diffusion of the ammonium. A number of diffusion procedures for determining the ammonium of a Kjeldahl digest have been described. A microdiffusion technique was developed by Conway (1962) primarily for ammonium and urea determination, and extended later to other compounds (amides, bromides, glucose). The diffusion is run in a microdiffusion apparatus (the "standard unit"), described in Fig. 1 and commercially available from Falcon Plastics (USA). The method depends on the gaseous diffusion of a volatile substance (ammonium) from the outer chamber into the inner chamber which contains an absorbing liquid. These analyses can be carried out using very small amounts of material with a high degree of accuracy, the

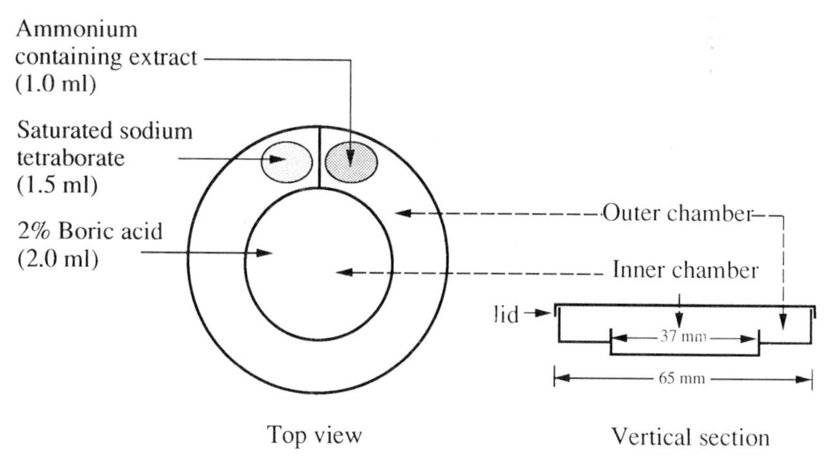

Fig. 1. Plan and principle of use of the microdiffusion dish used in the microdiffusion analysis of nitrogen (commercially available from Falcon Plastics, Oxnard, CA, USA).

technique is simple and a minimum of apparatus is required. The microdiffusion technique is also of value in considerably diminishing the work involved in large serial determinations of total nitrogen.

2. *Procedure*

Step 1: digestion

- Weigh out 30–100 mg of dry powdered material and transfer to a mineralization tube (Pyrex).
- Add 3 ml of nitrogen-free concentrated sulphuric acid and 100 mg of selenium reagent (97:1.5:1.5 w/w/w, sodium sulphate: copper sulphate: selenium).
- Heat gently on a digestion stand until fumes of sulphuric acid are freely evolved and then heat more strongly until digest has become clear or slightly green in colour. Blanks of the reagent alone should be run at the same time.

Step 2: diffusion

- Dilute the cooled digest in distilled water (usually 10 or 25 ml) and withdraw an aliquot for diffusion in the standard unit.
- Deliver, as described in Fig. 1:
  1.0 ml or less of the digest in the outer chamber;
  1.5 ml of saturated sodium tetraborate adjusted with HCl to pH 10 in the outer chamber;
  2.0 ml of 1% boric acid in the inner chamber.
- Cover with the lid and gently rock the unit to mix the digest and the tetraborate solution.
- Run the diffusion for about 16 h to ensure complete diffusion of the ammonium. The ammonium content can now be determined in the boric-HCl solution (see Section III.A)

B. Insoluble and soluble nitrogen

The soluble nitrogen fraction includes ammonium, nitrate and nitrite, urea, purines and pyrimidines and their derivatives, nucleosides and nucleotides, amino acids and their amides. The insoluble nitrogen fraction includes proteins and peptides.

1. *Principle*

Insoluble and soluble fractions can be easily separated and recovered by using a modified extraction method described by Bieleski and Turner (1966) (Fig. 2) and widely used in the study of nitrogen metabolism in plant or fungal tissues.

2. *Procedure*

- Grind 30–100 mg freeze-dried material in 3 ml of extraction medium consisting of methanol, methylene chloride and water (12:5:3; v/v/v).
- To the ground material add 5 ml of extraction medium and mix thoroughly. The homogenate and several rinses are combined and centrifuged at 10 000 g for 20 min. The pellet is extracted twice again and the supernatants pooled (now about 20 ml).

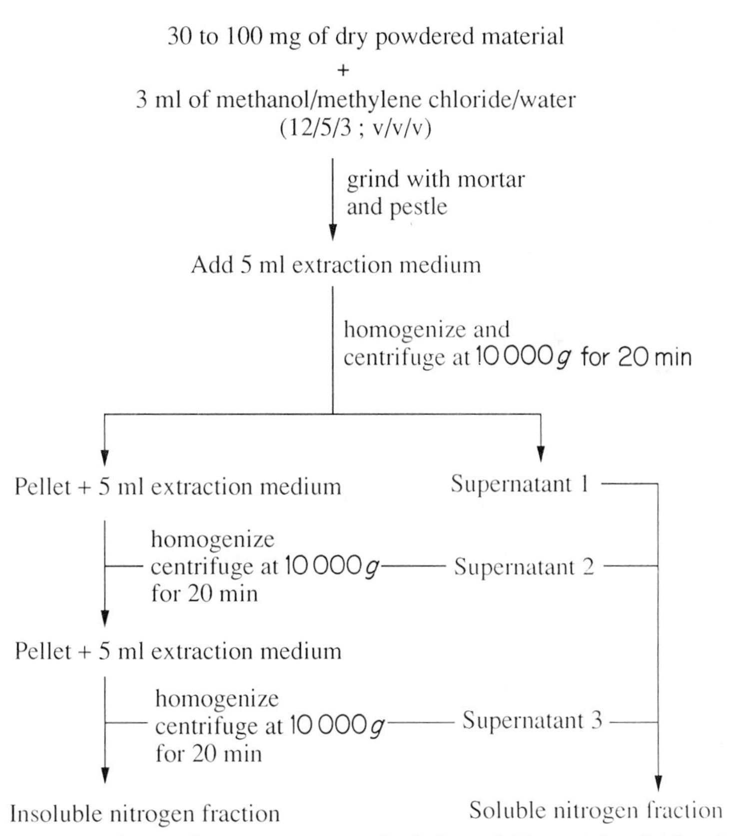

Fig. 2. Extraction and recovery protocol of the soluble and insoluble nitrogen fractions according to Bieleski and Turner (1966).

- The pellet contains the insoluble nitrogen compounds and can be dried in an oven to determine the total nitrogen of the insoluble nitrogen fraction after digestion and diffusion. The pooled supernatants contain the soluble nitrogen compounds (amino acids and related compounds) and can be dried under vacuum in a rotary evaporator (at 30 °C) for total nitrogen of the soluble nitrogen fraction (after digestion and diffusion) or amino acid determinations.

C. Extraction of proteins

Once released from its native environment, an intracellular protein will be subject to many inactivating conditions. Thus when root or fungal tissues are disrupted, several substances such as phenolic compounds may interact with proteins. As a result, protein denaturation and enzyme inactivation are to be expected. The use of thiol compounds (2-mercaptoethanol, dithiothreitol) or phenolic adsorbents like polyvinylpyrrolidone (soluble or insoluble form) may be included in the extraction medium, either separately or in different combinations to avoid the formation of complexes between proteins and phenolic compounds (Loomis *et al.*, 1979; Pitel and Cheliak, 1985; Weimar and Rothe, 1987). Mercaptoethanol (5–20 mM) and dithiothreitol (1–5 mM) should be added to the buffer immediately prior to use. However the protective action of 2-mercaptoethanol is lost within 24 h since it oxidizes rapidly. Dithiothreitol lasts longer and is therefore more suitable for use in storage buffers (Harris, 1989).

To facilitate solubilization of tissues, various non-ionic detergents (Nonidet P-40, Triton X-100 or Tween 20) may be added to the basic extraction medium. Findlay (1990) has reviewed a selection from the wide range of detergents and their properties.

The presence of tissue proteases that are liberated during the extraction procedure requires the addition of an inhibitor of proteases such as phenylmethylsulphonyl fluoride (PMSF) (0.1–1 mM) or tetrathionate to avoid protein degradation. It is also essential to run the extraction procedure at a temperature of 4 °C at which the proteases are inactive.

More elaborate precautions must be taken when considering enzymatic proteins (Loomis, 1974; Haissig and Schipper, 1975). Thus, apart from the inclusion of detergents, thiol compounds or protease inhibitors, the addition of glycerol (usually 10–50%, v/v) may be useful to preserve the activity of protein during the extraction procedure but also during subsequent fractionation and storage (Findlay, 1990). Many enzymes are also stabilized by including their cofactors or substrates in buffers.

However, the cofactor may be removed from its enzyme during purification by dialysis or gel permeation chromatography with a resultant loss of activity. Since some enzymes are metal-ion dependent, such substances may be added to the extraction buffer. Conversely the presence of metal ions can also inactivate sulphydryl groups, therefore a complexing agent, such as EDTA (0.1–1 mM) should be added to all buffers. Table I illustrates the role of a few protective agents on the activity of two nitrogen-metabolizing enzymes extracted from ectomycorrhizal roots of spruce.

When isolating one particular enzyme of interest from tissues it is essential to determine the optimal conditions of pH and the protective agent concentrations that give the optimal activity. Nevertheless in order to simplify the extraction procedure when isolating various enzymes at the same time, it is often convenient to determine a common extraction medium.

The composition of an extraction medium used to isolate key enzymes of nitrogen metabolism (GS, GDHs, AAT, AlaAT) from ectomycorrhizal roots of spruce is:

Tris-HCl (pH 7.6)	100 mM
Glycerol	10% (v/v)
PVP	5% (w/v)
PVPP	10% (w/w)
DTT	1 mM
2-Mercaptoethanol	14 mM
$MgCl_2$	2 mM
PMSF	1 mM

TABLE I

Effects of various protective agents on the activity of aspartate aminotransferase (AAT) and NADP-glutamate dehydrogenase (NADP-GDH) extracted from spruce ectomycorrhiza

Extraction medium	Proteins (mg g^{-1} fresh wt)	NADP-GDH (nkat g^{-1} fresh wt)	AAT (nkat g^{-1} fresh wt)
Complete medium	1.75	6.7	18.6
PVP omitted	0.15	1.8	1.2
PVPP omitted	1.66	4.0	16.6
2-Mercaptoethanol omitted	1.69	2.8	9.3

The complete extraction medium contains: 100 mM Tris HCl, pH 7.6; 10% (v/v) glycerol; 5% (w/v) PVP; 10% (w/w) PVPP; 1 mM DTT; 14 mM 2-mercaptoethanol; 2 mM $MgCl_2$ and 1 mM PMSF.

III. Quantification of the different forms of nitrogen

A. Ammonium

The determination of ammonium is of great importance when studying the nitrogen metabolism of ectomycorrhiza since ammonium is one of the principal mineral forms of nitrogen found in forest soils and absorbed by root systems. Ammonium is also the breakdown product of nitrogen compounds and its determination is therefore used in biochemistry as a measure of the total nitrogen in plant tissues (see Section II.A), as well as the total nitrogen of the soluble and insoluble fractions (see Section II.B). The estimation of ammonium concentration is also used in the measurement of enzyme activities that release ammonium (glutaminase, asparaginase or urease). We describe a method of ammonium determination that is based on the Berthelot reaction according to Weatherburn (1967).

1. Principle

The method is based on the capacity of ammonium to produce a blue indophenol chromophore in the presence of an alkaline solution of phenol and hypochlorite. The maximum absorbance is observed at 625 nm.

2. Procedure

- To 0.15 ml of ammonium containing solution, add:
  2.5 ml of the phenol reagent (1.0% phenol, w/v; 0.05% sodium nitroprusside, w/v);
  2.5 ml of the hypochlorite reagent (2.5% sodium hypochlorite, v/v; 0.5% sodium hydroxide, w/v).
- Homogenize and heat at 37 °C for 15 min.
- Read the absorbance at 625 nm against an adequate blank.

B. Nitrite

While the nitrite content of plants is usually extremely low because of its toxicity, the determination of nitrite is currently used for the determination of the nitrate assimilating enzyme, nitrate reductase, nitrite being the product of the reaction.

1. Principle

The method of nitrite determination is based on the reaction between sulphanilamide and nitrous acid with the formation of the diazo compound which reacts with naphthylamine to form a red azo dye that absorbs at 540 nm.

2. Procedure

- To 2 ml of nitrite containing extract, add:
  0.5 ml of 1% sulphanilamide prepared in HCl 3 M and
  0.5 ml of 0.02% dichlorhydrate of N-1 naphthylethylenediamine
- After 20 min of incubation, read the absorbance at 540 nm against a blank.

C. Nitrate

The nitrate content of plants is extremely variable but may usually be detected in actively growing tissues receiving abundant nitrogen. Nitrate forms coloured compounds with several organic substances such as brucine, diphenylamine or phenoldisulphonic acid and most of these reactions have been proposed as a basis of a method of estimation. The phenoldisulphonic acid method is commonly used but shows interferences with nitrites and chlorides. A rapid colorimetric determination of nitrate in plant tissues was developed by Cataldo *et al.* (1975). It is free of interferences caused by the presence of ammonium, chlorides or nitrite.

1. Principle

The nitrates extracted from plant tissues are complexed with salicylic acid under highly acidic conditions to form a chromophore that absorbs at 410 nm in a basic solution. Absorbance is directly proportional to the amount of nitrate present and the colour is stable for at least 48 h.

2. Procedure

- To 0.2 ml of nitrate containing extract, add 0.8 ml of 5% (w/v) salicylic acid prepared in concentrated sulphuric acid and mix thoroughly.
- After 20 min of incubation at 20 °C, gently add 19 ml of 2 N sodium hydroxide and homogenize. A yellow dye appears.

- After cooling, read the absorbance at 410 nm against a blank where salicylic acid is omitted.

D. Urea

Urea is the degradation product of ureides, which may be of great quantitative importance as nitrogenous constituents of plants. Urea is also the end product of the urea cycle which has been found in mycorrhizal roots of *Pinus nigra* and *Corylus avellana* (Krupa *et al.* 1973).

The most satisfactory method for the determination of urea is based on its enzymic transformation to ammonia and carbon dioxide. It is therefore important to use a urease of the purest grade.

1. Principle

Urea is hydrolysed to carbon dioxide and ammonia by the action of urease. The ammonia content is determined by the colorimetric method of Berthelot (see Section III.A) and is proportional to the urea nitrogen concentration.

2. Procedure

The determination of urea may be simplified by using commercial kits for urea determination. Such kits are available from Sigma and include urease, urea nitrogen standard and the reagents (phenol nitroprusside and alkaline hypochlorite) for the colorimetric determination of ammonium. It is essential to run adequate controls (omitting urease) to eliminate the background due to existing ammonium.

E. Amino acids

1. Ion-exchange chromatography

Modern amino acid analysers are very reliable and are still based on ion-exchange separation and post-column derivatization with ninhydrin. The original method has been much modified with dramatic improvements in speed and sensitivity of the analysis due to improved ion-exchange resin (Williams, 1986). Thus analysis times have been drastically reduced from 24 to about 2–3 h for the complete separation of physiological extracts and less than 1 h for protein hydrolysates. Some of the general problems commonly encountered with the analysis of amino acids by automated ion-exchange chromatography have been fully discussed (Williams, 1986).

2. *High performance liquid chromatography*

More recently, methods employing pre-column derivatization combined with reversed-phase high performance liquid chromatography (HPLC) have become increasingly popular.

Such reagents as orthophthaldialdehyde (OPA) (Hill *et al.*, 1979), phenylisothiocyanate (PITC) (Heinrikson and Meredith, 1984), fluorenylmethyloxycarbonyl chloride (FMOC) (Einarsson *et al.*, 1983) and dansyl chloride have been used to achieve separations of protein hydrolysates as well as various biological samples. Figure 3a shows the reaction of OPA with a primary amine and a thiol (2-mercaptoethanol for instance), which is required as an auxiliary agent to form an unstable fluorescent molecule. Also known as Edman's reagent, PITC has been used in peptide and protein sequencing for decades. Phenylisothiocyanate reacts at alkaline pH with primary and secondary amino acids

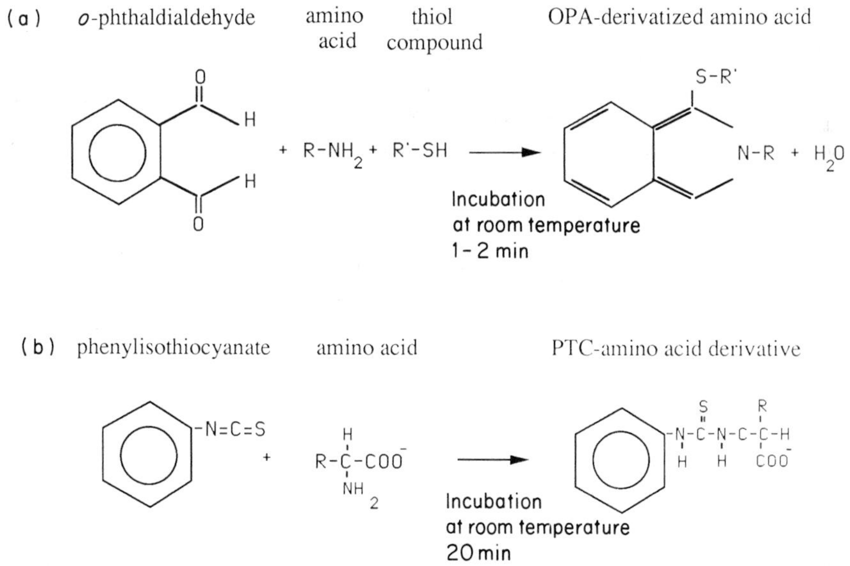

Fig. 3. Simplified reactions of amino acid derivatizations with *O*-phthaldialdehyde (a) and phenylisothiocyanate (b).

(Fig. 3b) to give phenylisothiocarbamoyl (PTC) amino acid derivatives, which absorb in the UV region. Each of the two methods described has its own particular advantages and disadvantages (Table II). If the determination of imino acids (proline and hydroxyproline) and cystine is not of particular importance, the OPA method is frequently selected because of its reduced analysis time. Although the derivatization procedure with PITC is relatively time-consuming, more than 40 samples can be derivatized simultaneously by using a vacuum centrifuge fitted with a cooling trap. Because of the stability of the PTC derivatives, the PITC pre-derivatization method can be fully automated with the aid of an autoinjector. An HPLC system consisting simply of an injector, pump, mixing valve, UV monitor (PITC and OPA) or fluorometer (OPA) and an integrator can give reproducible results and both methods have been successfully used for measuring amino acid contents of ectomycorrhizal tissues.

Derivatization procedure with PITC:

- Place 10 μl of standard or filtered sample in an Eppendorf vial and properly dry the sample using a vacuum centrifuge fitted with a cooling trap under 70 millitorr for about 20 min.
- Add 10 μl of drying solution consisting of 2:2:1 ethanol–triethylamine–sodium acetate, mix and repeat the drying step for 30 min.
- Add 20 μl of derivatization reagent to each re-dried sample. The derivatization reagent consists of 7:1:1:1 solution of ethanol:triethylamine:water:phenylisothiocyanate (PITC) and must be freshly prepared. Mix by vortexing for a few seconds and let stand for 20 min at room temperature (20–25 °C). Repeat the drying step,

TABLE II
Characteristics of derivatization methods with OPA and PITC

	OPA	PITC
Sample derivatization	Simple	Complex
Automation of derivatization reaction	Yes	No
Derivative stability	No	Yes
Detection of secondary amines	No	Yes
Detection method	Fluorescence or UV	UV
Detection limit	50 pmol (UV) 5 pmol (Fluor.)	10 pmol

allowing the sample to dry thoroughly for 30–45 min to remove all traces of PITC.

- Reconstitute the sample in 100 μl of diluent solution consisting of 0.071% (w/v) disodium hydrogen phosphate (Na_2HPO_4) and 5% (v/v) acetonitrile adjusted to pH 7.4 with 10% phosphoric acid. The samples are now ready for HPLC analysis. At room temperature, samples dissolved in the diluent provide accurate results for up to 12 h, allowing the automation of analysis from the derivatization step. Refrigerated samples will remain stable for up to 3 days. For long-term storage, if diluent is not added to the dried samples, they can be stored in the freezer for several months.

Figure 4 shows a typical chromatogram of spruce ectomycorrhiza PTC derivative amino acids.

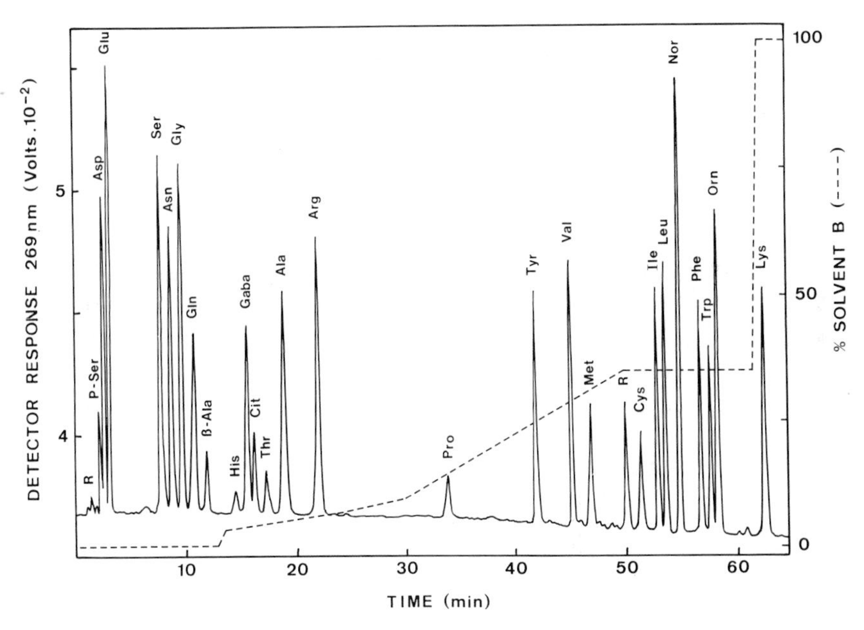

Fig. 4. Reverse-phase HPLC analysis of PITC-derivatized amino acids from spruce ectomycorrhiza on a Novapak C18 column, 3.9 × 300 mm (Waters). PTC derivatives were separated at a column temperature of 46 °C with a gradient of solvent A (acetonitrile–water, 2.5: 97.5 (v/v) containing 70 mM sodium acetate) and solvent B (acetonitrile–water–methanol, 45:40:15 (v/v/v)). The absorbance of the column eluent was monitored at 269 nm. R, reagent; Nor, norleucine as internal standard.

F. Proteins

A number of methods are available for the estimation of protein concentration. Some of these methods, such as Kjeldahl derivative methods, are not protein specific. Other methods have greater specificity for protein but were originally developed for animal tissues and physiological fluids, where protein concentrations are high and interfering components are low. In plants, protein nitrogen is in low concentrations (1–4%) and phenolics are ubiquitous (Loomis *et al.*, 1979). These compounds interfere with many of the assays that are based on the reduction of proteins (i.e. copper reduction: Lowry *et al.*, 1951; silver binding: Krystal *et al.*, 1985).

The Lowry procedure (Lowry *et al.*, 1951), based on the biuret reaction, is one of the most frequently cited in life science articles. However, a variety of substances have been found to interfere with this procedure. Such compounds include amino acids, buffers, detergents, sugars, lipids, salts and sulphydryl reagents. While modified procedures designed to overcome interferences by specific substances have been described, the two-step nature of the method complicates the assay and presents problems for its automation. Here we describe the Bradford assay (Bradford, 1976), which is widely used for plant protein determinations.

1. *Principle*

The assay depends upon the binding of Coomassie brilliant blue G-250 dye to protein to produce a complex absorbing at 595 nm. The dye will not react with free amino acids, low molecular weight polypeptides or many other nitrogen-containing or protein-like molecules (Compton and Jones, 1985; Jones *et al.*, 1989). The method is a rapid, simple one-step procedure that is adaptable to a micro-method which allows for the detection of less than 1.0 μg of protein per ml. Commercial preparations of the reagent are available from Biorad or Pierce.

2. *Procedure*

- An aliquot (usually 10–100 μl) of the protein containing extract is made up to 500 μl with a potassium phosphate buffer (pH 7.6).
- Add 500 μl of Coomassie Protein assay reagent (Pierce—ready to use) and mix immediately.
- Read the absorbance within 10 min at 595 nm against a blank consisting of buffer and the reagent. The protein concentration is

determined using a standard curve ranging from 1–25 $\mu g\,ml^{-1}$ bovine serum albumin (BSA).

IV. Use of enzyme inhibitors

It is necessary to use enzyme inhibitors with great care but provided that they are reasonably specific and the results are critically assessed, they can be employed effectively.

As these molecules block specific reactions in intact cells, the consequence of their action is an intracellular accumulation of precursor metabolites and a decrease in equality of the synthesized products. It is always advisable to limit the use of the inhibitors to short-term experiments because of the possibility of secondary indirect effects. Moreover, because of the importance of amino acids in nucleic acid and protein synthesis, prolonged blockage of their synthesis or utilization leads to inhibition of protein synthesis and to a decrease of cell growth.

A. Glutamine synthetase inhibitors

The most common inhibitor is methionine-S-sulphoximine (MSX or MSO) which has been shown to be an irreversible inhibitor of glutamine synthetase (GS) (Ronzio *et al.*, 1969; Meister, 1969). In the presence of ATP and metal ions, the MSX becomes tightly bound to the enzyme in the form of MSX-phosphate, in close analogy to the formation of enzyme-bound glutamyl phosphate as an intermediate of GS-catalysed synthesis of glutamine from glutamate.

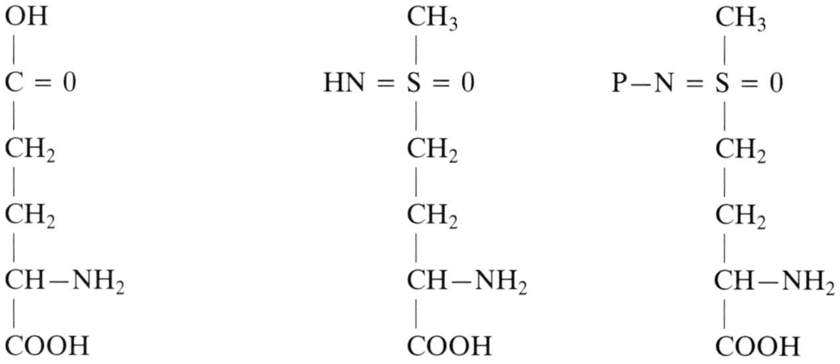

OH	CH_3	CH_3
|	|	|
C = 0	HN = S = 0	P—N = S = 0
|	|	|
CH_2	CH_2	CH_2
|	|	|
CH_2	CH_2	CH_2
|	|	|
CH—NH_2	CH—NH_2	CH—NH_2
|	|	|
COOH	COOH	COOH
L-Glutamic acid	Methionine sulphoximine	A possible structure for MSX-phosphate

MSX-phosphate is bound to the enzyme and can be removed by heating to 100 °C. It is therefore an extremely powerful tool in *in vivo* studies since it cannot be exchanged away from the enzyme once reacted. However its binding may be hindered by high concentrations of glutamate and prevented by the presence of both ammonium and glutamate. MSX has been reported to have certain other effects on plant tissues, e.g. in methionine uptake (Meins and Abrams, 1972), and to some extent can inhibit glutamate synthase at relatively high concentrations (Miflin and Lea, 1975). MSX has been shown by several authors to have no effect on GDH (Brenchley, 1973; Probyn and Lewis, 1978). It is usually used at a concentration ranging from 1 to 5 mM. Figure 5, redrawn from Martin *et al.* (1986), shows an example where MSX was added to a medium containing excised beech–*Lactarius* sp. ectomycorrhiza.

When MSX was included in the medium, the free glutamine pool and other free amino acids diminished sharply while the ammonium pool increased in size. This is consistent with the occurrence of the GS-GOGAT pathway of ammonium assimilation in beech ectomycorrhiza. GDH activity was not able to sustain NH_4^+ assimilation when GS was inhibited by MSX. An initial small increase in concentration of glutamine and other amino acids was probably due to a lag in GS inhibition by MSX.

In addition to MSX, two other molecules, which can also be regarded as structural analogues to glutamate, are potent inhibitors of GS. One of these inhibitors is L-phosphinothricin (PPT), which is the active component of the herbicides Glufosinate (Hoechst) and Biolaphos (Meiji Seika) and is produced by some *Streptomyces viridochromogenes* strains (Bayer *et al.*, 1972; Johnson *et al.*, 1990). The other one is Tabtoxinine-β-lactam (T-β-L) which irreversibly inactivates GS (Langston-Unkefer *et al.*, 1984).

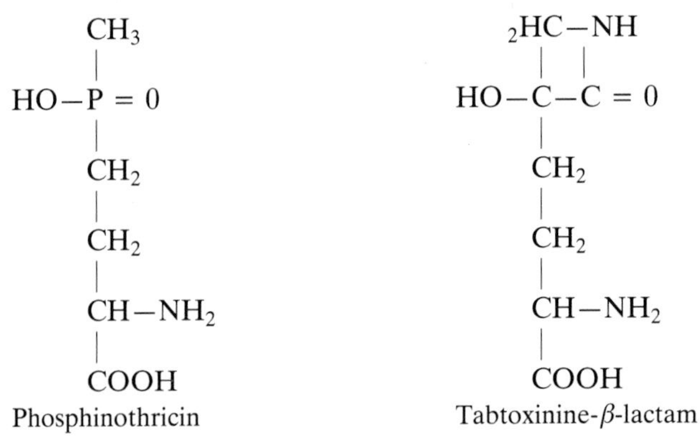

Phosphinothricin Tabtoxinine-β-lactam

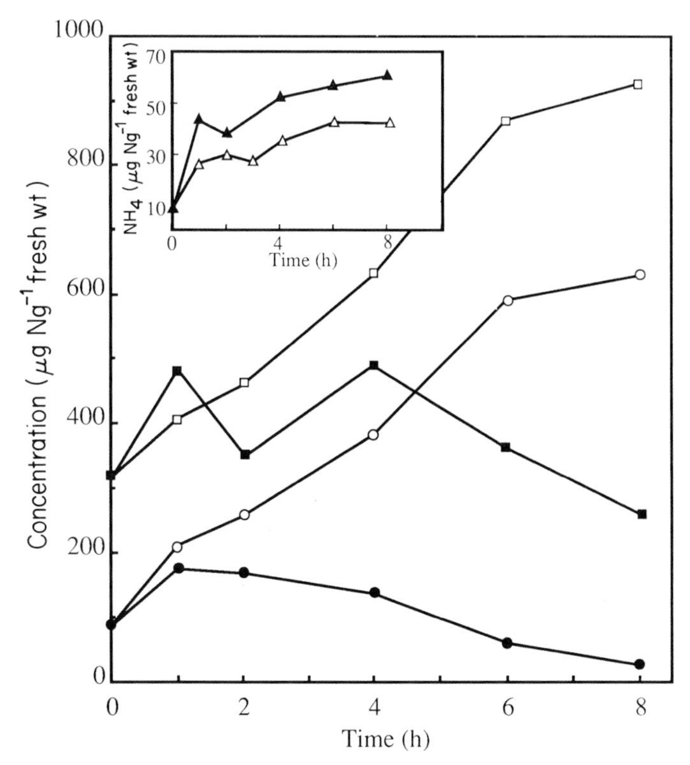

Fig. 5. Effect of ammonium feeding and 1 mM MSX on the levels of intracellular ammonium (inset: ▲, △), total free amino acid (■, □) and glutamine (●, ○) in ectomycorrhiza of *Fagus sylvatica*. Open symbols, without MSX; closed symbols, with MSX.

Whole detached ectomycorrhiza collected from a forest soil were placed in an Erlenmeyer flask containing 100 mM glucose, 7 mM KH_2PO_4, 2.5 mM $MgSO_4$ and 2.5 mM $(NH_4)_2SO_4$ in 0.5 l (pH 5.5). Streptomycin (0.04%) and tifomycin (0.4%) were added to prevent bacterial growth. Air was continuously bubbled through the incubation medium and incubations were performed at 20 °C. Redrawn from Martin *et al.* (1986).

These inhibitors have already been used to inactivate chloroplast and cytosolic forms of GS from higher plants and the enzyme of some non-mycorrhizal fungi at a concentration of 5 mM, but to our knowledge have not yet been used in mycorrhizal association studies.

B. Glutamate synthase inhibitors

Albizzine (L-2-amino-3-ureidopropionic acid), azaserine (*O*-diazoacetyl-L-serine) and DON (6-diazo-5-oxo-L-norleucine) are inhibitors of a wide range of glutamine-using enzymes in which the amide-amino group is transferred. Because of the structural similarity between these molecules and glutamine they compete for binding to the active site of glutamate synthase (GOGAT). However, once bound, the diazo-acetyl group of the analogue becomes irreversibly attached and cannot be displaced. Neither albizzine, azaserine, nor DON have any known effect on GDH (Miflin and Lea, 1975).

		N^-	N^-
		‖	‖
		N^+	N^+
		‖	‖
NH_2	NH_2	CH	CH
|	|	|	|
C = O	C = O	C = O	C = O
|	|	|	|
CH_2	NH	O	CH_2
|	|	|	|
CH_2	CH_2	CH_2	CH_2
|	|	|	|
$CH{-}NH_2$	$CH{-}NH_2$	$CH{-}NH_2$	$CH{-}NH_2$
|	|	|	|
COOH	COOH	COOH	COOH
Glutamine	Albizzine	Azaserine	DON

Albizzine is usually used at a concentration of 2.5 mM while azaserine, which is assumed to enter the cells more easily, is added at concentrations ranging from 1 to 2 mM. DON has not been used in mycorrhizal association studies.

Azaserine was used to investigate the route of ammonium assimilation in the mycorrhizal fungus *Pisolithus tinctorius* (Kershaw and Stewart, 1989). After 18 days of growth in Melin–Norkrans medium containing 1 mM ammonium, nitrogen was depleted and the nitrogen-starved mycelia were transferred to flasks of fresh medium containing 1 mM ammonium (control) (Fig. 6A), or 1 mM ammonium with 1 mM azaserine (Fig. 6B). Rapid ammonium assimilation was shown in the control by a marked increase in the glutamate and glutamine pools after 2 h in the fresh medium (Fig. 6A). Glutamate levels remained constant after 2 h but the glutamine concentration continued to increase up to 6 h, indicating glutamine as the primary product of assimilated ammonia.

Inhibition of glutamate synthase by azaserine blocked the transfer of amide nitrogen from glutamine to glutamate and the size of the glutamate pool did not increase over the time of the experiment (Fig.

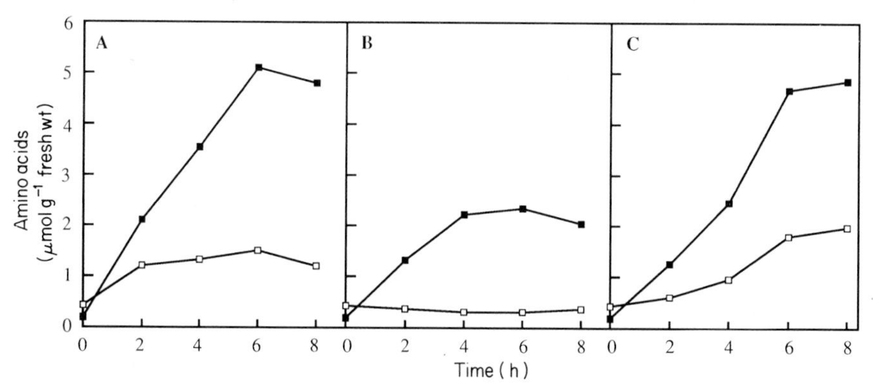

Fig. 6. Assimilation of ammonia into glutamate (□) and glutamine (■) by N-starved mycelia of *Pisolithus tinctorius*. (A) Control; (B) glutamate synthase inhibition by azaserine (1 mM); (C) aminotransferase inhibtion by aminooxyacetate (0.2 mM). See details in the text. Redrawn from Kershaw and Stewart (1989).

6B). Thus in this fungus GDH played no significant role and in the presence of azaserine, the glutamine pool also decreased over the last 4 h of the experiment.

However, very frequently, the role played by the fungal glutamate synthase remains difficult to demonstrate, even in the mycorrhizal associations. Indeed, in *Picea excelsa–Hebeloma* sp. mycorrhiza, albizzine added in the growth medium did not modify the time course of ^{15}N incorporation from $^{15}NH_4^+$ into glutamate and glutamine (Martin and Botton, 1991).

C. Aminotransferase inhibitors

The carbonyl-binding reagent aminooxyacetate (AOA) is structurally related to the aminotransferase substrate and has a particularly strong affinity for this kind of enzyme.

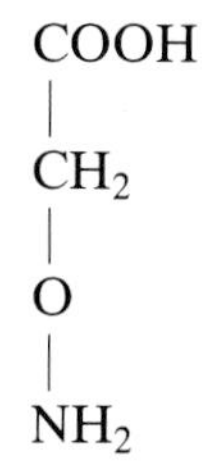

Aminooxyacetate (carboxymethoxylamine)
(The commercially available molecule is stabilized as hemihydroxychloride)

As the K_i is low, AOA is useful as a fairly specific inhibitor of transaminases in metabolic studies. It can be employed as a tool to determine whether a transamination step is involved in a particular sequence of reactions (Givan, 1980). Figure 6C shows that inhibition of aminotransferases by aminooxyacetate in *Pisolithus tinctorius* led to an accumulation of glutamate after 6 h, suggesting that glutamate is an important source of nitrogen for the anabolism of nitrogenous metabolites.

Smith *et al.* (1977) have reassessed the use of AOA and state that pyruvate and acetaldehyde may reverse AOA inhibition of aspartate and alanine aminotransferases, even after preincubating the enzymes with AOA. They recommend 2-amino-4-methoxy-*trans*-but-3-enoic acid as a preferable alternative to AOA for inhibiting aspartate aminotransferase, but point out that alanine aminotransferase (as well as glutamate dehydrogenase) is relatively insensitive to this compound.

Numerous investigators have also examined the effects of carbonyl-reacting compounds which are especially antagonistic to pyridoxal-5′-phosphate (PLP), the cofactor involved in the reaction of transamination. These compounds include hydroxylamine, cyanide and isonicotinic acid hydrazide (isoniazid), the latter being known to interact with the aldehyde group of the cofactor.

This compound was used to demonstrate the pyridoxal phosphate requirement in the aspartate aminotransferase of *Cenococcum geophilum* (Khalid *et al.*, 1988). In the absence of exogenous pyridoxal-5′-phosphate, the aspartate aminotransferase activity was not entirely suppressed, suggesting that the enzyme could partly retain the cofactor (Table III). Indeed, the activity of the purified enzyme was considerably inhibited by the addition of isoniazid, indicating that the coenzyme moiety was loosely bound to the apoenzyme of this aminotransferase.

D. Quantification of inhibitors in cell extracts

Methionine sulphoximine and albizzine, because of their structural analogies respectively with glutamate and glutamine, were both detected as PTC derivatives (Fig. 7B), as are other common amino acids. Extracts from MSX- or albizzine-treated spruce mycorrhiza contained high levels of inhibitors (Fig. 7C and D) that were not detected in untreated controls (Fig. 7A). Albizzine, when added to standard amino acids, was eluted between phosphoethanolamine and serine, and MSX detected between β-alanine and histidine. Coelutions of albizzine with phosphoethanolamine, and MSX with β-alanine might have occurred in some cases where albizzine or MSX were found at high concentrations

TABLE III
Effects of pyridoxal 5′-phosphate and isonicotinic acid hydrazide on the aspartate aminotransferase activity of *Cenococcum geophilum*

	Transaminase activity	
	(nkat mg^{-1} protein)	(%)
+ PLP	3200	100
− PLP	560	17.5
+ PLP + INAH	1440	45
− PLP + INAH	168	5.2

Pyridoxal 5′-phophate (PLP) and isonicotinic acid hydrazide (INAH) were supplied at a final concentration of 80 mM and 10 mM, respectively. The latter compound was incubated for 30 min at 20 °C in the reaction mixture before starting the reaction by addition of α-ketoglutarate. Assays were carried out with a coupled reaction at pH 7.6 (from Khalid *et al.*, 1988).

in amino acid extracts. However, by changing the running conditions it was easily demonstrated that phosphoethanolamine was not present in spruce extracts while β-alanine was found to be present in constant amounts in all cases. Results showed that MSX and albizzine rapidly accumulated in the tissues (Fig. 7C and D). Thus, after a 90 min incubation period, MSX concentration was about 2 $\mu mol\,g^{-1}$ fresh weight, and albizzine 1.7 $\mu mol\,g^{-1}$ fresh weight. Maximum levels were found after 3 h of incubation.

V. Protein purification methods

A. Purposes of the purification and knowledge of the protein under investigation

Before considering practical aspects, it is essential to define the purpose of the purification. If for example, the biological activity of the protein must be retained for use in enzyme assays, only a small amount of protein may be required. It may not need to be absolutely pure, but must be free of interfering compounds. The same is true for use in kinetic studies. Where the structure of a protein is under investigation, however, it must be available in larger amounts and of higher purity, especially for structure–function studies. This is also true when the protein is used for antibody production.

A thorough review of existing knowledge of the composition and properties of the protein in question is important. If it is a known

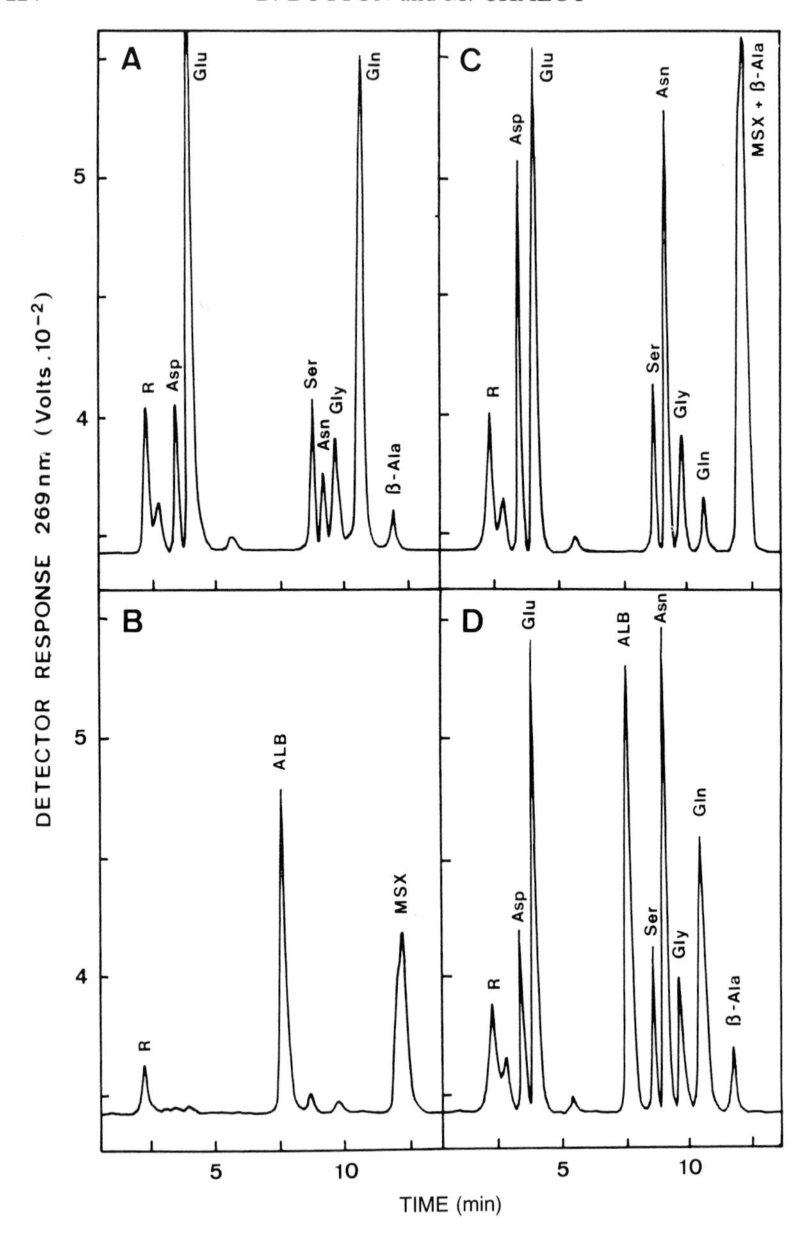

Fig. 7. HPLC-chromatograms showing elution of Methionine sulphoximine (MSX) and Albizzine (ALB) in control standards (B), in MSX treated (C) and ALB treated (D) spruce ectomycorrhiza extracts compared with an untreated control extract (A). The retention times of ALB and MSX were 7.6 and

lipoprotein or glycoprotein, for example, its properties will be markedly different from those of proteins lacking covalently bound lipids or sugars. Such properties can be valuable separation factors. Knowledge of the protein's native environment, whether it is intracellular, extracellular, soluble or membrane bound, independent or attached to other structural elements, will greatly influence the choice of buffer systems.

How stable is the molecule? Is it heat or acid labile? Are its molecular weight and isoelectric point (pHi) known? Is it sensitive to air, metal ions or proteases? Are there any known stabilizing compounds, or cofactors required? The answers to all such questions will certainly help the experimentalist to select the most efficient extraction and fractionation methods and preserve the activity of the protein.

It is also worthwhile reviewing the nature of the impurities present in the source. Quantities of inhibitors or proteases may be present and must be removed. This is also the case with polyphenols, very often present to a large extent when the plant partner of the mycorrhiza is involved. The early stages in a purification scheme are best aimed at removing specific contaminants which may interfere with later stages or cause loss of activity.

B. Extraction methods

In many cases mycorrhizal samples can be ground in a mortar with acid-washed sand. The grinding can be carried out in liquid nitrogen to fix active molecules. An alternative method is to use a Waring blender (Polytron type) but mechanical cell disruption may cause local overheating within the extract and so bring about protein denaturation. It is therefore, important to keep the lysate chilled at all times and several short treatments are preferable to a single prolonged one.

Cell disintegration by ultrasound can be used as a complementary method, especially when aggregated structures such as rhizomorphs are difficult to homogenize by classical methods. Indeed, good results have been obtained in the Ascomycete *Sphaerostilbe repens*, the rhizomorphs of which were particularly resistant to cell disruption (Botton and Bonaly, 1982).

It is noteworthy that some enzymes can be altered by sonic disintegration. However, the use of ultrasound is useful for disintegrating organelles such as mitochondria. As an example, purified mitochondria of

---

12.3 min, respectively. R, reagent. The spruce–*Hebeloma* sp. ectomycorrhiza collected from a nursery were incubated in the presence of 5 mM ammonium in Erlenmeyer flasks. Amino acid assays were performed after 3 h incubation.

the ectomycorrhizal fungus *Cenococcum geophilum* were successfully disintegrated by six 15-s periods at 20 kHz with 50% rest time (Khalid, 1988).

The use of vibration mills such as the Braun shaker has proved to be useful for disintegrating *C. geophilum* mycelia for purification of the cell walls (Mangin *et al.*, 1986). Glass beads of 0.4 mm diameter were included in the cell suspension with a high ratio of particles to micro-organism (2/3) and the treatment lasted 10 min to disrupt the cells completely. Heat is generated by this method and its effect can be suppressed by a CO_2 cooling treatment. The instrument has been used for disintegrating a wide variety of cells including yeasts.

C. Differential solubility

Differential solubility techniques are generally applied at the start of a separation when the protein of interest constitutes less than 1% of the total sample protein. They have a high capacity but low resolving power, and are used to eliminate the gross impurities from the sample. There are several useful techniques in this category including: heat treatment, isoelectric focusing, polyethylene glycol precipitation, salt fractionation, etc.

Salt fractionation is probably the most common technique in this category. Ammonium sulphate is most widely used for two reasons: first, it is highly soluble in water; and second, it has a stabilizing effect upon proteins. The solubility of proteins as a function of ammonium sulphate concentration follows a convex curve and Dixon's nomogram gives the quantity of solid ammonium sulphate to add to the solution to achieve desired saturations (Dixon, 1953). At low ionic strengths, addition of ammonium sulphate promotes the hydration of proteins—a salting-in effect. As the concentration of ammonium sulphate increases, the proteins become less soluble—a salting-out effect which culminates in the precipitation of the proteins. The concentration at which each species precipitates differs and varies with pH, temperature and protein concentration. If this technique is to be used regularly, conditions must be identical to ensure reproducibility. This is particularly important with regard to protein concentration, where variations from experiment to experiment can significantly affect recovery.

It is worth noting that proteins precipitated in this way retain their native conformation and can be redissolved without appreciable loss of activity. Thus, the technique can be used for precipitating either the bulk contaminants or the active protein fraction itself, whichever is most convenient. An ammonium sulphate precipitate is frequently the best form in which to store a protein either during purification or when pure.

1. Procedure of ammonium sulphate fractionation

Choose an amount of ammonium sulphate to be added that will precipitate the maximum amount of protein without precipitating a significant portion of the enzyme which is to be purified (for example 30% as ammonium sulphate in solution).

Add the solid ammonium sulphate slowly to the protein fraction with gentle stirring (magnetic stirrer). Allow the mixture to remain at 0 °C for 20 min to assure maximum precipitation. Remove the precipitated protein by centrifugation (40 000 *g*, 20 min).

To the supernatant fluid add solid ammonium sulphate in quantity to precipitate most of the enzyme, but no more protein than necessary (for example from 30 to 80% saturation). Allow the mixture to remain at 0 °C for 20 min. Collect the precipitated protein by centrifuging as before, and dissolve it in a buffer.

Figure 8 shows an example of purification of aspartate aminotransferase (AAT) and NAD-dependent malate dehydrogenase (NAD-MDH), in *C. geophilum* by ammonium sulphate fractionation. The crude extract was precipitated between 50 and 95% saturation. Maximum AAT was recovered at a concentration of 80% in ammonium sulphate, and NAD-MDH was recovered at 75% while maximum amount of total protein precipitation occurred between 50 and 75%. The 75–95% ammonium sulphate fraction was recovered for further purification of the AAT. At this stage the enzyme was purified 12-fold with a yield of 82% (Table IV).

2. Removal of salts and exchange of buffer

If the next step in purification is a gel filtration, the concentrated extract can be applied directly to the column. However, in the case where ion-exchange chromatography must be used, any remaining ammonium sulphate must be removed from the sample. Moreover the sample must be equilibrated with a buffer of the proper ionic strength and pH for the chromatographic step.

Dialysis can be carried out with a dialysis tubing (Union Carbide no. 8) previously heated in $NaHCO_3$–EDTA and rinsed. After having transferred the sample to the tube and tied the ends, place the dialysis tube in several litres of buffer. Allow the dialysis to proceed with gentle stirring (magnetic stirrer) for 10–20 h at 1–4 °C.

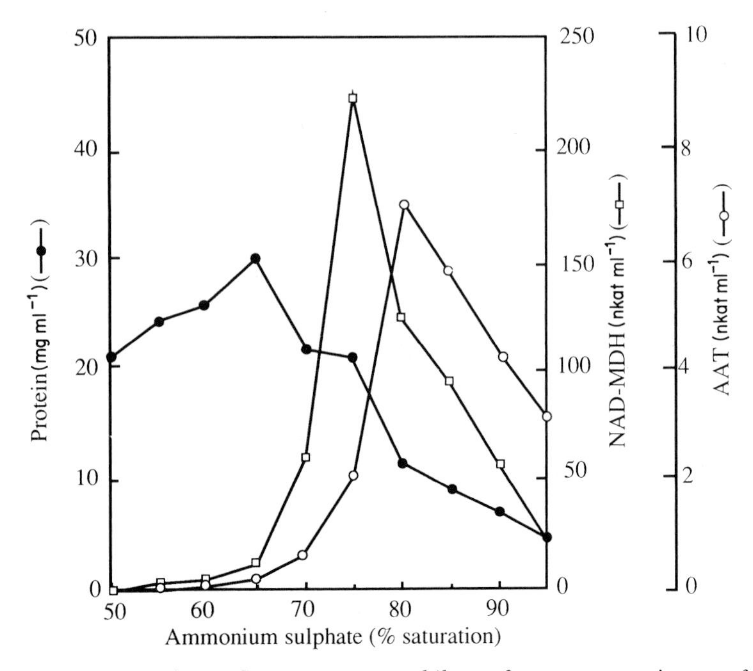

Fig. 8. Recovery from *Cenococcum geophilum* of aspartate aminotransferase, NAD-malate dehydrogenase and protein using ammonium sulphate fractionation. The crude extract containing 40 mg protein was precipitated between 50 and 95% saturation by increasing steps of 5% saturation. Proteins were estimated according to Bradford (1976). Adapted from Khalid *et al.* (1988).

Desalting can also be carried out by gel filtration, also known as "molecular-sieve chromatography". The term refers not only to the removal of salts, but also to the removal of other low molecular weight compounds from solutions of macromolecules. Since the introduction of a cross-linked dextran, commercially known as Sephadex, gel filtration has been developed extensively in both analytical and preparative methods for proteins.

Separations by this technique depend on the fact that molecules larger than the largest pores of the granules, i.e. above the exclusion limit, cannot penetrate the gel particles and therefore pass through the bed in the liquid phase outside the particles. They are thus eluted first. Smaller molecules, however, penetrate the gel particles to a varying extent depending on their size and shape, the particles then acting as a molecular sieve. Molecules are therefore eluted in the order of decreasing molecular size.

TABLE IV

Purification of aspartate aminotransferase from *Cenococcum geophilum*. The starting material was 83 g (fresh weight) homogenized with approximately 200 ml of Tris buffer (from Khalid *et al.*, 1988)

Purification step	Volume (ml)	Total protein (mg)	Total activity (nkat)	Specific activity (nkat mg^{-1} protein)	Yield (%)	Purification (fold)
Crude extract	164	430	3188	7.4	100	1
Ammonium sulphate (75–95% sat.)	12	29	2640	91	82	12
Phenyl sepharose CL 4B	19	2.1	1710	814	53	110
DEAE-cellulose	11	0.5	924	1848	30	250
Sephadex G-200	3.6	0.13	421	3238	14	438

In desalting, all proteins in a mixture are to be separated from relatively low molecular weight solutes (< 5000) and are totally excluded from the pore matrix. Suitable media include Sephadex G-25, or G-50 (Pharmacia-LKB) and Biogels P-6 or P-10 (Bio-Rad). Pre-packed Sephadex (G-25) columns are available from Pharmacia-LKB (PD-10 columns) specifically for this purpose.

Determine the void volume of the column by using Blue Dextran (molecular weight 2 000 000) which is excluded from the matrix.

This method is only applicable to small volumes and the maximum sample volume should not exceed 25–30% of the volume of the column to ensure adequate resolution between the protein and salt. If only the void volume is known, avoid sample volumes greater than half the void volume.

Once the void volume of buffer has passed, collect the sample up to a volume of 1.5 or 2 times the volume of the void volume.

A quicker, alternative method for desalting or buffer exchange is diafiltration which uses stirred cells available to cover the range 1–400 ml (e.g. Amicon, Filtron, Sartorius). Several types of membranes are available with a molecular weight cut-off from 500 to 1 000 000. Apply the minimum pressure of nitrogen (or possibly compressed air) required to give an acceptable flow rate (usually less than 75 p.s.i.). When the desired concentration has been achieved turn off the pressure and open the pressure relief valve. Continue stirring for 5 min to resuspend proteins absorbed to the membrane.

This procedure, unlike dialysis and gel filtration, allows concentration of the sample.

D. Chromatographic methods

1. Ion-exchange chromatography

Ion-exchange is the separation of proteins on the basis of their charge. Proteins carry both positively and negatively charged groups on their surfaces, due largely to the presence of side chains of acidic and basic amino acids. The net charge on a protein depends on the relative numbers of positively and negatively charged groups; this varies with pH. The pH at which a protein has an equal number of positive and negative groups is termed its isoelectric point (pHi).

A wide variety of matrices and functional groups is available. Matrices are derivatized with positively-charged groups (anion-exchangers) for the adsorption of anionic proteins, or with negatively-charged groups (cation-exchangers) for the adsorption of cationic proteins.

As proteins are anionic or cationic, depending on whether the buffer pH is above or below the isoelectric point, respectively, a choice of anionic or cationic matrix must be made. In practice the decision is sometimes restricted by the pH stability of the protein. If a protein is more stable above its pHi, then an anion-exchanger should be chosen. Conversely if the protein is more stable below its pHi, then a cation-exchanger should be chosen.

The most commonly used functionalities are the weak inorganic groups. The diethylaminoethyl (DEAE) group ($-C_2H_4NH^+(C_2H_5)_2$) is usually used in anion-exchange while the carboxymethyl (CM) group ($-CH_2COO^-$) is frequently used in cation-exchange. Strong ion-exchangers are becoming more popular, notably the Sepharose Fast Flow packings, based on sulphomethyl ($-CH_2SO_3^-$) and triethyaminoethyl ($-C_2H_4N^+(C_2H_5)_3$) functionalities for strong cation- and anion-exchange, respectively.

In protein purification, the pHi of the molecule to be purified is generally unknown, but anion-exchange functionalities (e.g. DEAE) are most frequently used since proteins of pHi below 7 are more common.

Procedures for purification of glutamine synthetase (GS) and NADP-glutamate dehydrogenase (NADP-GDH) of the ectomycorrhizal fungus *Laccaria bicolor* are examples where an anion-exchanger (DEAE-Trisacryl) has been used (Fig. 9.). The crude extract was homogenized in 100 mM Tris-HCl buffer pH 7.6 (see Section II.C). The volume was applied to a DEAE-Trisacryl column (Pharmacia-LKB) (170 × 150 mm) equilibrated with 50 mM Tris-HCl buffer at pH 7.6 including 10% glycerol, 1 mM DTT and 14 mM mercaptoethanol. After washing the column with the same buffer until all unabsorbed proteins were removed, NADP-GDH and GS were eluted with a linear gradient from 0 to 0.3 M NaCl in Tris-HCl buffer. After this step the NADP-GDH was purified 13.4-fold with a yield of 87.3% and GS was purified 5.7-fold with a yield of 65%.

2. *Hydrophobic interaction chromatography*

Most proteins have on their accessible surfaces small hydrophobic patches through which they can be induced to associate with hydrophobic substituents on insoluble supports.

The resins are generally dextran-based (Sepharoses or Superoses) or agaroses and are derivatized with uncharged moieties (hexyl, octyl, phenyl), through which interaction with the protein occurs.

The choice of the hydrophobic ligand is of considerable importance in determining the ease of subsequent protein elution. Phenyl Sepharose,

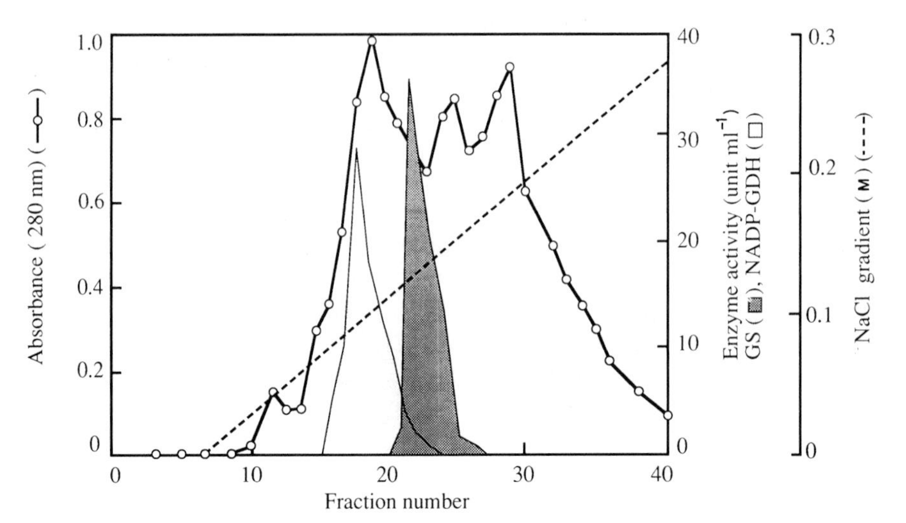

Fig. 9. Elution profiles of glutamine synthetase (GS), NADP-glutamate dehydrogenase (NADP-GDH), and proteins, of *Laccaria bicolor* from a DEAE-Trisacryl column. The column was loaded with 100 ml of crude extract containing 20 mg of proteins. Proteins were eluted with a linear gradient from 0 to 0.3 M NaCl in Tris-HCl buffer. The flow rate was 60 ml h^{-1} and 2.0 ml fractions were collected. One unit of NADP-GDH activity corresponds to 1 nmol of product formed per second. One unit of GS activity corresponds to 1 μmol of product formed per min $\times 10^{-3}$. From A. Brun, M. Chalot and B. Botton (unpubl. res.).

for example is less hydrophobic than Octyl Sepharose. Extremely hydrophobic proteins (e.g. membrane proteins) may absorb too strongly on Octyl Sepharose and require very strong eluting conditions. Phenyl Sepharose should be used for these molecules.

A significant step in purification of the aspartate aminotransferase in the ectomycorrhizal fungus *C. geophilum* was carried out using Phenyl Sepharose CL-4B, Pharmacia (Fig. 10) (Khalid *et al.*, 1988). The column (100 × 13 mm) was equilibrated with 50 mM sodium phosphate buffer (pH 7.5) containing 1 M ammonium sulphate and loaded with proteins precipitated from 75 to 95% saturation in ammonium sulphate. After washing away unbound material with the same buffer, bound proteins were recovered by decreasing the ionic strength using the salting-out effect with a linear gradient of sodium phosphate buffer (pH 7.5) from 50 mM to 5 mM and a linear gradient of ammonium sulphate from 1 M to 0 M. A major peak of enzyme activity was eluted at 0.3 M ammonium sulphate and a minor peak was detected at a salt concentration of 0.1 M. After ammonium sulphate fractionation and hydrophobic interaction

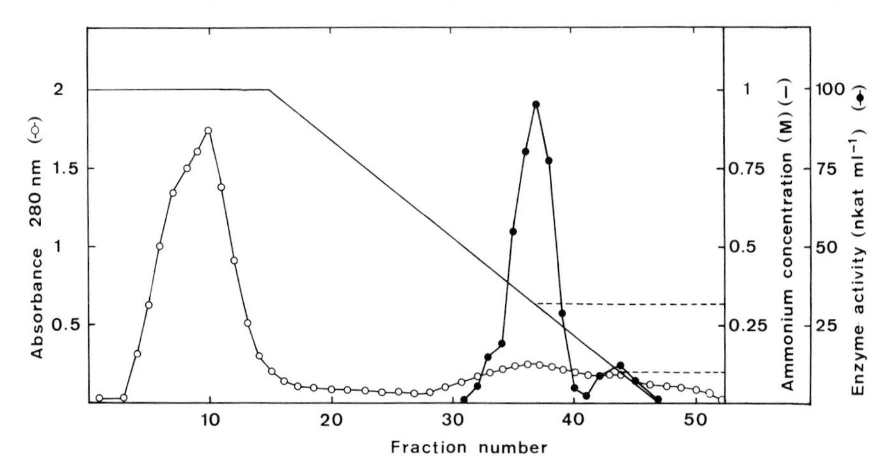

Fig. 10. Protein elution pattern and related aspartate aminotransferase activity obtained from *C. geophilum* extract, after hydrophobic interaction chromatography on phenyl Sepharose CL-4B. The column was loaded with 29 mg of proteins which were eluted with a linear gradient of phosphate buffer (pH 7.5) including ammonium sulphate. Phosphate buffer concentration ranged from 50 to 5 mM and ammonium sulphate concentration ranged from 1 M to 0 M. The flow rate was 20 ml h^{-1} and 2.5 ml fractions were collected. From Khalid *et al.* (1988).

chromatography, the AAT of *C. geophilum* was purified 110-fold with 53% of initial total activity (Table IV).

3. *Affinity chromatography*

Proteins carry out their biological functions through one or more binding activities and consequently contain binding sites for interaction with other biomolecules, called ligands. Suitable protein and ligand couples for affinity chromatography are antigen–antibody, hormone–receptor, glycoprotein–lectin or enzyme–substrate/cofactor/effector.

Affinity chromatography was used to purify the NADP-GDH of *C. geophilum* (Dell *et al.*, 1989) and *L. bicolor* by using 2′,5′-ADP-Sepharose (Pharmacia). In this matrix, 2′,5′-ADP is a highly dehydrogenase-specific ligand. The extract from *L. bicolor* collected from a DEAE-Trisacryl column (see Section V.D.1) was applied to the column, equilibrated with 50 mM Tris-HCl, pH 7.6, including 10% glycerol, 1 mM DTT and 14 mM mercaptoethanol. After washing the column with the same buffer, the NADP-GDH was eluted by a linear gradient of NADPH from 0 to 0.5 mM included in the buffer (Fig. 11). After such a two-step

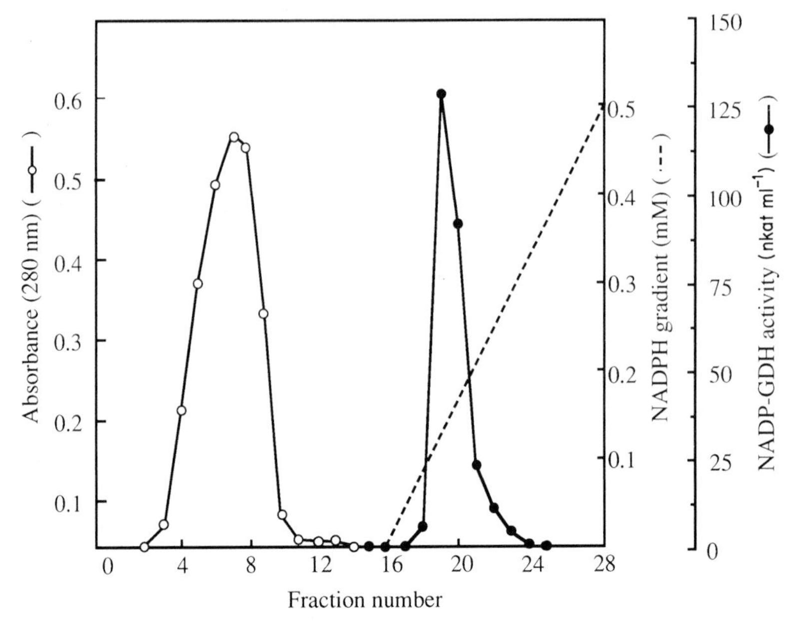

Fig. 11. Elution profile of NADP-glutamate dehydrogenase of *Laccaria bicolor* by using a 2′,5′-ADP-Sepharose column. The column was loaded with 8 mg of proteins which were eluted with a linear gradient of NADPH from 0 to 0.5 mM. As NADPH absorbs at 280 nm proteins were only detected during the washing of the column. The flow rate was 10 ml h^{-1} and 1.5 ml fractions were collected. From M. Chalot, A. Brun and B. Botton (unpubl. res.).

procedure including DEAE-Trisacryl and 2′,5′-ADP-Sepharose, the NADP-GDH was purified 173-fold with a yield of 54.5%.

4. *Gel filtration chromatography*

The basic principle of gel filtration is that molecules are partitioned between solvent and a stationary phase of defined porosity. The separation process is carried out using a porous gel matrix (in bead form) packed in a column and surrounded by solvent. The molecules are eluted in order of decreasing size as the smaller molecules can enter the matrix pores and hence move more slowly through the column, while the larger molecules are excluded from the stationary phase and hence elute first from the column.

The elution volume of a particular molecule is dependent on the fraction of the stationary phase available to it for diffusion; this is represented by the constant K_{av}, given by

$$K_{av} = \frac{V_e - V_0}{V_t - V_0}$$

where V_e is the elution volume of sample, V_0 is the void or exclusion volume, and V_t is the total bed volume.

The most commonly used supports are gels consisting of cross-linked polyacrylamide, agarose, dextran or combinations of these. For resolution of different proteins, a matrix of the correct fractionation range should be chosen such that the molecules do not elute in either the void or the total volume. This information can be found in the technical bulletins available from the gel manufacturers.

In addition to desalting, gel filtration is often used as one of the final steps of protein purification, due to its relatively low capacity. For example, as indicated in Table IV, the aspartate aminotransferase of *C. geophilum* was efficiently brought to electrophoretic homogeneity by gel filtration using Sepadex G-200 (Khalid *et al.*, 1988).

Gel filtration is also a valuable alternative to SDS-PAGE for the determination of molecular weights of proteins. The elution volumes of globular proteins are mainly determined by their hydrodynamic radius which is related to their molecular weight; thus the elution volume is an approximately linear function of the logarithm of the molecular weight.

Figure 12 illustrates the determination of the molecular weight of the aspartate aminotransferase in *C. geophilum*. The most active fractions collected by DEAE-cellulose chromatography were applied to the top of a Sephadex G-200 column (465 × 16 mm) equilibrated with 50 mM potassium phosphate buffer, pH 7.5, containing 0.1 M NaCl. Proteins were eluted with the same buffer at a flow rate of 5 ml h^{-1} and 1.2 ml fractions were collected. The column was calibrated with the following protein markers: ferritin (440 kDa), catalase (232 kDa), aldolase (158 kDa), lysozyme (143 kDa) and bovine serum albumin (67 kDa). Blue dextran 2000 was used to mark the front of elution. The constant K_{av} of each protein marker is plotted against the logarithm of the molecular weight and the value obtained with the aminotransferase is used to obtain the molecular weight from the calibration graph (155 kDa in this example) (Fig. 12).

5. *Choosing and combining techniques*

There are no hard and fast rules that one can apply in deciding the order in which one should use the techniques available. Indeed, there

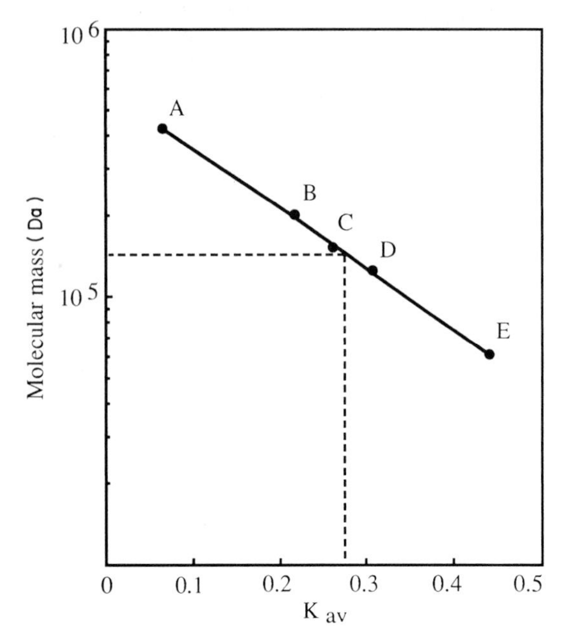

Fig. 12. Calibration graph and molecular mass estimation of the purified native aspartate aminotransferase of the ectomycorrhizal fungus *C. geophilum*, by using gel filtration. Elution profiles of molecular weight standards on a Sephadex G-200 column at a flow rate of 5 ml h^{-1} in 50 mM phosphate buffer (pH 7.5) containing 0.1 M NaCl. Gel was calibrated with: A, ferritin (440 kDa); B, catalase (232 kDa); C, aldolase (158 kDa); D, lysozyme (143 kDa); E, bovine serum albumin (67 kDa). The dashed line marks the mobility and the corresponding molecular mass of the aspartate aminotransferase. Redrawn from Khalid (1988).

are general considerations of capacity, recovery, resolving power and selectivity.

In general, the broad, less specific techniques of protein precipitation tend to be used in the first one or two steps of a protocol, followed by the high resolution steps of ion-exchange chromatography, affinity chromatography and gel filtration. Ion-exchange is often used early on for its high capacity and affinity chromatography later for its high specificity. Gel filtration is frequently used in the later stages as a final, polishing step to remove the last traces of impurity.

Results from total protein assays, when combined with those from specific assays, provide information on the degree of purification achieved by each step and the specific activity of the protein of interest.

The following two parameters must be determined at each step of purification.

$$\text{Specific activity} = \frac{\text{Protein of interest (units)}}{\text{Total protein (mg)}}$$

$$\text{Degree of purification (fold)} = \frac{\text{Specific activity at Step 2}}{\text{Specific activity at Step 1}}$$

Table IV summarizes the purification to electrophoretic homogeneity of the aspartate aminotransferase of *C. geophilum*. In the final step, the enzyme was purified 438-fold with 14% of initial total activity and a specific activity of 3238 nkat mg^{-1} protein. This sequence is by no means definitive or suitable for all cases. What is important is that the steps complement one another in their selectivities and that the active fraction from each step is, as far as possible, a suitable sample for the following step.

E. Analytical electrophoretic methods

Electrophoresis encompasses all operations in which charged molecules migrate in electric fields through solutions. Specific forms of electrophoresis and numerous support media have been developed over the last decades. This chapter only describes in detail the practicalities of one- and two- dimensional polyacrylamide gel electrophoresis of proteins and specifically some results obtained with mycorrhizal partners (further details can be found in Hames and Rickwood, 1990).

1. One-dimensional polyacrylamide gel electrophoresis

Polyacrylamide gel results from the polymerization of acrylamide monomer into long chains and the cross-linking of these by bifunctional compounds such as *N,N'*-methylene bisacrylamide (usually abbreviated to bisacrylamide) reacting with free functional groups at chain termini. Polymerization of acrylamide is initiated by the addition of either ammonium persulphate or riboflavin which requires light to initiate polymerization. In addition, *N,N,N',N'*-tetramethylethylenediamine (TEMED), or, less commonly, 3-dimethylamino-propionitrile (DMAPN) are added as accelerators of the polymerization process. Oxygen inhibits polymerization and so gel mixtures are usually degassed prior to use.

Discontinuous buffer systems have the major advantage that relatively large volumes of dilute protein samples can be applied to the gels with good resolution. The reason for this is that the proteins are concentrated into extremely narrow zones (or stacks) during migration through the large-pore stacking gel prior to their separation during electrophoresis in the small-pore resolving gel.

Composition of the gels used in non-denaturing conditions is given in Table V for the very common concentrations of 6 and 7% in acrylamide. Composition of the stacking gel is given in Table VI.

One-dimensional gel electrophoresis is used not only to resolve protein components from complex mixtures, but also to follow the purification step of a particular molecule. Another aim of this technique is to estimate molecular masses of the proteins.

(a) Molecular mass estimation of native proteins. During the electrophoresis of native proteins in polyacrylamide gel, separation takes place according to both size and charge differences of the molecules. The mobility of a given protein in a series of gels of different acrylamide concentration allows construction of a Ferguson plot, that is a plot of $\log_{10}$ relative mobility (R_f) versus gel concentration (% T, i.e. percentage acrylamide plus bisacrylamide). Since the Ferguson plot relates to mobility during electrophoresis when only the gel pore size (as deter-

TABLE V

Recipe for the preparation of non-denaturing resolving gels at 6 and 7% of acrylamide. The columns represent volumes (μl) of the various reagents required to make 20 ml of gel mixture

	Final acrylamide concentration 6%	7%
Acrylamide-bisacrylamide (30:0.8)	3984	4648
Tris-HCl buffer 1 M, pH 8.8	7480	7480
1.5% ammonium persulphate	1000	1000
Distilled water	7516	6852
TEMED	20	20

TABLE VI

Recipe for the preparation of non-denaturing stacking gels. The figures represent volumes (μl) of the various reagents required to make about 8 ml of gel mixture

Acrylamide–bisacrylamide (10:2.5)	2400
Tris-HCl buffer 1 M, pH 6.8	1000
1.5% ammonium persulphate	400
Distilled water	4200
TEMED	20

mined by % T) is varying, then the slope of the Ferguson plot, K_R, is a measure of the retardation of the protein by the gel, that is, K_R is a retardation coefficient which can be related to molecular size. Hedrick and Smith (1968) found that there is a linear relationship between K_R and the molecular mass of native proteins so that, by first using a series of standard native proteins of known molecular mass to construct a plot of K_R against molecular mass, one can determine the molecular mass of any sample protein simply by determining its K_R and then referring to the standard curve.

It is much simpler to use a single slab polyacrylamide gel with a gradient of acrylamide from the top to the bottom of the gel. During electrophoresis in gradient gels, proteins migrate until the decreasing pore size impedes further progress and once this "pore limit" is reached, the protein banding pattern does not change appreciably with time.

As an example, the molecular weight of an alkaline protease of the ectomycorrhizal fungus *C. geophilum* was estimated by using the gradipore gels PAA 4/30 Pharmacia (Fig. 13). This gel has an acrylamide concentration ranging from 4 to 30% and fractionates proteins over an approximate molecular weight range of 50 000–2 000 000. Larger proteins cannot enter the gel and smaller proteins eventually migrate from the gel.

The resolution of native proteins in polyacrylamide gradient gels enables a much larger range of protein sizes to be fractionated than gels of uniform concentration, and the protein bands are sharpened by the gel gradient.

After pre-equilibrium of the gel for 20 min at 70 V without any sample, 20 μg of protease were applied to the top of the gel. 70 V were maintained for 20 min until the proteins moved into the gel and then a constant voltage of 150 V was applied for 16 h in 90 mM Tris, 80 mM boric acid and 2.5 mM Na_2EDTA at pH 8.4. Gels were calibrated with standard molecular weight proteins: thyroglobulin (669 kDa), ferritin (440 kDa), catalase (232 kDa), lactate dehydrogenase (140 kDa) and BSA (67 kDa) (Fig. 13A). There exists a linear relationship between the molecular mass of the protein expressed on a logarithmic scale and its mobility (Fig. 13B). According to this method the molecular weight of the protease was estimated to be 94 kDa.

(b) Electrophoresis of denatured proteins. The vast majority of studies employing electrophoresis of proteins in polyacrylamide gel use a buffer system designed to dissociate all proteins into their individual polypeptide subunits. The most common dissociating agent used is the ionic detergent, sodium dodecyl sulphate (SDS).

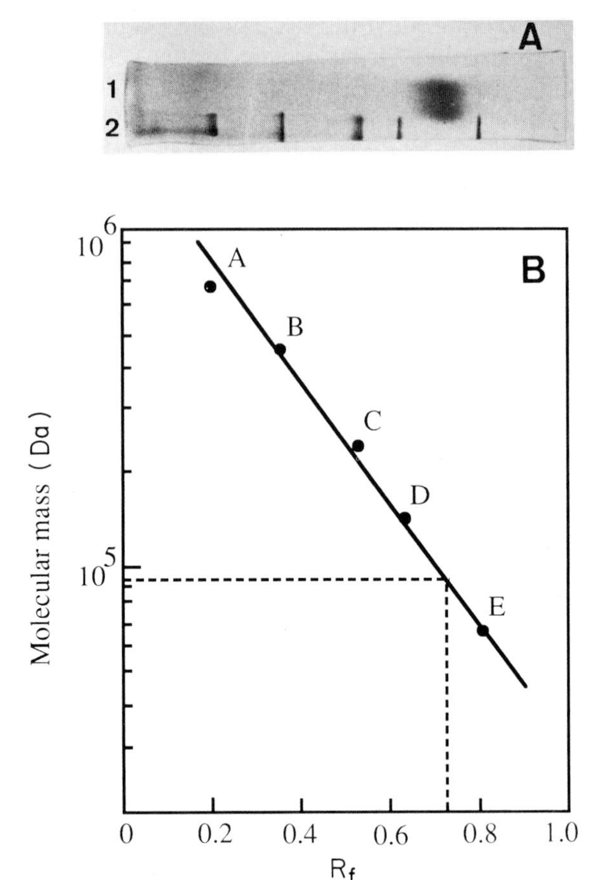

Fig. 13. Molecular mass estimation of a native protease of *C. geophilum* by using a concentration gradient gel (Gradipore PAA 4/30 Pharmacia). (A) Electrophoretic pattern of the purified protease (lane 1) and of standard molecular weight proteins (lane 2). Gel was calibrated with: A, thyroglobulin (669 kDa); B, ferritin (440 kDa); C, catalase (232 kDa); D, lactate dehydrogenase (140 kDa); and E, bovine serum albumin (67 kDa). Staining with Coomassie blue. (B) Calibration curve of $\log_{10}$ polypeptide molecular mass versus distance of migration. The dashed line marks the mobility and the corresponding molecular mass of the protease. From K. El Badaoui and B. Botton (unpubl. res.).

The protein mixture is denatured by heating at 100 °C in the presence of 2.5% SDS and 5% mercaptoethanol (to cleave disulphide bonds). Under these conditions, polypeptides bind SDS in a constant weight ratio (1.4 g of SDS per gram of polypeptide). The intrinsic charges of the polypeptide are insignificant compared to the negative charges provided by the bound detergent, so that the SDS-polypeptide complexes have essentially identical charge densities and migrate in polyacrylamide gels of the correct porosity strictly according to polypeptide size.

Under these conditions, a plot of $\log_{10}$ polypeptide molecular mass versus relative mobility (R_f) reveals a straight-line relationship (Weber and Osborn, 1969).

Polyacrylamide gel electrophoresis in the presence of SDS (SDS-PAGE) implies that SDS must also be present in the gel. Composition of a resolving gel with 15% acrylamide and of a stacking gel with 5% acrylamide are given in Tables VII and VIII, respectively.

Although in the past polypeptide molecular mass estimation using SDS-PAGE has usually been carried out in gels of uniform acrylamide concentration, concentration gradient gels are now often used for this purpose.

TABLE VII

Recipe for the preparation of SDS resolving gels at 15% acrylamide. The figures represent volumes (μl) of the various reagents required to make 20 ml of gel mixture

30% Acrylamide	11 360
1% Bisacrylamide	1 969
Tris-HCl buffer 1.5 M, pH 8.8	6 134
10% ammonium persulphate	114
20% SDS	114
Distilled water	303
TEMED	6

TABLE VIII

Recipe for the preparation of SDS stacking gels. The figures represent volumes (μl) of the various reagents required to make 10 ml of gel mixture

30% Acrylamide	1670
1% Bisacrylamide	2600
Tris-HCl buffer 0.5 M, pH 8.8	2500
10% ammonium persulphate	50
20% SDS	50
Distilled water	3125
TEMED	5

2. *Two-dimensional polyacrylamide gel electrophoresis*

The commonest two-dimensional electrophoresis method for analysing mixtures of polypeptides is to separate the proteins in a first dimension on the basis of charge by isoelectric focusing and then to separate the polypeptides in the second dimension in the presence of SDS (SDS-PAGE) which gives a separation primarily on the basis of molecular mass of the polypeptides.

Because of the wide range of applications and techniques, only the methodologies already used successfully in mycorrhiza analysis (Hilbert and Martin, 1988) will be reported.

(a) Preparation of the sample. Material is powdered in liquid nitrogen and extracted in cold acetone (−20 °C) containing 10% trichloracetic acid, 0.07% β-mercaptoethanol, then precipitated at −20 °C for 45 min (Zivy, 1986). After centrifugation, pellets are washed with 0.07% β-mercaptoethanol in cold acetone. After a new centrifugation the β-mercaptoethanol solution is discarded and pellets are dried under vacuum in order to eliminate acetone. They are resuspended in the O'Farrell (1975) lysis buffer (30 $\mu l\, mg^{-1}$ dry weight pellet) to solubilize the proteins and centrifuged at 10 000 *g* for 3 min.

The O'Farrell lysis buffer is composed of: 9.5 M urea, 2% Nonidet P40, 0.5% dithiothreitol, 1.6% ampholytes pH 5–8, 0.4% ampholytes pH 3.5–10.

Supernatants are stored at −70 °C for further analysis.

(b) Preparation of gels for the first dimension. Rod gels are prepared in capillary tubes (150 × 1 mm I.D.) and are composed of: 0.46% acrylamide, 0.03% bisacrylamide, 9.2 M urea, 0.4% Nonidet P40, 3.2% ampholytes pH 5–8, 1.6% ampholytes pH 3.5–10, 0.01% ammonium persulphate, 0.7% TEMED.

After polymerization for 1 h the top of the capillary tube is filled with the O'Farrell lysis buffer. A pre-electrophoresis is carried out for 1 h in order to establish the pH gradient (1 W for 6 gels; voltage limit 1200 V). The electrode solutions are 10 mM H_3PO_4 (anode buffer) and 20 mM NaOH (cathode buffer) (Hilbert and Botton, 1986).

Thirty μl of protein samples (150–300 μg) are loaded on the basic end of the tube gel after having removed the buffer.

Gels are run for 17.5 h in the same conditions as above followed by 0.5 h at 1500 V. They are extracted by pressure with a 62.5 mM Tris-HCl buffer (pH 6.8) containing 2.3% SDS, 10% glycerol and 0.015% bromophenol blue. This equilibrating buffer is also designed for removing ampholytes.

(c) Preparation of gels for the second dimension. The second dimension is electrophoresis in 15% SDS polyacrylamide slab gels as described previously. The gel used for the first dimension is loaded between the glass plates. Several experiments have shown that neither stacking gels nor sealing gels of agarose are required to obtain the best results. The electrode buffer is that recommended by Laemmli (1970): 25 mM Tris, 192 mM glycine, 0.1% SDS. With a slab gel of 15 × 15 cm, electrophoresis is run for 10 min at 1 W gel^{-1} then at 6 W gel^{-1} until the end of the migration. Proteins are silver-stained as in Blum *et al.* (1987).

Such a protocol was applied to resolve proteins of ectomycorrhiza (Hilbert and Martin, 1988). These authors showed that in the *Pisolithus-Eucalyptus* association, approximately 50% of the fungal polypeptides and 80% of the plant polypeptides disappeared during the development of the ectomycorrhiza. Moreover, ten polypeptides called ectomycorrhizins were ectomycorrhiza-specific in that they were present in ectomycorrhiza, but not in free-living mycelia nor in non-infected roots.

A promising alternative has been developed by Pharmacia that consists of running the first dimension on immobilized pH gradient gels supported on plastic films (Immobiline dry-strip gels). The gel used for the second dimension is very often of uniform concentration, but for complex samples better results can be obtained by using a gradient gel. Pre-cast gels with a gradient ranging from 8 to 18% in acrylamide are available (Pharmacia).

VI. Localization and quantification of proteins by immunological methods

Immunological procedures can be valuable aids in the identification, localization and quantification of proteins.

The basis of these procedures is the antigen–antibody reaction, which is highly specific and of high sensitivity and reproducibility. The primary reaction of antibody with antigen sometimes detected by immunoprecipitation is not always visible, but it can be made so by linking the antibody to labels that can be detected with great sensitivity. Indeed, numerous molecules such as fluorochromes, radioisotopes, enzymes or solid particles can be linked to an antibody without deleterious effects upon its specific ability to combine with the homologous antigen.

The purpose of this chapter is to describe some immunological procedures already used in mycorrhiza to study protein synthesis as well as tissue and intracellular locations of a few enzymes.

A. Antibody production

Numerous antibodies against enzymes involved in nitrogen metabolism have been obtained from ectomycorrhizal fungi (NADP-GDH, GS, aminotransferases). Although there are many possible variations, a typical routine procedure is as follows:

Due to variation between animals, three rabbits (or at least two) are useful for each antigen. They are immunized by two subcutaneous and two intramuscular primary injections. The band of polyacrylamide gel exhibiting the enzyme activity is cut out and homogenized in a Potter–Elvehjem homogenizer with phosphate buffer 10 mM, pH 7.0 containing 0.9% NaCl.

One injection (500 μl emulsion) is composed of 250 μl polyacrylamide gel suspension containing the enzymatically active fraction (100–250 μg protein) and 250 μl Complete Freund's Adjuvant.

Two intravenous booster injections of antigen are given one month later at two-day intervals. For this purpose the protein is extracted from the polyacrylamide gel usually by electroelution (Botton *et al.*, 1987), diafiltered on Amicon PM 10 and resuspended in phosphate buffer 10 mM pH 7.5 with 0.9% NaCl. Each rabbit receives 100–250 μg of enzyme in 500 μl of buffer by injection.

Two intramuscular booster injections at 15-day intervals with the same protocol as that used for primary injections are also satisfactory and allow using smaller amounts of antigens.

Bleedings are taken from the marginal ear vein 15 days after the last injection. The blood is kept for 12 h at 4 °C, then the immune serum is separated by centrifugation. The immunoglobulins, precipitated by ammonium sulphate (35% saturation) are collected by centrifugation, dissolved in borate buffer 50 mM, pH 8.1 containing 0.9% NaCl, then dialysed against the same buffer. This preparation is sterilized by filtration on a Millipore membrane and kept frozen at −25 °C until use.

The specificity of the immune serum must be checked by the intermediate gel technique with crossed immunoelectrophoresis (Bøg-Hansen, 1990), or at least by the Ouchterlony double diffusion test: a crude extract, or the proteins concentrated by ammonium precipitation (85% saturation) should give only one sharp precitation band in the presence of antibodies (Fig. 14).

B. The application of fluorescent antibody techniques

The fluorescent antibody technique combines the unique specificity of an immunological reaction with the high sensitivity obtained by fluor-

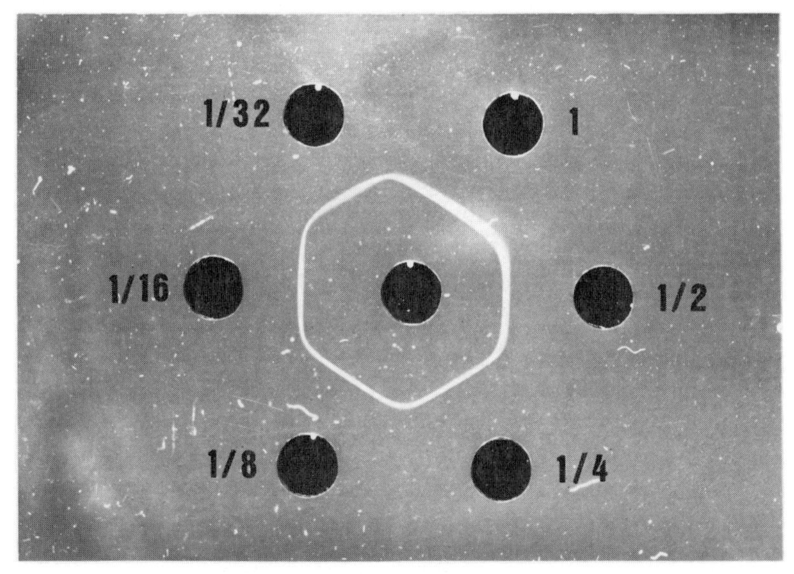

Fig. 14. Ouchterlony double diffusion test with antibodies raised against the aspartate aminotransferase of *C. geophilum*. The antigen extract concentrated by ammonium sulphate precipitation at 95% saturation was deposited in the central well. The outer wells contained different dilutions of the antiserum (1–1/32). Precipitation lines were observed by examining the slide on a black background without any staining.

escence emission such that the site of interaction between the antigen and labelled antibody can be observed by fluorescence microscopy.

The derivatives of two fluorochromes, fluorescein and rhodamine, are the commonest, and, as they provide a good colour contrast, are eminently suitable for the simultaneous detection of two antigens. Visualization of a fluorescent marker is achieved by exposure of the specimen to excitation light of a wavelength maximally absorbed by the fluorochrome. This results in the emission of high intensity visible light detected by fluorescence microscopy.

The method given below refers to the use of fluorescein isothiocyanate (FITC)-labelled antibodies to localize antigens (fungal NADP-dependent glutamate dehydrogenase) in ectomycorrhizal tissues of the association spruce–*Hebeloma* sp. (B. Dell and B. Botton, unpubl. res.).

Mycorrhiza are fixed for 3 h in 0.1 M sodium phosphate buffer, pH 7.2, containing 2% paraformaldehyde. After dehydration in ethanol and propylene oxide series, material is embedded in Epon 812 and polymerized at 60 °C for three days.

Ultra-thin sections are placed on slides and resin is eliminated with several washes of sodium methoxide-benzene (1:1), methanol-benzene (1:1) and acetone. After rehydration and several washes in PBS (phosphate buffer saline) the sections are brought in contact for 60 h at 4 °C with the immune serum diluted in PBS (dilution usually between 1:600 and 1:5000).

The immune serum is removed gently and the sections are rinsed by several changes of PBS, then incubated for 45 min at 20°C in the presence of FITC-labelled sheep anti-rabbit antiserum diluted in PBS. Sections are rinsed with PBS, stained with Evans blue (1×10^{-4} w/v in veronal buffer 0.1 M, pH 7.3), and mounted in buffered glycerol (glycerol–PBS, 66%, v/v). Slides are examined by fluorescence microscopy at 490 nm.

Specific controls including use of PBS or pre-immune serum instead of specific serum are performed to assess labelling specificity.

C. The application of gold-labelled antibody techniques

Antibodies labelled with gold particles of uniform and designed size (range 3–20 nm) are particularly valuable for the demonstration of antigenic determinants. Gold particles, easily visible in electron microscopy, are localized at the site in the cell where homologous antigenic structures are situated.

Detailed electron microscopy techniques cannot be given here, and standard works of reference should be consulted for these. In brief, cells should be fixed, embedded and sectioned, prior to staining. As an example, the routine method which was used to localize the fungal aspartate aminotransferase in ectomycorrhizal associations and in the fungal partners (Chalot *et al.*, 1990) is described here.

Tissues (either free-living mycelia or mycorrhiza) are fixed with 1% glutaraldehyde in 0.1 M sodium phosphate buffer, pH 7.2, at 0 °C and post-fixed with 1% osmium tetroxide in sodium cacodylate buffer. Dehydration and embedding are performed as indicated previously. Ultra-thin sections are mounted on nickel grids and processed for cytochemical labelling. After restoration of antigenicity with sodium methoxide and washing with water, sections are incubated with 1% (w.v) BSA, 10% (v/v) normal sheep serum in 0.1 M PBS, pH 7.4, for 30 min to block non-specific antibody sticking. Grids are then transferred to a drop of anti-AAT antiserum diluted 1/10 (v/v) with 1% BSA in PBS buffer for 2 h. Sections are rinsed with 0.1% BSA in PBS buffer to remove unbound antibody excess and then incubated for 1 h with goat anti-rabbit immunoglobulin G complexed with colloidal gold particles (10 nm) purchased from Jansen Pharmaceutica (Beerse, Belgium). Removal of non-specific labelling is achieved by transferring the grids to 0.1% BSA in PBS buffer and washing with water. Grids are post-stained with uranyl acetate and viewed in an electron microscope. Specific controls must be performed at different steps of the procedure (Rhode *et al.*, 1988).

Such a protocol applied to free-living *C. geophilum* cells showed aspartate aminotransferase to be located in the cytoplasm and, to a greater extent, in the peripheral region of the cell (Fig. 15).

D. Quantification of proteins by rocket immunoelectrophoresis and enzyme-linked immunosorbent assay

These immunological methods can be used to detect and quantify individual proteins in complex mixtures such as crude extracts, tissue homogenates and culture media, all without prior purification.

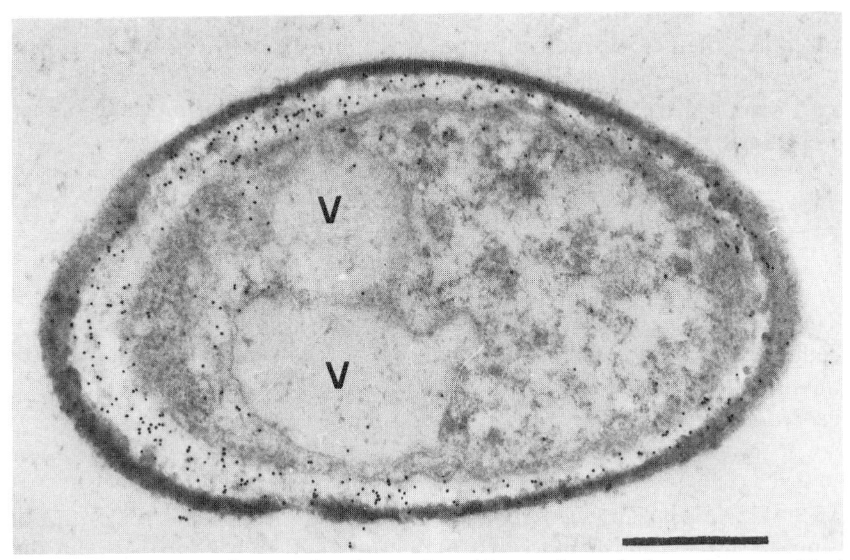

Fig. 15. Electron micrograph of an ultra-thin section of the ectomycorrhizal fungus *C. geophilum*. The mycelium was treated with gold-labelled specific anti-aspartate aminotransferase immunoglobulins. Dark dots are gold particles indicating localization of the enzyme, V, vacuole; bar, 0.5 μm. From Chalot *et al.* (1990).

1. Rocket immunoelectrophoresis

When electrophoresis of an antigen is performed in an agarose gel containing the corresponding antibody, a long rocket-like immunoprecipitate develops. The height of the rocket is linearly correlated with the amount of antigen. This method, originally developed by Laurell (1966), is still a very useful alternative to other procedures. Only if the antigen in question is of low molecular mass (below 30 kDa) or if it exhibits extensive electrophoretic heterogeneity, should single radial immunodiffusion or ELISA (enzyme-linked immunosorbent assay) be considered.

A routine procedure was developed for assaying NADP-glutamate dehydrogenase in the ascomycete *Sphaerostilbe repens* (Botton *et al.*, 1987). Rocket immunoelectrophoresis was performed in 1.5% agarose (Indubiose A 37, Reactifs IBF) dissolved in sodium barbital (20 mM) and barbital (4.3 mM), pH 8.6. Immune serum was incorporated into the agarose maintained at 50 °C, at a concentration of 0.04%.

The gel was cast in slabs of 120 × 120 × 2 mm and wells (3 mm diameter) contained 5 μl of antigen. Electrophoresis was carried out at 83 V and 10 mA for 18 h. Slabs were water-cooled at 15 °C. Plates were rinsed overnight in 0.85% NaCl so as to eliminate unprecipitated proteins, then after dehydration, precipitin lines were stained with Coomassie blue and measured.

A standard curve was constructed with a quantity of enzyme in each well ranging from 50 to 500 ng.

2. Enzyme-linked immunosorbent assay (ELISA)

This method has been optimized as a competitive immunoassay procedure (Martin *et al.*, 1983) and has been successfully applied to quantify aspartate aminotransferase in the ectomycorrhizal fungus *C. geophilum* (Fig. 16).

Step 1: adsorption of the antigen to polystyrene. About 30 ng (250 μl) of purified antigen in 50 mM $NaHCO_3$ buffer (pH 9.8) is incubated in the polystyrene wells for 2.5 h at 28 °C. After incubation, the solution can be saved for re-use. The plate is then washed three times with phosphate-buffer saline (pH. 7.0) containing 0.9% NaCl and 0.05% Tween 80 (PBS-T).

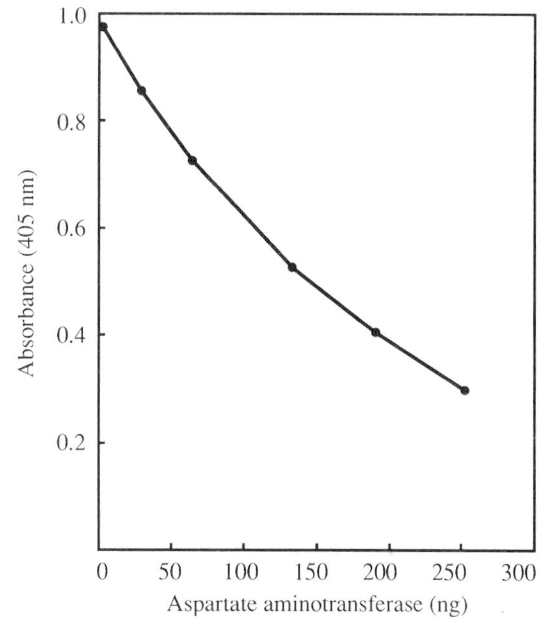

Fig. 16. Standard curve for the immunoassay of the aspartate aminotransferase (AAT) in *C. geophilum*. Polystyrene wells coated with purified fungal AAT at 120 ng ml^{-1} were incubated with specific antibodies for 2.5 h at 28 °C and with ALP-IgG for 1.5 h at 28 °C. Soluble purified AAT was incubated with specific antibodies for 1 h before performing the assay. Redrawn from Khalid (1988).

Step 2: competitive immunoassay procedure. To each of the antigen-coated wells is added 250 μl of the antibodies diluted in PBS-T and the wells are kept at 28 °C for 3 h. At the end of the incubation period, the solution is removed and the wells are rinsed with PBS-T. Then 250 μl of alkaline phosphatase conjugated to goat anti-rabbit immunoglobulins (ALP-IgG) at 1:300 dilution in PBS-T are added and incubated again at 28 °C for 1.5 h. Wells are washed three times. For the assay of surface-bound alkaline phosphatase activity, 250 μl of a 1 mg ml^{-1} solution of *p*-nitrophenylphosphate (PNPP) in 100 mM glycine-NaOH buffer (pH 10.3) containing 1 mM $MgCl_2$ and 1 mM $ZnCl_2$ freshly prepared, are added to each cup. The reaction is allowed to proceed at 28 °C for 1.5 h. Incubation is stopped by adding 100 μl of 2 N NaOH. A 200 μl aliquot is transferred to a spectrophotometer cuvette containing 800 μl of $NaHCO_3$ buffer and readings are taken at 405 nm against a blank containing the buffer. The amount of yellow colour developed per unit of time is dependent on the concentration of bound ALP-IgG and is in turn a measure of the extent of antigen–antibody interaction.

The concentration of antigen standards is determined by mixing of increasing amounts of purified antigen with a fixed concentration of antibodies and a pre-incubation for 1 h at 28 °C prior to addition to the antigen-sensitized polystyrene wells. Increasing concentrations of soluble antigen prevent proportional specific antibody binding by the immobilized antigen. A standard curve is thus generated yielding the inhibition of the PNPP hydrolysis as a function of soluble antigen concentration (Fig. 16).

Samples containing the antigen, pre-incubated in the same way with specific antibodies, can then be assayed by the extent of inhibition of the phosphatase activity when compared to that obtained by a standard amount of antigen solution.

This technique allowed us to demonstrate that the synthesis of the aspartate aminotransferase in *C. geophilum* mycelia varied with the age of the colony and with the nature of the nitrogen source (Khalid, 1988).

VII. Conclusions

Early research on mycorrhiza focused on identification and organization of the partners at the tissue and cellular levels. More recently the attention of mycorrhizologists has been turning strongly towards understanding the physiological functions of both the host plant and symbiont. In this context the study of nitrogen metabolism forms a large part of the topics investigated because of the central position of nitrogen in the nutrition and growth of the partners.

The study of nitrogen metabolism is inevitably complex and involves a wide variety of biological and physiological techniques: electron microscopy, radioisotope tracing, enzymology, immunology, molecular biology, etc. It is not possible or appropriate in this text to cover all the many disciplines and principles used in the study of nitrogen metabolism and only a few techniques encountered frequently in the field have been described.

In the future, there is no doubt that molecular cloning and analysis of structural genes will be powerful tools to obtain a better understanding of nitrogen metabolism and the techniques involved will also be diversified. The application of these techniques should lead to exciting new discoveries in the area of nitrogen metabolism. These may eventually enable us to improve the efficiency of the symbiosis and thus increase its effectiveness in agricultural production.

References

Bayer, E., Gugel, K. H., Hagele, K., Hagemaier, H., Jessipo, W. S., Konig, W. A. and Zaner, Z. (1972). *Helv. Chim. Acta* **55**, 224–239.

Bieleski, R. L. and Turner, N. A. (1966). *Anal. Biochem.* **17**, 278–293

Blum, H., Beier, H. and Gross, H. J. (1987). *Electrophoresis* **8**, 93–99.

Bøg-Hansen, T. C. (1990). In *Gel Electrophoresis of Protein. A Practical Approach* (B. D. Hames and D. Rickwood, eds), pp. 273–300. IRL Press, Oxford and New York.

Botton, B. and Bonaly, R. (1982) *Arch. Microbiol.* **131**, 291–297.

Botton, B., Msatef, Y. and Godbillon, G. (1987) *J. Plant Physiol.* **128**, 109–119.

Bradford, M. M. (1976) *Anal. Biochem.* **72**, 248–254.

Brenchley, J. E. (1973). *J. Bacteriol.* **114**, 666–673.

Cataldo, D. A., Haroon, M., Schrader, L. E. and Youngs, V. L. (1975). *Commun. Soil Sci. Plant Anal.* **6**, 71–80.

Chalot, M., Brun, A., Kahlid, A., Dell, B., Rohr, R. and Botton, B. (1990). *Can. J. Bot.* **68**, 1756–1762.

Compton, S. J. and Jones, C. G. (1985) *Anal. Biochem.* **151**, 369–374.

Conway, E. J. (1962). *Microdiffusion Analysis and Volumetric Error*, 5th ed. Crosby Lockwood & Son Ltd., London. 467 pp.

Dell, B., Botton, B., Martin, F. and Le Tacon, F. (1989). *New Phytol.* **111**, 683–692.

Dixon, M. (1953). *Methods in Enzymology*, Vol. 2, pp. 76–77. Academic Press, New York.

Einarsson, S., Josefsson, B., and Lagerkvist, S. (1983) *J. Chromatogr.* **282**, 609–618.

Findlay, J. B. C. (1990). In *Protein Purification Applications. A Practical Approach* (E. L. V. Harris and S. Angal, eds), pp. 59–82. IRL Press, Oxford and New York.

Givan, C. V. (1980). In *The Biochemistry of Plants* (B. J. Miflin, ed.), Vol. 5, pp. 329–357. Academic Press, New York.

Guiraud, G. and Fardeau, J. C. (1977). *Ann. Agron.* **28**, 329–333.

Haissig, B. E. and Schipper, J. R. (1975) *Phytochemistry* **14**, 345–349.

Hames, B. D. and Rickwood, D. (1990). In *Gel Electrophoresis of Proteins. A Practical Approach* (B. D. Hames and D. Rickwood, eds), pp. 1–272. IRL Press, Oxford and New York.

Harris, E. L. V. (1989). In *Protein Purification Methods. A Practical Approach* (E. L. V. Harris and S. Angal, eds), pp. 51–64. IRL Press, Oxford and New York.

Hedrick, J. L. and Smith, A. J. (1968). *Arch. Biochem. Biophys.* **126**, 155–164.

Heinrikson, R. L. and Meredith, S. C. (1984) *Anal. Biochem.* **136**, 65–74.

Hilbert, J. L. and Botton, B. (1986). *Physiol. Plant.* **68**, 403–409.

Hilbert, J. L. and Martin, F. (1988). *New Phytol.* **110**, 339–346.

Hill, W. D., Walters, F. H., Wilson, T. D. and Stuart, J. D. (1979) *Anal. Chem.* **51**, 1338–1341.

Johnson, C. R., Boettcher, B. R., Cherpeck, R. E. and Dolson, M. G. (1990) *Biorgan. Chem.* **18**, 154–159.

Jones, C. G., Hare, J. D. and Compton, S. J. (1989). *J. Chem. Ecol.* **15**, 979–992.

Kershaw, J. L. and Stewart, G. R. (1989). *Ann. Sci. For.* **46** (Suppl)., 706–710.

Khalid, A. (1988). Purification et caractérisation de l'aspartate aminotransférase du Champignon ectomycorhizien *Cenococcum geophilum*. PhD Thesis, University of Nancy I, 137 pp.
Khalid, A., Boukroute, A., Botton, B. and Martin, F. (1988). *Plant Physiol. Biochem.* **26**, 17–28.
Krystal, G., MacDonald, C., Munt, B. and Ashwell, S. (1985) *Anal. Biochem.* **148**, 451–460.
Krupa, S., Fontana, A., Palenzona, M. (1973) *Physiol. Plant.* **28**, 1–6.
Laemmli, U. K. (1970) *Nature* **227**, 680–685.
Langston-Unkefer, P. L., Macy, P. A. and Durbin, R. D. (1984) *Plant Physiol.* **76**, 71–74.
Laurell, C. B. (1966). *Anal. Biochem.* **15**, 45–52.
Loomis, W. D. (1974). In *Biomembranes*, Part A, *Methods in Enzymology*, Vol. 31, pp. 528–545. Academic Press, New York.
Loomis, W. D., Lile, J. D., Sandstrom, R. P. and Burbott, A. J. (1979). *Phytochemistry* **18**, 1049–1054.
Lowry, O. H., Rosebrough, N. J., Farr, A. L. and Randall, R. J. (1951). *J. Biol. Chem.* **193**, 265–275.
Mangin, F., Bonaly, R., Botton, B. and Martin, F. (1986). In *Physiological and Genetical Aspects of Mycorrhizae*, Proceedings of the 1st European Symposium on Mycorrhiza (V. Gianinazzi-Pearson and S. Gianinazzi, eds), pp. 451–456. INRA Press, Paris.
Martin, F. and Botton, B. (1991). In *Advances in Plant Pathology* (D. S. Ingram and P. H. Williams, series eds), Vol. 8: *Mycorrhiza: A Synthesis* (I. C. Tommerup, ed.). Academic Press, London.
Martin, F., Botton, B. and Msatef, Y., (1983). *Plant Physiol.* **72**, 398–401.
Martin, F., Stewart, G. R., Genetet, I. and Le Tacon, F. (1986). *New Phytol.* **102**, 85–94.
Meins, Jr. and Abrams, M. L. (1972). *Biochim. Biophys. Acta* **266**, 307–311.
Meister, A. (1969). *Biochem. J.* **8**, 1066–1075.
Miflin, B. J. and Lea, P. J. (1975) *Biochem. J.* **149**, 403–409.
O'Farrell, P. H. (1975). *J. Biol. Chem.* **250**, 4007–4021.
Paech, K. and Tracey, M. V. (1956). *Modern Methoden der Pflanzen Analyse*, Vol. 1, pp. 479–483. Springer-Verlag, Berlin.
Pitel, J. A. and Cheliak, W. M. (1985). *Can. J. Bot.* **64**, 39–44.
Probyn, T. A. and Lewis, O. A. M. (1978) *New Phytol.* **81**, 519–526.
Rhode, M., Gerberding, H., Mund, T. and Kohring, G. W. (1988). *Methods in Microbiology* (F. Mayer, ed.), Vol. 20, pp. 175–210. Academic Press, London.
Ronzio, A. R., Rowe, W. B. and Meister, A. (1969). *Biochem. J.* **8**, 1066–1075.
Smith, S. M., Briggs, S., Triebwasser, K. C. and Freddland, R. A. (1977). *Biochem. J.* **162**, 453–455.
Weber, K. and Osborn, M. (1969). *J. Biol. Chem.* **224**, 4406–4010.
Weimar, M. and Rothe, G. M. (1987). *Physiol. Plant.* **69**, 692–698.
Weatherburn, M. W. (1967). *Anal. Chem.* **39**, 971–974.
Williams, A. P. (1986). *J. Chromatogr.* **373**, 175–190.
Zivy, M. (1986). In *Recent Progress in Two-dimensional Electrophoresis* (M. M. Galteau and G. Siest, eds), pp. 69–72. University of Nancy Press, Nancy.

9

Ectomycorrhizal DNA: Isolation, RFLPs and Probe Hybridization

PAUL T. RYGIEWICZ

US Environmental Protection Agency, Environmental Research Laboratory, Corvallis, OR 97333, USA and Department of Forest Science, Oregon State University, Corvallis, OR 97331, USA

JOHN L. ARMSTRONG

US Environmental Protection Agency, Environmental Research Laboratory, Corvallis, OR 97333, USA

I. Introduction

We are in a new era for research in terrestrial ecology. As with diagnostics and research in medicine and forensics, molecular biology is transforming the way we pose hypotheses, plan experiments, and conduct research in rhizosphere ecology. Many soil-borne organisms are morphologically indistinguishable, can change morphology from one sexual state to another, or are not culturable *in vitro*. These properties hinder our understanding of rhizosphere structure and function. Contemporary, molecular-based methods to isolate DNA from soil organisms will enhance our ability to track, identify and quantify individuals. The purpose of this chapter is to present a primer of basic methods that have direct application to recovery of DNA from mycorrhizal fungi in soils.

The methods are organized into three broad categories: cell lysis, DNA purification, and DNA analysis. Major difficulties typically arise during cell lysis and during the early steps in extraction when DNA is still associated with soil constituents (e.g. humic and tannin-like substances). Therefore, these steps are given greater attention. Our approach is to present considerable detail for techniques that may vary among mycorrhizal fungi and mycorrhiza and may need modification for successful application. When necessary, protocols have been redesigned so that they can be used with a small number of mycorrhizal root tips. Since purified DNA behaves similarly regardless of its source, we present fewer alternatives for purifying DNA. As a logical thread for presenting molecular-based techniques for analysing DNA, we focus on producing and interpreting restriction fragment length polymorphisms (RFLPs). An exhaustive description of techniques for all possible analyses and manipulations of isolated DNA is beyond the scope of this chapter. Those investigators needing more information on theory, skills and applications can refer to Darbre (1988), Davis *et al.* (1986), Maniatis *et al.* (1982), Perbal (1988) or Sambrook *et al.* (1989). Kirby's (1990) treatise on DNA fingerprinting may be valuable for a more thorough discussion of RFLPs.

II. Cell lysis and DNA isolation

A. Carpophores and mycelia from pure culture

1. Overview

The methods presented here are intended as guidelines for lysis and extraction of DNA from a group of organisms whose DNAs are not well studied. In this section, we present methods of extraction with several options to provide examples of procedures that have been used with mycorrhizal fungi or related organisms.

All DNA extractions must begin with cell disruption and lysis. Because of the chemical and physical nature of the cell walls of mycorrhiza and mycorrhizal fungi, these steps require the most modification of existing methods to optimize yield and quality of the resultant DNA. Available cell lysis techniques for fungi generally are based on the cetyltrimethylammonium bromide (CTAB) method for plant DNA extraction (Murray and Thompson, 1980) or the sodium dodecyl sulphate (SDS) method used for several filamentous fungi (Leach *et al.*, 1986; Raeder and Broda, 1985). Cell lysis is enhanced by combining chemical treatments with mechanical, tissue-disrupting methods. However, a compromise must be reached between DNA yield and size, since the same forces which break open cell walls also shear DNA. Many DNA isolation procedures begin with lyophilized fungi since dry tissue can be efficiently disrupted with minimal shearing of the unhydrated DNA. Expected DNA yields will be variable, depending on the fungal species and growth stage.

2. *CTAB-based methods*

CTAB-based protocols utilize the selective salt-concentration-dependent precipitation of nucleic acids by the detergent CTAB. The two protocols described here are derived from the procedures of Armstrong *et al.* (1989) and Rogers *et al.* (1989), and are modifications of the CTAB isolation methods of Murray and Thompson (1980) and Taylor and Powell (1982). The methods yield up to 2 mg of purified, high molecular weight (approximately 25 kb) DNA per gram of lyophilized mycelia. The two procedures employ different equipment, making it possible for researchers to customize protocols to their laboratories.

The procedures can be used for direct extraction of DNA from carpophores. However, researchers need to consider that contaminating DNAs from other organisms associated with the carpophores may affect subsequent analyses and lead to false inferences. For example, less specific DNA probes might hybridize with both the contaminating and carpophore DNAs (see Section IV.D). If contaminant-free carpophore DNA is desired, make isolations from sterile carpophore tissue onto an appropriate growth medium. Harvest mycelia by vacuum filtration and store frozen. Lyophilize and then grind tissue to a very fine powder with a mortar and pestle at room temperature after added liquid nitrogen has evaporated from the mortar, or in the presence of crushed dry ice. If dry ice is used for grinding, the powder is stored frozen in an open container until the dry ice is sublimated. Store pulverized hyphae in a freezer until used (storage for at least one year is possible).

(a) Procedure of Armstrong et al. (1989):

1. Add approximately 20 μl *low CTAB–high salt* buffer per mg dry wt of mycelial powder [*low CTAB–high salt* buffer: 50 mM Tris-HCl* (pH 8.0), 0.7 M NaCl, 10 mM EDTA[†], 1.0% 2-mercaptoethanol (v/v), 1.0% CTAB; add CTAB last, heat in water bath at 60 ± 2 °C]. Keep CTAB above 15 °C at all times.
2. Heat in water bath for 1 h or longer at 50–60 °C with occasional gentle mixing until buffer becomes viscous.
3. Add an equal volume of chloroform:isoamyl alcohol (24:1). Cap tube and emulsify with moderate shaking.
4. Centrifuge (10 min, 25 ± 2 °C, 20000*g*; or at maximum speed in a microfuge).
5. Transfer the aqueous (upper) phase to a clean tube.
6. Add 0.1 volume *high CTAB – high salt* buffer [10% CTAB, 0.7 M NaCl].
7. Repeat Steps 3–5. Evaluate cloudiness of the supernatant. If supernatant is cloudy (due to presence of protein), repeat Steps 3–5 again. Otherwise go to Step 8.
8. Add 1.0 volume *low CTAB – no salt* buffer [1% CTAB, 50 mM Tris-HCl (pH 8.0), 10 mM EDTA]. Incubate for 30 min at room temperature. For convenience, incubate overnight.
9. Centrifuge (5 min, 25 ± 2 °C, 2000 *g*; or for 1–3 min at maximum speed in a microfuge).
10. Dry nucleic acid pellet (under vacuum or with lyophilizer) and resuspend in TE [10 mM Tris-HCl (pH 8.0), 1 mM EDTA]. Purify further according to procedures described in Section III.

(b) Procedure of Rogers et al. (1989):

1. Lyse as follows:
   (a) Dry tissue: 10–15 μl 1 × CTAB buffer per mg tissue [1 × CTAB buffer: 1.0% CTAB, 50 mM Tris-HCl (pH 8.0), 0.7 M NaCl, 10 mM EDTA, 0.5% PVP* (MW 40000)],
   (b) Fresh tissue: 1–2 μl 2 × CTAB buffer per mg tissue.
2. Shake vigorously to hydrate tissue. Heat to 65 ± 2 °C for up to 60 min. Lysis may occur with a shorter incubation time and can be determined empirically.
3. Add an equal volume of chloroform:isoamyl alcohol (24:1). Emulsify by shaking.

*Tris-HCl: tris (hydroxymethyl) aminomethane hydrochloride.
[†]EDTA: ethylenediaminetetraacetic acid.
*PVP: polyvinylpyrrolidene.

4. Centrifuge (10 min, 25 ± 2 °C, 11000 *g*; or for 1 min at maximum speed in microfuge).
5. Collect the top phase.
6. Add 0.1 volume 10% CTAB buffer [10% CTAB, 0.7 M NaCl] warmed to 65 ± 2 °C in a water bath.
7. Repeat Steps 3–4. If top phase is turbid, continue centrifugation.
8. Transfer top phase to new tube. With small amounts of tissue ($<$ 1–2 mg dry wt, or $<$ 25 mg fresh wt), add 25 μg yeast tRNA to improve nucleic acid precipitation (see Section II.C.).
9. Add 1–1.5 volumes 1 × CTAB buffer, mix gently and thoroughly. If precipitate is not apparent after addition of CTAB buffer, place sample on ice for 30 min.
10. Centrifuge (5 min, 25 ± 2 °C, 11000 *g*; or for 1–3 min at maximum speed in a microfuge).
11. Dry nucleic acid pellet (under vacuum or with lyophilizer) and resuspend in TE. Purify further according to procedures described in Section III.

3. *SDS-based methods*

An SDS-based cell lysis technique (Leach *et al.*, 1986; Raeder and Broda, 1985) has been used to isolate DNA from several non-mycorrhizal basidiomycetous and ascomycetous fungi. Martin (F. Martin *et al.*, 1991) modified the SDS cell lysis procedure of Kafatos *et al.* (1979) to extract DNA from mycorrhizal fungi.

(a) Procedure of Raeder and Broda (1985):

1. Lyse powdered mycelia with approximately 10 μl extraction buffer per mg dry wt of tissue [extraction buffer: 200 mM Tris-HCl (pH 8.5), 250 mM NaCl, 25 mM EDTA, 0.5% SDS].
2. Add 7 μl phenol/extraction buffer per mg of starting material [redistilled phenol that has been pH-equilibrated and buffer-saturated with 1 volume of extraction buffer and stored under buffer in freezer]. Shake until homogeneous as determined by visual inspection.
3. Add 3 μl chloroform per mg of starting material and shake.
4. Centrifuge (1 h, 4 ± 2 °C, 13000 *g*).
5. Recover the aqueous (upper) phase, add 0.54 volume isopropanol, incubate in freezer for 30 min, recover pellet by centrifugation.
6. Dry nucleic acid pellet (under vacuum or with lyophilizer) and resuspend in TE. Purify further according to procedures described in Section III.

(b) Procedure of Leach et al. (1986):

1. Lyse powdered mycelia with 35–140 μl LETS buffer per mg dry wt tissue [LETS buffer: 0.1 M LiCl, 10 mM EDTA, 10 mM Tris-HCl (pH 8.0), 0.5% SDS].
2. Cover tube with parafilm and mix with a tube mixer (e.g. Vortexer 2™*, VWR Scientific, San Francisco, CA, USA) at top speed for 1–2 min. Glass beads (< 0.5 mm diameter) may be added to aid in tissue disruption.
3. Add an equal volume of phenol–chloroform–isoamyl alcohol (25:24:1, v:v:v). Mix with a tube mixer at medium speed for 20–30 s. [Phenol is previously prepared: melt redistilled phenol crystals by heating in water bath (65–70 °C); add 8-hydroxyquinoline to 0.1% final concentration; extract several times with equal volumes of 1 M Tris-HCl (pH 8.0) until the pH is above 7.6; add 0.1 M Tris-HCl (pH 8.0) containing 0.2% 2-mercaptoethanol. Store frozen until needed.]
4. Centrifuge (5 min, 25 ± 2 °C 11000 *g*; or for 1 min at maximum speed in microfuge).
5. Transfer the top (aqueous) phase to a clean tube.
6. Add 2 volumes of 100% ethanol (or 95% ethanol), swirl thoroughly, and place sample in freezer (or on dry ice) for 30–60 min.
7. Centrifuge (15 min, 4 ± 2 °C, 11000 *g*; or for 15 min at maximum speed in a microfuge).

8 Dry nucleic acid pellet (under vacuum or with lyophilizer) and resuspend in TE. Purify further according to procedures described in Section III.

(c) Procedure of Martin et al. (1991)

1. Lyse powdered mycelia with approximately 10 μl lysis buffer per mg dry wt tissue [lysis buffer: 20 mM Tris · HCl (pH 8.0), 250 mM NaCl, 0.5% SDS, 50 mM EDTA]. Add 2 μl proteinase K stock solution per mg of starting material [proteinase K stock solution: 1 mg ml^{-1} water; store in freezer; see Section III.B for preparation of proteinase K].
2. Incubate 4–6 h at 55 ± 2 °C. Gently mix.
3. Centrifuge (30 min, 4 ± 2 °C, 32000 *g*).
4. Transfer the top phase to a clean tube.
5. Extract sequentially with 1 volume each of redistilled phenol–

*Mention of trade names or commercial products does not constitute endorsement.

chloroform–iso-amyl alcohol (24:24:2, v:v:v) and chloroform-isoamyl alcohol (24:1, v:v). To prepare phenol, see Step 3 of Leach *et al*. (1986) procedure in Section II.A.3.
6. Centrifuge (15 min, 4 ± 2 °C, 7500 *g*).
7. Repeat Steps 4–6.
8. Add 0.1 volume 3 M Na acetate and 2.5 volumes 100% ethanol, mix thoroughly.
9. Centrifuge (15 min, 4 ± 2 °C, 7500 *g*).
10. Dry nucleic acid pellet (under vacuum or with lyophilizer) and resuspend in TE. Purify further according to procedures described in Section III.

B. Mycorrhiza

1. Overview

Isolation of DNA from mycorrhiza poses a formidable task. Protocols must be designed to accommodate small amounts of starting material, since the dry weight of an individual mycorrhiza may be only a few mg. Pooling samples from several mycorrhiza may suffice to achieve sufficient dry weight, but these samples may contain DNA from more than one mycobiont. We present procedures for isolating DNA from one mycorrhizal root tip. However, to extract enough DNA to produce an RFLP (see Section IV.D), one needs 10–15 mg fresh weight (2–3 mycorrhizal root tips). Protocols must permit removal of DNA from the complex chemicals associated with the host and soil (e.g. polysaccharides and tannins). Plant contaminants can impede extraction of high molecular weight DNA (Murray and Thompson, 1980) and interfere with restriction nucleases. Polysaccharide-like contaminants can lead to incomplete digestion of DNA by restriction endonucleases and can cause anomalous reassociation kinetics (Merlo and Kemp, 1976).

2. Methods

The CTAB method of Rogers *et al*. (1989) (Section II.A.2.) was designed to extract DNA from field samples of mycorrhiza. We modified the CTAB procedure of Armstrong *et al*. (1989) (Section II.A.2.) to extract DNA from mycorrhiza receiving minimal treatment to remove adhering soil. This protocol is a modification of the method developed by Wagner *et al*. (1987) and Saghai-Maroof *et al*. (1984), who also based their methods on Murray and Thompson (1980).

(a) Modified CTAB procedure for mycorrhiza:

1. Harvest mycorrhiza from soil. Gently shake mycorrhiza to remove adhering soil, or wash gently with sterile distilled water if soil is fine. Store frozen in capped, glass vials.
2. While keeping sample frozen, grind 15–200 mg fresh weight in the glass vial with a plastic pestle (Pellet Pestle™, Kontes, Vineland, NJ, USA). Prevent thawing of tissue by periodically immersing vial in liquid nitrogen during grinding.
3. Immediately add powdered sample to 1 ml ice-cold extraction buffer [50 mM Tris-HCl (pH 8.0), 50 mM EDTA, 0.5% 2-mercaptoethanol (v/v), 0.1% bovine serum albumin (BSA) 0.35 M sorbitol, 10% polyethylene glycol 4000, 0.1% spermine tetrachloride, 0.1% spermidine trihydrochloride, 0.1 M diethyldithiocarbamic acid (DEDTC), 0.1 M PVP; buffer must be refrigerated and used within one week of preparation]. Homogenize for 1 min with a mechanical tissue homogenizer, especially if the mycorrhiza are thick or woody. If minimum capacity of available homogenizer is more than 1 ml, scale up the volume and quantity of tissue.
4. Filter through one layer of fine-mesh nylon cloth (e.g. MiraCloth™, Calbiochem Corp., La Jolla, CA, USA) and collect in a plastic or siliconized glass tube (Sambrook *et al.*, 1989). Rinse cloth with 500 μl of extraction buffer.
5. Centrifuge (10 min, 4 ±2 °C, 10 000 *g*).
6. Discard supernatant. Resuspend pellet in 500 μl ice-cold wash buffer [50 mM Tris-HCl (pH 8.0), 50 mM EDTA, 0.5% 2-mercaptoethanol, 0.1 M DEDTC, 0.1 M PVP, 0.35 M sorbitol; buffer must be used within one week of preparation, refrigerate]. Transfer to a microfuge tube. Keep on ice.
7. Add 100 μl 5% *n*-lauroyl sarcosine and shake vigorously. If doing multiple samples, keep them on ice until *n*-lauroyl sarcosine is added to all samples.
8. Incubate at room temperature for 10 min.
9. Add 90 μl 5 M NaCl, and shake vigorously.
10. Add 60 μl CTAB solution [8.6% CTAB, 0.7 M NaCl], and shake vigorously.
11. Incubate (60 ± 2 °C, 10 min).
12. Add 1.5 volume of chloroform/octanol (24:1), and shake vigorously.
13. Centrifuge (10 min, 25 ± 2 °C, maximum speed in a microfuge).
14. Transfer upper phase to a new microfuge tube.
15. Concentrate by alcohol precipitation (Section III. D).

16. Centrifuge (5 min, 25 ± 2 °C, 11 000 *g*; or for 5 min at maximum speed in a microfuge).
17. Dry nucleic acid pellet (under vacuum or with lyophilizer) and resuspend in TE. Purify further according to procedures described in Section III.

Porteous and Armstrong (1991) developed a bulk DNA extraction method which yields up to 200 μg DNA per gram of soil. The procedure has not been used with mycorrhiza. It offers promise as a means to reduce chemical constituents associated with field samples, but must be tested to determine its suitability as an additional procedure for mycorrhiza.

(b) Procedure of Porteous and Armstrong (1991):

1. Add 6 ml mixing buffer per gram of sieved soil [mixing buffer: 0.5 M sorbitol, 15% polyethylene glycol 4000, 2% DEDTC, 100 mM EDTA, 50 mM Tris-HCl (pH 8.0); prepare just prior to use]. Mix with a tube mixer for 1 min at room temperature.
2. Add 500 mg polyvinylpolypyrrolidone (PVPP) and mix with a tube mixer.
3. Add 100 μl lysozyme solution [50 mg ml^{-1}; prepared just prior to use] and 120 μl Novozym 234 (Novo Biolabs, Bagsvaerd, Denmark) solution [50 mg ml^{-1}; prepared just prior to use]. Mix with a tube mixer for 15 s at room temperature, and incubate on ice for 1–2 h.
4. Add 3.8 ml lysis buffer [4% SDS, 100 mM EDTA, 500 μg proteinase K ml^{-1}, 50 mM Tris-HCl (pH 8.0); prepared just prior to use], mix by slowly inverting and return to ice for 16 h.
5. Centrifuge (5 min, 4 ± 2 °C, 5000 *g*) to separate nucleic acids (aqueous phase) from PVPP, humic compounds, soil particles and other debris. Store aqueous phase in sterile tube on ice.
6. Add 3 ml wash buffer [50 mM Tris-HCl (pH 8.0), 100 mM EDTA] to pellet, invert, mix with a tube mixer for 2 s, then centrifuge and pool aqueous phase with aqueous phase from Step 5.
7. Repeat Step 6.
8. Centrifuge (5 min, 4 ± 2 °C, 15 000 *g*) and transfer supernatant to sterile tube and hold at room temperature.
9. Add 5 M potassium acetate to a final concentration of 0.5 M, incubate on ice for 1–2 h.
10. Centrifuge (10 min, 4 ± 2 °C, 15 000 *g*).
11. Add 2 volumes 95% ethanol (room temperature), mix and centrifuge (10 min, 10–15 °C, 15 000 *g*).

12. Dry nucleic acid pellet (under vacuum or with lyophilizer) and resuspend in TE. Purify further according to procedures described in Section III.

The lysis procedures we describe for mycorrhizal fungi and mycorrhiza do not depend on cell wall-degrading enzymes (e.g. chitinase, cellulase, hemicellulase; Peberdy, 1989). Researchers may want to add these enzymes to improve cell lysis. However, the available enzyme preparations are typically contaminated with nucleases which degrade DNA after its release from cells. The nuclease activity can be inhibited with an increased EDTA concentration (e.g. 100 mM EDTA) in the extraction buffer, as was done by Porteous and Armstrong. These investigators (Porteous and Armstrong, 1990) also used pulverized glass (Geneclean®; Section III.F) to separate soil chemicals from the partially purified DNA at Step 10.

C. Comments

Depending on the type of sample and use of the DNA, additional procedures may be required to increase DNA purity. Some uses generally may not require highly purified DNA (e.g. amplification by the polymerase chain reaction; Innis *et al.*, 1990), so the DNA may not need extensive purification. However, endonuclease activity increases as DNA purity increases. The high concentration of phenolics in some species of fungi (e.g. *Pisolithus tinctorius* and *Paxillus involutus*) may require purification using CsCl density gradient ultracentrifugation (Section III.E.) or another procedure (Section III.F.) before the DNA is usable. Katterman and Shattuck (1983) overcame the effects of secondary host and soil-borne chemicals while extracting DNA from nuclei of *Gossypium* samples from the field by including a high concentration of glucose in the maceration buffer (citrate buffer, Triton X-100, and 1 M glucose). The glucose was a reducing agent to suppress formation or activity of oxidized phenolic groups.

As cells lyse and DNA is released, an increase in solution viscosity is indicative of the presence of DNA. The optimal buffer:tissue ratio permits the suspension to be mixed to homogeneity (visual check) by gentle shaking. If the suspension does not move freely in the tube during the shaking, add more buffer. Our CTAB procedure (Section II.B.2.) yields high amounts of DNA when the first CTAB extraction buffer:tissue ratio is 10–20 $\mu l \, mg^{-1}$ dry wt tissue. Some experimentation may be needed to find the optimum buffer:tissue ratio for various types of mycorrhiza. Generally, from 10 to more than 500 mg powder can be extracted as long as the buffer:tissue ratio is maintained. Depending on the use of the extracted DNA, it can be treated with RNase and proteinase K at various steps during the extraction procedure (see Sections III.B and III.C).

Many polysaccharides are insoluble at the salt concentrations used in the CTAB procedures to precipitate nucleic acids, possibly requiring additional effort to eliminate these contaminants if DNA is isolated from host and mycobiont. Differential precipitation with high acetate salts selectively removes many carbohydrates with a high pectin content. When isolating DNA from small amounts of tissue (i.e. expected DNA yield is approximately 10 ng ml^{-1}), the alcohol precipitation can be done at lower temperatures (− 80 °C) for longer times (overnight), or a carrier nucleic acid (e.g. 25 μg tRNA mg^{-1} dry wt tissue) can be added during alcohol precipitation.

The requirement of sorbitol in the extraction and wash buffers of our CTAB procedure to isolate total DNA from mycorrhiza has not been investigated. The original Wagner *et al.* (1987) and Saghai-Maroof *et al.* (1984) protocols used sorbitol as an osmotic stabilizer to isolate chloroplast DNA from needles of coniferous species.

Additional protocols utilizing CTAB for isolating DNA from filamentous fungi have been developed by Jahnke and Bahnweg (1986) and Manicom *et al.* (1987). Readers may consider the work of Verma (1988) who studied *Neurospora crassa* and used an alternative way to lyse cells with guanidine hydrochloride and guanidinium isothiocyanate.

III. Techniques to purify DNA

A. Overview

After cell lysis and deproteinization by extraction with phenol, pheno–chloroform, isoamyl alcohol or chloroform (e.g. Step 3 in Leach *et al.* procedure, Section II.A.3.), the DNA should be essentially protein-free. The absence of a denatured protein layer at the aqueous–organic interface is diagnostic of protein-free DNA. This can be achieved after repeated deproteinizations, followed by dialysis (Sambrook *et al.*, 1989) to remove the chemicals. Other options to purify DNA are CsCl density gradient ultracentrifugation (Section III.E), Geneclean® (Section III.F.) and chromatography (Malliaros, 1988; Merion, 1988, 1989). Digestion with ribonuclease (RNase) is a convenient means to obtain RNA-free DNA.

During the cell lysis procedures, DNA is not degraded as the constituents of the lysis solutions inactivate nucleases. However, once isolated, DNA is susceptible to degradation by nucleases associated with the skin. In general, use of sterile glassware and solutions and aseptic handling of the DNA extract is recommended. Gloves should be worn whenever one expects to touch surfaces which will contact the extract (e.g. during preparation of dialysis tubing).

B. Proteinase K

Residual protein can be removed with 100 μg proteinase K ml^{-1} for 30 min at 37 ± 2 °C, followed by extraction with redistilled phenol–chloroform–isoamyl alcohol (25:24:1, v:v:v) to remove the proteinase K. A stock solution can be prepared at 10 mg ml^{-1} in TE and stored frozen. DNase contaminants in proteinase K are eliminated by incubating the stock solution at 37 ± 2 °C for 30 min.

C. Ribonuclease

Treat the protein-free DNA with RNase that is free of DNase [10 mg RNase A ml^{-1} in 10 mM Tris-HCl (pH 7.5) and 15 mM NaCl] at 50–100 μ ml^{-1} for 30–60 min at 37 ± 2 °C in a water bath. Inactivate DNase by boiling RNase stock solution for 15 min, cool slowly to room temperature and store in freezer.

For RFLP analyses, there is no reason to remove the RNase. However, if subsequent uses involve RNA (e.g. RNA–DNA hybridization, *in vitro*; Sambrook *et al.*, 1989), the RNase is eliminated with proteinase K.

D. Alcohol precipitation

Two options are available to concentrate DNA by alcohol precipitation: ethanol and isopropanol. The more commonly used is 95% ethanol. Alcohol precipitation depends on salt in solution: 2.5 M sodium or potassium acetate (pH 5.2) at a final concentration of 0.25 M, 7.5 M ammonium acetate at a final concentration of 2.5 M, or 5 M NaCl at 0.1 M (residual Na^+ may interfere with restriction endonuclease activity). Next, ethanol (2 volumes) or isopropanol (1.1 volume) is added, the solution is mixed and chilled (60 min to overnight) at − 20 °C or colder, if a low DNA yield is expected. Collect precipitate by centrifugation (30 min, 4 ± 2 °C, 12 000*g*), dry under vacuum or by lyophilization, and resuspend in TE.

E. Caesium chloride density gradient ultracentrifugation

Separation of DNA from protein and RNA can be accomplished with CsCl gradients which separate molecules by density (Sambrook *et al.* 1989). Density of protein is approximately 1.2 g ml^{-1}, RNA is greater than 1.8 g ml^{-1}, and DNA is about 1.5–1.7 g ml^{-1}, depending on the amount of ethidium bromide intercalated in the DNA. A good final density of CsCl solution is 1.57–1.62 g ml^{-1}, but it may need adjustment depending on the base composition of the DNA. Ethidium bromide is a potent mutagen and is moderately toxic. Researchers should wear gloves when handling the dye. To inactivate ethidium bromide, refer to Sambrook *et al.* (1989).

One procedure that works well for fungal DNA is:

1. Prepare solutions according to Murray and Thompson (1980):
   (a) Solution 1—1.0 M CsCl (density = 1.12 g ml^{-1}), 50 mM Tris-HCl (pH 8.0), 10 mM EDTA, 200 μg ethidium bromide ml^{-1}
   (b) Solution 2—6.6 M CsCl (density = 1.82 g ml^{-1}), 50 mM Tris-HCl (pH 8.0), 10 mM EDTA, 0.1% *n*-lauroyl sarcosine
   (c) Solution 3—4.5 M CsCl (density = 1.55 g ml^{-1}), 50 mM Tris-HCl (pH 8.0), 10 mM EDTA, 0.1% *n*-lauroyl sarcosine, 100 μg ethidium bromide ml^{-1}.
2. Dissolve DNA in 0.5 volume CsCl Solution 1 (1.0 volume is the volume of the ultracentrifuge tube). If necessary, heat at 60 ± 2 °C in a water bath until pellet dissolves.
3. Add 0.5 volume of CsCl Solution 2.
4. Add CsCl solution 1, or CsCl crystals to adjust density to 1.58 ± 0.01 g ml^{-1}.
5. Transfer to ultracentrifuge tubes and fill with CsCl Solution 3.
6. Centrifuge (22 h, 15 ± 2 °C, 338 000 *g*).
7. Collect DNA band with a hypodermic syringe under long-wave ultraviolet light (use protective eyeglasses).
8. Remove ethidium bromide by repeated extractions with water-saturated *n*-butanol or isoamyl alcohol until solution is no longer pink by visual check. Then, extract two more times.
9. Dialyse overnight against a large volume of TE.
10. Precipitate DNA with alcohol (95% ethanol) and vacuum-dry or lyophilize.

F. Geneclean®

Geneclean® (BIO 101, Inc., La Jolla, CA, USA) is a good substitute for alcohol precipitation as a means to purify DNA. It is pulverized glass and binds single-stranded DNA and double-stranded DNA without binding DNA contaminants. The procedure is simple, rapid and yields highly purified DNA. It is especially useful for removing DNA from agarose gels, desalting DNA, separating DNA from RNA and proteins, and for removing phenol, chloroform or ether from DNA. Geneclean® has been used to purify DNA from bulk soil prior to restriction enzyme analysis (Porteous and Armstrong, 1991; Section II.B.2, Step 10), has nearly eliminated electroelution of DNA from agarose gels, and offers promise for DNA extraction from mycorrhiza.

IV. Techniques to analyse DNA

A. Estimating DNA concentration

1. Ultraviolet spectrophotometry

Ultraviolet spectrophotometry at 260 nm is a common method to determine DNA concentration. For estimates of concentration in the linear portion of the curve, readings should be between 0.1 and 1.0 absorbance units (1 cm path length). A value of 1 corresponds to approximately 50 $\mu g\,ml^{-1}$ for double-stranded DNA and 40 $\mu g\,ml^{-1}$ for single-stranded DNA or RNA. The following formulae can be used directly if the light path through the cuvette is 1 cm:

native, double-stranded DNA (dsDNA), $mg\,ml^{-1} = A_{260} \div 20$;
denatured, single-stranded DNA (ssDNA), $mg\,ml^{-1} = A_{260} \div 23$

Spectrophotometry also permits detection of contaminating materials such as proteins and phenol, which absorb light at 280 nm. Readings are made at 260 nm and 280 nm. For preparations of pure DNA or RNA, the A_{260}/A_{280} is between 1.8 and 2.0. A lower ratio indicates contamination.

2. Fluorimetry

Fluorimetry can be used to measure ng to μg amounts of DNA (Cesarone *et al.*, 1979; Labarca and Paigen, 1980). The technique is based on the enhancement of bis-benzimidazole (Hoechst 33258) fluorescence when bound to DNA. In the presence of DNA, Hoechst 33258 has a peak excitation at 365 nm, and an emission peak at 458 nm. Without DNA the excitation peak is 356 nm and the emission peak is 492 nm. The fluorescence enhancement is due to preferential binding of the dye to A–T-rich regions, and is highly specific for DNA. The dye binds more strongly to dsDNA than to ssDNA, but does not appear to intercalate. RNA enhances the fluorescence to a lesser degree. Binding of the dye to DNA is unaffected by most buffers used to extract DNA, by high salt (up to 3 M NaCl), or by low levels of detergents. However, SDS above 0.01% should not be used, and one must assess the potential interferences of the various buffers described in the cell lysis procedures presented herein. RNA does not compete with the binding of DNA, so Hoechst 33258 is very good for estimating DNA in crude extracts which contain RNA. With some dye concentration adjustments, the assay is linear from 10 ng DNA ml^{-1} to 10 μg DNA ml^{-1}.

3. Ethidium bromide-stained DNA in agarose gels

A commonly used method for estimating DNA concentration is to compare band intensity of unknown DNAs in agarose gels containing ethidium bromide (see Section IV.B) with band intensities of DNA standards. The standards must be the same conformation (e.g. linear, circular, supercoiled) and approximately the same molecular weight as the sample DNA. The method is especially useful for small quantities of DNA in low concentrations.

B. Electrophoresis

Electrophoresis in agarose gels is one of the most useful tools for characterizing DNA. We present here some general information and useful hints on agarose gel electrophoresis. Refer to Andrews (1986) for a more detailed discussion.

Several factors affect the rate of DNA migration through agarose: conformation, length, charge and weight. As the charge:length ratio is fixed, and the molecular weight:length ratio is constant, mobility through agarose is related primarily to molecular weight. All linear, non-supercoiled dsDNA moves through the electrophoretic field in agarose at a rate that is inversely proportional to $\log_{10}$ of the molecular weight. Since the molecular weight of DNA is directly proportional to its length, its size can be calculated from the distance it migrates compared with the migration of molecules of known sizes. Include a set of standards whose size range brackets that of the fragments of interest. Accurate estimates of size may require standards in adjacent lanes. Standards are produced by digesting a well-characterized DNA (e.g. bacteriophage λ or SV40) with an endonuclease that yields discrete fragments of known sizes.

Agarose concentration also affects rate of movement since pore size and number are related to concentration. Useful concentrations range from 0.4 to 2.5%. Larger molecules (60 kb) are conveniently separated in 0.3% agarose, while up to 2.0% agarose can be used to separate fragments from 0.1 to 3 kb. Composite polyacrylamide–agarose gels and 1–2% agarose gels separate fragments less than 5 kb. Gel concentrations of 0.3–0.7% are most suitable for molecules ranging from 4 to 40 kb. As the pore space of polyacrylamide gels is smaller than that of agarose gels, the former can be used to separate lower molecular weight nucleic acids. Since separations can be done among fragments varying by one base pair, polyacrylamide is used to sequence nucleic acids.

Temperature affects nucleic acid movement. The electric current generates heat in the gel and temperature gradients form across its thickness. Mobility is greater in the warmer centre than near the cooler surfaces of the gel, therefore forming curved mobility zones. Heat distortions are minimized by reducing gel thickness or voltage, but heat effects are more pronounced as electrophoresis progresses.

Both dsDNA and ssDNA are visualized with ethidium bromide, but binding of the dye is greater with dsDNA so the extinction coefficient favours native DNA. Low concentrations of ethidium bromide can be used, permitting detection of as little as 2 ng of native DNA. If one includes 0.5 μg ethidium bromide ml^{-1} in the electrophoresis buffer and the gel during the run, progress of the separation can be observed by ultraviolet illumination. Since ethidium bromide stains DNA and RNA, this procedure can be used for both. To inactivate ethidium bromide (a powerful mutagen) refer to Sambrook *et al*. (1989).

A variety of electrophoresis buffers are available (Sambrook *et al*., 1989) which permit high resolution of DNA fragments (used usually at an ionic strength between 0.03 and 0.1). Tris–acetate–EDTA (TAE) is often used, but its low buffering capacity makes it ineffective during long periods of electrophoresis [1 × TAE (pH 7.5–7.8): 0.04 M Tris base, 1 mM EDTA (pH 7.8), 1.14 ml glacial acetic acid litre^{-1}]. Tris–borate–EDTA is a better buffer and is not depleted as quickly. Three additional advantages of Tris–borate–EDTA are the greater mobility of DNA fragments, the higher resolution of distinct bands when fragments are about 500 bases and smaller, and Tris–borate–EDTA can be recycled (4–5 runs).

Since agarose gels have very low, and nucleic acids have high, absorbances in ultraviolet light (260 nm), quantitative estimates of nucleic acids can be made by direct densitometry of unstained gels. As little as 0.05 μg can be detected in this way.

Adjustments may be necessary for the amount of DNA loaded into the well of the agarose gel. For example, too much DNA can result in smearing of the bands, leading to ambiguity in fragment identification. On the other hand, too few target sequences available for hybridization with probe DNA may result in a signal which is too low for detection, even if the lanes are overloaded.

As a suggested starting point for electrophoresis after endonuclease digestion of DNA, the following procedure may be used to produce RFLPs:

1. Add 0.7% agarose to TAE running buffer [1 × TAE (pH 7.5–7.8): 0.04 M Tris base, 1 mM EDTA (pH 7.8), 1.14 ml glacial acetic acid litre^{-1} containing 0.5 μg ethidium bromide ml^{-1} (prepared as 10 mg ml^{-1} stock]. Prepare enough agarose to make a 4–5 mm thick gel.
2. Microwave soution at medium power for 1–3 min, or alternatively boil in water bath, until clear and free of agarose particles. If necessary add water to restore to its original volume. Mix.
3. Cool agarose to 55 ± 2 °C and pour into a clean gel plate with ends taped. Insert electrophoresis comb near end. Allow gel to solidify.
4. Remove tape and place plate into the electrophoresis unit. Add TAE until buffer covers gel. Remove comb.
5. Load nucleic acid samples (10 ng to 2 μg DNA) in TE containing 0.25 volume of loading buffer [50% glycerol, 0.4% bromophenol blue, 5 mM. Tris-HCl (pH 7.0)]. Connect gel unit to power source, negative lead at well end of gel. Set power unit to 50–125 V constant and electrophorese for 1–3 h (lower voltages may be used for longer times). Monitor movement of bromophenol blue dye to assess progress of the electrophoresis. To determine separation of fragments, interrupt the electrophoresis and examine the gel while it is still on the gel plate with a short-wave ultraviolet light source. As fragment separation varies with the DNA and restriction enzyme, time of electrophoresis is determined empirically.
6. Destain gel in water for 15–30 min in shaking bath at room temperature. Photograph gel under ultraviolet light.

C. Endonucleases and restriction fragment length polymorphisms

Analysis of DNA using restriction fragment length polymorphisms (RFLPs) is recognized as a powerful technique for comparing DNA fragments containing specific nucleotide sequences (Anderson *et al.*, 1987; Loftus *et al.*, 1988; Summerbell *et al.*, 1989). Restriction fragments are produced by digesting DNA with restriction endonucleases that recognize specific sequences in the dsDNA. The restriction fragments are then separated electrophoretically into RFLPs. RFLPs are frequently used to develop a restriction enzyme map using different combinations of enzymes; to identify regions of the genome that encode for specific genes by hybridizing with probes containing specific gene sequences; to determine relatedness among individuals within or between populations, and between parent and offspring; to estimate gene copy number; and to diagnose disease.

Caution is advisable when interpreting RFLPs. Since an RFLP is an array of DNA fragments of different sizes, two similarly sized fragments with different nucleotide sequences are not electrophoretically distinguishable. Based on probability alone, between 20 and 40% of RFLP fragments from two unrelated isolates will be similar in size. DNA hybridization increases sensitivity to discriminate between similar sized fragments, if the nucleotide sequence of the probe is unique to one of the two fragments. Point mutations or methylation of bases within a restriction site may hinder the recognition of the site by the enzyme, leading to ambiguity in determining relatedness between isolates which are otherwise isogenic.

The first step in producing RFLPs is to digest DNA until no further fragmentation occurs. The manufacturer's product information sheet accompanying each enzyme should be consulted to determine the amount of enzyme and buffer added. Since DNA purity influences enzyme activity, the DNA:enzyme ratio and incubation period should be determined empirically. Complete digestion can be assured by longer incubation of DNA with nuclease and/or addition of more restriction enzyme after an initial incubation period. Resulting fragments, which may vary from a few nucleotide pairs to about 25 kb, are analysed by gel electrophoresis. Observation and quantitative measurement of fragments are the same as for intact nucleic acids. Refer to Section IV.B to determine fragment size.

The class of restriction endonucleases used to produce RFLPs are cleavage-site specific, and with a specific DNA will form characteristic specific fragments. Fragments have single-stranded tails on the 5′ and 3′ ends or have blunt ends. Restriction sites are usually 4–6 consecutive bases. Wherever a base within a recognition sequence is protected (e.g. methylated), the endonuclease will not cleave the site. By using a variety of enzymes, either singly or in combination, it is possible to create different RFLPs. Selection of endonucleases will depend on the experiment being done and the specific organism under investigation. Consult Armstrong *et al*. (1989) and Rogers *et al*. (1989) for enzymes used with DNA from mycorrhizal fungi and mycorrhiza.

A general procedure for digestion of DNA by endonucleases is:

1. Prepare reaction mixture:
   (a) Predetermined volume of H_2O to bring final volume to 10 μl
   (b) 1 μl BSA (stock is 1 mg ml)
   (c) 1 μl of 10 × endonuclease buffer
   (d) 0.5–2 μg DNA in 1–7 μl TE
   (e) 1 μl (5–10 units) enzyme.
2. Incubate for 2 h at 37 ± 2 °C (to ensure complete digestion, we suggest adding another μl of enzyme and incubating an additional 1 h).

D. DNA hybridization and related techniques

Ultraviolet light can be used to observe the large number of fragments typical of RFLPs. If a highly qualitative comparison is desired, this approach is suitable. Major differences between two patterns will be observable. However, as the number of fragments increases, so does the difficulty in comparing specific bands, especially if the bands are faint. For greater sensitivity and specificity, hybridization of RFLPs is performed with specific DNA sequences (probes) that may encode specific genes to reduce the number of bands needed for analysis (Sambrook *et al.*, 1989).

DNA hybridization is the association of ssDNA molecules with nucleotide sequences containing a high degree of homology. The probe sequence is labelled to detect its association with target DNA. Hybridization conditions are optimized for maximal hybrid formation between all sequences independent of the degree of homology. Next, the hybrids are incubated under "stringent" conditions that select for more stable hybrids (more homologous). Hybrids are visualized by autoradiography when probes are labelled with ^{35}S or ^{32}P.

Several theoretical issues must be considered for reaction conditions and probe selection. One is stringency, i.e. the incubation conditions determining the degree of sequence homology necessary to maintain a stable hybrid. At lower temperatures and higher salt concentrations, hybridization occurs between DNA fragments with less homology that can lead to erroneous conclusions on similarity between fragments. A second issue concerns the spectrum of specificity in the hybridization reaction that depends on the choice of probe. For example, if an RFLP of *Laccaria bicolor* chromosomal DNA is probed with the entire genome of *L. bicolor*, the autoradiogram should be identical to the RFLP observed with ultraviolet light. However, if the probe encodes a specific gene isolated from *L. bicolor*, only one or a few bands in the RFLP of *L. bicolor* chromosomal DNA will be visualized on the autoradiogram. Thus, careful choice of probes permits comparison of DNAs at differing levels of rigor.

Several steps are necessary for visualization of hybrids: RFLPs are transferred to a solid matrix (e.g. nitrocellulose or nylon filters); the probe is treated with either radioactive or non-radioactive labels; the hybridization reaction is performed; and an autoradiogram is prepared or visualization reaction is done.

1. Southern transfer

The method to transfer the RFLP to nitrocellulose paper is:

1. Soak gel in 0.25 M HCl in a glass tray with mild agitation for 15 min (low pH cleaves DNA by depurination so high molecular weight fragments will pass quickly through the gel).
2. Rinse gel in distilled water for 15 min.
3. Soak gel in 0.5 N NaOH–1.5 M NaCl for 10 min to denature the DNA.
4. Repeat Step 3 two times (30 min each).
5. Wash two times (30 min each) in 1 M Tris-HCl (pH 8.0)–1.5 M NaCl to neutralize excess NaOH.
6. Place gel on Whatman 3MM paper which is layered over a glass plate raised above 10 × SSC [1 × SSC: 0.15 M NaCl, 0.015 M tri-sodium citrate (pH 7.0)] in a glass tray (diagram of apparatus; Darbre, 1986).
7. Cut a sheet of nitrocellulose paper (BA85, Schleicher & Schuell, Keene, NH, USA) to the size of the gel. Wet in 2 × SSC for 5 min. To aid in orienting the lanes, mark cut sheet with indelible ink.
8. Place the nitrocellulose on top of the gel. Overlay with several pieces of absorbent paper wetted in 10 × SSC. Cover with 10–12 cm of paper towels or household napkins and put a 0.5 kg weight on the top.
9. Incubate overnight to transfer DNA.
10. Remove nitrocellulose, rinse in 6 × SSC for 5 min, and bake for 2 h at 80 ± 2 °C in a vacuum oven.
11. Store under vacuum until used for hybridization.

2. Selecting and preparing probes

The factors that must be considered for selecting and constructing DNA probes (Keller and Manak, 1989) extend well beyond the scope of this chapter. To illustrate some of the criteria for choosing probes, we consider several research topics about mycorrhizal fungi: relatedness between isolates, species or other taxonomic levels; identification; and detection in the environment. To study these topics, one needs to select probes of varying specificity. At one extreme, there are less specific probes appropriate for studies with organisms that are not closely related. Examples of these probes are the prokaryotic- and eukaryotic-specific oligonucleotides derived from rRNA sequences (Giovannoni *et al.*, 1988), total genomic DNA (Cooper *et al.*, 1987), total and non-conserved regions of rRNA genes (Wu *et al* 1983; Specht *et al.*, 1984; Armstrong *et al.*, 1989) and mitochondrial genes (Camougrand *et al.*, 1988; Bruns *et al.*, 1989; Förster *et al.*, 1989). At the other extreme, probes with high specificity can distinguish closely related organisms, and are prepared from the organisms under investigation. Examples of these probes are the conserved regions of rRNA genes (DeLong *et al.* 1989) and genes encoding for proteases, lignases, cellulases or pigment production. Within rRNA genes of all organisms, there are regions containing a spectrum of variability in nucleotide sequence, that are characteristic of organisms at a particular taxonomic level (Noller, 1984; Sogin *et al.,* 1986). This property of rRNA lends itself to phylogenetic analyses because sequences can be identified for probes, ranging from species-specific (perhaps isolate-specific) to those which hybridize with DNA from all organisms.

The impact of environmental biotechnology has led to considerable efforts by molecular ecologists to produce probes which can be applied to environmental issues. One example is the development of high-specificity probes to track genetically engineered fungi released into the environment (Steffan and Atlas, 1988; Dickman and Partridge, 1989). Currently, the most common approach for marking fungi is transformation with genes conferring antibiotic resistance. The *Escherichia coli* hygromycin B resistance gene has been introduced into many fungi, including *Laccaria bicolor* (Barret *et al.*,1990), *Colletotrichum trifolii* (Dickman, 1988), *Cochliobolus heterostrophus* (Turgeon *et al.*, 1987), and *Ustilago maydis* (Wang *et al.*, 1988). Dickman (1988) also transformed *C. trifolii* with the β-tubulin gene of *Neurospora crassa*, which confers resistance to benomyl. Such marked strains could be used to investigate the environmental issue of movement of introduced genes into indigenous soil fungi. To accomplish this, one could use two DNA probes: one characteristic of the organism and the other specific to the introduced gene. Use of specific probes is sufficiently sophisticated for use with environmental samples, such as sediments and soils. Steffan and Atlas (1988) were able to track a herbicide-degrading strain of *Pseudomonas cepacia* introduced into sediments by using a probe specific for the gene. Dickman and Partridge (1989) studied the persistence and stability of transformed antibiotic resistance genes in *C. trifolii* that were recovered from soil supporting growth of corn plants.

During the early stages of molecular ecology research, investigators may want to rely on others for probes. As the skills become more familiar, other options for producing probes are gene cloning (Sambrook *et al.*, 1989), *in vitro* synthesis of oligonucleotides based on sequences obtained from databases (GenBank®, IntelliGenetics, Inc., Mountain View, CA, USA), and purchase from commercial firms.

Labelling of DNA can be accomplished in a variety of ways. One of the most popular methods is nick translation. DNase I forms 3′ hydroxyl ends (nicks) in the DNA. DNA polymerase I catalyses addition of the appropriate labelled deoxynucleotide triphosphate (dNTP; e.g. ^{32}P-dNTP) into each nick. The exonuclease activity of polymerase I continues to excise deoxynucleotide monophosphates and new labelled dNTPs are inserted into the nicked region. If high specific activities (greater than 10^8 cpm μg^{-1}) are required, no unlabelled dNTPs are added to the reaction mixture and all four dNTPs are labelled. For general analysis of RFLPs, only one labelled dNTP is used.

The nick translation procedure to prepare a radiolabelled probe with a specific activity in excess of 10^8 cpm μg^{-1} is as follows:

1. After all constituents (except DNA polymerase I) have thawed on ice, mix the following:
   (a) 5.0 μl of 10 × nick translation buffer (NTB) [1 × NTB: 0.5 M Tris-HCl (pH 7.2), 0.1 M $MgSO_4.H_2O$, 1 mM dithiothreitol, 0.5 mg BSA m^{-1}]
   (b) 1.0 μg DNA in water or TE
   (c) 1 μl of each unlabelled dNTP from 1 mM stock solutions (all nucleotides except for the labelled ones)
   (d) 50–75 μCi [α-^{32}P]dNTP (50–100 pmol)
   (e) Water to 48 μl final volume (volume can be scaled between 10 and 100 μl if constituents are kept in same proportions)
2. Keep on ice.
3. Make a 10 000-fold dilution of DNase I in 1 × NTB–50% glycerol. Then, dilute 10-fold in 1 × NTB. Dilution of the DNase depends on the specific activity of the enzyme. The final concentration of enzyme must be empirically determined so that about 30% of the label is incorporated. After nick translation, the DNA molecules should be 0.4–0.8 kb.

4. Add 1 μl diluted DNase and 1 μl polymerase I.
5. Incubate at 15 ± 2 °C for 60 min.
6. Stop reaction with 5 μl 0.5 M EDTA.
7. Separate unincorporated dNTPs with column (1 ml plastic pipette) packed with Sephadex G-50, or with a disposable spun column (Nu-Clean TM D50, International Biotechnologies, Inc., New Haven, CT, USA).
8. Store the labelled probe in freezer.

A second method yielding more efficient incorporation of label is random-primer labelling (Sambrook *et al.*, 1989). Hexanucleotides are used to initiate synthesis of the complementary strand of ssDNA with concomitant incorporation of up to four of the dNTPs being radiolabelled. The specific activity of the labelled DNA is usually 10–50 times greater than that achieved with nick translation.

Non-radioactive probes can be prepared using nick translation to incorporate biotin-labelled analogues of dNTPs, permitting colorimetric visualization of hybrids (Leary *et al.*, 1983).

3. *Hybridization and autoradiography*

Conditions for hybridization are chosen to maximize hydrogen bond formation between probe and target DNAs. Then, the less stable associations between heterologous sequences are dissociated by incubation at a higher temperature (e.g. 65–68 °C) and lower ionic strength. The higher the incubation temperature allowing base pairing, the greater the complementarity between strands of the hybrids. If formamide (e.g. 50%) is added, the incubation temperature can be lowered (e.g. 42 °C). For lower stringency of base pairing, formamide concentration is reduced (e.g. 30%). Overnight incubations for hybrid formation are long enough to ensure complete hybridization. The shortest time for complete hybridization can be calculated from the quantity and size of probe, and reaction mixture volumes (Sambrook *et al.*, 1989).

The following protocol for hybridizing probe to DNA on nitrocellulose paper can be applied to high and low stringency conditions, depending on the formamide concentration in the pre-hybridization and hybridization solutions. In both cases, the remaining steps are the same:

1. Moisten baked nitrocellulose papers (Section IV.D.1; Step 10) for 5 min in 6 × SSC [1 × SSC: 0.15 M NCl, 0.015 M trisodium citrate (pH 7.0)].
2. Wash for 1 h at 42 ± 1 °C in solution containing 50 mM Tris-HCl (pH 8.0), 1 M NaCl, 1 mM EDTA, 0.1% SDS.

3. Place nitrocellulose papers in heat-sealable bags with 2–5 ml of pre-hybridization solution [deionized formamide (v/v) (stringency: high, 50%; low, 30%), 5 × Denhardt's solution, 5 × SSPE, 0.1% SDS, salmon sperm or calf thymus DNA at 100 $\mu g\,ml^{-1}$]. Store in freezer. Use 0.2 ml liquid per cm^2 of paper; up to 3 papers can be put into a single bag. To prepare the pre-hybridization solution, use aliquots of the following stocks:
   (a) Deionized formamide (mix 5 g Dowex XG8 resin with 50 ml formamide, stir for 30 min, filter through Whatman No. 1 filter paper, store in freezer)
   (b) 50 × Denhardt's solution: 1% Ficoll, 1% PVP, 1% BSA (sterilize by filtering; store in freezer)
   (c) 20 × SSPE: 3 M NaCl, 0.2 M $NaH_2PO_4.H_2O$ (pH 7.4), 0.022 M EDTA (sterilize by autoclaving)
   (d) Salmon sperm or calf thymus DNA at 10 $mg\,ml^{-1}$ (shear by forcing solution several times through a hypodermic needle; denature by boiling for 5 min and cool quickly on ice).
4. Lay bag on a paper towel and force the air bubbles out with a pipette. Seal bag and incubate for 4–6 h at 42 ± 1 °C.
5. Heat a needle and make a small hole in the top of a microfuge tube (to allow release of gas), which contains the labelled probe. Denature the probe by boiling for 5 min and cool on ice. Cut open a corner of the bag containing the nitrocellulose papers and introduce the probe to make the hybridization solution. Do not let air enter bag. If bubbles form, push them out by moving a pipette along the side of the bag. Seal and put into a second bag.
6. Incubate bag overnight at 42 ± 1 °C with gentle agitation.
7. Remove nitrocellulose and discard hybridization solution (or store frozen for future use).
8. Wash nitrocellulose 4 times in 100–300 ml wash solutions (first wash, 5 min, 25 ± 2 °C, 2 × SSC–0.5% SDS; second, 15 min, 25 ± 2 °C, 2 × SSC–0.5% SDS; third, 2 h, 42 ± 2 °C, 0.1 SSC–0.1% SDS; and fourth, 30 min–1 h, 42 ± 2 °C, 0.1 × SSC–0.5% SDS).
9. Air-dry nitrocellulose.

The probe can be reused several times. Just before re-use, heat at 90 ± 2 °C for 5 min and cool on ice. Then, instead of adding the probe to the bag in Step 5, substitute the pre-hybridization solution with this previously used hybridization solution.

Autoradiography is done with or without intensifier screens. High specific activity of the radiolabelled probe yields a visual signal within hours. However, in general, film needs exposure for as long as 48–96 h. Note orientation of nitrocellulose when placing film on paper. Wrap the nitrocellulose in plastic film to protect the cassette from contamination by radioisotopes and to prevent the paper from sticking to the film.

V. Conclusion

The new hybrid science of molecular ecology has arisen from the merging of molecular biology and ecology. The combined philosophy and methods of both disciplines offer a powerful tool for new vistas in the study of mycorrhiza ecology. Since methods for handling and analysing nucleic acids are now routine, they are rapidly becoming available to a diverse research community. One of the major breakthroughs that relates to rhizosphere ecology is the ability to identify and quantify organisms present in very small numbers at a variety of taxonomic levels.

This chapter presents basic methods directly applicable to recovery of DNA from mycorrhizal fungi in soils. DNA extraction from soil organisms poses a formidable task. Thus, we detail the steps in DNA extraction that might pose difficulties for researchers, e.g. cell lysis and other steps when soil constituents and secondary plant products still contaminate the extract. The variety of techniques provides investigators with a basis for designing procedures specific to their individual needs. Most of the protocols were tailored to accommodate small amounts of starting material, since individual mycorrhiza may weigh only a few mg dry weight. It is a widely accepted fact that purified DNA behaves similarly regardless of source, and basic methodology texts for handling purified DNA are plentiful. Therefore, we present fewer options and less detail for manipulating purified DNA. These include methods for quantifying, endonuclease digesting, and separating DNA molecules; and techniques related to hybridizing RFLPs with probes.

The phenomenon of specific base pairing (hybridization) is one of the fundamental properties of nucleic acids upon which many of the procedures of molecular biology are based. These procedures are now being applied to many soil eukaryotes. Although most of the methods have not yet been applied to mycorrhizal fungi, they have been used with other fungi and should be directly transferable. For example, hybridization permitted the direct probing of crushed nodules for detection of *Frankia* collected from a single alder stand (Simonet *et al.*, 1988). Subtraction hybridization was used to isolate strain-specific DNA sequences of *Rhizobium loti* for use as probes (Bjourson and Cooper,

1988), and stage- and cell-specific genes from *Aspergillus nidulans* (Timberlake, 1986). RFLPs were analysed for relatedness among such organisms as *Agaricus bisporus* (Loftus *et al*., 1988); *A. brunnescens* (Summerbell *et al*., 1989); *Armillaria mellea* (Anderson *et al*., 1987); *Frankia* spp. (Lalonde *et al*., 1988); and *Bremia lactucae* (Hulbert and Michelmore, 1988). Hybridization was used to correlate the genetic and physical map of the mitochondrial genome of *Agaricus brunnescens* (Hintz *et al*., 1988) and the rRNA genes of *Coprinus* spp. (Cassidy *et al*., 1984, Cassidy and Pukkila, 1987). Cassidy *et al*. (1984) determined the copy number of rDNA genes in *C. cinereus*.

A limitation of RFLPs is its relatively low sensitivity (the quantity of DNA needed to perform the analysis). A very recent development that has far-reaching potential for molecular ecology is the use of the polymerase chain reaction (PCR; Innis *et al*., 1990). PCR is an *in vitro* method of amplifying exponentially a region of DNA between two segments of known sequence. The reaction depends on a thermo-stable polymerase and site-specific oligonucleotide primers. New DNA is synthesized during alternating denaturation and re-annealing of the specific sequence being amplified. Lee and Taylor (1990) amplified rDNA sequences from a single spore of *Neurospora tetrasperma*. PCR was also used for phylogenetic studies of fungi (Bruns *et al*., 1989; White *et al*., 1990) Researchers working on mycorrhiza can benefit greatly from PCR, as nucleic acids from one mycorrhiza can be analysed after amplification (Gardes *et al*., 1991), provided adequate measures are taken to control for effects of contaminating DNAs.

References

Anderson, J. B., Petsche, D. M. and Smith, M. L. (1987) *Mycologia*. **79**, 69–76.

Andrews, A. T. (1986) *Electrophoresis: Theory, Techniques, and Biochemical and Clinical Applications*, 2nd edn. Clarendon Press, Oxford. 452 pp.

Armstrong, J. L., Fowles, N. L. and Rygiewicz, P. T. (1989). *Plant Soil* **116**, 1–7.

Barrett, V., Lemke, P. A. and Dixon, R. K. (1990). *Appl. Microbiol. Biotechnol*. **33**, 313–316.

Bjourson, A. J. and Cooper, J. E. (1988). *Applied. Environ. Microbiol.* **54**, 2852–2855.

Bruns, T. D., Vogel, R., White, T. J. and Palmer, J. D. (1989). *Nature (Lond.)* **339**, 140–142.

Camougrand, N., Mila, B., Velours, G., Lazowska, J. and Guérin, M. (1988). *Curr. Genet.* **13**, 445–449.

Cassidy, J. R. and Pukkila, P. J. (1987). *Curr. Genet.* **12**, 33–36.
Cassidy, J. R., Moore, D., Lu, B. C. and Pukkila, P. J. (1984). *Curr. Genet.* **8**, 607–613.
Cesarone, C. F., Bolognesi, C. and Santi, L. (1979). *Anal. Biochem.* **100**, 188–197.
Cooper, J. E., Bjourson, A. J. and Thompson, J. K. (1987). *Appl. Environ. Microbiol.* **53**, 1705–1707.
Darbre, P. D. (1988). *Introduction to Practical Molecular Biology*. John Wiley and Sons, New York. 117 pp.
Davis, L. G., Dibner, M. D. and Battey, J. F. (1986). *Basic Methods in Molecular Biology*. Elsevier, New York. 388 pp.
DeLong, E. F., Wickham, G. S. and Pace, N. R. (1989). *Science* **243**, 1360–1363.
Dickman, M. B. (1988). *Curr. Genet.* **14**, 241–246.
Dickman, M. B. and Partridge, J. E. (1989). *Theoret. Appl. Genet.* **77**, 535–539.
Förster, H., Kinscherf, T. G., Leong, S. A. and Maxwell, D. P. (1989). *Can. J. Bot.* **67**, 529–537.
Gardes, M., White, T. J., Fortin, J. A., Bruns, T. D. and Taylor, J. W. (1991). *Can. J. Bot.* **69**, 180–190.
Giovannoni, S. J., DeLong, E. F. Olsen, G. J. and Pace, N. R. (1988). *J. Bacteriol.* **170**, 720–726.
Hintz, W. E. A., Anderson, J. B. and Horgen, P. A. (1988). *Curr. Genet.* **14**, 43–49.
Hulbert, S. H. and Michelmore, R. W. (1988). *Mol. Plant–Microbe Interact.* **1**, 17–24.
Innis, M. A., Gelfand, D. H. Sninsky, J. J. and White T. J. (eds) (1990). *PCR Protocols: A Guide to Methods and Applications*. Academic Press, London. 482 pp.
Jahnke, K.-D., and Bahnweg, G. (1986). *Br. Mycol. Soc.* **87**, 175–191.
Kafatos, F. C., Jones, C. W. and Efstratiadis, A. (1979). *Nucl. Acids Res.* **7**, 1541–1552.
Katterman, F. R. H. and Shattuck, V. I. (1983). *Prep. Biochem.* **13**, 347–359.
Keller, G. H. and Manak, M. M. (1989). *DNA Probes*. Stockton Press, New York. 256 pp.
Kirby, L. T. (1990) *DNA Fingerprinting: An Introduction*. Stockton Press, New York. 256 pp.
Labarca, C. and Paigen, K. (1980). *Anal. Biochem.* **102**, 344–352.
Lalonde, M., Simon, L., Bousquet, J. and Séguin, (1988). In *Nitrogen Fixation: Hundred Years After, Proceedings of the 7th International Congress on [triple-bond] Nitrogen Fixation,* Köln, West Germany (H. Bothe, F. J. de Bruin and W. E. Newton, eds), pp. 671–691. Gustav Fischer, Stuttgart.
Leach, J., Finkelstein, D. B. and Rambosek, J. A. (1986). *Fungal Genet. Newslett.* **33**, 32–33.
Leary, J. J., Brigati, D. J. and Ward, D. C. (1983). *Proc. Natl. Acad. Sci. USA* **80**, 4045–4049.
Lee, S. B. and Taylor, J. W. (1990) In *PCR Protocols: A Guide to Methods and Applications* (M. A. Innis, D. H. Gelfand, J. J. Sninsky and T. J. White, eds), pp. 282–287. Academic Press, London.
Loftus, M. G., Moore, D. and Elliot, T. J. (1988). *Theoret. Appl. Genet.* **76**, 712–718.

Malliaros, D. (1988). *Anal. Biochem.* **169** 121–131.

Maniatis T., Fritsch, F. E. and Sambrook, J. (1982). *Molecular Cloning. A Laboratory Manual*. Cold Spring Harbor Laboratory, Cold Spring Harbor, New York. 545 pp.

Manicom, B. Q., Bar-Joseph, M., Rosner, A., Vigodsky-Haas, H. and Dotze, J. M. (1987). *Phytopathology* **77**, 669–672.

Martin, F., Zaiou, M., Le Tacon, F. and Rygiewicz, P. T. (1991). *Ann. Sci. For.* **48**, 297–305.

Merion, M. (1988). *Biotechniques* **6**, 246–251.

Merion, M. (1989). *Biotechniques* **7**, 60–67.

Merlo, D. J. and Kemp, J. D. (1976). *Plant Physiol*. **58**, 100–106.

Murray, M. G. and Thompson, W. F. (1980). *Nucl. Acids Res*. **8**, 4321–4325.

Noller, H. F. (1984). *Ann. Rev. Biochem.* **53**, 119–162.

Peberdy, J. F. (1989). *Mycol. Res.* **93**, 1–20.

Perbal, B. (1988). *A Practical Guide to Molecular Cloning*, 2nd edn. John Wiley and Sons, New York. 811 pp.

Porteous, L. A. and Armstrong, J. L. (1991). *Curr. Microbiol.* **22**, 345–348.

Raeder, U. and Broda, P. (1985). *Lett. Appl. Microbiol.* **1**, 17–20.

Rogers, S. O., Rehner, S., Bledsoe, C. S., Mueller, G. J. and Ammirati, J. F. (1989). *Can. J. Bot.* **67**, 1235–1243.

Saghai-Maroof, M. A., Soliman, K. M., Jorgensen, R. A. and Allard, R. W. (1984). *Proc. Nat. Acad. Sci. USA* **81**, 8014–8018.

Sambrook, J., Fritsch, F. E. and Maniatis, T. (1989). *Molecular Cloning. A Laboratory Manual.* 2nd edn. Cold Spring Harbor Laboratory, Cold Spring Harbor, NY. 1200 pp.

Simonet, P., Li, N. T. du Cros, E. T. and Bardin, R. (1988). *Appl. Environ. Microbiol.* **54**, 2500–2503.

Sogin, M. L.. Elwood, H. J. and Gunderson, J. H. (1986). *Proc. Nat. Acad. Sci. USA* **83**, 1383–1387.

Specht, C. A., Novotny, C. P. and Ullrich, R. C. (1984). *Curr. Genet.* **8**, 219–222,

Steffan, R. J. and Atlas, R. M. (1988). *Appl. Environ. Microbiol.* **54**, 2185–2191.

Summerbell, R.C., Castle, A. J., Horgen, P. A. and Anderson, J. B. (1989). *Genetics* **123**, 293–300.

Taylor, B. and Powell, A. (1982). *Focus* **4**, 4–6.

Timberlake, W. E. (1986). In *Biology and Molecular Biology of Plant–Pathogen Interactions* (J. Bailer, ed.) Vol. 3, pp. 343–359. Springer-Verlag, Berlin.

Turgeon, B. G., Garber, R. C. and Yoder, O. C. (1987). *Mol. Cell. Biol.* **7**, 3297–3305.

Verma, M. (1988). *BioTechniques* **6**, 848–853.

Wagner, D.B., Furnier, G. R., Saghai-Maroof, M. A. Williams, S. M., Dancik, B. P. and Allard, R. W. (1987). *Proc. Nat. Acad. Sci. USA* **84**, 2097–2100.

Wang, J., Holden, D. W. and Leong, S. A. (1988). *Proc. Nat. Acad. Sci. USA* **85**, 865–869.

White, T. J., Bruns, T., Lee, S. and Taylor, J. (1990). In *PCR Protocols: A Guide to Methods and Applications* (M. A. Innis, D, H. Gelfand, J. J. Sninsky and T. J. White, eds), pp. 315–322. Academic Press, London.

Wu, M. M. J., Cassidy, J. R. and Pukkila, P. J. (1983). *Curr. Genet.* **7**, 385–392.

10

Procedures and Prospects for DNA-Mediated Transformation of Ectomycorrhizal Fungi

P. A. LEMKE and V. BARRETT

Department of Botany and Microbiology, Auburn University, Auburn, AL 36849 USA

R. K. DIXON

Environmental Research Laboratory, Environmental Protection Agency, Corvallis, OR 97333, USA

I. Background considerations

Mycorrhizal fungi represent an integral component of the root–soil ecosystem or rhizosphere (Harley and Smith, 1983; Curl and Truelove, 1986) and, if genetically altered to further benefit the plant, offer potential to improve the establishment and yield of many crop plants. Certainly, genetic manipulation of mycorrhizal fungi, through studies involving genes for agronomically important traits or for improved

METHODS IN MICROBIOLOGY
VOLUME 23 ISBN 0-12-521523-1

Copyright © 1991 by Academic Press Limited
All rights of reproduction in any form reserved

symbiosis, has the potential to improve agricultural and forest production world-wide. Despite their ubiquity in the rhizosphere and their importance to plants, mycorrhizal fungi have unfortunately not been amenable to detailed genetic study. Only more recently, owing to the development of procedures for DNA-mediated transformation of filamentous fungi, has the prospect for research with mycorrhizal fungi been extended to include molecular genetic manipulations.

A. Limitations to conventional genetic study

The extent of mutualistic interdependence between a mycorrhizal fungus and a plant root system, while of overall benefit to both partners, limits the ability to study the fungus as an experimental or genetic system. The endomycorrhiza or vesicular-arbuscular mycorrhiza involve a limited number of species of zygomycetous fungi (Schneck, 1982). These fungi do sporulate during vegetative or mitotic growth, often in profusion, but they are developmentally dependent biotrophs with no known sexual or meiotic spore states (Harley and Smith, 1983). The vesicular-arbuscular mycorrhizal fungi, as obligate symbionts with no known sexual phase, thus offer little or no opportunity for conventional or other genetic investigation. By contrast, the ectomycorrhiza collectively involve more than 1000 species of mainly basidiomycetous and ascomycetous fungi (Trappe, 1962; Zak, 1973), most with an ability to grow saprophytically. Many ectomycorrhizal fungi thus can grow vegetatively, often quite well, in pure culture, but in general they do not sporulate in the vegetative state (Hutchinson, 1989) and are rarely able to complete a sexual cycle under standard laboratory conditions (Kropp, 1988). Their sexual or meiotic spores, when produced, germinate asynchronously and often only at low frequency (Fries, 1983; Kropp, 1988). Such intrinsic limitations for sexual competence are the principal reason why ectomycorrhizal fungi have not been examined extensively through conventional genetic methods. To date, only very few ectomycorrhizal basidiomycetes, species principally from the genera *Laccaria* and *Hebeloma*, have been examined critically through breeding experiments.

B. Fungal transformation systems

DNA-mediated transformation extends the potential for genetic manipulation, even among established genetic systems, and provides access to genetic experimentation in fungi less amenable to study by conventional genetic analysis. Transgenic introduction of genes and/or their control mechanisms has proven to be important in current research involving several filamentous fungi (Timberlake and Marshall, 1989; Leong and Berka, 1991). There is presently a variety of transformation/selection systems available and in theory such systems are applicable to any fungus that can be grown *in vitro*.

Fungal transformation typically involves use of regenerative protoplasts, identification of selectable marker genes, and stable integration of the transforming DNA into the chromosome of the organism (Fincham, 1989; Ballance, 1991). Both polyethylene glycol (4000–8000 mol. wt. PEG) and calcium ions (10–50 mM $CaCl_2$) are important ingredients of a protoplast transformation mixture. Vectors developed for fungal transformation are typically unable to replicate autonomously in the target fungus, since the transformation event is expected to be integrative and, once integrated, this transforming DNA is stably inherited as a chromosomal or genomic insert.

Various fungal species have been transformed following incubation of protoplasts with DNA that encodes genes for selectable traits, such as required metabolite synthesis (Yelton *et al.*, 1984), antibiotic or pesticide resistance (Punt *et al.*, 1987; Avalos *et al.*, 1989; Henson *et al.*, 1989; van Engelenberg *et al.*, 1989), or utilization of some novel nitrogen source (e.g. acetamide) (Tilburn *et al.*, 1983; Kelly and Hynes, 1985; Geisen and Leistner, 1989). Transformants are selected by virtue of some requisite function provided by the introduced gene allowing growth against a background of stasis or non-growth. Transformation frequencies are generally in the order of 10–100 integrative transformants per μg added DNA. Expression of the introduced gene depends variously on the number and position of inserts and upon appropriate control mechanisms (e.g. promoter and terminator signals, respectively, at the 5′ and 3′ ends of the introduced gene).

A number of basidiomycetes have been transformed using genes that complement auxotrophic mutations. Tryptophan-requiring auxotrophs of both *Coprinus cinereus* and *Schizophyllum commune* were transformed to prototrophy using homologous *trpC* genes and regulator signals (Binninger *et al.*, 1986; Munroz-Rivas *et al.*, 1986). An adenine auxotroph of *Phanerochaete chrysosporium* was transformed with an adenine biosynthetic enzyme gene from *S. commune* (Alic *et al.*, 1989). Transformants of the basidiomycetous yeast, *Rhodosporidium toruloides*, were selected for the ability to use phenylalanine as the sole nitrogen source by virtue of the expression of the homologous phenylalanine ammonia-lyase gene (Tully and Gilbert, 1985). More recently, an ectomycorrhizal basidiomycete, *Laccaria laccata*, has been transformed with a bacterial-derived gene conferring resistance to the aminoglycoside antibiotic hygromycin B (Barrett *et al.*, 1990). This result provides the first evidence for transformation of a mycorrhizal fungus.

These studies on transformation of basidiomycetes illustrate several features typical of fungal transformation systems. Transforming DNA is integrated into high molecular weight (chromosomal) DNA and stably maintained within the fungal genome. There is some interspecific expression of genes, as *P. chrysosporium* will express the *S. commune ade* gene (Alic *et al.*, 1989). This represents a conservative example of heterologous gene expression involving fungi that belong to the same taxonomic class. The successful transformation of *L. laccata* employed promoter and terminator signals derived from an unrelated fungus, *Aspergillus nidulans*. This result is especially fortuitous, as it indicates that control signals of ascomycetous origin, and perhaps as well from other eukaryotic sources, are able to function in a basidiomycete.

The extent of heterologous promoter recognition among filamentous fungi (Turgeon *et al.*, 1987; Saunders *et al.*, 1989) and the sequence requirements for transcription termination and polyadenylation of messenger RNA transcripts in these organisms are virtually unknown (Ballance, 1991).

C. Prospects for molecular genetic manipulation

Results with the *L. laccata* system for transformation encourage research to transform this fungus as well as other ectomycorrhizal fungi with specific genes useful to improve their symbiotic benefit to plants.

The intention of such research would be to introduce useful genes into the root-associated fungus, rather than into the host plant genome. It is anticipated that relevant gene expression at the level of the fungus on the surface of the root will exert more direct influence on rhizoplane activities. Moreover, since the growth of the mycorrhizal fungus extends beyond the immediate root zone, it may be possible to influence the total rhizosphere and surrounding soil more generally through transgenic functions introduced into the fungus. Candidate genes from plant or microbial sources for these transformation studies may involve genes encoding production of phytohormones that stimulate root and mycorrhizal development, genes encoding metabolites that foster osmotolerance or disease resistance, genes for substances repellent to insect or nematode predation, genes for resistance to pesticides, or genes encoding enzymes, such as phosphatase or nitrate reductase, that influence nutrient availability or alleviate various edaphic stress factors.

Transgenic fungal inocula may effectively substitute for fertilizers and/or pesticides during forest seedling production, providing a biotechnological alternative to expensive chemical treatment in the control of biotic and abiotic stress phenomena.

In addition to engineering novel gene activities into an ectomycorrhizal fungus, it should be possible to disrupt by transformation undesirable genes resident in the fungal genome. Procedurally, this has been accomplished with *A. nidulans* by homologous recombination using a defective gene or one that is interrupted by some selectable marker gene insert (Miller *et al.*, 1985).

The introduction of novel genes or the elimination of unwanted genes in ectomycorrhizal fungi by transformation is an area of research that promises to provide genetically superior fungal inocula for improved practices for forest nursery production and/or tree seedling outplanting. Success of this approach will depend on interdisciplinary research efforts and not just upon molecular genetic manipulations in the laboratory. The preparation of transgenic inocula in sufficient quantity and of high quality and a careful assessment of ectomycorrhizal formation following inoculation must accompany transformation initiatives. These initiatives must also address, as part of an evaluation for overall benefit, aspects of risk assessment as related to release of genetically engineered organisms to the environment (Russell and Gruber, 1987; Tiedje *et al.*, 1989).

II. Protoplasts and vectors for transformation

Prerequisite to the development of a transformation system is the preparation of propagules from a cell line in sufficient numbers and with acquired competence for uptake of the added DNA and, ultimately, for efficient phenotypic expression of some rare transforming event. Also important in the development of a transformation system is the design of the vector or added DNA so as to contain at least one selectable marker gene for recognition and recovery of this rare transforming event, The introduced gene(s) that are targeted for expression by transformants should be bracketed with control signals appropriate to the cell line under investigation.

Regenerative protoplasts or cell wall-less propagules, derived experimentally from either intact fungal cells or their mitotic spore states, are commonly employed in the development of fungal transformation systems. Moreover, vectors for such experiments are generally designed for integrative, rather than replicative, transformation. These vectors often exhibit a wide latitude for heterologous gene expression (Fincham, 1989; Timberlake and Marshall, 1989).

A. Protoplast formation

Protoplasts have been liberated *in vitro* from a wide variety of fungi (Davis, 1985; Peberdy and Ferenczy, 1985), including several ectomycorrhizal basidiomycetes (Kropp and Fortin, 1986; Hebraud and Fevre, 1988; Barrett *et al.*, 1989). The absence of mitotic spores from ectomycorrhizal species requires that mycelial fragments serve as starting material for protoplast formation with these fungi.

Commercially available enzyme mixtures, such as Novozyme 234 or Caylase C3, are generally effective for protoplast generation with fungi. These lytic enzyme preparations include principally β-1,3 and β-1,4 glucanases, α-1,3 and α-1,4 glucanases, as well as chitinase and glucuronidase activities. Preparations may also contain proteinase activities detrimental to stability of liberated protoplasts. Care must be taken in evaluating incubation in the enzyme mixtures for efficiency in yielding stable and regenerative protoplasts.

Optimal conditions for protoplast formation and maintenance, involving enzyme concentration/exposure time, culture age, pH and osmotic conditions of buffers/media, vary with individual species. Once the cell wall is removed, an osmotic stabilizer, such as 0.6 M KCl, 1.2 M sorbitol or 0.5 M mannitol, must be present both in maintenance buffers and in regeneration media, at least until some resynthesis of cell wall material has occurred (Tilburn *et al.*, 1983; Peberdy and Ferenczy, 1985; Barrett *et al.*, 1989). Regeneration rates also vary and among ectomycorrhizal fungi are low, often only 1–5%, with yields of the order of 10^7 viable protoplast regenerates per gram fresh weight of mycelia (Kropp and Fortin, 1986; Hebraud and Fevre, 1988; Barrett *et al.*, 1989).

The ability to form and regenerate large numbers ($> 10^6$) of protoplasts is an incentive for attempting transformation with any fungus, provided a suitable vector employing a selectable marker gene is available. To date, a few species of the genera *Laccaria* and *Hebeloma* are the only ectomycorrhizal fungi among several evaluated for protoplast yield/regeneration with potential for such experimentation (Kropp and Fortin, 1986; Hebraud and Fevre, 1988; Barrett *et al.*, 1989). The only vectors presently available for such experimentation, however, are those developed for studies involving other filamentous fungi, primarily ascomycetous species.

B. Representative vectors

The vector pAN7-1, developed originally for studies with *Aspergillus nidulans*, is representative of the type of vector used in fungal transformation studies (Punt *et al.*, 1987). This vector was adopted by Barrett and co-workers (1990) for the successful transformation of *L. laccata*. Transformation was based on positive selection for resistance to hygromycin B (HmB) using the *Escherichia coli* aminocyclitol phosphotransferase (*hpt*) gene bracketed by an *Aspergillus nidulans* glyceraldehyde-3-phosphate dehydrogenase (*gpd*) promoter and the transcription terminator region of the *A. nidulans* tryptophan synthetase (*trpC*) gene.

The pAN7-1 is 6.7 kbp in size and, in addition to the *hpt* gene and adjacent control signals, contains sequences derived from the bacterial plasmid pBR322. The latter sequences are required for replication and recovery of pAN7-1 DNA from *E. coli* clones and include a bacterial origin of replication (*ori*) and a selectable marker gene for ampicillin (Ap) resistance (Maniatis *et al.*, 1982). The pAN7-1 DNA obtained from *E. coli* clones is normally purified by caesium chloride/ethidium bromide density gradient centrifugation for use in transformation experiments. A detailed map of pAN7-1 and related plasmids is provided by Punt and co-authors (1987).

The successful application of pAN7-1 to *L. laccata* transformation encourages the development of comparable vectors either with other selectable marker genes or with reporter genes or genes for still other traits. Reporter genes that can be easily and quantitatively assayed for function could be effectively adopted to study promoter specificity/activity in transformed *L. laccata*. Two reporter genes of promise in this regard are the *E. coli uidA* gene, coding for β-glucuronidase (GUS) activity (Jefferson, 1987; Jefferson *et al.*, 1987; Roberts *et al.*, 1989), and the bacterial-derived A-B *lux* fusion gene, coding for luciferase activity (Shaw and Kado, 1986; Carmi *et al.*, 1987). Such genes might be introduced onto a vector in tandem with a selectable marker gene or placed upon a separate vector, the latter suitable for a cotransformation experiment involving a second vector carrying a dominant selectable marker gene. Cotransformation frequencies, using equal molar ratios of selectable and unselectable vectors, have proven to be remarkably high (e.g. 20–44%), and procedurally cotransformation may be generally possible among filamentous fungi (Roberts *et al.*, 1989).

III. Transformation protocols

While transformation protocols for filamentous fungi generally employ protoplasts and the use of a PEG–$CaCl_2$ regimen for DNA uptake, procedural details vary for each transformation system. Details that relate to the optimization of DNA-mediated transformation of *L.*

laccata protoplasts are discussed below, and alternative procedures for transformation of this fungal system and perhaps other species of ectomycorrhizal fungi are described.

A. Polyethylene glycol–calcium chloride treatment

The initial mycelium of *L. laccata*, grown in MMN medium (Marx, 1969), is macerated in a blender for 30 s and added to a 250 ml flask with fresh medium (equal vol.) and allowed to grow for 3–4 days. Resultant mycelia are recovered by centrifugation and washed once with MMC (Barrett *et al.*, 1990). The packed volume of mycelia is usually 2–4 ml. This mycelial pellet is suspended in 15 ml enzyme mix containing 5–7 mg ml^{-1} Novozyme 234 and 1 mg ml^{-1} bovine serum albumin (BSA) in MMC. This enzyme mix is filter sterilized before addition to the mycelia. The mycelia are incubated in the enzyme mix for 2–4 h at 31 °C with gentle shaking (125 rpm).

Novozyme 234 is a lysing enzyme preparation from *Trichoderma harzianum* sold by Novo Industries. Comparable enzyme preparations from Sigma Chemical or Cayla also work well and may be substituted.

Liberated protoplasts are separated from hyphae by filtering the slurry through cotton. A wad of cotton is tamped into the bottom of a 10 ml plastic syringe. The cotton and the syringe are autoclaved before use. The cotton forms a loose filter 1 ml deep. Wet the cotton with MMC to make sure it stays in place and pour the protoplast/hyphal slurry through, letting the clear solution drip through by gravity. Centrifuge the filtrate in a table-top centrifuge at 2500 rpm to pellet the protoplasts. Wash the protoplasts twice in MMC by successive centrifugations.

Resuspend the protoplast pellet in MMC and count the protoplasts on a haematocytometer. A pellet resuspended in 1 ml usually contains $1–2 \times 10^8$ protoplasts. If there are more, dilute with MMC; if there are fewer, pellet again and resuspend to achieve $1–2 \times 10^{-8}$ protoplasts ml^{-1}.

In a sterile microfuge tube, incubate 10–50 μg of pAN7-1 DNA with 50 μg heparin for 10–15 min. Add 100 μl ($1–2 \times 10^{-7}$) protoplasts and 50 μl PEG solution (Yelton *et al.*, 1984). Mix well. Incubate on ice 30–45 min. While the protoplasts are incubating, add 1 ml of the PEG solution to a sterile 15 ml tube. When the protoplast incubation period is over, transfer the protoplasts to the 15 ml tube. This step facilitates handling because the PEG solution is difficult to pipette and it is more expedient to transfer the protoplasts into the 1 ml of PEG than to pipette the PEG into the microfuge tube. Mix well and continue the incubation for 10–15 min at 22 °C.

Wash the protoplasts by centrifugation with regeneration medium (Kitamoto *et al.*, 1988) modified to contain only 0.4 M sucrose. The usual regeneràtion concentration of 0.6 M is probably too high because of the high concentration of PEG. Resuspend the protoplasts in regeneration medium containing 0.6 M sucrose and allow the protoplasts to recover in this liquid medium for 5–7 days. This allows the protoplasts to regenerate and permits early expression of the *hpt* gene in transformants before plating.

Plate the regenerates on MMN medium containing 2% agar and 200 $\mu g\,ml^{-1}$ hygromycin B (HmB). Plates are incubated at 22 °C in the dark and observed after at least 1–2 weeks.

By this protocol between 5 and 50 putative transformants are recovered. Transfer colonies that grow on 200 $\mu g\,ml^{-1}$ HmB to fresh plates with 200 $\mu\,ml^{-1}$ and then to MMN agar medium containing higher levels of the drug. Many of the isolates fail to grow at higher concentrations of the antibiotic. Isolates that grow on 500 and 1000 $\mu g\,ml^{-1}$ are selected for further study. Hygromycin B is also a mutagen and some isolates recovered on 1000 $\mu g\,ml^{1}$ of HmB contain no plasmid sequences when examined by Southern (1980) or dot blot (Mohr, 1989) analyses of fungal DNA probed with ^{32}P-labelled pAN7-1.

Genomic DNA is isolated from selected colonies by growing gram fresh weight quantities of mycelium in MMN medium with HmB or by growing colonies on 0.45 μm filters (Millipore, nitrocellulose) placed on top of agar medium in a Petri dish. The colonies peel off the nitrocellulose and are free from agar contamination. The mycelium must be frozen and lyophilized, and is then very sensitive to lysis in sodium dodecyl sulphate (SDS). Use 1% SDS, 0.05 M EDTA pH 8, 25 $\mu g\,ml^{-1}$ proteinase to lyse the cells.

Either the procedure of Zolan and Pukkila (1986) is followed or extractor columns (Molecular Biosystems, Inc.) are used to purify the fungal DNA. The DNA usually forms a flocculant rather than a spoolable precipitate in cold ethanol, but it seems to be of high molecular weight upon analysis by electrophoresis in 0.8% agarose gels (Barrett et al., 1990).

1. Media and buffers

MMN medium (Marx, 1969)

0.3%	malt extract
1.0%	dextrose

0.0025%	$(NH_4)_2HPO_4$
0.05%	KH_2PO_4 (pH 5.6)
0.015%	$MgSO_4$
0.005%	$CaCl_2$
0.0025%	NaCl
25 μg litre^{-1}	thiamine-HCl
1.2 μg litre^{-1}	$FeCl_3$

MMC buffer (Barrett *et al.*, 1990)

0.5 M	mannitol
50.0 mM	maleic acid (pH 5.4)
50.0 mM	$CaCl_2$

Regeneration medium (Kitamoto *et al.*, 1988)

0.1%	peptone
0.1%	soytone
0.1%	KH_2PO_4
0.03%	$MgSO_4$
0.05%	yeast extract
0.6 M	sucrose

PEG solution (Yelton *et al.*, 1984)

60%	polyethelene glycol 4000
10 mM	$CaCl_2$
10 mM	Tris (pH 7.4)

B. Other procedures

The development of an efficient transformation protocol for *L. laccata* suggests that other ectomycorrhizal fungi can be transformed, provided sufficient numbers of viable protoplasts can be formed. Incubation of intact hyphae in lithium ion-containing buffers (Dhawale *et al.*, 1984; Binninger *et al.*, 1986; Bej *et al.*, 1989) has provided an alternative, although not always efficient, means of introducing DNA into fungal cells. This or other procedures may prove feasible for transforming species of ectomycorrhizal fungi not suited for efficient protoplast formation.

One strategy for transformation used recently with fungal systems is electroporation (Delmore, 1989; Richey *et al.*, 1989; Goldman *et al.*,

1990). Electroporation is non-destructive permeabilization of biological membranes in short-duration, high-amplitude electric fields. This technique may provide a method of introducing DNA into species of fungi not amenable to or inefficient for transformation by the traditional calcium ion and PEG treatments. Yet another technique, a ballistic process involving DNA bound to microprojectiles (Sanford, 1988), can introduce DNA into cells by high velocity propulsion in a partial vacuum. This procedure has been effective in transformation studies involving mammalian cells as well as intact plant cells (Sanford, 1988, 1990) and might be especially useful if adapted to fungal systems where cell wall removal and efficient protoplast generation have not been successful.

IV. Concluding remarks

Current research with ectomycorrhizal fungi involves the use of protoplasts as starting material for genetic manipulation through DNA-mediated transformation (Barrett *et al.*, 1990). Transformation procedures are expected to extend more traditional approaches to genetic study and thus enhance the potential for selective breeding and strain improvement of these fungi. Through the use of protoplasts and procedures of molecular genetics it should be possible to decipher the complex genetic behaviour of ectomycorrhizal formation and to identify traits which affect that symbiosis. The goal of such research is to develop improved fungal strains for use as inocula in sylvicultural practices related to tree seedling production, forest productivity and reafforestation. To accomplish this goal it will be necessary to obtain fundamental information on relevant genes and their expression in order to carry out genetic manipulations leading to improved ectomycorrhiza. In considering control strategies to improve symbiosis, it will be important to avoid non-target effects that could disrupt the balance among rhizosphere components. The overall intention of the research is to optimize symbiosis through genetic manipulation of the fungal component without impairment of components of rhizosphere ecology that may be supportive of that symbiosis.

Regardless of the specific gene(s) introduced, and perhaps in some cases impaired, by transformation, transformed phenotypes of ectomycorrhizal fungi should be stable due to integration of the foreign DNA into the fungal chromosome. This stability adds to the desirability of developing transgenic fungi as inocula to improve symbioses in order to ameliorate stress factors that adversely affect plants.

References

Alic, M., Kornegay, J. R., Pribnow, D. and Gold, M. H. (1989). *Appl. Environ. Microbiol.* **55**, 406–411.

Avalos, J., Geever, R. F. and Case, M. E. (1989). *Curr. Genet.* **16**, 369–372

Ballance, D. J. (1991). In *Molecular Industrial Mycology* (S.A. Leong and R. M. Berka, eds), pp. 1–29. Marcel Dekker, New York.

Barrett, V., Lemke, P. A. and Dixon, R. K. (1989). *Appl. Microbiol. Biotechnol*. **30**, 381–387.

Barrett, V., Dixon, R. K. and Lemke, P. A. (1990). *Appl. Microbiol. Biotechnol*. **33**, 313–316.

Bej, A. K. and Perlin, M. H. (1989). *Gene* **80**, 171–176.

Binninger, D. M., Skrzynia, C., Pukkila, P. J. and Casselton, L. A. (1986). *EMBO J*. **6**, 835–840.

Carmi, O. A., Stewart, G. S. A. B., Ulitzer, S. and Kuhn, J. (1987). *Bacteriol.* **169**, 2165–2170.

Curl, E. A. and Truelove, B. (1986). *The Rhizosphere*. Springer-Verlag, Berlin.

Davis, B. (1985). In *Fungel Protoplasts* (J. F. Peberdy and L. Ferenczy, eds), pp. 45–71. Marcel Dekker, New York.

Delmore, E. (1989). *Appl. Environ. Microbiol.* **55**, 2242–2246.

Dhawale, S. S., Paietta, J. V. and Marzluf, G. A. (1984). *Curr. Genet*. **8**, 77–79.

Fincham, J. R. S. (1989). *Microbiol. Rev*. **53**, 148–170.

Fries, N. (1983). *Mycologia* **75**, 221–227.

Giesen, R. and Leistner, L. (1989). *Curr. Genet.* **15**, 307–309.

Goldman, G. H., Van Montagu, M. and Herrera-Estrella, A. (1990). *Curr. Genet.* **17**, 169–174.

Harley, J. L. and Smith, S. E. (1983). *Mycorrhizal Symbiosis.* Academic Press, New York.

Hebraud, M. and Fevre, M. (1988). *Can. J. Microbiol.* **34**, 157–161.

Henson, J. M., Blake, N. K. and Pilgeram, A. L. (1989). *Curr. Genet*. **14**, 113–118.

Hutchinson, L. J. (1989). *Mycologia* **81**, 587–594.

Jefferson, R. A. (1987). *Plant Mol. Biol. Rep.* **5**, 387–405.

Jefferson, R. A., Kavanagh, A. A. and Bevan, M. W. (1987). *EMBO J.* **6**, 3901–3907.

Kelly, J. M. and Hynes, M. J. (1985). *EMBO J*. **4**, 475–479.

Kitamoto, Y., Mori, N., Yamamoto, M., Ohiwa, T. and Ichikawa, Y. (1988). *Appl. Microbiol. Biotechnol.* **28**, 445–450.

Kropp, B. R. (1988). In *Canadian Workshop on Mycorrhizae in Forestry* (M. Lalonde and Y. Piché, eds), pp. 131–133. Université Laval, Québec, Canada.

Kropp, B. R. and Fortin, J. A. (1986). *Can. J. Bot.* **64**, 1224–1226.

Leong, S. A. and Berka, R. M. (Eds) (1991). *Molecular Industrial Mycology*. Marcel Dekker, New York.

Maniatis, T., Fritsch, E. F. and Sambrook, J. (1982). *Molecular Cloning: A Laboratory Manual*. Cold Spring Harbour, New York.

Marx, D. H. (1969). *Phytopathology* **59**, 153–163.

Miller, B. L., Miller, K. Y. and Timberlake, W. E. (1985). *Mol. Cell. Biol.*, **5**, 1714–1721.

Mohr, G. (1989). *Appl. Microbiol. Biotechnical.* **30**, 371–374.
Munroz-Rivas, A. M., Specht, C. A., Drummond, B. J., Froelinger, E., Novotny, C. P. and Ullrich, R. C. (1986). *Mol. Gen. Genet.* **205**, 103–106.
Peberdy, J. F. and Ferenczy, L. (Eds), (1985). *Fungal Protoplasts: Applications in Biochemistry and Genetics*. Marcel Dekker, New York.
Punt, P. J., Oliver, R. P., Dingemanse, M. A., Powels, P. H. and van Hondel, C. A. M. J. J. (1987). *Gene* **56**, 117–124.
Richey, M. G., Marek, E. T., Schardl, C. L. and Smith, D. A. (1989). *Phytopathology* **79**, 844–847.
Roberts, L. N., Oliver, R. P., Punt, P. J. and van den Hondel, C. A. M. J. J. (1989). *Curr. Genet.* **15**, 177–180.
Russel, M. and Gruber, M. (1987). *Science* **236**, 286–290.
Sanford, J. C. (1988). *Trends Biotechnol.* **6**, 299–309.
Sanford, J. C. (1990). *Physiol. Plant* **79**, 206–209.
Saunders, G., Pickett, T. M., Tiute, M. F. and Ward, M. (1989). *Trends Biotechnol.* **7**, 283–287.
Schneck, N. C. (Ed), (1982). *Methods and Principles of Mycorrhizal Research*. American Phytopathological Society, St Paul, MN.
Shaw, J. J. and Kado, C. I. (1986). *Biotechnology* **4**, 560–564.
Southern, E. (1980). *Meth. Enzymol.* **69**, 152–179.
Tiedje, J. M., Colwell, R. K., Grossman, Y. L., Hodson, R. E., Lenski, R. E., Mack, R. N. and Regan, P. J. (1989). *Ecology* **70**, 298–315.
Tilburn, J., Scazzocchio, C., Taylor, C. G., Zabicky-Zissman, J. H., Mockington, R. A. and Davies, R. W. (1983). *Gene* **26**, 205–221.
Timberlake, W. E. and Marshall, M. A. (1989). *Science* **244**, 1313–1317.
Trappe, J. M. (1962). *Bot. Rev.* **28**, 538–606.
Tully, M. and Gilbert, H. (1985). *Gene* **36**, 235–240.
Turgeon, B. G., Garber, R. C. and Yoder, O. C. (1987). *Mol. Cell. Biol.* **7**, 3297–3305.
Van Engelenberg, R., Smit, R., Godsen, T., van den Broek, I.I. and Tudzynski, P. (1989). *Appl. Microbiol. Biotechnol.* **30**, 364–370.
Yelton, M. M., Hamer, J. E. and Timberlake, W. E. (1984). *Proc. Natl. Acad. Sci. USA* **81**, 1470–1474.
Zak, B. (1973). In *Ectomycorrhizae, Their Ecology and Physiology* (G. C. Marks and T. T. Kozlowski, eds), pp. 43–78. Academic Press, New York.
Zolan, M. E. and Pukkila, P. J. (1986). *Mol. Cell. Biol.* **6**, 195–200.

11

Principles of Use of Radioisotopes in Mycorrhizal Studies

P. B. TINKER

Natural Environment Research Council, Terrestrial and Freshwater Science Directorate, Polaris House, North Star Avenue, Swindon, SN2 1EU, UK

M. J. JONES and D. M. DURALL

Natural Environment Research Council, Plant Mycorrhizal Unit, Department of Plant Sciences, Parks Road, Oxford, OX1 3PF, UK

I. Introduction

The purpose of this chapter is to describe the place of the use of radioisotopes in studies of mycorrhizal symbioses. It is not intended to consider the details of the use of these techniques at bench level, but the great majority of the papers which report such work are referred to here, and the details of the techniques can be obtained by referring to these original publications.

A symbiotic association is by definition more complex than a single organism, and a particular component of this complexity is the fluxes of material between the two partners in the symbiosis. It is no accident

METHODS IN MICROBIOLOGY
VOLUME 23 ISBN 0-12-521523-1

Copyright © 1991 by Academic Press Limited
All rights of reproduction in any form reserved

that much of the use of radioisotopes has been directed towards understanding these interchanges, because the interface region in mycorrhiza is almost impossible to access directly. This applies both to the vesicular-arbuscular mycorrhiza, where it is believed that the most active part of the interface is in the arbuscules, and to ectomycorrhiza, where it is believed to be in the Hartig net region. In symbiotic associations there are few general techniques for investigating this interface region or the interfacial fluxes. One of them is the "inhibition" technique developed by Drew and Smith (1967) which uses radioactive isotopes to identify the compounds which are being exchanged between the two symbionts in lichens. It is rarely possible to apply this technique to mycorrhizal systems, because of the difficulty of physical access to the interface. Work with a few ectomycorrhiza has used physical separation to determine phosphate and carbon uptake, transfer and storage (Harley and Brierley, 1955; Lewis and Harley, 1965). Since the roots have been excised from the plant, however, these experiments have the disadvantage of not dealing with the entire symbiotic organism. In the vesicular-arbuscular mycorrhiza and most of the ectomycorrhiza, physical separation is extremely difficult. Thus, most information about the interchange between the symbiotic partners has been obtained by direct measurement of the transfer of radioisotopes: mineral nutrients from the fungus to the host plant, and carbon from the host to the fungus. Our knowledge of the function of the mycorrhizal symbiosis, in terms of the transfer and allocation of materials, has been enormously advanced by the use of radioisotopes. Without these, we would be largely confined to descriptive studies.

II. Carbon allocation and fluxes

It was believed from a very early stage that the fungal symbiont in mycorrhizal associations received its carbon nutrition from the host plant (Frank, 1885). Most of the ectomycorrhiza and ericoid mycorrhizal fungi have the potential to meet their carbon needs from external carbon sources when grown non-symbiotically in the laboratory. Nevertheless, it is generally agreed that they receive the majority of their carbon nutrition from the host plant while grown symbiotically. Since vesicular-arbuscular mycorrhiza have not yet been grown in pure culture it has been assumed that they receive their carbon nutrition solely from the host plant. In contrast, under natural conditions, the Orchidaceae are obligate symbionts with mycorrhizal fungi, and during the stage when orchids lack green leaves, the net movement of carbon compounds is from substrate through the fungus and into the host (Hadley and Purves, 1974; Purves and Hadley, 1975).

Direct proof of carbon transfer from host plant to fungus in both vesicular-arbuscular and ectomycorrhiza has been obtained by the use of ^{14}C (Melin and Nilson, 1957; Cox *et al.*, 1975; Cox and Tinker, 1976; Ho and Trappe, 1973) The simplest way of showing transfer is by allowing fixation of ^{14}C-labelled CO_2, and then using autoradiography to prove the presence of ^{14}C in various parts of the fungus. The work by Cox *et al.* (1975) was notable for showing the very high radioactivity concentrated in the vesicles of the vesicular-arbuscular mycorrhiza. This type of study can give some information about the speed of transfer, and it is obvious that significant movement takes place within a few hours. Recently, dynamic measurements of carbon translocation to roots in both non-mycorrhizal and mycorrhizal plants have been made with ^{11}C (Fares *et al.*, 1988; Wang *et al.*, 1989). Using this method, the translocation of nanomoles of carbon from shoot to root can be detected. These researchers suggest that the speed of carbon transport may be measured more precisely with ^{11}C than is possible with ^{14}C. They found that ^{11}C was translocated from the labelling point to the roots at a speed of 0.51–0.98 $cm\,min^{-1}$, though these results could depend upon the sensitivity of the detection system. One drawback of using ^{11}C is that because of its short half-life (20.3 min) experiments must be performed in the vicinity of a reactor.

Studies using ^{14}C and ^{11}C cannot prove that all fungal carbon comes from the higher plant, but this is assumed to be so for the vesicular-arbuscular mycorrhiza because of its obligate symbiotic nature. For the ectomycorrhiza, this cannot be assumed. Todd (1979) showed that ectomycorrhizal fungi in symbiotic association can utilize carbon from ^{14}C-labelled organic matter substrates, but it is not known whether this provides a substantial source of carbon for the symbiotic association. Thus, further experiments are needed to determine the amount of carbon transfer from soil organic matter via hyphae to ectomycorrhizal roots.

Carbon-14 has also been used to quantify carbon flow within the extra-matrical hyphal network. Miller *et al.* (1989) examined portions of hyphae growing in a separate compartment but still in symbiosis with the host. Thus, they were able to express ^{14}C in extra-matrical hyphae as a specific activity and subsequently to calculate the activity for the entire hyphal network.

In soil, the mycelial network of a particular mycorrhizal fungus can be connected directly to the roots of two or more plants, thus forming hyphal links between their mycorrhizal roots (Newman, 1988). Carbon transfers via hyphal links have been demonstrated using ^{14}C autoradiography (Finlay and Read, 1986a). If carbon transfers via these links are prevalent in the field then this would have a major effect on carbon cycling as well as on total ecosystem functioning (Newman, 1988).

Radioactive isotopes of carbon have been crucial in estimating the carbon cost to the plant of a mycorrhizal association. Field studies, using non-isotopic techniques, have suggested that very large fractions, up to 50%, of total carbon fixed by ectomycorrhizal trees is eventually utilized by the mycorrhizal fungus (Fogel and Hunt, 1979). If this were true, it would have important implications for our understanding of the total carbon budget of forests, the efficiency with which dry matter is formed, and the inputs of carbon into the soil ecosystem. Given the high variability of field conditions, most researchers have preferred to estimate the cost to the plant of a mycorrhizal association by using ^{14}C-labelled plants grown under controlled conditions.

Such studies have been carried out for ectomycorrhiza (Reid *et al.*, 1983) and vesicular-arbuscular mycorrhiza (Pang and Paul, 1980; Kucey and Paul, 1982; Snellgrove *et al.*, 1982; Harris *et al.*, 1985). All of these studies have used the pulse-chase technique. This approach involves labelling the host by giving a short pulse of $^{14}CO_2$. This is followed by a chase period where the plant is exposed to a $^{12}CO_2$ environment in which ample time is allowed for the transfer of ^{14}C to the fungal symbiont. All fixed ^{14}C, including that respired from shoots and roots and that deposited in the soil, is measured.

The overall results of these studies are sufficiently consistent to allow some conclusions to be drawn for vesicular-arbuscular mycorrhizal plants. All vesicular-arbuscular mycorrhiza studies show a difference of 5–10% between mycorrhizal and non-mycorrhizal plants in the percentage of the total carbon fixed which is allocated to the below-ground system. This appears to be a realistic estimate for the cost to these plants of the mycorrhizal symbiosis. Nevertheless, it is difficult to extrapolate these results to the field situation because these ^{14}C techniques are only suitable for use with small, often young plants, in somewhat artificial circumstances.

III. Carbon loss into soil

The total input of carbon into soil from a root system is very large, and is one of the basic supports of the soil biological systems. This input, referred to as rhizodeposition (Rovira *et al.*, 1979), consists of carbon passed via roots into the soil by root and microbial respiration, root exudation, sloughing off of roots, death of roots, etc. This is very difficult to measure directly in the field, since root death, root exudation and turnover rates are not easily nor accurately determined. Less ambiguous information on the plants' input of carbon to soil via roots has been obtained by the use of continuous ^{14}C labelling in sealed environmental chambers (Helal and Sauerbeck, 1983; Whipps and Lynch, 1985). These studies have fairly consistently shown that percentages of the order of 20–40% of total carbon fixed is passed into the soil by way of rhizodeposition. The evidence that a larger fraction of the total fixed plant carbon is directed below ground in mycorrhizal plants suggests that total carbon inputs into soil should be larger for these. There is some suggestion from ^{14}C studies, such as those of Snellgrove *et al.* (1982), that more labelled carbon is indeed found in the soil when the plant is mycorrhizal, though the effect is too small to place much reliance upon. It would certainly be expected that the external fungal network, together with spores and sporocarps of vesicular-arbuscular mycorrhiza and ectomycorrhiza, respectively, should result in a higher final deposition rate. Another point of importance is the suggestion from Graham *et al.* (1981) that the carbon from the host used by the mycorrhizal fungus is provided simply by the normal exudation processes. However, estimates of carbon loss to soil, although few in number, do not support this. Based on several non-radioisotopic studies on non-mycorrhizal roots, estimates range from 0.024–1.7% of fixed carbon lost as soluble exudates to soil (Ratnayake *et al.*, 1978; Johnson *et al.*, 1982; Lambers, 1987). These values are at least five times less than those values presented in the previous section, in which the cost to the plant of vesicular-arbuscular mycorrhizal symbiosis was 5–10% of fixed carbon. In a study focusing on mycorrhizal cost using ^{14}C, Snellgrove *et al.* (1982) found no difference between mycorrhizal and non-mycorrhizal plants in terms of ^{14}C-labelled soluble compounds exuded to the soil, but did find a difference between insoluble compounds. There has been a suggestion that a root system that is mycorrhizal has less exudation than that which is non-mycorrhizal but there are no data available to support this. Unfortunately no detailed studies appear to have been made of either exudation or total loss rates of carbon from mycorrhizal versus non-mycorrhizal plants, at a level of detail which would allow us to say whether the formation of the symbiosis alters the total input of carbon to the soil. What is clear is that when detailed studies are performed, the use of radioisotopes will help us to substantiate and quantify carbon deposition to soil by mycorrhizal plants.

IV. Mineral nutrient uptake

A. Mechanism of nutritional benefit by mycorrhiza

It was long suspected that the very visible benefit given to plants by mycorrhizal infection was due to the supply of additional mineral nutrients. To prove this, it was necessary to show that the fungus could absorb mineral nutrients from the growth medium, transport them through its hyphae, and transfer them to the higher plant. The more general approach was also to prove that the mycorrhizal root system was more efficient than the non-mycorrhizal one. i.e. in the same soil it absorbed nutrients at a higher rate per unit amount of root.

Radioisotope techniques have been essential in establishing many of these facts. Absorption, translocation and transfer of mineral nutrients by the external hyphae of mycorrhiza were originally demonstrated in a series of papers by Melin and Nilsson (1950, 1952, 1953, 1955). The authors showed that isotopic phosphorus, nitrogen and calcium could be translocated through the hyphae of ectomycorrhizal fungi to their hosts. The basic system was simple, with the fungal hyphae being led over a barrier from a nutrient medium to the root system. An isotope was then added to the medium, and was subsequently detected in the host. The experiments showed the potential for uptake and transfer, but they were not quantitative and so could not be related to any specified level of improvement in plant nutrition.

The most direct measurements of the latter type were made by Pearson and Tinker (1975), and Cooper and Tinker (1978). In these studies, radioisotopes were supplied to vesicular-arbuscular mycorrhizal fungal hyphae which had grown over a barrier into agar, with the host plant growing in a soil–agar mix on the other side of the barrier. An important aspect of this system was that the entire shoot could be inserted into a whole-plant radioactivity counter. This allowed the non-destructive observation of the rate at which radioactivity appeared in the shoot. With these systems, it was shown firstly that ^{32}P, ^{35}S and ^{65}Zn were taken up by vesicular-arbuscular mycorrhizal hyphae and transferred into the host. Additionally, the rates of uptake were measured, and it was shown that these were in the ratio $P > S > Zn$, as would be expected from the known composition of higher plants. By counting the hyphae crossing the barrier, and measuring their cross-section, estimates could be made of the minimum flux of these nutrients through the hyphae towards the plant. Finally, it was shown that the transfer rate was increased when the host plant was transpiring, which suggested a mass flow component within the fungal hypha, and that cytochalasin, a chemical which inhibits cytoplasmic streaming, brought the uptake to a halt.

Harley and co-workers, (Harley and McCready, 1950, 1952; Harley and Brierley, 1954), showed that excised beech mycorrhiza absorbed more ^{32}P than did non-mycorrhizal roots. Under conditions of continuous supply of phosphorus, 90% of this phosphorus was retained in the fungal mantle. This was measured by excising the mantle from the "core" of root tissue using an ophthalmic scalpel. When the roots were placed in a phosphate-free medium, however, transfer rates from the sheath to the core increased. These basic conclusions with respect to transfer of phosphorus from the fungus to the plant were confirmed by Morrison (1957) using intact plants. He placed mycorrhizal and non-mycorrhizal pine seedlings, which had previously been fed with ^{32}P, into a phosphorus-free medium. He then monitored ^{32}P levels in the shoots over 15 days, using a method similar to that of Pearson and Tinker (1975). He found that, while ^{32}P levels in shoots of non-mycorrhizal plants remained constant, those of mycorrhizal plants increased steadily with time, indicating transfer of ^{32}P from the roots. Thus, in a whole range of studies, radioisotopes have been essential in establishing that mycorrhizal hyphae can absorb mineral nutrients from the external medium and transfer them to the host plant.

The greater efficacy of vesicular-arbuscular mycorrhizal root systems in absorbing phosphorus from the soil was established by Sanders and Tinker (1971, 1973). They showed that inflow (uptake rate per unit length of root), was between three and four times higher for a mycorrhizal root, and that this inflow was much larger than could have arisen from the maximum rate of diffusion of phosphorus to the root surface, as shown by diffusion calculations. Radiotracers were not required in these initial studies, however they have been very useful in distinguishing between various interpretations of the results. By labelling the soil with ^{32}P, it was possible to show that the specific activity of the phosphorus absorbed was equal to that of the soil solution, and similar in both mycorrhizal and non-mycorrhizal plants (Sanders and Tinker 1971). Uptake was therefore from the same isotopically-exchangeable pool (Larsen, 1967). A whole series of subsequent experiments have broadly confirmed the fact that both mycorrhizal and non-mycorrhizal plants obtain their phosphorus supplies from the same isotopically-labelled pool of soil phosphorus (Hayman and Mosse, 1972; Mosse *et al.*, 1973; Powell, 1975; Pichot and Binh, 1976; Owusu-Bennoah and Wild, 1980; Gianinazzi-Pearson *et al.*, 1981). It is important to understand the precise implications of this technique. It is not able to detect shifts between the sorbed and solution phase phosphate which form the two components of the isotopically-exchangeable pool (Tinker, 1975), but it does show that phosphorus has not been "solubilized" from organic phosphates or from mineral phosphates that were not in isotopic equilibrium with the soil solution. If this had occurred, the specific activity of phosphorus in the plant would have been lower than that in the soil solution. Similar concepts can also be used to compare the efficiency of use of different fertilizer types by mycorrhizal and non-mycorrhizal plants.

Other types of experiments have used ^{32}P to support the theory that the increase in phosphorus uptake by mycorrhizal plants is due to absorption of phosphorus by the hyphae from beyond the depletion zones which develop around roots. In most soils. phosphorus is very poorly mobile. Thus any ^{32}P applied is rapidly sorbed to soil solids and so can be used as a marker in experiments investigating the location of phosphorus uptake relative to the root. Using this technique, Hattingh *et al.* (1973) and Rhodes and Gerdemann (1975) clearly confirmed that external vesicular-arbuscular mycorrhizal mycelia can absorb phosphate up to 3–7 cm away from the root. In an ectomycorrhizal system, ^{32}P applied to the cut end of a rhizomorph was transported 40 cm to the root (Finlay and Read, 1986b). In the latter case, autoradiography of the root system growing along a perspex plate was used to visualize transport.

These varied studies with radioisotopes have therefore given rise to the general conclusion that the mycorrhizal function is due to direct uptake by hyphae at a distance from the root, and not by some chemical modification of the soil. The results of Bolan *et al.* (1984) cannot be explained on this basis, but no other convincing explanation has been advanced for them. An experiment by Owusu-Bennoah and Wild (1979), which determined the depletion zone around roots by autoradiographic techniques with ^{32}P, showed that mycorrhizal roots had a slightly wider depletion zone than a non-mycorrhizal root. This would not be expected if the additional inflow to the mycorrhizal root arises from a widely dispersed mycorrhizal network, with the host root continuing to function in the same way as before, as was assumed by Sanders and Tinker (1973) and Sanders *et al.* (1977). More information is obviously required on the distribution of external hyphae in soil. The general conclusion on mycorrhizal function therefore still stands, whilst emphasizing that there is further need for research on the general nutritional properties of mycorrhiza.

B. Other ^{32}P studies

Due to its rapid immobilization in soil, ^{32}P has been used to investigate the localization of phosphorus uptake relative to the distribution of the root systems of species competing for nutrients (Goodman and Collison, 1981; Caldwell *et al.*, 1985; Perry *et al.*, 1989). Perry *et al.* (1989) found that the competitive ability of two tree species to extract ^{32}P from different soil depths was affected by the species and numbers of ectomycorrhizal fungi present on their root systems. Langlois and Fortin (1984), also using field-grown roots, found that pine mycorrhiza varied in their ^{32}P uptake capacity in a predictable way through the year.

One of the most interesting questions regarding mycorrhizal functioning in the field is the importance of hyphal connections between plants. Radioisotopes have been particularly useful in investigating transfer between plants. When two plants are grown in the same soil and ^{32}P is applied to one plant, radioactivity is soon detected in the second plant (Chiariello *et al.*, 1982; Whittingham and Read, 1982; Ritz and Newman, 1984). The amount of ^{32}P transferred to the second plant is often higher when the plants are mycorrhizal and this has been used as evidence for mineral nutrient transfer via mycorrhizal connections (Whittingham and Read, 1982). There are, however, many problems in interpreting this type of result. For example, the presence of radioisotope in the receiver plant should not necessarily be interpreted as confirming net movement of the element concerned. It could merely be due to isotopic exchange between phosphorus pools in the two organisms (Newman, 1988). Alternatively, some exchange processes can be better explained as occurring through loss into the soil by the donor plant, and re-absorption by the recipient plant (Newman and Ritz. 1986). However, the results of other experiments indicate that phosphorus flow between the symbionts is only unidirectional, being from the fungus to the host (Finlay and Read, 1986b). Readers are referred to Newman (1988) for a full treatment of this topic.

C. Other mineral nutrients

Although research efforts have focused primarily on the uptake of phosphorus using ^{32}P, other isotopes such as ^{45}Ca, ^{35}S, and ^{65}Zn have been used as tracers for the uptake of the analogous ions by mycorrhizal plants (Morrison, 1962; Cooper and Tinker, 1978; Swaminathan and Verma, 1979; Ascon-Aguilar *et al.*, 1986). The uptake of still other radioactive elements has been studied specifically because there can be raised levels of these elements in the environment as a result of nuclear activity. Thus vesicular-arbuscular mycorrhizal plants have been shown to absorb more ^{134}Cs, ^{137}Cs, ^{60}Co and ^{90}Sr than non-mycorrhizal plants (Jackson *et al.*, 1973; McGraw *et al.*, 1979; Rogers and Williams 1986). The results of these experiments must, however, be interpreted carefully. If the mycorrhizal plants are larger than the contrasted non-mycorrhizal plants, an increase in the total uptake of any mineral element would not be surprising.

In plant nutrition studies, radioisotopes are often used to interpret the mechanisms of uptake. Inhibitors or low temperatures are applied to determine if, for example, the uptake (or in the case of mycorrhiza, the transfer) of the element requires an input of metabolic energy. This has been found to be the case for ^{32}P transfer between the fungal mantle and the root tissue in excised beech mycorrhizae (Harley and Brierley, 1955). Cooper and Tinker (1978) using ^{32}P, ^{35}S and ^{65}Zn found suggestions of a lag phase in nutrient transfer between a vesicular-arbuscular mycorrhizal fungus and its host, an initial period during which uptake was extremely slow and which could not be explained by considerations of isotopic dilution along the root up into the shoot. It was therefore suggested that there was a need for the induction of some enzyme associated with uptake or transfer, but this matter was not settled. Jones *et al.* (1988) used ^{63}Ni to show that nickel-tolerant mycorrhizal birch exhibited different patterns of metabolic and non-metabolic nickel uptake from those shown by non-mycorrhizal birch. These patterns suggested a mechanism for their differences in tolerance.

Rygiewicz and Bledsoe (1984), using ^{86}Rb as a tracer for potassium, investigated why more potassium was retained in mycorrhizal than in non-mycorrhizal conifer roots. They performed a compartmental analysis and found that inward fluxes across the tonoplast were greater, and that effluxes out of the vacuole were smaller, in the fungal cells of the mycorrhiza than in root cells.

V. Metabolism

Radioisotopes can give greater insight into cellular processes. For example, they can be used to follow the metabolism of certain labelled compounds. but these types of studies have only rarely been performed on mycorrhizal plants. In one of the few examples we are aware of, Miura and Hall (1973) showed that ectomycorrhizal fungi may be able to synthesize cytokinins. They supplied an axenic culture of *Rhizopogon roseolus* with [8-^{14}C]-N^6-2-isopentenyl adenosine and found that the fungus was able to convert this compound to *trans*-ribosylzeatin, an active cytokinin. This is a step in the same pathway as that used in higher plants for cytokinin biosynthesis.

Graham *et al.* (1981) used the common technique of estimating membrane leakiness by measuring ^{86}Rb efflux. They hypothesized that the increased release of soluble carbohydrates and amino acids caused by "leaky" membranes in low phosphorus roots was the reason for the higher levels of mycorrhizal infection in these roots.

VI. Conclusions

In conclusion, radioisotopes have been crucial in establishing some basic principles of mycorrhizal functioning. They can give clear, quantitative results in situations where other types of experiments are forced to rely on tenuous assumptions and conjecture. Most of the studies cited here have been involved in verifying and quantifying the exchange of carbon and mineral nutrients between symbionts. Useful directions for future work include the identification of the individual compounds exchanged. Phosphorus-32 could be used in conjunction with a dissection technique to determine the types of phosphorus compounds transferred from fungus to host. Additionally, more use could be made of techniques developed by plant physiologists and biochemists to study the biochemical pathways of labelled compounds. These techniques could be used to determine whether the functioning of a certain metabolic pathway differs between mycorrhizal and non-mycorrhizal plants. In these ways, the use of radioisotope experiments could be imaginatively expanded to elucidate further more detailed aspects of mycorrhizal physiology.

References

Azcon-Aguilar, C., Gianinazzi-Pearson, V., Fardeau, J. C. and Gianinazzi, S. (1986). *Plant and Soil*, **96**, 3–16.
Bolan, N. S., Robson, A. D., Barrow N. J. and Aylmore L. A. G. (1984). *Biol. Biochem.* **16**, 299–304.
Caldwell, M. M., Eissenstat, D. M., Richards, J. H., and Allen, M. F. (1985). *Science* **229**, 384–386.
Chiariello, N., Hickman, J. C. and Mooney, H. A. (1982). *Science* **217**, 941–943.
Cooper, K. M. and Tinker, P. B. (1978). *New Phytol*. **81**, 43–52.
Cox, G. and Tinker, P. B. (1976). *New Phytol.* **77**, 371–378.

Cox, G., Sanders, F. E., Tinker, P. B. and Wild, J. A. (1975). In *Endomycorrhizas* (F. E. Sanders, B. Mosse and P. B. Tinker, eds) pp. 297–312. Academic Press, London.
Drew. E. A. and Smith D. C. (1967) *New Phytol.* **66**, 379–388.
Fares, Y., Goeschl. J. D., Magnuson. C. E., Scheld, H. W. and Strain, B. R. (1988). *J. Radioanal. Nucl. Chem.* **124**, 103–122.
Finlay, R. and Read, D. J. (1986a). *New Phytol.* **103**, 145–156.
Finlay, R. D. and Read, D. J. (1986b). *New Phytol.* **103**, 157–163.
Fogel. R. and Hunt, G. (1979). *Can. J. For. Res.* **9**, 245–256.
Frank, B. (1885). *Deutsch. Bot. Gesellschaft.* **3**, 128–145 (translation by J. M. Trappe). In *6th North American Conference on Mycorrhizae*, Bend, Oregon 1984 (R. Molina, ed.) pp. 18–25.
Gianinazzi-Pearson, V., Fardeau, J. C., Asimi, S. and Gianinazzi, S. (1981). *Physiol. Veg.* **19**, 33–43.
Goodman, P. J. and Collison. M. (1981). *Ann. Appl. Biol.* **98**, 499–506.
Graham, J. H., Leonard, R. T. and Menge, J. A. (1981). *Plant Physiol.* **68**, 548–552.
Hadley, G. and Purves, S. (1974). *New Phytol.* **73**, 475–482.
Harley, J. L. and Brierley, J. K. (1954). *New Phytol.* **53**, 240–252.
Harley, J. L. and Brierley, J. K. (1955). *New Phytol.* **54**, 296–301.
Harley, J. L. and McCready, C. C.(1950). *New Phytol.* **49**, 388–397.
Harley, J. L. and McCready, C. C. (1952). *New Phytol.* **51**, 56–64.
Harris, D., Pacovsky, R. S. and Paul, E. A. (1985). *New Phytol.* **101**, 427–440.
Hattingh, M. J., Gray, L. E. and Gerdemann, J. W. (1973). *Soil Sci.* **116**, 383–387.
Hayman, D. S. and Mosse, B. (1972). *New Phytol.* **71**, 41–47.
Helal, H. M. and Sauerbeck, D. R. (1983). *Soil Biol.* **15**, 223–225.
Ho, I. and Trappe, J. M. (1973). *Nature* **244**, 30–31.
Jackson, N. E., Miller, R. H. and Franklin, R. E. (1973). *Soil Biol. Biochem.* **5**, 205–212.
Johnson, C. R., Graham, J. H., Leonard, R. T. and Menge, J. A. (1982). *New Phytol.* **90**, 665–669.
Jones, M. D., Dainty, J. and Hutchinson, T. C. (1988). *Can. J. Bot.* **66**, 934–940.
Kucey, R. M. N. and Paul, E. A. (1982). *Soil Biol. Biochem.* **14**, 407–412.
Lambers, H. (1987). *Root Development and Function* (P. J. Gregory, J. V. Lake and D. A. Rose, eds), pp. 125–145. Cambridge University Press, Cambridge.
Langlois, C. G., and Fortin, J. A. (1984). *Can. J. For. Res.* **14**, 412–415.
Larsen, S. (1967). *Adv. Agronomy* **19**, 151.
Lewis, D. H. and Harley, J. L. (1965). *New Phytol.* **64**, 256–269.
McGraw, A. C., Gamble, J. F. and Schenck, N. C. (1979). *Phytopathol* **69**, 1038.
Melin, E. and Nilsson, H. (1950). *Physiol. Plant.* **3**, 88–92.
Melin, E. and Nilsson, H. (1952). *Svensk. Bot. Tidskr.* **46**, 281–285.
Melin, E. and Nilsson, H. (1953). *Nature* **171**, 134.
Melin, E. and Nilsson, H. (1955). *Svensk. Bot. Tidskr.* **49**, 119–122.
Melin, E. and Nilsson, H. (1957). *Svensk. Bot. Tidskr.* **51**, 166–186.
Miller, S. L. Durall, D. M. and Rygiewicz, P. T. (1989). *Tree Physiol.* **5**, 239–249.
Miura, G. and Hall, R. H. (1973). *Plant Physiol.* **51**, 563–569.

Morrison, T. M. (1957). *Nature* **179**, 907–908.
Morrison, T. M. (1962). *New Phytol*. **61**, 21–27.
Mosse, B., Hayman, D. S. and Arnold, D. J. (1973). *New Phytol*. **72**, 809–815.
Newman, E. I. (1988). *Adv. Ecol. Res.* **18**, 243–270.
Newman, E. I. and Ritz. K. (1986). *New Phytol*. **104**, 77–87.
Owusu-Bennoah, E. and Wild, A. (1979). *New Phytol*. **82**, 133–140.
Owusu-Bennoah, E. and Wild, A. (1980). *Plant and Soil* **54**, 233–242.
Pang, P.C. and Paul, E. A. (1980). *Can. J. Soil. Sci.* **60**, 241–250.
Pearson, V. and Tinker, B. H. (1975). In *Endomycorrhizas* (F. E. Sanders, B. Mosse, and P. B Tinker, eds), pp. 277–287. Academic Press, London.
Perry, D. A. Margolis, H., Choquette, C., Molina, R. and Trappe, J. M. (1989). *New Phytol*. **112**, 501–511.
Pichot, J. and Binh, T. (1976). *Agron. Trop.* 31, 375–378.
Powell, C. L. (1975). *New Phytol*. 75, 563–566.
Purves, S. and Hadley, G. (1975). In *Endomycorrhizas* (F. E. Sanders, B. Mosse, and P. B. Tinker, eds), pp. 175–194. Academic Press, London.
Ratnayaka, M., Leonard, R. T. and Menge, J. A. (1978). *New Phytol*. **81**, 543–552.
Reid, C. P. P., Kidd, F. A. and Ekwebelam, S. A. (1983). *Plant and Soil* **71**, 415–432.
Rhodes, L. H. and Gerdemann, J. W. (1975). *New Phytol*. **75**, 555–561.
Ritz, K. and Newman, E. I. (1984). *Oikos* **43**, 138–142.
Rogers, R. D. and Williams, S. E. (1986). *Soil Biol. Biochem.* **18**, 371–376.
Rovira, A. D. (1979). In *The Soil–Root Interface* (J. L. Harley and R. Scott Russell, eds), pp. 145–160. Academic Press, London.
Rygiewicz, P. T. and Bledsoe, C. S. (1984). *Plant Physiol.* **76**, 918–923.
Sanders, F. E. and Tinker, P. B. (1971). *Nature* **233**, 278–279.
Sanders, F. E., Tinker, P. B. (1973). *Pestic. Sci.* **4**, 385–395.
Sanders, F. E., Tinker, P. B., Black, R. L. B. and Palmerley, S. M. (1977). *New Phytol*. **78**, 257–268.
Snellgrove, R. C., Splittstoesser, W. E., Stribley, D. P. and Tinker, P. B. (1982). *New Phytol*. **92**, 75–87.
Swaminathan, K. and Verma, B. C. (1979). *New Phytol*. **82**, 481–487.
Tinker, P. B. H. (1975). In *Endomycorrhizas* (F. E. Sanders, B. Mosse, and P. B. Tinker, eds.), pp. 353–371. Academic Press, London.
Todd, A. W. (1979). Abstr. *4th North American Conference on Mycorrhizae*, Fort Collins, Co.
Wang, G. M., Coleman, D. C., Freckman, D. W., Dyer, M. I., McNaurghton, S. J., Acra, M. A. and Goeschl, J. D., (1989). *New Phytol*. **112**, 489–493.
Whipps, J. M. and Lynch, J. M. (1985). *Ann. Proc. Phytochem. Soc. Eur.* **26**, 59–71.
Whittingham, J. and Read, D. J. (1982). *New Phytol*. **90**, 277–284.

12

Techniques for Studying the Functional Aspects of Rhizomorphs of Wood-rotting Fungi: Some Possible Applications to Ectomycorrhiza

D. H. JENNINGS

Department of Genetics and Microbiology, The University, PO Box 147, Liverpool L69 3BX, UK

I. Introduction

Rhizomorphs are long linear multi-hyphal structures produced by higher fungi, principally members of the Basidiomycotina (but also members of the Ascomycotina), for exploring the substratum in which the fungus is living. In particular, rhizomorphs allow a fungus to move from one

METHODS IN MICROBIOLOGY
VOLUME 23 ISBN 0-12-521523-1

Copyright © 1991 by Academic Press Limited
All rights of reproduction in any form reserved

source of combined carbon to another, an example being the movement from one piece of timber to another within the forest floor.

In recent years, the term rhizomorph has been used by some (Garrett, 1960) to apply only to the linear structure of *Armillaria*, which produces an apex which is clearly defined morphologically. It has been believed that the apex develops in a manner similar to a root, namely through the activity of a meristematic region (Garrett, 1960; Motta, 1969, 1971). Examination of the rhizomorph of *Armillaria* by scanning and transmission electron microscopy suggests otherwise (Granlund *et al.*, 1984; Rayner *et al.*, 1985; Cairney *et al.*, 1988b). It is now believed (Rayner *et al.*, 1985) that the rhizomorph of *Armillaria* is only one extreme of the continuum exhibited by growing mycelia capable of producing linear organs (Fig. 1). That continuum extends from the almost completely undifferentiated mycelial front of *Serpula lacrymans* (Fig. 2a) through increasing aggregation of hyphae, as exemplified by *Phanerochaete velutina* (Fig. 2b) and *Phallus impudicus* (Fig. 2c), through to *Armillaria* (Fig. 2d). On the basis of the similarity of mature linear organs produced by members of Basidiomycotina, Cairney *et al.* (1988b, 1989) have argued for the use of the term "rhizomorph" for all long multi-hyphal linear organs. The historical precedence of this term rules out those introduced at a later date, namely syrrotia (Falck, 1912) and cord (Rayner and Todd, 1979), and probably also strand when used mycologically. Cairney *et al.* (1989) suggested that the terms "simple" and "complex" might be used to describe the degree of differentiation of the internal structure encountered within the rhizomorph, while "apically diffuse" (e.g. *S. lacrymans*), "apically spreading" (e.g. *P. impudicus*), "apically branched" and "apically dominant" (e.g. *Armillaria*) might be used to describe that part of the mycelial system which brings about the extension of the rhizomorph (Rayner *et al.*, 1985). There is as yet no general acceptance of this terminology. Nevertheless it is the most appropriate, given our present understanding of the manner of growth and structure of the linear organs of members of the Basidiomycotina.

It might be argued that the less well developed structures which can be produced by fungi forming mycorrhiza should not be termed rhizomorphs. But for convenience it is inappropriate to follow this line of thought. In any case, there are very strong similarities between the manner of development of such rhizomorphs and the early stages of development of the rhizomorphs of *S. lacrymans* (Butler, 1958; Agerer, 1988).

That said, this chapter is concerned with the techniques which can be used to study how rhizomorphs of wood-rotting fungi function. In this

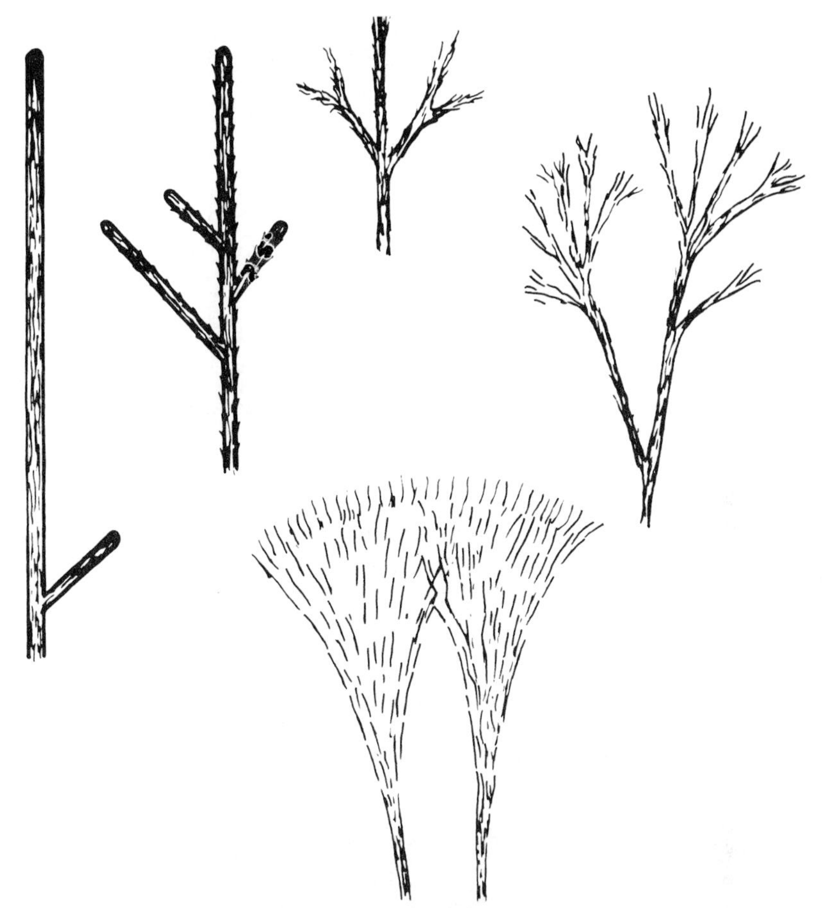

Fig. 1. Diagram illustrating the spectrum of mycelial outgrowth patterns resulting in the production of rhizomorphs. There is a continuum from mycelium without any apical coherence, with rhizomorphs being produced only in the older parts (bottom centre), through increasing apical control, until there is strongly polarized outgrowth (far left) (based on figure from Rayner *et al.*, 1985).

context, the term rhizomorph is very apposite, since by highlighting the root-like morphology of the fungal linear organs, the investigator is encouraged to think of rhizomorphs in experimental terms as being like roots. Thus the higher plant physiological literature contains a source of ideas and techniques for studying how rhizomorphs function and the investigator needs to keep this in mind.

Though this chapter is concerned with rhizomorphs of wood decay fungi, where relevant there is reference to mycorrhizal rhizomorphs.

Fig. 2(a). Mycelium of *Serpula lacrymans* growing from colonized spruce blocks over cement (showing white patches of efflorescence as a result of mycelial activity). No apical control at the mycelial front and development of rhizomorphs in the older parts of the mycelium in the bottom portion of the figure.

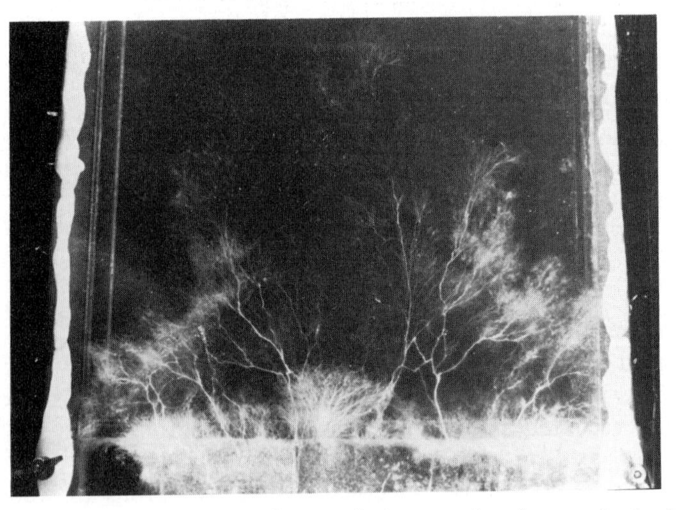

Fig. 2(b). Mycelium of *Phanerochaete velutina* growing from colonized beech blocks onto a Perspex platform in a sterile chamber.

Fig. 2(c). Mycelium of *Phallus impudicus* growing as in (b) showing tendency of mycelium towards polarized growth.

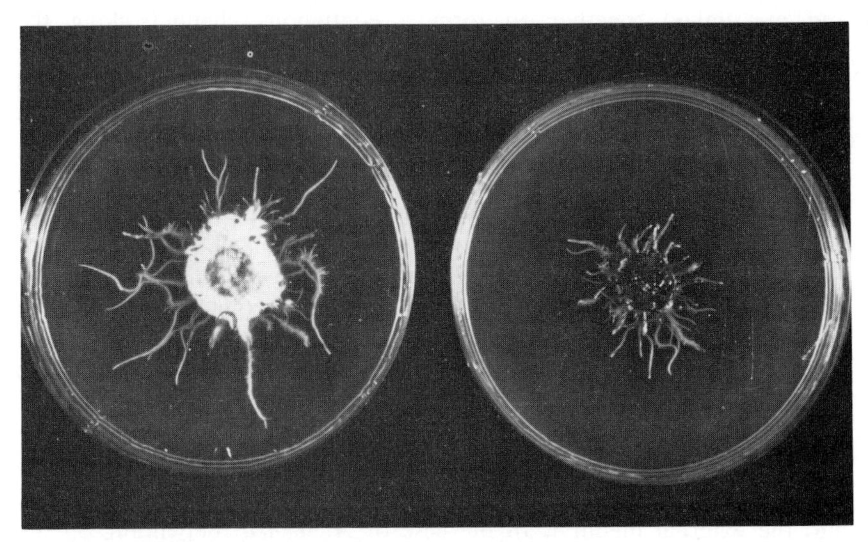

Fig. 2(d). Mycelium of *Armillaria mellea sensu lato* growing on agar showing rhizomorphs extending beyond the undifferentiated colony margin.

But, as yet, it is not clear how far the methods described here are applicable to these latter rhizomorphs. Nevertheless, there is a commonality of function between them and rhizomorphs of the wood decay fungi. Thus the characteristics being measured by the methods described here apply also to mycorrhizal rhizomorphs. It is only when we have quantitative information about such characteristics that we will have a true picture as to how these rhizomorphs function.

II. Salient features of morphology and anatomy

Rhizomorphs of wood-rotting fungi are never more than a few millimetres in diameter but can extend over many metres. Branching can occur and fruit bodies may be formed from them. When mature, rhizomorphs of wood decay fungi, for the most part, exhibit a similar internal structure (Townsend, 1954; Cairney *et al.*, 1989), being composed of an outer region surrounding an inner region containing both large and small diameter hyphae. Invariably, the outer region contains many thick-walled or fibre hyphae, the inner region large diameter hyphae at least 5 mm but up to around 35 mm across [for a description of the possible hyphal types, the reader should consult Hornung and Jennings (1981)]. These large diameter hyphae, in most instances more correctly described as vessel hyphae, are frequently devoid of contents and are linked together to form a continuous channel along the rhizomorph (Cairney *et al.*, 1988b). There is convincing evidence for *Armillaria mellea sensu lato* and *Phallus impudicus* that the vessel hyphae are the main channels for the flow of water along the rhizomorph (Eamus *et al.*, 1985).

There are species which are much less differentiated than just described (Townsend, 1954). They can be completely undifferentiated, e.g. *Hymenogaster luteus* or *Mutinus caninus*, or show zonation due to arrangement of similar hyphae, e.g. *Collybia platyphilla*. Lack of clearly defined hyphal channels does not preclude such rhizomorphs from acting as translocatory organs, though due to the presence of septa, the rate of translocation will be lower than in rhizomorphs in which the channels are unobstructed.

Less well-established in many instances, particularly in quantitative terms, is the outer surface of the more highly differentiated rhizomorphs. The matter of concern is the extent to which hyphae extend into the external medium. In the case of *A. mellea*, depending on the conditions of growth, particularly the humidity and degree of aeration, the outer surface may appear smooth with a covering of mucilage or be covered to a greater or lesser extent with so-called fringing hyphae which can explore the substratum for nutrients (Cairney *et al.*, 1988b).

Rhizomorphs of mycorrhiza show a comparable diversity of structure to those of wood-rotting fungi, although examples showing the degree of differentiation possible amongst the latter fungi appear to be absent from the former group (Foster, 1981; Agerer, 1986, 1988; Fox, 1987). Where larger diameter hyphae have been shown to be present in rhizomorphs of mycorrhiza, it has been found that cross-walls can be absent (Brownlee *et al.*, 1983; Agerer, 1988) and in certain instances are clearly vessel hyphae (Fox, 1987).

III. Growth and preparation of material for experimental purposes

A. Field-growth material

In the case of rhizomorphs of wood-rotting fungi, it is frequently possible to collect material from the field. Under such circumstances, there is often little problem about finding enough material, although several square metres of leaf litter must be uncovered. In practice, there is little difference between collection of rhizomorphs and collection for experimental purposes of ectomycorrhizal roots of a forest tree such as beech. Collections of rhizomorphs should preferably be made under a uniform stand of trees to minimize gathering material from more than one species. Identification of material can present a problem, so tracing a rhizomorph back to a fruit body is recommended whenever possible.

B. Laboratory-grown material

1. Decaying wood

Decaying wood is particularly appropriate for obtaining rhizomorphs of *A. mellea*. The best material is fallen timber 30 cm or more in diameter, clearly covered by mature black rhizomorphs. This covering should be stripped off and pieces of timber around 100 cm long incubated in a water-saturated atmosphere. After a period of time, rhizomorphs grow out of the wood and can be harvested for experimental purposes. Rhizomorphs with cream-coloured apices should be used; those with dark brown apices should be discarded. To my knowledge, this procedure has not been used with other fungi.

2. *Inoculated wood*

Figures 2a, b and c illustrate how rhizomorphs might be produced from a wood inoculum such that they grow over a non-nutrient surface, in this instance Perspex. Section VI indicates how the capability to produce rhizomorphs in this manner can be used for studying the translocation of nutrients along them. Rhizomorphs can be harvested from such systems for experimental treatment or chemical analysis.

3. *Agar*

Mycelium growing on nutrient agar may develop rhizomorphs (Fig. 2d). The anatomy of the rhizomorphs can be investigated, but otherwise they cannot be used to much advantage experimentally. The exception to this is when the rhizomorphs can be induced to grow into agar, as is the case for *A. mellea*, when they can be several centimetres in length. This is possible because oxygen can diffuse to the apex from the air above the agar via the central space within the rhizomorph (Smith and Griffin, 1971). Experimentally, tubes (6–12 cm i.d.) are filled with nutrient agar to within 20 cm of the top (Granlund *et al.*, 1985; Cairney *et al.*, 1988a). The concentration of agar used (< 1.5% w/v) should be such that the gel is semi-liquid. The tubes should be inoculated from agar plate cultures with plugs which have diameters slightly larger than the tube to minimize sinking. Growth at 20 °C gives rhizomorphs 15–25 cm in length after 8–15 weeks.

IV. Water relations

A. Introduction

All the evidence that is available indicates that translocation of solution through both stranded, i.e. producing rhizomorphs, and unstranded mycelium of *S. lacrymans* occurs by the pressure-driven flow of solution (Jennings, 1987). The pressure is generated osmotically. Essentially, the volume flow of water across the outer membrane of the mycelium growing on the food source, as the result of a water potential gradient across that membrane, generates the pressure which is dissipated as a flow of solution through the mycelium, most particularly in the growing front. In mature rhizomorphs the flow of solution almost certainly takes place in the vessel hyphae (Eamus *et al.*, 1985).

It is highly likely that translocation in all rhizomorphs occurs by the same mechanism, irrespective of the degree of their internal complexity.

The circumstantial evidence is strongly in favour of the mechanism for rhizomorphs of *A. mellea* (Jennings, 1987). The data for water movement in mycorrhizal rhizomorphs produced by *Suillus bovinus* between pine seedlings (Brownlee *et al.*, 1983) are in keeping with the above view. But compelling evidence of the probable ubiquity of pressure-driven translocation of solutions in all rhizomorphs comes from the investigations on translocation within mycelium of *S. lacrymans* (Jennings, 1987), which indicate that the process is the same irrespective of the degree of differentiation of the mycelium. In the very young mycelium, the larger diameter hyphae are clearly a very important route for translocation (Brownlee and Jennings, 1982b) and there is indirect evidence that such hyphae are of the "armbone" kind in terms of their shape (Eamus *et al.*, 1985; Hornung and Jennings, 1981).

B. Determination of turgor potential

If we accept that translocation is likely to be driven osmotically, then determination of the turgor potential (pressure) gradient will give both a measure of the driving force and the direction of water flow. The turgor potential (ψ_p) is related to the water potential (ψ) as follows:

$$\psi_p = \psi - \psi_s,$$

where ψ_s is the solute potential. While ψ_p within a hypha involved in translocation has not yet been determined directly, it can be determined readily for a tissue by thermocouple psychrometry. It can be assumed that the turgor potential on a tissue-averaged basis by this technique gives a value for a translocating hypha. It would be difficult to envisage a system where the conducting channels had a lower turgor than the surrounding tissue (Eamus and Jennings, 1984).

ψ and ψ_s are determined directly and ψ_p determined by difference. Samples of *c.* 0.1 g are excised from the mycelium or rhizomorph and rapidly sealed in the sample chamber of the psychrometer. ψ is determined, then the samples are frozen in liquid nitrogen, thawed, when ψ_p is reduced to zero, and ψ_s then determined. Thompson *et al.* (1985) used a Wescor MR-33 T microvoltmeter coupled to a C-52 or L-51 sample chamber, the latter modified to function as a C-52 chamber by the use of a silicone rubber base with centre well. The base is produced by taking a 1 cm × 1 cm square of rubber of 1 mm thickness and punching a 3 mm diameter hole out of the centre. A square of similar size but without the hole is bonded onto this. Mycelium is placed in the well and inserted into the L-51 body, such that the well lies directly below the thermocouple. The thermocouple and the mounting are lowered into position and pressed down firmly on the silicone rubber to form the roof of the well. All measurements are made in the dew-point mode after calibration with a series of sodium chloride solutions of known water potential.

C. Water flow through rhizomorphs

1. Determination of volume flow and hydraulic conductivity

If the rhizomorph can be attached to a reservoir of water, the volume of which can be monitored, then any change in volume will give a measure of flow of water through the mycelium. This procedure was used by Thompson *et al.* (1985) and a diagram of the apparatus is given in Fig. 3. Thompson *et al.* were able to determine the velocity of water flow from a knowledge of the total cross-sectional area of the hyphae within the mycelium. The value obtained was higher than velocities of translocation determined with radiotracers (Brownlee and Jennings, 1982a). This difference might be due to the following: (1) evaporation from hyphae; (2) movement of water from translocating hyphae into those which are not translocating; (3) presence of hyphae which allow translocation of water but not solutes; or (4) inaccuracies in the determination of hyphal cross-sectional area. These possible errors are relevant to any determination of the kind described. It should be noted that, because of the manner of its determination, the value for the velocity of water moving through the mycelium (cm s^{-1}) is formally equivalent to the flux of water through the mycelium ($\text{cm}^3\,\text{cm}^{-2}\,\text{s}^{-1}$).

It might be argued that the information obtained from such a system is of little value to any consideration of natural situations. This is not so. If the determination of water uptake is combined with determination of the turgor gradient within the mycelium, as was done in the investigation of Thompson *et al.* (1985), it is possible to calculate the hydraulic conductivity (L_p) of the mycelium (Jennings, 1987), which is given by

$$J_\text{v} = L_\text{p} \cdot \frac{\text{d}P}{\text{d}l},$$

where J_v is the water flow ($\text{cm}^3\,\text{cm}^{-2}\,\text{s}^{-1}$) and $\text{d}P/\text{d}l$ is the pressure gradient (Pa cm^{-1}). L_p has units of $\text{cm}^2\,\text{Pa}^{-1}\,\text{s}^{-1}$ and is thus a measure of the resistance of a channel to the pressure-driven flow of water. Thus, given that the turgor pressure gradient can be determined—and this is very easy, certainly in the laboratory where one could be using model systems or microcosms—knowledge of L_p for a rhizomorph allows an assessment of the water flow through it.

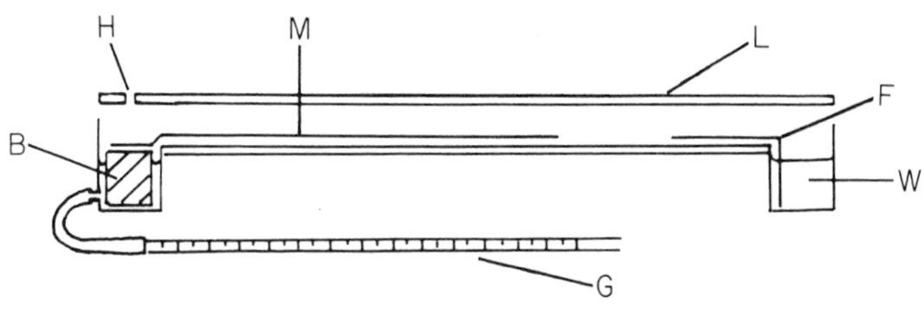

Fig. 3. Vertical longitudinal section through Perspex cabinet used for measuring water uptake as well as linear extension of mycelium of *Serpula lacrymans* (Thompson *et al.*, 1985). Key: B, wood block; F, filter paper; G, graduated pipette slanting upwards to the side of the cabinet; H, hole for addition of solution; L, Perspex lid; M, mycelium; W, sterile water.

If the rhizomorph is robust, L_p can be determined readily by determining the throughput of water under the driving force of a known applied pressure. This has been achieved for rhizomorphs of *A. mellea* and *P. impudicus* (Eamus *et al.*, 1985). Samples of material were cut under water into 1.5 cm lengths. One end was sealed into a plastic tube which tapered from approximately 5.0 mm to 0.5 mm internally. The free end of the tube was attached to a water suction pump via an on-line vacuum gauge. The free end of the rhizomorph was inserted into a 3.0 mm i.d. glass U-tube filled with water to a pre-determined mark. Suction was applied at a known, but not pre-determined, stable level for a known period of time, approximately 60 min. The volume of water taken up was determined by adding water with a Gilson micropipette until the original volume was restored. Calculation of L_p requires the determination of the cross-sectional area for water movement. In the case of rhizomorphs in which vessel hyphae are well developed, it is the cross-sectional area of these hyphae which should be determined, since there is clear evidence that these hyphae are the channels for the pressure-driven flow of water (Eamus *et al.*, 1985). The necessary measurements can be readily made with the light microscope using thin sections of the rhizomorph. Where there is no obvious differentiation into vessel hyphae, there is a problem. At the very least, the total cross-sectional area of the rhizomorph can be used to obtain a minimum value for L_p.

Tritiated water (^3H_2O) has been used to demonstrate the movement of water through rhizomorphs to mycorrhizal seedlings of pine (Duddridge *et al.*, 1980). Three such seedlings were transferred to Perspex chambers (15 cm × 17 cm × 1 cm) with detachable sides and base, containing non-sterile peat. When the peat had been extensively exploited by rhizomorphs with many on the wall of the chamber, a shallow

Perspex cup (1 cm diameter, 1 mm deep) containing ^3H_2O was attached to the chamber wall with anhydrous lanolin. After about an hour, the plants and rhizomorphs could be analysed for radioactivity. The procedure is a very good way of demonstrating that water is moving through rhizomorphs, although success with the procedure will depend on (1) restriction of movement of ^3H_2O as a vapour in the medium external to the rhizomorph, as will be the case in a medium such as peat, and (2) the absence of a major air space which would allow very rapid diffusion of ^3H_2O in the vapour phase. The procedure allows only an assessment of the velocity of water movement—volume flow cannot be determined directly from the data obtained. Nevertheless, if the cross-sectional area for water flow can be assessed, the rate of volume flow can be estimated.

V. Nutrient absorption

A. Introduction

A considerable amount is known about the mechanisms which underlie nutrient absorption by fungi (Jennings, 1990b). While it would be of considerable interest to know how rhizomorphs absorb specific ions and molecules, at present there is a greater need to know what ions and molecules can be absorbed and for any particular solute what is its flux (mol $cm^{-2} s^{-1}$) into the rhizomorph under conditions which are ecologically relevant. Equally, there is a need to know the extent to which rhizomorphs are able to degrade insoluble material to produce molecules or ions which then can be absorbed.

Conceptually, there is little difficulty in determining the flux of a nutrient into a rhizomorph, provided one has a suitable assay for the nutrient, which could include the use of a radioisotope, and the area of the absorbing surface can be determined. It is the determination of the latter which can present problems. Suggestions are given below for determining in the laboratory the absorbing surface area, but there must be concern about the effectiveness of any method devised in the laboratory for transference to the field situation in which hyphae ramify out of rhizomorphs into the surrounding medium in the manner demonstrated by Cairney *et al.* (1988b). Such hyphae will be important, particularly in the absorption of non-mobile elements such as phosphorus. As yet no assessment has been made of the extent to which hyphae do ramify from a rhizomorph in the soil. Equally flux values determined in the laboratory will only have ecological relevance if they can be related to values for nutrient concentrations in the field and pH of that part of the soil from where absorption is taking place. It is not appropriate to discuss here how such values can be obtained; the matter is discussed at length by Jennings (1990b). Nevertheless, the investigator needs to be aware of the problems.

B. Laboratory and field studies

Clipson *et al.* (1987) and Cairney *et al.* (1988a) give details about uptake of phosphate by mycelial and rhizomorph material of several members of the Basidiomycotina. Mycelium was grown in stationary liquid culture, the agar plug inoculum removed before measurements of uptake commenced; rhizomorphs were either collected from the field or in the case of *A. mellea* grown in tubes or from fallen timber as described above. The procedures used were no different from those used extensively by others investigating phosphate uptake by fungi (Beever and Burns, 1980). Nevertheless one matter needs attention, namely the amount of phosphate remaining in the free space, which can be defined as that space in the rhizomorph readily accessible to the external solution (Jennings, 1986). Cairney *et al.* (1988a), studying phosphate uptake into rhizomorphs of *A. mellea*, placed the tissue after incubation in ^{32}P-phosphate in 25 cm^3 unlabelled KH_2PO_4 at the same concentration as the experimental solution at 4 °C and agitated for 15 min. Preliminary experiments had shown that loss of label into the external medium was complete after 6 min; 15 min ensured all label was removed from the free space of the rhizomorph prior to determination of ^{32}P in the non-free space, presumably within the protoplasm.

In the experiments of Cairney *et al.* (1988a), 50% of the phosphate absorbed by rhizomorphs over a 2 h period was found in the free space. No attempt was made to delineate the free space; there is clearly a need to do this. Also there is a need to establish the characteristics of the free space of a rhizomorph such as that of *A. mellea* with respect to cations. It may well be that significant binding may take place, especially in view of the large amount of wall material which can be present. Such binding could confound any considerations of the availability of ions for translocation within the rhizomorph which is based on the content (mol g^{-1}) of the ion.

At its simplest, the flux of a solute into a rhizomorph can be expressed in terms of a surface area calculated on the basis of a perfect cylinder. The high value for the free space of the rhizomorphs of *A. mellea* determined with ^{32}P-phosphate suggests that a flux value so calculated could be a considerable overestimate. Thus there is a need for procedures leading to a proper estimate of the surface area over which absorption of a solute takes place. One approach is via stereology, i.e. the estimation of surface areas of membranes of cellular compartments from micrographs (Wiebel, 1979; Mary and Rigaut, 1986), one which has been demonstrated to be successful with mycelial pellets of the marine fungus *Dendryphiella salina* (Clipson *et al.*, 1989). Another approach, somewhat indirect, would be through a detailed analysis of the free space, using solutes which do not penetrate into the protoplast or bind to cell walls. Essentially, the procedures behind this approach are those which have been used to determine the osmotic volume of fungal hyphae and cells (Jennings, 1990a; Jennings and Garrill, 1991).

Using nuclear magnetic resonance (NMR) spectroscopy it is possible to study the partitioning of phosphorus between various pools in rhizomorphs which occurs whilst uptake is taking place, as has been done by Cairney *et al.* (1988a) with rhizomorphs of *A. mellea*. Success with the technique depends on maintaining the mycelium in a well-aerated condition within the NMR machine. The value of results obtained by this technique can only be suggested at present. It is better than destructive methods of analysis because it gives information about uptake into the vacuole. In terms of present considerations, importantly NMR allows an assessment of the relationship between uptake of a nutrient and its translocation. Thus sequestration of phosphate in the vacuole might reduce the amount of phosphorus translocated (Cairney *et al.*, 1988a).

Though not concerned with uptake *per se* significant information about the relationship between nutrient uptake and growth can be obtained from studying rates of growth of rhizomorphs from an inoculum over an appropriate substratum in a glass tube (Fig. 4). The rate of growth is readily determined by the position of the mycelial front after increasing periods of time. Most experiments have used soil or other substrata (Butler, 1958; Thompson and Rayner, 1983), but the opportunity exists for using substrata which can be charged up with individual nutrients. Jennings (1990c) reported the use of spillage granules (BDH plc). While there is considerable holding capacity, there is some nutrient movement from granule to granule. However other possibilities exist, e.g. ion-exchange materials, and these need to be investigated.

To date, there has been only one report on uptake of a nutrient by rhizomorphs in the field (Clipson *et al.*, 1987). Though unique, the method generated kinetic data for phosphate uptake by rhizomorphs of *Mutinus caninus* which were similar to those obtained in the laboratory. Rhizomorphs were exposed from the litter layer in a stand of beech

Fig. 4. Experimental system used to investigate the effect of nutrients on the extension of rhizomorph-forming mycelium (Jennings, 1990c). Key: A, stopper covered in tin foil; B, glass tube; C, wood block infected with mycelium of *Phanerochaete velutina*: D, moist filter paper; E, plastic spacer; F, fungal mycelium; G, "spillage granules". Tube on left: granules loaded with water and mycelium grown throughout the tube; tube on right: granules loaded with 50 mM $(NH_4)_2SO_4$ with extending mycelial front readily visible.

(*Fagus sylvatica*) over a 10 cm length and about 20 cm^3 of soil removed from beneath the rhizomorph in order to insert the vessel containing the solution from which uptake was to take place (Fig. 5). The vessel was a cylinder of Perspex glued onto a base of the same material with two 0.5 cm grooves on each side of the rim; the volume of the vessel was 14 cm^3. The rhizomorph was lowered into the two grooves and sealed with a water-impermeable mastic so that 22 mm was exposed to the

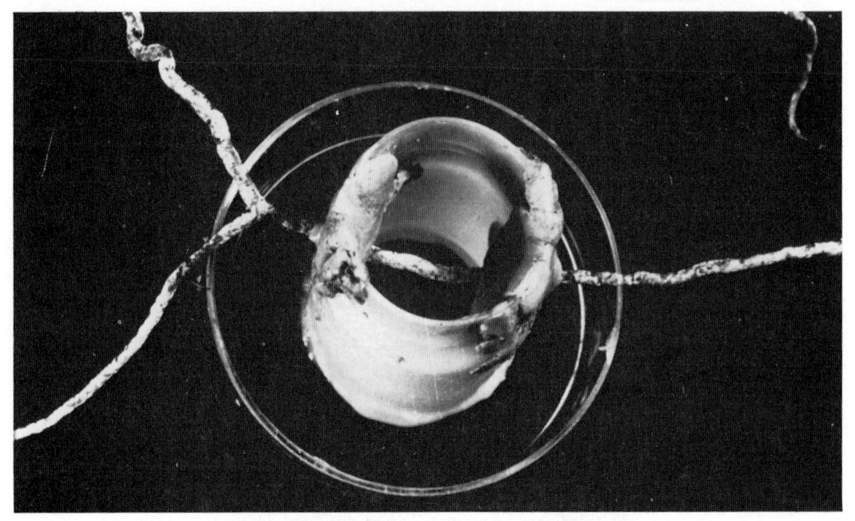

Fig. 5. Vessel used for investigating the uptake of nutrients by rhizomorphs of *Mutinus mutinus* in the field. The rhizomorph is fixed in position by mastic. For further details, see text (Clipson *et al.*, 1987).

solution, which was added to the vessel until it was full. In the experiments of Clipson *et al.* (1987), uptake of phosphate was determined over a 4–30 h period, using ^{32}P; the vessel was surrounded by polystyrene to minimize temperature fluctuations. At the appropriate time, the external solution was removed and replaced by phosphate-free solution at 4 °C, such that the part of the rhizomorph in the vessel was given a 10 min wash. Then the vessel and mastic were removed and the rhizomorph excavated, the labelled section separated, and the radioactivity in it determined. Radioactivity in other parts of the rhizomorph could also be determined, to establish the extent of translocation of the isotope.

VI. Translocation

Translocation—the movement of nutrients along a rhizomorph—can be characterized in two ways: either as a velocity (cm s^{-1}), usually by using a radiotracer; or as a flux (g or mol cm^{-2} s^{-1}). The former value provides information relevant to elucidating the mechanism (Brownlee and Jennings, 1982a) but, unless the specific activity of the radioisotope is known, does not provide much information which is relevant ecologically. A flux value provides information of value for comparing the effectiveness of rhizomorphs in moving nutrients, although for such purposes it is necessary that the cross-sectional area (cm^2) used in calculating the flux be that of the channels involved in translocation such as the vessel hyphae (Section IV.C). A large diameter rhizomorph is not necessarily more effective in translocation than one of smaller diameter. Ecologically, the amount moved per unit time (g or mol s^{-1}) may be the important characteristic to describe translocation along a rhizomorph in relation to its growth, or to its ability to supply nutrients to the host if it is connected to an ectomycorrhizal sheath.

The simplest method of determining the rate of translocation is to allow the mycelium to grow over a non-nutrient surface for a known period of time and harvest the mycelium which was produced within the period. The dry weight and/or content of a particular nutrient can be determined. Given the cross-sectional area of the pathway of translocation, a dry matter ($g\,cm^{-2}\,s^{-1}$) or nutrient ($mol\,cm^{-2}\,s^{-1}$) flux can be calculated. Dry matter fluxes of 1.1×10^{-5} for rhizomorphs of *A. mellea* and $1.96–3.4 \times 10^{-5}$ $g\,cm^{-2}\,s^{-1}$ for mycelium of *S. lacrymans* have been obtained from such a procedure.

There are two published reports on the determination of the velocity of translocation in mycelium of rhizomorphs of wood-rotting fungi. One is concerned with mycelium of *S. lacrymans* (Brownlee and Jennings, 1982a). The experimental set-up (Fig. 6) would be applicable to those fungi whose rhizomorphs develop well behind the mycelial front. The

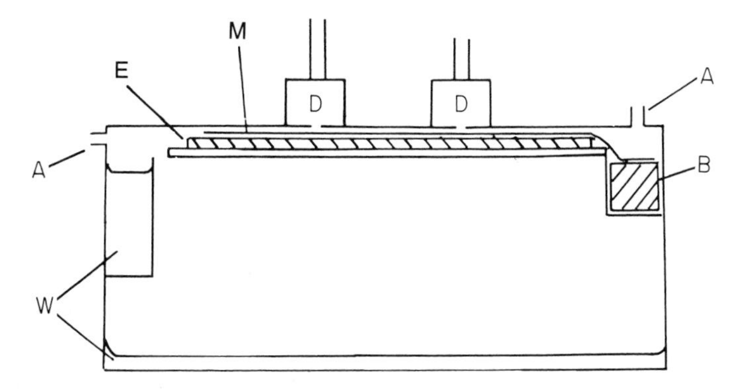

Fig. 6. Vertical longitudinal section through Perspex cabinet used for direct monitoring of radioactivity at set points along mycelium of *Serpula lacrymans* (Brownlee and Jennings, 1982a). Key: A, adaptor for gas inlet or outlet: B, wood block; D, radioactivity detector; E, plaster substratum; M, mycelium; W, water.

other study was concerned with rhizomorphs of *A. mellea* (Fig. 7). The procedure in both instances was essentially the same, in that a radioactive nutrient was supplied to the mycelium and the time determined for the isotope to move between two Geiger counters set at a known distance apart. From the specific activity of the U-[^{14}C]-glucose supplied, the flux of carbohydrate was found to be between 20 and 400

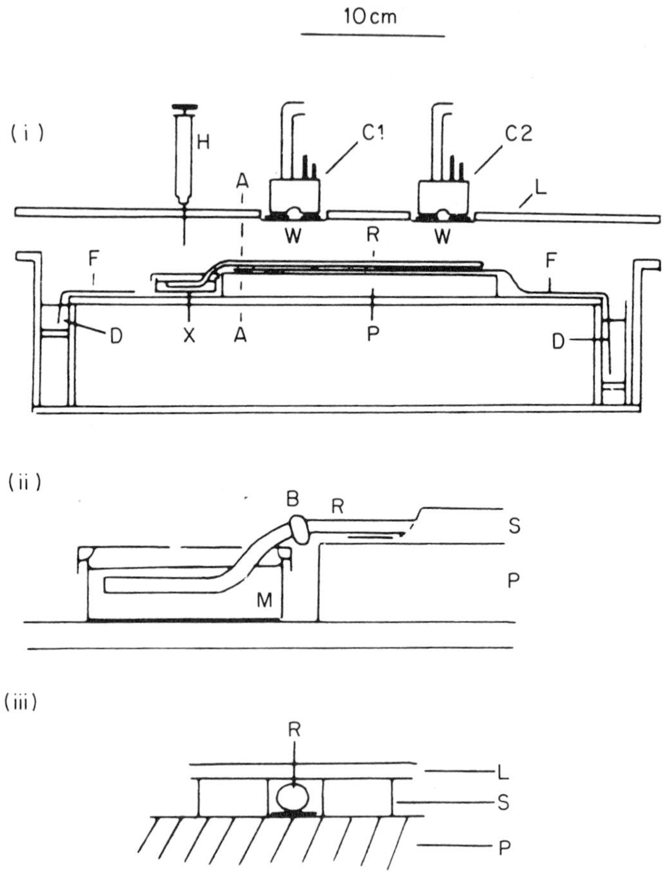

Fig. 7. Experimental cabinet used for continuous monitoring of radioactivity at set points along a rhizomorph of *Armillaria mellea sensu lato*. (i) Diagram of complete cabinet with lid raised; (ii) detail showing the relationship between rhizomorph and medium supplying radioactivity; (iii) section A–A through lid (in place), rhizomorph and Perspex supports. Key: B, lanolin barrier; C1, counter 1; C2, counter 2; D, reservoir containing water; F, filter paper; H, hypodermic syringe containing labelled compound; L, lid; M, bathing medium; P, Perspex support; R, rhizomorph; S, microscope slides; W, thin window; X, Petri dish (Granlund *et al.*, 1985).

times lower than would have been expected from knowledge of the flux of dry matter, assuming it was all in the form of carbohydrate. While dry matter must come from sources other than carbohydrate, it will be the major one. The data suggest that there is dilution of label as it is translocated, probably as exchange with non-labelled material, but also through loss as carbon dioxide if the carbohydrate is being translocated through living protoplasm and capable of being metabolized.

Essentially the same procedure as just described has been used by Finlay and Read (1985a,b) for determining the velocity of ^{14}C and ^{32}P movement through rhizomorphs of *Pisolithus tinctorius* and *Suillus bovinus* forming mycorrhiza with seedlings of *Pinus* sp. Carbon-14 was fed to the seedlings while ^{32}P was fed to mycelium via small dishes or presented directly to peat in front of advancing mycelium. A Geiger tube was placed in a series of pre-determined positions at known intervals of time to determine velocities. It was argued that transpiration had no effect on translocation. However, that conclusion came from indirect evidence; the rate of transpiration of the seedlings was not measured nor was there any information about the extent of dilution of the isotope as it moved through the rhizomorph.

The matter of isotopic exchange and/or dilution as a labelled compound is translocated along a rhizomorph demands further exploration. Until there are data for the extent of such exchange or dilution, the use of isotopes to obtain translocation flux values will lead to underestimates. One possible way to obtain relevant data would be to perfuse excised portions of rhizomorphs with solutions of known composition using procedures similar to those described above for determining hydraulic conductivity (Section IV.C). Change in the composition (both chemical and radioactive) after passage of the solution through the rhizomorph could be determined.

VII. Conclusions

Rhizomorphs of wood-rotting fungi are mostly robust and easy to handle, but not always so. However rhizomorphs of mycorrhiza are both fragile and not necessarily susceptible to a varied experimental attack. Read and his colleagues (as exemplified by Finlay and Read, 1986b) have demonstrated very elegantly that mycorrhizal rhizomorphs can link mycelium on separate plants; thus the rhizomorphs can channel nutrients, particularly combined carbon, from one plant to another, as well as both exploring the soil for other sources of nutrients and enhancing water uptake. To date, much of what is known about the functioning of rhizomorphs, both non-mycorrhizal and mycorrhizal, is qualitative in nature. We now need quantitative information. For mycorrhizal rhizomorphs it may well be that the most productive approach will be a combination of mathematical modelling and experiment. The power of this approach has been well demonstrated by analysis of the effect of vesicular-arbuscular mycorrhiza on the phosphate uptake and growth of their higher plant hosts (Jennings, 1986). But modelling can only yield results of value if the characteristics of the system under consideration are defined with some degree of precision. The techniques and approaches described here could help that definition with respect to the rhizomorphs of the mycorrhiza.

When considering the functioning of the total higher plant–fungus system with respect to nutrient uptake, there is a need to take into account water flow through the system. While the previous sections have addressed some of the features of the water relations of rhizomorphs, there are some features that have not been considered specifically. In particular, we know nothing as yet of the hydraulic conductivity to water flow when there are negative pressures (tension/suction) within the conducting channels within the rhizomorph, as will be the case when there is a transpiration pull exerted by the higher plant. In this particular instance, as indeed will be the case when considering the total higher plant–fungus system, the investigator needs to consult the literature from higher green plant physiology with respect to the theoretical issues which need to be addressed and the approaches which can be used from that particular area of biology.

References

Agerer, R. (1986). *Mycotaxon* **27**, 1–59.

Agerer, R. (1988). *Nova Hedwigia* **47**, 311–334.

Beever R. and Burns, D. J. W. (1980). *Adv. Bot. Res.* **8**, 127–219.

Brownlee, C. and Jennings, D. H. (1982a). *Trans. Br. Mycol. Soc.* **79**, 143–148.

Brownlee, C. and Jennings, D. H. (1982b). *Trans. Br. Mycol. Soc.* **79**, 401–407.

Brownlee, C., Duddridge, J. A., Malibari, A. and Reed, D. J. (1983). *Plant Soil* **71**, 433–443.

Butler, G. M. (1958). *Ann. Bot.* **21**, 523–537.

Cairney, J. W. G., Jennings, D. H., Ratcliffe, R. G. and Southon, T. E. (1988a). *New Phytol.* **109**, 327–333.

Cairney, J. W. G., Jennings, D. H. and Veltkamp, C. J. (1988b). *Nova Hedwigia* **46**, 1–25.

Cairney, J. W. G., Jennings, D. H. and Veltkamp, C. J. (1989). *Can. J. Bot.* **67**, 2266–2271.

Clipson, N. J. W., Cairney, J. W. G. and Jennings, D. H. (1987). *New Phytol.* **105**, 449–457.
Clipson, N. J. W., Jennings, D. H. and Smith, J. L. (1989). *New Phytol.* **113**, 21–27.
Duddridge, J. A., Malibari, A. and Read, D. J. (1980). *Nature* **287**, 834–836.
Eamus, D. and Jennings, D. H. (1984). *J. Exp. Bot.* **35**, 1782–1786.
Eamus, D,, Thompson, W., Cairney, J. W. G. and Jennings, D. H. (1985). *J. Exp. Bot.* **36**, 1110–1116.
Falck, R. (1912). In *Hausschwammforschungen* (A. Moller, ed.), Vol. 6, pp. 1–405. Gustav Fischer, Jena.
Finlay, R. D. and Read, D. J. (1986a). *New Phytol.* **103**, 143–156.
Finlay, R. D. and Read, D. J. (1986b). *New Phytol.* **103**, 157–165.
Foster, R. C. (1981). *New Phytol.* **88**, 705–712.
Fox, F. M. (1987). *Trans. Br. Mycol. Soc.* **89**, 551–560.
Garrett, S. D. (1960). In *Plant Pathology* (J. G. Horsfall and A. E. Diamond, eds), pp. 23–56. Academic Press, New York and London.
Granlund, H. I., Jennings, D. H. and Veltkamp, K. (1984). *Nova Hedwigia* **39**, 85–99.
Granlund, H. I., Jennings, D. H. and Thompson, W. (1985). *Trans. Br. Mycol. Soc.* **84**, 111–119.
Hornung, U. and Jennings, D. H. (1981). *Nova Hedwigia* **34**, 101–126.
Jennings, D. H. (1986). In *Plant Physiology: A Treatise* (F. C. Steward, ed.) Vol. VII, pp. 225–379. Academic Press, Orlando, FL.
Jennings, D. H. (1987). *Biol Rev.* **62**, 215–243.
Jennings, D. H. (1990a). In *Microbiology of Extreme Environment* (C. Edwards, ed.), pp. 117–146. Open University Press, Milton Keynes.
Jennings. D. H. (1990b). *Proc. Ind. Nat. Acad. Sci. Plant Sci.* **100**, 153–164.
Jennings, D. H. (1990c). In *Nutrient Cycling in Terrestrial Ecosystems : Field Methods, Application and Interpretation* (A. F. Harrison, P. Ineson and O. W. Heal, eds) pp. 233–245. Elsevier Applied Science, London and New York.
Jennings, D. H. and Garrill, A. (1991). In *The Isolation and Study of Marine Fungi* (E. B. Gareth Jones, ed.). Academic Press, London (in press).
Mary, J. Y. and Rigaut, J. P. (1986). *Quantitative Image Analysis in Cancer Cytology and Histology*. Elsevier Science Publishers, Amsterdam.
Motta, J. J. (1969). *Am. J. Bot.* **56**, 610–619.
Motta, J. J. (1971). *Am. J. Bot.* **58**, 80–87.
Rayner, A. D. M. and Todd, N. K. (1979). *Adv. Bot. Res.* **7**, 333–420.
Rayner, A. D. M., Powell, K. A., Thompson, W. and Jennings, D. H. (1985). In *Developmental Biology of Higher Fungi* (D. Moore, L. A. Casselton, D. A. Wood and J. C. Frankland, eds), pp. 249–279. Cambridge University Press, Cambridge.
Smith, A. M. and Griffin, D. M. (1971). *Austral. J. Biol.* **24**, 231–262.
Thompson, W. and Rayner, A. D. M. (1983). *Trans. Br. Mycol. Soc.* **81**, 333–345.
Thompson, W., Eamus, D. and Jennings, D. H. (1985). *Trans. Br. Mycol. Soc.* **84**, 601–608.
Townsend, B. A. (1954). *Trans. Br. Mycol. Soc.* **37**, 222–233.
Wiebel, E. R. (1979). *Stereological Methods*, Vol. I: *Practical Methods for Biological Morphometry*. Academic Press, London.

13

Structural and Ontogenic Study of Ectomycorrhizal Rhizomorphs

JOHN W. G. CAIRNEY

Department of Soil Science, Waite Agricultural Research Institute, University of Adelaide, Glen Osmond, SA 5064, Australia

1. Introduction

The vegetative mycelium of ectomycorrhizal fungi comprises the hyphal mantle and the extramatrical mycelium (where present) that extends from the mantle surface, often for a considerable distance, into the surrounding soil. Extramatrical mycelium is known to absorb nutrients and water from the soil (Skinner and Bowen, 1974; Brownlee *et al.*, 1983) and to facilitate their translocation to the mantle, from where they are transferred across the symbiotic interface to the root surface. Extramatrical mycelium also acts as an effective inoculum, infecting roots on the same or on other plants (Read, 1984).

In common with a number of saprotrophic basidiomycetes, the extramatrical mycelium of some ectomycorrhizal species aggregates in soil to form linear mycelial organs. Depending on the species, linear mycelial organs can be loose, undifferentiated hyphal aggregates, or they can be differentiated to a greater or lesser extent (rhizomorphs). The major advantages of differentiation to form rhizomorphs are

METHODS IN MICROBIOLOGY
VOLUME 23 ISBN 0-12-521523-1

Copyright © 1991 by Academic Press Limited
All rights of reproduction in any form reserved

thought to be protection of hyphae from environmental fluctuations and antagonistic organisms, and an increase in inoculum potential mediated by more efficient translocation of resources from carbon-rich food bases (roots or woody substrate) towards the growing front (Thompson, 1984).

From the information currently available, the structure of mature rhizomorphs appears to be intraspecifically consistent (Cairney, 1991b; Cairney *et al.*, 1991) and there is evidence of strong structural similarities at the genus level (Cairney, 1990, 1991a). Rhizomorph structure is likely, therefore, to be an important taxonomic criterion, particularly in the absence of basidiomata. Indeed, several workers routinely include details of rhizomorph structure in their descriptive characterization of ectomycorrhiza (see Agerer, 1987). The aim of this chapter is to rationalize the nomenclature relating to linear mycelial organs, and to highlight techniques which have proven useful, or are of potential value, in the investigation of rhizomorph structure and development.

II. Nomenclature

Although a rather confused area, due to the use of a number of terms by different authors, attempts have recently been made to clarify the terminology relating to linear mycelial organs. In the ectomycorrhizal context the terms used over the years have included "hyphal bundle" (e.g. Masui, 1926), "mycelial cord" (e.g. Dell *et al.*, 1989), "rhizomorph" (e.g. Agerer, 1988) and "strand" (e.g. Foster, 1981). It is probably fair to say, however, that "strand", either unqualified (e.g. Garbaye and Bowen, 1989), or qualified by "fungal" (e.g. Ashford and Allaway, 1985), "hyphal" (e.g. Duddridge and Read, 1984), "mycelial" (e.g. Fox, 1987) or "mycorrhizal" (e.g. Dighton and Mason, 1985), has been favoured by the majority of ectomycorrhizal workers in recent years. This appears to be founded (Bowen, 1973; Read, 1984) on the separation of rhizomorphs and strands on the basis of the presumed differentiated meristem in rhizomorphs of *Armillaria mellea sensu lato* (Garrett, 1970).

Rayner and Todd (1979), however, in discussing linear mycelial organs produced by saprotrophic basidiomycetes, argued that "strand" describes a single filament—thus a single hypha would constitute a "mycelial strand". "Mycelial cord" was proposed as a more appropriate term, and gained wide acceptance in this context (e.g. Thompson, 1984). Our understanding of linear mycelial organs has been further advanced by Rayner *et al.* (1985), who demonstrated that all linear mycelial organs, including those produced by *A. mellea sensu lato*, differentiate behind a front of normal apically-extending hyphae. While still advocating the use of both "rhizomorph" and "mycelial cord", these authors suggested the adoption of qualifying terms such as "apically dominant", "apically spreading" or "apically diffuse" to describe the outgrowth patterns at the growing front. (See Fig. 1, Chapter 12, Jennings, this volume.)

More recently, Cairney *et al.* (1989) highlighted structural similarities common to all differentiated linear mycelial organs irrespective of the degree of hyphal aggregation at their growing front, stressing that only a single, qualified term is required in their description. "Rhizomorph", qualified as described by Rayner *et al.* (1985) along with a term, such as "simple" or "complex" (see Fig. 1), to describe the degree of internal differentiation attained in the mature organ, was thus regarded as most appropriate.

Although generally smaller and composed of fewer hyphae, the differentiated linear mycelial organs associated with ectomycorrhiza

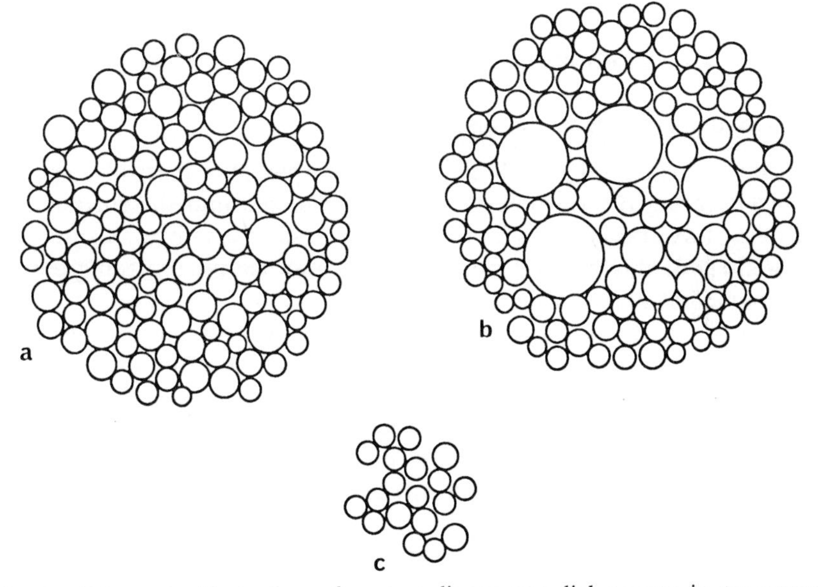

Fig. 1. Schematic illustration of mature linear mycelial organs in transverse section showing relative hyphal arrangement and size. (a) "Simple" rhizomorph (little differentiation between cortex and medulla; large diameter "vessel" hyphae poorly defined from other hyphae). (b) "Complex" rhizomorph (clear distinction between cortex and medulla; distinct large diameter "vessel" hyphae in medulla). (c) Undifferentiated mycelial "cord" (loose aggregate of hyphae of similar dimensions; large diameter "vessel" hyphae are absent). Based on Cairney *et al.* (1989, 1991).

should be regarded as both structurally and functionally equivalent to those produced by saprotrophic basidiomycetes (Cairney, 1991b; Cairney *et al.*, 1991). The continued use of "strand" in the ectomycorrhizal context cannot be justified and will only serve to confound such similarities. It is essential, therefore, that we adopt a common nomenclature, and I encourage ectomycorrhizal workers to use "rhizomorph", qualified as outlined above, in describing all differentiated linear mycelial organs produced by ectomycorrhizal fungi.

The undifferentiated linear hyphal aggregates (see Fig. 1) which are associated with the mycorrhiza of certain species (e.g. *Dermocybe crocea*; (Uhl and Agerer, 1987) are neither structurally nor functionally equivalent to the differentiated rhizomorphs so far discussed (Cairney, 1991b; Cairney *et al.*, 1991). Although I have recently suggested that either "cord" or "strand" might be appropriate here, "cord" (*sensu* Rayner and Todd, 1979) probably best describes these undifferentiated aggregates.

III. Study of rhizomorph structure and development

A. Obtaining experimental material

The simplest method for obtaining rhizomorphs of a particular species involves tracing the rhizomorph from a basidioma into the surrounding soil (e.g. Fox, 1987). Whilst providing material that is likely to be fully differentiated, this method provides little information regarding earlier developmental stages of the organ.

In order to investigate earlier stages of development, rhizomorphs must be produced *in vitro*. This will only occur where the extramatrical mycelium of rhizomorph-forming species is allowed to grow from the roots of a host plant onto a medium which is essentially non-nutritive in terms of organic carbon. The growth pouch technique (Piché and Fortin, 1982) provides such an environment, and has been used successfully to study rhizomorph structure in several species by Samson and Fortin (1988). Although, to date, not exploited in the study of rhizomorph ontogeny, this technique shows great promise, particularly for the study of the early developmental stages close to the rhizomorph growing front. Pieces of the paper wick underlying the relevant region of the mycelium can be excised and viewed, for example using scanning electron microscopy, without disturbing the constituent hyphae.

An alternative method involves allowing rhizomorphs to develop in soil, in either Petri dishes (Agerer, 1988) or Perspex root chambers

(Finlay and Read, 1986; Skinner and Bowen, 1974). Moist peat is a suitable substrate and incubating the chambers at a slight angle encourages production of rhizomorphs on the lower soil surface (Fig. 2). It is important to use non-sterile soil, particularly when studying rhizomorph development, since this can be adversely affected by soil sterilization (Thompson and Rayner, 1983). Study of the early stages of rhizomorph development may be limited by this technique as material must be removed from the substrate prior to microscopy. Modified root chambers, such as the root-mycocosm of Rygiewicz *et al.* (1988), in which the extramatrical mycelium can grow from the soil onto an alternative substrate such as glass fibre filter paper, might prove invaluable here. Like the wick from the growth pouch, sections of the filter paper could be removed and processed directly for scanning electron microscopy. Alternatively, and potentially much more useful, the glass filter paper might be replaced by glass cover-slips which could be removed and viewed directly under the light microscope. A version of this technique has been used successfully in the study of developing rhizomorphs of *Serpula lacrimans* (Brownlee and Jennings, 1982).

B. Rhizomorph structure

Before discussing methods used in the study of fungal rhizomorphs, it is important to stress that rhizomorphs differentiate progressively behind the growing front (see Section II). The position of the material studied relative to that growing front will influence the structure observed. Furthermore, neither distance behind the growing front nor rhizomorph diameter can be relied upon as absolute indicators of developmental maturity (Cairney *et al.*, 1989). It is essential, therefore, that the mature condition of the rhizomorph is ascertained prior to description of internal structure. This can only be achieved through the study of a number of rhizomorphs of the species under investigation. The failure of some authors to ensure this probably explains the intraspecific inconsistancies in rhizomorph structure that occasionally arise in the literature (Cairney *et al.*, 1988, 1989).

1. Transverse sections (non-embedded material)

Light microscopy, scanning electron microscopy (SEM) and transmission electron microscopy (TEM) have all been used in the study of rhizomorph structure. In terms of simply demonstrating the arrangement and size of hyphae within the rhizomorph, transverse sections are easiest to interpret and provide most information. The simplest method here is to

Fig. 2. *Paxillus involutus* (Batsch) Fr., ectomycorrhizal with Sitka spruce (*Picea sitchensis* Bong. Carr), growing in a Perspex root chamber with a moist peat substrate. The chamber was angled during incubation to encourage rhizomorph growth on the lower soil surface. Bar, 10 cm.

view unstained sections (10–20 μm thick), cut either by hand or with the aid of a cryomicrotome, using Nomarski interference contrast microscopy (e.g. Kammenbauer *et al.*, 1989).

Transverse sections of rhizomorphs have also been successfully viewed using SEM, and provide more information with regard to hyphal wall thickness, extracellular material and the three-dimensional arrangement of hyphae within the organ (e.g. Giovannetti and Fontana, 1986; Cairney *et al.*, 1989). Preparation of rhizomorph material for SEM is relatively simple and involves dehydrating either unfixed material (Eamus *et al.*, 1985) or material fixed in glutaraldehyde (Fox, 1987) through an ethanol or acetone series prior to critical-point drying. Some workers include an osmium tetroxide post-fixation step before dehydration (Giovannetti and Fontana, 1986). In order to obtain "clean" faces in transverse section, I have found it useful to glue pieces of rhizomorph (*c.* 0.5 cm in length) vertically onto stubs following critical-point drying, and then to cut transverse sections with a fresh double-sided razor blade. Enhanced retention of extracellular material and better tissue preservation might be achieved using cryofixation as an alternative to fixation/dehydration/critical-point drying (Alexander *et al.*, 1987). Material for SEM is finally sputter-coated with gold prior to viewing. SEM has the additional advantage of allowing external surface characteristics to be studied along with internal anatomy.

2. Transverse sections (embedded material)

The alternative method for obtaining transverse sections involves embedding rhizomorphs in resin and viewing either semi-thin sections under the light microscope or thin sections using TEM. A number of preparation schedules have been utilized (e.g. Foster, 1981; Brownlee *et al.*, 1983; Fox, 1987; Cairney *et al.*, 1988), all following the normal TEM methodology of chemical fixation, post-fixation in osmium tetroxide (not required for light microscopy), solvent dehydration and resin-embedding. Transverse sections viewed under the light microscope do not provide much additional information *per se*. However, if stained with a variety of histochemical reagents, in the manner already used in sclerotia (Bullock *et al.*, 1980) or stromata (Kohn and Grenville, 1989), they could provide invaluable information relating to rhizomorph function and division of labour amongst constituent hyphae.

TEM can reveal still further detailed information regarding cellular contents, wall thickness and extracellular material (e.g. Foster, 1981; Brownlee *et al.*, 1983; Fox, 1987). This technique also provides the potential for using molecular probes to investigate differences in the chemistry of cell walls and extracellular material (e.g. Bonfante-Fasolo *et al.*, 1987) that might relate to functional differentiation between hyphal types.

3. Longitudinal sections

Longitudinal sections in themselves provide little information regarding the structure of mature rhizomorphs. When considered along with transverse sections, however, longitudinal sections at the light microscope (e.g. Brand, 1989), SEM (e.g. Cairney *et al.*, 1988) or TEM level (Brownlee *et al.*, 1983) can greatly enhance our understanding of the degree of hyphal branching, anastomosis and interweaving within the rhizomorph.

C. Rhizomorph ontogeny

It is well established that the mode of apical extension of a rhizomorph does not necessarily reflect the degree of differentiation attained by the mature organ (Butler, 1966; Cairney *et al.*, 1989). Ontogenic patterns exhibited by particular species are therefore of great interest. Differentiation can be studied as a series of transverse/longitudinal sections taken along the rhizomorph length (Foster, 1981; Cairney *et al.*, 1988). Again, both SEM and TEM are particularly suitable here, with differential staining at the TEM level revealing important changes in hyphal characteristics as the rhizomorph develops (Foster, 1981). This, however, reveals little concerning the actual patterns of hyphal growth resulting in aggregation within the growing front. Careful observation of this region with the light microscope (Agerer, 1988) or SEM (Hornung and Jennings, 1981; Jennings and Watkinson, 1982) is invaluable in the study of early ontogenic events. Histochemical staining of light microscope preparations can permit early identification of chemical differences in hyphae which might be correlated with morphological/functional differentiation (Butler, 1958; Hornung and Jennings, 1981). Labelling with molecular probes (see Section III.B.2) during these early stages of development is also likely to yield valuable information in this respect.

IV. Conclusions

As with all studies of structure and development, a number of techniques are of potential value in the study of rhizomorphs. The techniques selected will ultimately depend on the nature of the investigation and the facilities available to the individual researcher. However, integrated studies of individual species, incorporating a number of the techniques outlined above, are likely to provide the most significant advances in our understanding of rhizomorph structure and ontogeny, and how this relates to the function of these important organs.

References

Agerer, R. (1987). *Colour Atlas of Ectomycorrhizae*. Einhorn Verlag, Schwäbisch Gmünd.

Agerer, R. (1988). *Nova Hedwigia* **47**, 311–334.

Alexander, C., Jones, D. and McHardy, W. J. (1987). *New Phytol*. **105**, 613–617.

Ashford, A. E. and Allaway, W. G. (1985). *New Phytol*. **100**, 595–612.

Bonfante-Fasolo, P., Perotto, S., Testa, B. and Faccio, A. *Protoplasma* **139**, 25–35.

Bowen, G. D. (1973). In *Ectomycorrhizae* (G. C. Marks and T. T. Kozlowski, eds), pp.151–201. Academic Press, New York.

Brand, F. (1989). *Nova Hedwigia* **48**, 469–483.

Brownlee, C. and Jennings, D. H. (1982). *Trans. Brit. Mycol. Soc.* **79**, 401–407.

Brownlee, C., Duddridge, J. A., Malibari, A. and Read, D. J. (1983). *Plant and Soil* **71**, 433–443.

Bullock, S., Ashford, A. E. and Willetts, H. J. (1980). *Protoplasma* **104**, 333–351.

Butler, G. M. (1958). *Ann. Bot.* **22**, 219–236.

Butler, G. M. (1966). In *The Fungi*, (G. C. Ainsworth and A. S. Sussman, eds), Vol. II, pp. 83–112. Academic Press, New York.

Cairney, J. W. G. (1990). *Mycol. Res.* **94**, 117–119.

Cairney, J. W. G. (1991a). *Cryptogam. Bot.* (in press).

Cairney, J. W. G. (1991b). *The Mycologist* **5**, 5–10.

Cairney, J. W. G., Jennings, D. H. and Veltkamp, C. J. (1988). *Nova Hedwigia* **46**, 1–25.

Cairney, J. W. G., Jennings, D. H. and Veltkamp, C. J. (1989). *Can. J. Bot.* **67**, 2266–2271.

Cairney, J. W. G., Jennings, D. H. and Agerer, R. (1991). *Cryptogam. Bot.* (in press).

Dell, B., Botton, B., Martin, F. and Le Tacon, F. (1989). *New Phytol*. **111**, 683–692.

Dighton, J. and Mason, P. (1985). In *Developmental Biology of Higher Fungi* (D. Moore, L. A. Casselton, D. A. Wood and J. C. Frankland, eds), pp. 117–139. Cambridge University Press, Cambridge.

Duddridge, J. A. and Read, D. J. (1984). *New Phytol*. **96**, 565–573.

Eamus, D., Thompson, W., Cairney, J. W. G. and Jennings, D. H. (1985). *J. Exp. Bot.* **36**, 1110–1116.

Finlay, R. D. and Read, D. J. (1986). *New Phytol*. **103**, 143–156.

Foster, R. C. (1981). *New Phytol.* **88**, 705–712.
Fox, F. M. (1987). *Trans. Br. Mycol. Soc.* **89**, 551–560.
Garbaye, J. and Bowen, G. D. (1989). *New Phytol.* **112**, 383–388.
Garrett, S. D. (1970). *Pathogenic Root-infecting Fungi.* Cambridge University Press, London.
Giovannetti, G. and Fontana, A. (1986). In *Physiological and Genetical Aspects of Mycorrhizae* (V. Gianinazzi-Pearson and S. Gianinazzi, eds), pp. 641–645. INRA Press, Paris.
Hornung, U. and Jennings, D. H. (1981). *Nova Hedwigia* **34**, 101–126.
Jennings, L. and Watkinson, S. C. (1982). *Trans. Brit. Mycol. Soc.* **78**, 465–474.
Kammenbauer, H., Agerer, R. and Sanderman, H. (1989). *Trees* **3**, 78–84.
Kohn, L. M. and Grenville, D. J. (1989). *Can. J. Bot.* **67**, 371–393.
Masui, K. (1926). *Mem. Coll. Sci., Kyoto Imperial Univ. (B)* **2**, 161–187.
Piché, Y. and Fortin, J. A. (1982). *New Phytol.* **91**, 211–220.
Rayner, A. D. M. and Todd, N. K. (1979). *Adv. Bot. Res.* **7**, 333–420.
Rayner, A. D. M., Powell, K. A., Thompson, W. and Jennings, D. H. (1985). In *Developmental Biology of Higher Fungi* (D. Moore, L. A. Casselton, D. A. Wood and J. C. Frankland, eds), pp. 249–279. Cambridge University Press, Cambridge.
Read, D. J. (1984). In *The Ecology and Physiology of the Fungal Mycelium* (D. H. Jennings and A. D. M. Rayner, eds), pp. 215–240. Cambridge University Press, Cambridge.
Rygiewicz, P. T., Miller, S. T. and Durall, D. M. (1988). *Plant Soil* **109**, 281–284.
Samson, J. and Fortin, J. A. (1988). *Mycologia* **80**, 382–392.
Skinner, M. F. and Bowen, G. D. (1974). *Soil Biol. Biochem.* **6**, 53–56.
Thompson, W. (1984). In *The Ecology and Physiology of the Fungal Mycelium* (D. H. Jennings and A. D. M. Rayner, eds), pp. 185–214. Cambridge University Press, Cambridge.
Thompson, W. and Rayner, A. D. M. (1983). *Trans. Br. Mycol. Soc.* **81**, 333–345.
Uhl, M. and Agerer, R. (1987). *Nova Hedwigia* **45**, 509–527.

14

Fine-structural Analysis of Mycorrhizal Fungi and Root Systems: Negative Staining and Cryoelectron Microscopic Techniques

F. MAYER, M. MADKOUR and A. NOLTE

Institut für Mikrobiologie der Georg-August-Universität Göttingen, D-3400 Göttingen, Germany

A. VARMA

School of Life Sciences, Jawaharlal Nehru University, New Delhi 110067, India

---

METHODS IN MICROBIOLOGY
VOLUME 23 ISBN 0-12-521523-1

Copyright © 1991 by Academic Press Limited
All rights of reproduction in any form reserved

I. Introduction

Visualization of fine-structural details in cells interacting in mycorrhizal systems is of major importance; structure–function relationships can only be understood when the structural organization is known. Electron microscopic techniques which allow a reliable preservation of the *in vivo* conditions and high resolution of structural aspects should be favoured. Nevertheless, conventional techniques should also be used to complement other findings. Knowledge of the spatial distribution of macromolecular components in tissues and cells is another prerequisite for a better understanding of mycorrhizal interactions. Many biological systems other than mycorrhiza require the same reliability of cellular and molecular preservation when their structural aspects are to be analysed. The recent progress in the development of preparation techniques designed for other biological samples besides mycorrhiza should also be exploited for the investigation of mycorrhizal systems.

This chapter looks at several different approaches that can be used for fine-structural analysis. These are the negative staining technique, not commonly used in mycorrhiza research; the preparation of frozen-hydrated samples for high resolution transmission electron microscopy; the freeze substitution of material subsequently cut for transmission electron microscopy; the preparation of ultra-thin cryosections; and immunoelectron microscopic techniques. Some potentially useful combinations of such approaches which have not yet been used in studies of mycorrhiza are here described.

A wealth of information has been accumulated by application of conventional transmission electron microscopy for the analysis of ultra-thin sections prepared by the usual steps, i.e. chemical fixation, dehydration, embedding in resin, heat-polymerization of the resin, sectioning, and post-staining. In combination with this approach, studies have been made of the cytochemical localization of protein and/or polysaccharide (Gianinazzi-Pearson *et al.*, 1981) and of enzyme activities by visualization of specific precipitates (Lacaze, 1983) or by indirect immunogold labelling of enzyme (Straker *et al.*, 1989; Spanu *et al.*, 1989). By use of very similar sample preparation techniques, but combined with the application of scanning transmission electron microscopy and atomic absorption spectrometry, granular inclusion bodies have been investigated with respect to their chemical composition (Vare, 1990). Conventional scanning electron microscopy often reveals well-preserved structures when sample preparation is done with freeze-drying or freeze-substitution. Certain disadvantages of this procedure can be overcome by cryofixation, where specimens are frozen in liquid nitrogen and viewed in the frozen-hydrated state (Alexander *et al.*, 1987; Jones *et al.*, 1987). By a combination of cryoultramicrotomy and a labelling technique making use of ferritin linked to wheat-germ agglutinin, the presence of *N*-acetylglucosamine in the fungal cell wall has been demonstrated by Bonfante-Fasolo (1982) in a study of the cell wall architectures in a vesicular-arbuscular mycorrhizal association. This study concluded that cryoultramicrotomy is a powerful tool for the elucidation of macromolecular arrangements.

Amongst the approaches described in this chapter that have not been extensively applied in studies of mycorrhiza is that of negative staining, a procedure often applied in other areas. Negative staining is appropriate for very small cells such as bacteria and for cell components of sizes down to macromolecules such as enzyme complexes. However, the negative staining technique does not lend itself to good structural preservation due to the air-drying which constitutes the final step of the procedure prior to electron microscopic examination. Another approach not commonly used in mycorrhiza research is the preparation and analysis of "frozen-hydrated" samples; this technique avoids air–drying. It can be applied for the visualization of ultrastructural details at the macromolecular level and up to dimensions of viruses, membrane vesicles, isolated cell components, cells and tissues. The preparation procedure appears to be simple; however, in practice difficulties are encountered when good quality micrographs are needed. The reasons for this are two-fold: first, freezing ("vitrification") may not proceed without complications which may lead to the appearance of artefacts; and second, this kind of sample requires adequate instrumentation and careful treatment (i.e. low dose electron microscopy) during screening and image recording; focusing especially is a matter of experience, and usually the primary contrast obtained by imaging is low.

II. Negative staining

A. Principle

Negative staining is achieved by using solutions of heavy metal salts dissolved in water (Valentine *et al.*, 1968). The solution surrounds the specimen placed on a support film and penetrates to some extent into cavities and clefts of the sample. After air-drying, the remaining dried negative staining salt appears dark in conventional transmission electron microscopy; the sample proper is more electron transparent. Thus, a "negative" image is obtained by stain exclusion. A side effect of the procedure is that the stain, which is in tight contact with the sample and forms a "shell" around it, reduces severe radiation damage to the specimen caused by interaction of the electron beam with the sample. However, as mentioned above, air-drying may nevertheless cause flattening of the sample (Johannssen *et al.*, 1979). In addition, the final image formed from the sample is complex: it originates from both sides of the object. Therefore, image interpretation may be difficult.

B. Preparation of support films

1. Carbon support film

Carbon support films are prepared from the tips of two carbon rods brought in contact and heated up to sublimation temperature by an electric current (resistance evaporation), in a vacuum apparatus (Sleytr *et al.*, 1988). The carbon is deposited onto a piece of freshly cleaved mica (about 2 cm × 5 cm). The thickness of the carbon film may be estimated by visual inspection (a grey to dark-grey layer should be obtained); estimation of film thickness is simplified by placing a piece of white paper below the mica sheet. Insufficient or excessive thickness of the carbon film can be seen immediately by removing the paper from below the mica sheet. The mica covered by the carbon film is stored in a Petri dish.

2. Plastic film

A conventional plastic support film can be produced from a solution containing 0.5% (w/v) Formvar in water-free chloroform (Spiess and Mayer, 1976). A light-microscope glass slide is dipped into this solution and removed after a few seconds. Complete air-drying of the thin layer of solution covering the two faces of the slide produces two thin Formvar films still attached to the glass slide. They can be removed from the glass surface by slowly submerging the slide into distilled water; prior to this step, the edges of the slide should be scratched with a knife in order to remove the thin Formvar film along these edges, thus allowing the two films to float freely off the glass surfaces. Electron microscope copper or nickel grids (available in a number of mesh sizes) are then carefully placed on top of the floating films. The films covered with the grids are removed from the water surface by placing a piece of filter paper of appropriate size onto each floating film. As soon as the filter paper is completely wettened, it is removed, together with the adhering grids and films, from the water surface and dried on filter paper with the grids facing upwards. The filmed grids, still attached to the filter paper, are stored in a Petri dish. After one day, the Formvar films covering the grids can be stabilized by a thin layer of carbon deposited on the films by carbon sublimation as described above (Section II.B.1).

C. Preparation of negative staining solutions

Different stains (1–4%, w/v, aqueous solutions) may be used (Sleytr *et al.*, 1988). The conventional ones are:

- uranyl acetate, pH 4.2–4.8
- phosphotungstic acid neutralized with KOH or NaOH
- ammonium molybdate, pH 6.5–7.0

The solutions should be filtered to remove visible precipitates. They should be stored at room temperature; the uranyl acetate solution should be kept in a brown glass bottle. Note that uranyl acetate is radioactive.

D. Preparation of samples: mounting and negative staining

The sample in appropriate buffer (buffer systems other than phosphate buffer are preferred) should be prepared at concentrations of around 20–100 μg protein ml^{-1} in the case of proteins; when small cells are to be analysed, the suspension should be visibly turbid. Additions of sugar (up to 10%) or glycerol (up to 20%) are allowed. Detergents should be avoided.

1. Direct mounting on filmed grids

One droplet of the sample is placed directly onto a grid covered with a film of Formvar or carbon-reinforced Formvar (see above) (Lalucat,

1988). The components in the sample are allowed to sediment and adsorb for 1–3 min. The excess liquid is then removed by use of a piece of filter paper. Next, a droplet of buffer is placed onto the grid (washing) and removed with filter paper (this washing step may be omitted). Before the sample becomes completely dry, a drop of staining solution is deposited onto the grid and removed with filter paper after 30–60 s. The grid is now air-dried and is ready for electron microscopic inspection. The grids can be stored in a desiccator for several months.

2. *Floating technique*

The sample is placed in a small container (0.5–1 ml) or, as a drop, onto a hydrophobic surface such as Parafilm. Next, a piece, measuring 3 mm × 3 mm, is cut from carbon-coated mica (see Section II.B.1 above), and submerged, at an angle, into the solution or suspension containing the sample. This causes the carbon film to partially float off the mica surface; it should still have contact with a small area of the mica sheet to facilitate manipulation. After a few seconds to 1–2 min, depending on the sample concentration, the mica with the adhering carbon film is removed from the sample. The carbon film returns into its former position on the mica sheet, with the adhering sample components attached to the surface facing the mica. The carbon film can then be floated for a few seconds on a buffer surface (washing), and removed (this washing step may be omitted). The mica–carbon sandwich is now touched, for 1–2 s, to a piece of filter paper; make sure that the sample does not become completely dry. Finally, the carbon film is floated completely off the mica on the surface of the negative staining solution and left for a few seconds. The carbon film with adhering sample components is then removed from the staining solution by submerging a grid (without any support film) into the negative staining solution, positioning it below the floating film, and lifting it up. Thus, the floating film is attached to the grid. The grid (carbon-covered side facing upwards) is carefully touched to a piece of filter paper. Make sure that the staining solution is not completely sucked off the sample; a minimal amount of staining solution has to remain on the grid which is now air-dried, without further contact with filter paper, and is ready for electron microscopic inspection (Valentine *et al.*, 1968). Figure 1 shows the sequence of steps.

Neither of the variations of negative staining described here include chemical fixation. However, this kind of fixation may be applied either by using a sample pre-fixed with glutaraldehyde (1–2%, v/v), or by replacing the washing step mentioned for the floating technique by floating on the surface of a fixing solution (also glutaraldehyde at the concentrations given above).

E. Electron micrography

Negatively stained samples should be investigated with electron doses that are as low as possible. This is because the electron beam interacts

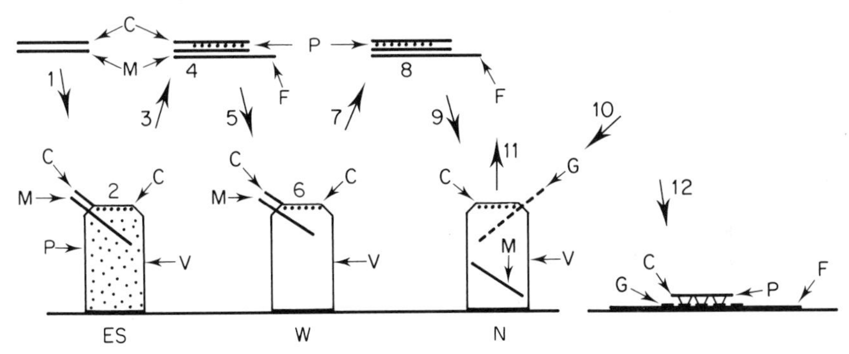

Fig. 1. The "negative staining" procedure (floating technique). A thin layer of carbon (C) deposited onto a clean mica (M) surface by vacuum heat sublimation is partially floated off on the surface of the solution or suspension containing the sample (ES) in a small vessel (V) (Step 1). The sample particles (P) adsorb to the lower surface of the carbon film (Step 2); adsorption times depend on the size of the sample particles (the larger the particles, the longer the time) and their concentration (the lower the concentration, the longer the time). Next the carbon-mica sandwich is removed from the sample suspension (Step 3); the sample particles are trapped between carbon and mica, and still surrounded by a small amount of buffer. The buffer is partially removed from the sandwich by a brief contact with filter paper (F) (Step 4), and the carbon film with the adhering sample particles is immediately (partially) floated off on the surface of a washing solution (W, buffer of low concentration or water, or a solution containing fixatives; see text) (Step 5). By this procedure, both buffer from the original sample suspension and remaining sample particles not attached to the carbon support film are removed (Step 6). Afterwards, the carbon–mica sandwich is lifted off from the washing solution (Step 7), and again partially dried on filter paper (Step 8), and the carbon film with the attached sample particles is completely floated off from the mica on the surface of the negative staining solution (N) (Step 9). The mica is no longer needed. Now the carbon support film with the sample particles still adhering to its lower side is picked up (Step 10) from underneath with an electron microscope grid (G), lifted off from the vessel (Step 11), and partially blotted dry on filter paper (Step 12). Care must be taken that a very small amount of negative staining salt solution remains surrounding the sample particles. The sample is finally air-dried and ready for electron microscopic inspection. The whole procedure (Steps 1 to 12) takes between 1 and 3 min.

with the biological components of the sample and with the staining salt. At high electron doses, the ultrastructure at the macromolecular level is partially lost due to breaks of bonds, loss of material and rearrangement of parts of macromolecules; microcrystals are formed in the negative staining salt. These effects taken together result in a loss of resolution.

III. Preparation and investigation of frozen-hydrated samples

A. Principle

Samples containing protein macromolecules, viruses, membrane vesicles or small cell components can be visualized, without chemical treatments such as fixation and staining, in the frozen-hydrated state (Dubochet and McDowall, 1981; Dubochet *et al.*, 1982; Dubochet, 1984; Müller, 1988). To achieve this, the sample has to be shock-frozen (vitrified) in an ultrathin layer of buffer spanning small holes in a support film. Thus, the sample components are surrounded by amorphous ice. Larger samples such as cells can also be prepared in the frozen-hydrated state (Dubochet and McDowall, 1981; Dubochet *et al.*, 1982).

B. Preparation of support films with small holes

Vigorous shaking of 50 ml of 0.2% (w/v) collodion–ethylacetate solution to which 0.33 ml of glycerol–water solution (86%, v/v) and 0.1 ml Lenor (fabric softener, Procter and Gamble Ltd) are added, produces a homogeneous suspension of small droplets. This suspension is used to deposit, by the conventional method (see Section II.B.2 above), a holey collodion film on grids. Note that collodion is used instead of Formvar. Because collodion is less stable than Formvar, for stabilization a thick layer of carbon should be evaporated onto the film (see Sections II.B.1 and II.B.2 above). This thick carbon layer does not interfere with the investigation of the sample; the sample which is analysed is contained in the thin layer of ice spanning the holes. For further stabilization, this support is finally heated at 180 °C for 10 min (Johannssen *et al.*, 1979). A modified version of the procedure is described by Lünsdorf and Spiess (1986).

The surface of the support films has to be rendered hydrophilic by dipping the grid, shortly prior to use, into 96% ethanol. Hydrophilicity will be retained for several hours. An alternative procedure is glow discharge (Dubochet *et al.*, 1971).

C. Preparation of frozen-hydrated samples

1. Samples for direct visualization (without sectioning)

The concentration of the sample components to be depicted has to be optimized. A useful rule is that protein macromolecules should be present at a concentration between 0.1 and 5 mg ml^{-1}.

A grid, covered with a hydrophilic holey film, is held vertically with forceps, submerged in the solution or suspension, and removed, again vertically. Next, under visual inspection, the liquid clinging to the grid is allowed to drain, forming a drop hanging at the lower edge of the grid, and leaving a very thin layer of liquid covering the upper half of the grid. The drop is removed with filter paper, and the sample is shock-frozen in liquid propane cooled with liquid nitrogen in an appropriate apparatus. Warming up of the sample has to be prevented during the period between shock-freezing and inspection in the pre-cooled electron microscope (Dubochet, 1984).

2. Preparation of ultra-thin cryosections

(a) Pre-treatment of the sample and freezing. Details are given in Section IV.B below.

(b) Ultra-thin sectioning and collection of sections. Details are given in Section V.C.2 below. The collection of the cryosections is done by picking up the sections adhering to the knife with untreated copper or nickel grids. Note that any warming up during this procedure and within the period of time between collection of sections and their visualization in the pre-cooled electron microscope must be avoided.

D. Electron microscopic investigation of frozen-hydrated samples

As mentioned above, the grid carrying the frozen-hydrated sample (vitrified small particles or cryosections) is investigated in an electron microscope equipped with a low-temperature stage. The total electron dose allowed should not exceed 3000 e nm^{-2}, including screening of the grid and focusing. This means that screening is usually done at very low magnification and low beam intensity and micrographs are taken at magnifications as low as possible (usually not higher than 20 000 ×). Focusing must not be done at the exact location of the detail of interest, but close to it. Electron microscopes are available which fulfil all necessary prerequisites to perform this kind of analysis. An inbuilt energy filter simplifies the observation and enhances the final contrast (Bauer, 1988).

IV. Freeze-substitution and preparation of ultra-thin sections

A. Principle

Although the omission of chemical fixation would be most desirable, protein and lipid fixation is usually necessary for morphological studies of cells and tissues. On the other hand, application of immunoelectron microscopic techniques to freeze-substituted samples requires that the concentrations of fixatives be as low as possible. This means abandoning high quality preservation and visibility of ultrastructural details. In any case, a compromise has to be found (Robards and Sleytr, 1985; Müller, 1988).

The technique of freeze-substitution combines shock-freezing, substitution of amorphous ice by methanol, embedding in low temperature resin, polymerization of the resin in ultraviolet light at low temperature, and conventional ultra-thin sectioning. After optimization, this approach not only gives reasonable preservation of cell ultrastructure, but also protects antigenic components in the cell from severe loss of antigenicity. This is important when immunoelectron microscopy is applied.

B. Pre-treatment of samples and freezing

1. Instrumentation

Equipment (an instrument for freezing, an instrument for freeze-substitution), available from various companies, is needed which allows rapid freezing of specimens followed by substitution of the ice in the sample by an organic solvent. This has to be done at temperatures which avoid damage by recrystallization of the frozen water in the sample. An automatically controlled apparatus for freeze-substitution is a major advantage.

2. Sample pre-treatment

Depending on the type of specimen (isolated cell components, cells, tissues), addition of fixatives such as glutaraldehyde (up to 30%, v/v) uranyl acetate (0.5–2%, w/v) or osmium tetroxide (0.5–1%, w/v), (Slot *et al.*, 1988) and of glycerol for cryoprotection (Robards and Sleytr, 1985; Müller, 1988) may be necessary. These additions should be applied at concentrations which, together with the buffer system, are in sum isotonic to the osmotic state of the sample. Optimization is necessary, especially when plant material is to be examined. A considerable amount of data in this respect has accumulated, derived in particular from the field of freeze-etching and freeze-fracturing (Müller, 1988).

3. *Freezing*

A drop of liquid containing the pre-treated sample is put on a copper grid; its volume should be as low as possible (around 1 mm^3); note that the most rapid freezing only takes place close to the surface of the specimen; inside, the cooling rate is considerably lower. Now the sample is quickly dipped into propane (which again is cooled by liquid nitrogen) by use of pre-cooled tweezers, removed, and immediately transferred to liquid nitrogen to avoid warming up (Dubochet, 1984).

C. Freeze-substitution, resin-embedding, and sectioning

1. *Substitution of the amorphous ice by methanol*

The frozen sample is transferred, without any warming up, to the freeze-substitution apparatus, and quickly submerged into the pre-cooled absolute methanol which has a temperature of −80 °C. Substitution is achieved by keeping the sample for a period of time (to be optimized; at least 18 h) at −80 °C in methanol; afterwards, the system is slowly warmed up (+ 1.5 °C h^{-1}) until −40 °C is reached.

2. *Low temperature resin-embedding and resin polymerization*

Now the sample is transferred into a 1:1 mixture of methanol and low temperature resin pre-cooled at −35 °C. This step is followed by the conventional treatment for low temperature embedded samples, i.e. the final replacement of any remaining methanol by resin (at −35 °C), ultraviolet (UV) polymerization (at −35 °C) for several days, and finally UV polymerization at room temperature for 1–3 days.

3. *Ultra-thin sectioning*

Ultra-thin sections are produced at room temperature by the routine procedure usually used for any kind of resin-embedded material. The resulting sections can either be post-stained as usual for morphological studies, or they can be used for immunoelectron microscopy.

V. Preparation of ultra-thin cryosections for morphological studies and immunoelectron microscopy

A. Principle

Cryotechniques are suited for good structural preservation combined with preservation of antigenicity of the sample components. However, usually it is not possible to obtain the best ultrastructural preservation as well as the best efficiency of immunolabelling. This is because chemical fixation is obligatory for the best structural preservation when cell and tissue components have to be depicted with high resolution (Robards and Sleytr, 1985; Müller, 1988; Slot *et al.*, 1988). The necessary concentrations of chemical fixatives may cause severe reduction of reactivity of antigenic sites with antibodies due to alterations of the exposed epitopes of the antigens. Thus, optimization of the procedure is necessary, and often this can only be achieved by trial and error. Below, certain basic data will be given for the start of such optimization experiments.

B. Pre-treatment of samples and freezing

The procedures described in IV.B can be followed for pre-treatment steps, such as fixation and cryoprotection, and for freezing.

C. Cryosectioning and treatment of sections

1. Instrumentation

Cryoultramicrotomes are available from a number of suppliers. Note that liquid nitrogen should be routinely available in reasonable amounts (up to 20 litres per day). The knife may be made from glass (Tokuyasu and Ukamara, 1959; Roberts, 1975; Griffiths *et al.*, 1983, 1984).

2. Mounting of the sample in the cryoultramicrotome, and optimization of sectioning conditions

The sample is attached to the pre-cooled sample holder in the cryoultramicrotome. As optimal freezing of the sample only occurs close to its surface, it should have a suitable shape (i.e. not too large, not too flat) for obtaining sections of optimal size from the very first few cuts. The temperature of the sample (i.e. the holder together with the sample), the knife, and the sectioning chamber can be selected separately in up-to-date cryoultramicrotomes. As a rule, the sample should have a temperature of around −85 °C; however, this temperature has to be optimized by inspection of the resulting sections. They should not fall apart but should be obtained in their full size. In most cases, the sections will be single; serial sectioning cannot be expected, although it is possible. The temperature of the knife should, for a start, also be set to a value close to −85 °C. Note that the sections may fall apart and crumble at sample temperatures below −85 °C because the sample may then be too brittle. Speed of cutting should be around 3 $mm\,s^{-1}$.

3. *Collection and post-treatment of sections*

The sections produced at these low temperatures cannot be collected on the surface of a conventional liquid such as water in a trough as in conventional ultra-thin sectioning procedures (Robards and Sleytr, 1985). The sections remain attached to the knife, close to its edge. They are removed from the knife and brought onto a support as follows (Robards and Sleytr, 1985; Slot *et al.*, 1988): a drop of highly concentrated (about 2.3 M) sucrose solution is filled into a loop formed from platinum wire. The loop needs to have a diameter very similar to that of a copper or nickel grid (i.e. 3 mm). The sucrose solution in the loop is now cooled in the instrument chamber, and brought in contact with the sections. Note that the sections adhering to the knife are usually rolled to variable extents. The sections are flattened by attachment to the surface of the pre-cooled ("sticky") drop of sucrose. Usually, the section will now remain in contact with the surface of the cooled drop when the drop is removed from the knife. Next, the drop with the adhering section is exposed to room temperature; the drop starts to melt, and the section is now "floating" on the surface of the melting drop. It is brought in contact with a filmed (Formvar, carbon reinforced) copper or nickel grid (depending on which future treatments will follow; nickel grids for immunoelectron microscopy), with the result that the sections are removed from the surface of the melting drop and finally attached, no longer rolled but flat, to the surface of the support film. The grid with the sections is now floated, sections downwards, on water to remove the sucrose. If the sections become dispersed by this procedure, 0.5–2% gelatin may be added to the 2.3 M sucrose solution used to form the drop.

The sections attached to the support film should be stabilized with methyl cellulose and stained with uranyl acetate as follows (Slot *et al.*, 1988): A 2% solution is made from 25 centipoise methyl cellulose (e.g. methocel^R, Fluka AG, Buchs, Switzerland). The powder is added to water pre-heated to 95 °C. Mixing is done with a magnetic stirrer at 95 °C for a few minutes and then the solution is put on ice. The solution is mixed for at least 4–8 h at 0–4 °C, and then left for a further 3–4 days at 4 °C. It is then centrifuged at high speed (e.g. 60 000 rpm in a Beckman 60Ti or 70Ti) for 90 min at 4 °C. The tubes are stored in the refrigerator for at least 3–4 weeks (!) without disturbing the pellet that appears. For each experiment small amounts are pipetted carefully from the surface layers. For negative staining the methyl cellulose solution is mixed with 2–4% aqueous uranyl acetate solution to give a final concentration of 0.1–0.4% uranyl acetate. Convenient amounts of components are 0.9 ml of the 2% methocel mixed with 0.1 ml of 3% uranyl acetate.

After preparation of this mixture, it is centrifuged, shortly prior to use, for 90 min at 60 000 rpm (see above). The grids are now put onto drops of the mixture (two drops in sequence, room temperature), and then for 10 min on the surface of a larger quantity of the mixture. Afterwards, the grids are removed from the mixture with a loop (3.5 mm diameter) of platinum wire. The drop covering the grid is sucked off with filter paper. The optimal thickness of the methyl cellulose–uranyl acetate layer can be determined by variation of the remaining amount of the mixture; to this end, each sample is checked by electron microscopy for best contrast and the absence of drying artefacts.

VI. Immunoelectron microscopic techniques

A. Principle

The spatial distribution of antigens in an ultra-thin sectioned sample can be analysed by application of a post-embedding detection system based on IgG antibodies coupled to ferritin (Wagner *et al.*, 1980) or to colloidal gold particles (Faulk and Taylor, 1971; Romanow *et al.*, 1974, 1975; Roth *et al.*, 1978, 1981; Bendayan *et al.*, 1980; Bendayan, 1982; Bendayan and Zollinger, 1983; Acker, 1988; Rohde *et al.*, 1988; Slot *et al.*, 1988). Pre-conditions are that the position of the antigens is not altered during the preparation procedure, that a sufficient preservation of the ultrastructure is retained, that the antigenicity of the components of interest is preserved, and that the controls indicate the specificity of the antibodies and low background labelling. Both polyclonal antisera and monoclonal antibody preparations may be used. Experience tells us that, due to higher efficiency of labelling, monospecific polyclonal antisera are best suited for this kind of approach.

A number of variations of the method can be used:

- Protein A–gold complexes are coupled to the antigen-specific IgG antibody.
- The specific (rabbit) antibody bound to the antigen on the surface of the section is marked by a labelling system consisting of a secondary (goat anti-rabbit) IgG antibody coupled to gold without protein A.
- The antigen is labelled using specific rabbit IgG antibody coupled to colloidal gold without protein A.

The sample preparation may be performed by conventional low temperature embedding (using Lowicryl resins) or by freeze-substitution or cryoultramicrotomy.

B. Preparation of marker systems (colloidal gold)

1. Principle

Colloidal gold particles are electron dense markers which can be coupled to certain proteins in general and especially to IgG antibodies. Thus, the site where the antibody is located on the surface of an ultra-thin section can be visualized by depicting the distribution of the gold particles relative to the biological structure. This holds true as long as no cross-reactivities with other sample components occur (Rohde *et al.*, 1988). Other marker systems, e.g. ferritin (Wagner *et al.*, 1980), are available. However, the electron density of ferritin is relatively low compared to that of colloidal gold. In addition, the gold system is more versatile because colloidal gold particles of different sizes can be prepared. Thus, simultaneous labelling of two (or even three) different antigens in a sample may be achieved (Freudenberg *et al.*, 1989).

2. Glassware and reagents

Even small amounts of contaminants on the surface of the glassware will interact negatively with the reduction process in the preparation of colloidal gold. All glassware has to be extremely carefully cleaned, rinsed repeatedly with distilled water, and siliconized. Siliconization is achieved as follows (Rohde *et al.*, 1988): Sigmacote (article no. SL-2, Sigma) as available from stock is poured into the tube; after shaking (to cover the complete inner surface of the tube) the solution is poured out, and the tube is dried at an adequate temperature (80–120 °C, for 1–2 h).

3. *Preparation of colloidal gold particles of specific sizes*

(a) Particle size between 8 and 10 nm. 1 ml of 1% tetrachloroauric acid (stock solution, can be stored at 4 °C for several months) and 1.5 ml of 0.2 M K_2CO_3 are mixed with 25 ml water on ice. While stirring, 1 ml of 0.7% aqueous sodium ascorbate is added quickly. This causes a change of the colour to brownish purple-red. Finally, the volume is adjusted with water to 100 ml and heated until a wine-red colour appears (Slot and Geuze, 1981).

(b) Particle size 15 nm. 1 ml of 1% tetrachloroauric acid (see above) is added to 100 ml water. After boiling, 4 ml 1% aqueous trisodium citrate are rapidly added. To complete the reduction process the mixture is boiled for another 15 min until a wine-red colour appears (Frens, 1973; Slot and Geuze, 1981).

(c) Particle sizes of 4, 6, 8 or 12 nm.

Solution 1: 1 ml of 1% tetrachloroauric acid in 79 ml water.

Solution 2 (reduction mixture): 4 ml 1% trisodium citrate and 0–2 ml of 1% tannic acid (note: tannic acid, available from Mallinckrodt, St Louis, USA, code no. 8835, should be used). The more tannic acid is added, the smaller is the gold particle size.

Procedure: The same volume 25 mM K_2CO_3 and 1% tannic acid are combined to adjust the pH. The total volume is adjusted to 20 ml with distilled water. Both solutions are heated to 60 °C. The reducing mixture is added quickly to Solution 1 while stirring. The reduction process is completed by boiling until the solution turns to the final wine-red colour (Slot and Geuze, 1981).

4. *Stabilization of colloidal gold*

Colloidal gold particles have a tendency towards agglomeration and fusion. This has to be prevented by a stabilization procedure (Horisberger *et al.*, 1975; Horisberger and Rosset, 1977; Rohde *et al.*, 1988).

An important surface characteristic of colloidal gold particles is their negative surface charge; therefore, they can interact with positively charged proteins. The stability of colloidal gold is maintained by electric repulsion. Addition of electrolytes to unprotected gold solutions immediately causes agglomeration of the gold particles. The proteins used to stabilize the particles bind to the gold by non-covalent electrostatic van der Waals' forces. To achieve this, appropriate conditions of concentrations and pH have to be used. For protein A, a pH between 6.0 and 6.9 is adequate; the pH for binding of IgG antibodies is around 8.9. Adjustment of pH is performed by addition, to the colloid gold sample, of 0.2 M K_2CO_3 for raising and 0.1 N acetic acid for lowering the pH value. For the determination of the minimum amount of protein A or IgG antibody necessary to stabilize a certain amount of colloidal gold, the following procedure can be applied: 0.1 ml of serial dilutions of the respective protein are added to constant volumes (1 ml) of colloidal gold and kept at room temperature for 15 min. To check the state of stabilization, 0.1 ml of 10% sodium chloride is added. Stabilized solutions retain their wine-red colour (blue colour would indicate agglomeration of the gold). In the experiment proper, the protein is used in 10% excess of the minimum stabilizing amount determined by the dilution series experiment. For the experiment the colloidal gold solution is added to the protein, which should be dissolved in distilled water. After 10 min, a 1% aqueous polyethylene glycol (PEG 20000) solution is added for further stabilization.

Ultracentrifugation is used for removal of non-complexed protein and insufficiently stabilized gold particles as follows:

for 2–4 nm gold: 110000 *g*, 4 °C, 1.5 h
6–8 nm gold: 80000 *g*, 4 °C, 45 min
12–15 nm gold: 30000 *g*, 4 °C, 20 min

The small dark pellet formed during centrifugation is discarded. The colourless supernatant, containing unbound protein, is also discarded. A loose, intensely red-coloured band has formed close to the bottom of the tube. This is the material of interest. It is collected and resuspended in buffer (e.g. phosphate-buffered saline containing 0.2 mg PEG ml^{-1}). The centrifugation procedure is repeated twice.

Detailed recipes are as follows:

(a) Protein A–gold complex. 10% excess of the optimal stabilizing amount of protein A is dissolved in 0.2 ml distilled water. The colloidal

gold, pH 6.9, is added to this solution, followed by incubation for 15 min at room temperature. 1% PEG is added. The mixture is centrifuged under conditions depending on the gold particle size required (see above). Finally, the protein A–gold complexes are suspended in phosphate-buffered saline, pH 6.9, containing 0.2 mg ml^{-1} PEG and sodium azide (0.01%), and stored at 4–6 °C.

(b) IgG antibody–gold complex. The IgG antibody (either rabbit IgG raised specifically against the antigen of interest, or goat anti-rabbit IgG) solution is dialysed against 2 mM borax–HCl buffer, pH 9.0. Protein aggregates are removed by centrifugation (100 000 × *g*, 4 °C, 1 h). The colloidal gold solution, adjusted to pH 9.0, is added to a 10% excess of the optimal stabilizing amount of IgG antibody. After 10 min, bovine serum albumin (BSA) in distilled water (pH adjusted to 9.0 with NaOH) is added to give a final concentration of 1%. After centrifugation (conditions depending on gold size, see above), the clear supernatant is discarded, and the IgG antibody–gold complexes are suspended in 20 mM Tris-buffered saline, pH 8.2, containing 1% BSA. This centrifugation is repeated twice. The complexes are finally suspended in Tris-buffered saline containing BSA and sodium azide (0.01%), and stored at 4–6 °C.

C. Pre-treatment of antisera/antibodies

1. *Affinity chromatography*

When pure monoclonal antibody preparations are used, no pre-treatment (besides dilution) is necessary. For the use of polyclonal antisera it is recommended that the antiserum is purified by affinity chromatography using protein A–sepharose CL-4 B (Rohde *et al.*, 1988). Protein A only binds to the Fc part of an IgG antibody, and to free Fc fragments. Contaminating protein, possibly causing background labelling, is not bound and, thus is no longer present in the sample which is obtained by elution from the column. Contaminating proteins interfering with the specific labelling of the antigens of interest would be free Fab fragments detached from the IgG part carrying the Fc component. Therefore, no masking of antigenic sites on the ultra-thin sections can be caused by free Fab fragments, competing with intact IgG antibodies for the antigenic sites. The free Fc parts present in the final sample do not interfere with the procedure as they do not bind to the antigen.

2. *Reduction of background labelling, and optimization of antibody concentration*

It is reasonable to assume, and in fact it has been shown, that the efficiency of labelling is dependent on the concentration of IgG antibodies in the labelling mix (Rohde *et al.*, 1988). The higher the concentration, the higher is the chance that labelling is quantitative with respect to the antigens exposed at the surface of the ultra-thin section. Therefore, dilution series have to be prepared and examined. Usually, dilution rates between 1:10 and 1:1000 are adequate for sufficient labelling. Fully quantitative labelling cannot be expected because, in parallel with specific labelling, background labelling is increased. The background can be significantly reduced by covering the surface of the ultra-thin section, prior to labelling, with a layer of ovalbumin or milk powder. This is done by incubation, for 5 min, of the sections attached to the support film at room temperature, first on a drop of adequate buffer (phosphate-buffered saline, PBS) containing 1% ovalbumin or 0.5% milk powder. Next, the grids are blotted, for a few seconds, on filter paper. They should not be allowed to dry afterwards. Immunolabelling must follow immediately.

D. Immunolabelling of ultra-thin sections

1. *Protein A–gold technique*

After blotting (see Section VI.C.2) the sections on the grids are incubated, by floating on the antibody solution, for various periods of time and at different temperatures. Make sure that evaporation is minimized (cover). The incubation time should be as short as possible to keep background labelling low. The sections are then rinsed in PBS to remove unbound antibody. This is done by spraying gently from a plastic bottle for 10–20 s and by floating the grids for 5 min each on several drops of PBS. For marking with colloidal gold, the sections are incubated on drops of a dilution series of the marker system at room temperature for 1–2 h. Afterwards the grids are rinsed in PBS and distilled water as described above, and post-stained with uranyl acetate (4%, 3–5 min) and lead citrate (15–30 s) before electron microscopic inspection.

2. *Antibody–gold techniques*

The primary (rabbit) IgG antibody, instead of labelling with the protein A system, may be labelled with a secondary (goat anti-rabbit) IgG antibody which is coupled, without protein A, to colloidal gold. The procedure is very similar to that described above (see Section VI.D.1), with one exception: additional application of ovalbumin or milk powder is advisable prior to incubation with the secondary antibody–gold complex.

A simplified version of the labelling technique uses antigen-specific (rabbit) IgG antibody directly coupled, without protein A, to colloidal gold. The procedure is very similar to that described for the first steps of the protein A–gold technique; instead of the antigen-specific rabbit IgG antibody not coupled to gold, such an antibody already coupled to gold is used.

3. Controls

Obvious and necessary control experiments (Rohde *et al.*, 1988) are:

(1) Incubation of the sections with the electron-dense marker alone, omitting the incubation with the specific antibody. This control identifies the non-specific binding of the marker to the resin.
(2) Allowing the specific antibody to react with the corresponding antigen prior to the incubation of the sections, followed by incubation with the marker verifying the specificity of the antigen –antibody reaction.
(3) Incubation of the sections with the specific antibody, followed by incubation with IgG antibody or protein A alone, and then by the IgG antibody–gold and protein A–gold complexes; this control verifies the interaction between the specific antibody and IgG antibody or the protein A.
(4) Incubation of the sections with the pre-immuno serum, purified in the same way as the specific antibody, followed by the marker system; this reveals non-specific binding of IgG antibody to the resin.
(5) Performing controls taking advantage of physiological conditions in which, for example, the enzyme to be localized is not expressed.

E. Electron micrography and evaluation of micrographs

Representative samples should be depicted in a number sufficient to apply statistical evaluation procedures. Note that the values obtained are not absolutely quantitative: the labelling system will usually not detect all antigens exposed at the surface of the section; antigens buried within the section cannot be detected, and the washing steps may remove part of the label or the marker system. In complex labelling systems such as those involving primary and secondary antibody, a 1:1 ratio of gold particles to antigenic sites cannot be expected. The position of a gold particle may very well not exactly coincide with that of the labelled antigen due to the fact that always at least one IgG antibody (length about 12 nm) is between the antigen and the gold particle. The antibody may be tipped over. Thus, labelling of antigens located at the inside of the cytoplasmic membrane may result in their (erroneous) localization at the outside of the membrane. Techniques (pre-embedding labelling) are available to avoid this phenomenon. In this approach, labelling is performed prior to embedding (Rohde *et al.*, 1988). However, penetration of the labelling system into cells is not possible without severe damage to the cell. Only surface-exposed antigens can be identified and localized by this technique without major trial-and-error procedures (Acker, 1988; Rohde *et al.*, 1984, 1985, 1988). Figure 2 presents, in a diagrammatic way, the situation encountered when gold particles of different sizes are used as markers.

VII. Conclusions

The combination of electron microscopic preparation, imaging and image evaluation procedures which involve steps aimed at reasonable preservation of structure, identification and localization of cellular macromolecular components and good resolution will considerably improve our understanding of interactions between organisms in mycorrhizal systems. Some of these approaches are complex and sophisticated, and their application requires expensive instrumentation. However, as this instrumentation becomes more widespread, and the users become more experienced, the techniques described here have the potential to generate new data which cannot be obtained by other methods.

Acknowledgements

This work was supported by the Deutsche Forschungsgemeinschaft, the Fonds der Chemischen Industrie and the Biotechnologie-Schwerpunkt "Grundlagen der Bioprozesstechnik".

Fig. 2. Scheme illustrating the principles of the protein A–gold and the antibody–gold technique. The situation before air-drying of the sections is depicted. Two different sizes of colloidal gold particles are given as examples (from Rohde *et al.*, 1988, with permission).

References

Acker, G. (1988). *Meth. Microbiol.* **20,** 147-174.

Alexander, C., Jones, D. and Mchardy, W. J. (1987). *New Phytol.* **105,** 613–618.

Bauer, R. (1988). *Meth. Microbiol.* **20,** 113-146.

Bendayan, M. (1982). *Biol. Cell* **43,** 153-156.

Bendayan, M. and Zollinger, M. (1983). *J. Histochem. Cytochem.* **31,** 101-109.

Bendayan, M., Roth, J., Perrelet, A. and Orci, L. (1980). *J. Histochem. Cytochem.* **28,** 149-160.

Bonfante-Fasolo, P. (1982). *Protoplasma* **111,** 113-120.

Dubochet, J. (1984). *Proceedings of the 8th European Conference on Electron Microscopy*, Budapest, Vol. 2, p. 1379.

Dubochet, J. and McDowall, A. W. (1981). *J. Microsc.* **124,** RP 3.

Dubochet, J., Ducommun, M., Zollinger, M. and Kellenberger, E. (1971). *J. Ultrastruct. Res.* **35,** 147-167.

Dubochet, J., Lepault, J., Freeman, R., Berryman, J. A. and Homo, J.-C. (1982). *J. Microsc.* **128,** 219.

Faulk, W. P. and Taylor, G. M. (1971). *Immunocytochemistry* **8,** 1081-1083.

Frens, G. (1973). *Nature (Phys. Sci.)* **241,** 20-22.

Freudenberg, W., Mayer, F. and Andreesen, J. R. (1989). *Arch. Microbiol.* **152,** 182-188.

Gianinazzi-Pearson, V., Morandi, D., Dexheimer, J. and Gianinazzi, S. (1981). *New Phytol.* **88,** 633-640.

Griffiths, G., Simons, K., Warren, G. and Tokuyasu, K. T. (1983). *Meth. Enzymol.* **96,** 466-483.

Griffiths, G., McDowall, A., Back, R. and Dubochet, J. (1984). *J. Ultrastruct. Res.* **89,** 65-78.

Hobot, J. A., Carlemalm, E., Villiger, W. and Kellenberger, E. (1984). *J. Bacteriol.* **160,** 143-152.

Hobot, J. A., Villiger, W., Escaig, J., Maeder, M., Ryter, A. and Kellenberger, E. (1985). *J. Bacteriol.* **162,** 960-971.

Horisberger, M. and Rosset, J. (1977). *J. Histochem. Cytochem.* **25,** 295-305.

Horisberger, M., Rosset, J. and Bauer, H. (1975). *Experientia* **31,** 1147-1148.

Johannssen, W., Schuette, H., Mayer, F. and Mayer, H. (1979). *J. Mol. Biol.* **134,** 707-726.

Jones, D., Mchardy, W. J. and Alexander, C. (1987). *Scanning Microsc.* **1,** 1423-1430.

Lacaze, B. (1983). *Can. J. Bot.* **61,** 1411-1414.

Lalucat, J. (1988). *Meth. Microbiol.* **20,** 79-90.

Lünsdorf, H., Spiess, E. (1986). *J. Microsc.* **144,** 211-213.

Müller, M. (1988). *Meth. Microbiol.* **20,** 1-28.

Robards, A. W. and Sleytr, U. B. (1985). *Practical Methods in Electron Microscopy* (A. M. Glauert, ed.), Vol. 10, pp. 461–499. Elsevier, Amsterdam, New York and Oxford.

Roberts, I. M. (1975). *J. Microsc.* **103,** 113-119.

Rohde, M., Mayer, F. and Meyer, O. (1984). *J. Biol. Chem.* **259,** 14 788-14 792.

Rohde, M., Mayer, F., Jacobitz, S. and Meyer, O. (1985). *FEMS Microbiol. Lett.* **28,** 141-144.

Rohde, M., Gerberding, H., Mund, Th. and Kohring, G.-W. (1988). *Meth. Microbiol.* **20,** 175-210.

Romanow, E. L., Stolinski, C. and Hughes-Jones, N. C. (1974). *Immunocytochemistry* **11,** 521-522.

Romanow, E. L., Stolinski, C. and Hughes-Jones, N. C. (1975). *Br. J. Haematol.* **30,** 507-516.

Roth, J., Bendayan, M. and Orci, L. (1978). *J. Histochem. Cytochem.* **26,** 1074-1081.

Roth, J., Bendayan, M., Carlemalin, E., Villiger, W. and Garavito, M. (1981). *J. Histochem. Cytochem.* **29**, 663–669.

Sleytr, U. B., Messner, P. and Pum, D. (1988). *Meth. Microbiol.* **20,** 29-60.

Slot, J. W. and Geuze, H. J. (1981). *J. Cell Biol.* **90,** 533-536.

Slot, J. W., Geuze, H. J. and Weerkamp, A. J. (1988). *Meth. Microbiol.* **20,** 211-236.

Spanu, P., Boller, T., Ludwig, A., Wiemken, A., Faccio, A. and Bonfante-Fasolo, P. (1989). *Planta (Berl.)* **177,** 447-455.

Spiess, E. and Mayer, F. (1976). *Inst. Wiss. Film (Göttingen, Germany): Elektronenmikroskopische Präparationsmethoden.*

Straker, C. J., Gianinazzi-Pearson, V., Gianinazzi, S., Cleyet-Marel, J. C. and Bousquet, N. (1989). *New Phytol.* **111,** 215-222.

Tokuyasu, K. T. and Ukamara, S. (1959). *J. Biophys. Biochem. Cytol.* **6,** 305-308.

Valentine, R. C., Shapiro, B. M. and Stadtman, E. R. (1968). *Biochemistry* **7,** 2143-2152.

Vare, H. (1990). *New Phytol.* **116,** 663-668.

Venable, J. H. and Coggeshall, R. (1965). *J. Cell Biol.* **25,** 407-408.

Wagner, B., Wagner, M., Kubin, V. and Ryc, M. (1980). *J. Gen. Microbiol.* **118,** 95-105.

15

Epifluorescent Microscopy for Identification of Ectomycorrhiza

PAVEL CUDLÍN

Institute of Landscape Ecology, Czechoslovak Academy of Sciences, Ceské Budějovice, Czechoslovakia

I. Introduction

Since the discovery of mycorrhiza, more than 100 years ago, scientists have been looking for a simple staining method to distinguish mycorrhizal structures—the fungal mantle and the Hartig net. Practically all the advanced methods of anatomical botany have been tested. Screening for vesicular-arbuscular mycorrhizal infection has provided important techniques for tree root examination. Plant roots have most often been fixed in FAA, more recently in glutaraldehyde (Alexander and Bigg, 1981). For staining of free-hand or microtome-cut sections of mycorrhiza, aniline blue, cotton blue, acid fuchsin, thionin, safranin, fast green, crystal violet and many other stains dissolved in water, alcohol or lactophenol have been used (Wilcox, 1982). Lactophenol has been recently replaced by lactic acid or lactoglycerin (Kormanik *et al.*, 1980). In the last 20 years several methods have been elaborated to extract tannins and clear roots with KOH, H_2O_2, chloral hydrate or sodium hypochlorite (Nemec, 1982). The most frequently used routine method

METHODS IN MICROBIOLOGY
VOLUME 23 ISBN 0-12-521523-1

Copyright © 1991 by Academic Press Limited
All rights of reproduction in any form reserved

for vesicular-arbuscular mycorrhiza study—KOH clearing and trypan blue staining (Phillips and Hayman, 1970)—provides good results with tree root sections, but it has drawbacks in that there is a considerable loss of sections during the procedure and the operations involved are very time-consuming (Cudlín, 1980). With the increasing need to study tissue and cell structure, methods for embedding roots and the techniques of thin and semi-thin sectioning have been developed. Paraffin embedding has recently been replaced by the use of synthetic resins, e.g. Historesin (Weiss, 1988). For semi-thin sections, several single or combined stain techniques have been developed: Conant quadruple stain—safranin, crystal violet, fast green and gold orange (Johansen, 1940); chlorazol black and Pianese III-B (Wilcox and Marsch, 1964); toluidine blue O in sodium borate (Massicotte *et al.*, 1989); new fuchsin and crystal violet (Blasius and Oberwinkler, 1989).

The last-mentioned procedure, together with observation of sections in lactic acid under the epifluorescent microscope (wavelength 340–380 and 450–490 nm) and Nomarski's phase contrast, has been used for identification of ectomycorrhizal types of forest trees in Germany (Agerer, 1985). An adequate time-saving method for fungal mantle and Hartig net distinction using epifluorescent microscopy of sections stained with cotton blue in lactoglycerin has been proposed (Cudlín *et al.*, 1990).

II. Sample preparation, staining and observation

A. Sample preparation

Roots can be collected at any time for identification of ectomycorrhizal types. Nevertheless, sampling of the roots situated below the mycorrhizal fungus sporocarp (the closer the better) is recommended, especially when a new experimental plot is studied. Tracing rhizomorphs and hyphae to sporocarp is particularly easy in rotten wood, peat moss or broken stones (Zak, 1973).

Roots need not be processed immediately in the laboratory. They can be stored, after careful washing in tap water and cleaning between two fine brushes, in a buffered solution of glutaraldehyde. This fixative consists of 120 ml 25% glutaraldehyde, 500 ml 0.1 M KH_2PO_4, 400 ml 0.1 M KOH and 110 ml distilled water (Alexander and Bigg, 1981). Only young, turgid mycorrhiza with morphological features sufficiently documented in the fresh state should be preserved (Agerer, 1985).

B. Staining

The following procedure is used for staining.

1. Leave clean root specimens in glutaraldehyde fixative for at least 48 h before staining.
2. Make cross- and longitudinal sections of different diameter in the range 10–50 μm by hand using razor blades or by using a freezing microtome. Break out several pieces of fungal mantle using a needle under a dissecting microscope.
3. Place sections into a drop of 0.5% solution of cotton blue in lactoglycerin (250 ml lactic acid, 250 ml glycerin and 300 ml distilled water) on a ground microscope slide and leave for 24–48 h. In the case of poor penetration of the stain into root cells, it is necessary to heat the slide to 80 °C for 1 min. Checking the staining process under a light microscope (without cover-slip) is recommended.
4. With the aid of a needle and a small scalpel transfer stained sections onto a new ground slide with a drop of 0.01% solution of aniline blue in lactoglycerin and cover with a cover-slip.

C. Observation

The best time for observation under the epifluorescence microscope is between 24 and 48 h after staining and lasts for several days, depending on the material. The optimal combination of excitation filter (wavelength 355–425 nm) and additive filter is found by trial. The epifluorescent effect increases the visible contrast of stained fungal tissues. It enables mantle structure and single hyphae in Hartig net to be distinguished in cross- and longitudinal sections. In mantle fragments the structure of the outer and inner surfaces can be clearly seen. For photographic documentation, Kodak 800 colour slide film is convenient.

III. Potential and limitations

The proposed staining process is one of the routine methods for screening ectomycorrhizal infection. Its principal advantages are as follows: (1) it allows the possibility of working with both almost fresh and long-time fixed material; (2) good distinction of single fungal hyphae is achieved, comparable with that obtained using stained semi-thin sections; (3) relatively thick hand-cut sections can be used; and (4) the process is not time-consuming and its cost is low. It has universal applicability, especially when a considerable number of specimens are to be examined, e.g. for mycorrhiza identification or for screening mycorrhiza infection in seedlings from an early stage.

On the other hand, the method described can serve only as a part of a complex set of morphologico-anatomical approaches to type identification or ecophysiological studies. The surface of relatively thick sections is not completely flat and uniformly in focus. Unlike other fluorescent staining methods, the nature of the fluorescent substances in stained fungal tissue is not known; changes in the colour of hyphae cannot be interpreted or related to other changes in their condition or state. The process is not suitable for permanent preparations due to colour changes taking place in the sections.

References

Agerer, R. (1985). *Colour Atlas of Ectomycorrhizae*. Einhorn-Verlag, Schwäbisch Gmünd.

Alexander, I. J. and Bigg, W. L. (1981). *Trans. Br. Mycol. Soc.* **77**, 425–429.

Blasius, D. and Oberwinkler, F. (1989). In *Forest Tree Physiology* (E. Dreyer *et al.*, eds) *Ann. Sci. For. Suppl.* **46**, 758–761.

Cudlín, P. (1980). Study on mycorrhizal fungus cultures and pesticide effect on mycorrhiza development of seedlings of *Pinus sylvestris* L. (in Czech). PhD Thesis, Institute of Landscape Ecology, Czechoslovak Academy of Sciences, Prague.

Cudlín, P., Jansen, A. E. and Mejstřík, V. (1990). In *Abstracts of 4th International Mycological Congress*, Universität Regensburg. 73 pp.

Johansen, D. A. (1940). *Plant Microtechnique*. McGraw-Hill, New York.

Kormanik, P. P., Bryan, W. C. and Schultz, R. C. (1980). *Can. J. Microbiol.* **26**, 536–538.

Massicotte, H. B., Peterson, R. L. and Melville, L. H. (1989). *Am. J. Bot.* **76**, 1654–1667.

Nemec, S. (1982). In *Methods and Principles of Mycorrhizal Research* (N. C. Schenck, ed.), pp. 23–27. The American Phytopathological Society, St Paul, MN.

Phillips, J. M. and Hayman, D. S. (1970). *Trans. Br. Mycol. Soc.* **55**, 158–161.

Weiss, M. (1988). Ektomykorrhizen von *Picea abies*. Synthese, Ontogenie und Reaktion auf Umweltschadstoffe. PhD Thesis, Department of Special Botany, University of Tübingen, Tübingen.

Wilcox, H. E. (1982). In *Methods and Principles of Mycorrhizal Research* (N. C. Schenck, ed.), pp. 103–113. The American Phytopathological Society, St Paul, MN.

Wilcox, H. E. and Marsh, L. G. (1964). *Stain Technol.* **39**, 81–86.

Zak, B. (1973). In *Ectomycorrhiza: Their Ecology and Physiology* (G. C. Marks and T. T. Kozlowski, eds), pp. 43–78. Academic Press, New York.

16

Electron Energy Loss Spectroscopy and Imaging Techniques for Subcellular Localization of Elements in Mycorrhiza

INGRID KOTTKE

Universität Tübingen, Institut für Botanik, Lehrstuhl Spezielle Botanik, Mykologie, Auf der Morgenstelle 1, D-7400 Tübingen, Germany

I. Introduction

Mycorrhizal plants obtain a considerable proportion of their nutrient ions, as well as some potentially toxic elements, by way of the fungal hyphae. Pathways for solute movement in mycorrhiza are therefore different from those in non-mycorrhizal roots. The structure and physiology of the fungus interacts with that of the root in a poorly understood manner. Identification of sites of localization of elements at the subcellular level, in cell organelles, cytoplasm or cell walls, would help to clarify the process involved in their uptake, transport and deposition or detoxification in mycorrhiza.

METHODS IN MICROBIOLOGY
VOLUME 23 ISBN 0-12-521523-1

Copyright © 1991 by Academic Press Limited
All rights of reproduction in any form reserved

Electron energy loss spectroscopy (EELS) and electron spectroscopic imaging (ESI) are techniques that have recently become available for general use in association with transmission electron microscopy. Theoretically they offer the possibility for localization and identification of all elements at a resolution which is otherwise unattainable. They also simultaneously yield information on the ultrastructure of the specimen.

EELS and ESI are based on the element-specific energy loss (ΔE) and the deflection of the inelastically scattered electrons of the electron beam passing the inner shells of the atoms in the specimen (Fig. 1). The EELS signal is higher than the X-ray signal by some orders of magnitude and it increases with decreasing atomic number. For the biologically important elements, which are mostly of lower atomic number, EELS therefore has some major advantages for microanalysis in comparison with energy dispersive X-ray analysis (EDXA).

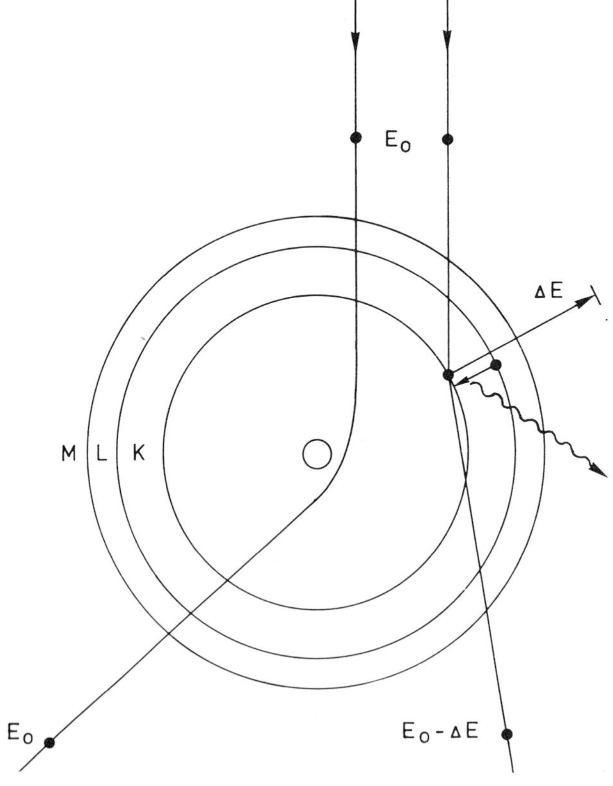

Fig. 1. Electron beam (E_0) passing an atom: electrons are scattered elastically (E_0) without energy loss and inelastically ($E_0 - \Delta E$) with element-specific energy loss (ΔE) (after Bauer, 1985).

II. Techniques

A. Equipment and facilities on the transmission electron microscope

The TEM902 Zeiss (80 kV) has been equipped with an imaging spectrometer of the prism–mirror–prism type and a slit system to separate and select the energy loss electrons desired for microanalysis (Fig. 2). For centring the deflected electrons with an energy of $E = E_0 - \Delta E$ on the optical axis, the accelerating voltage U_0 is increased specifically by

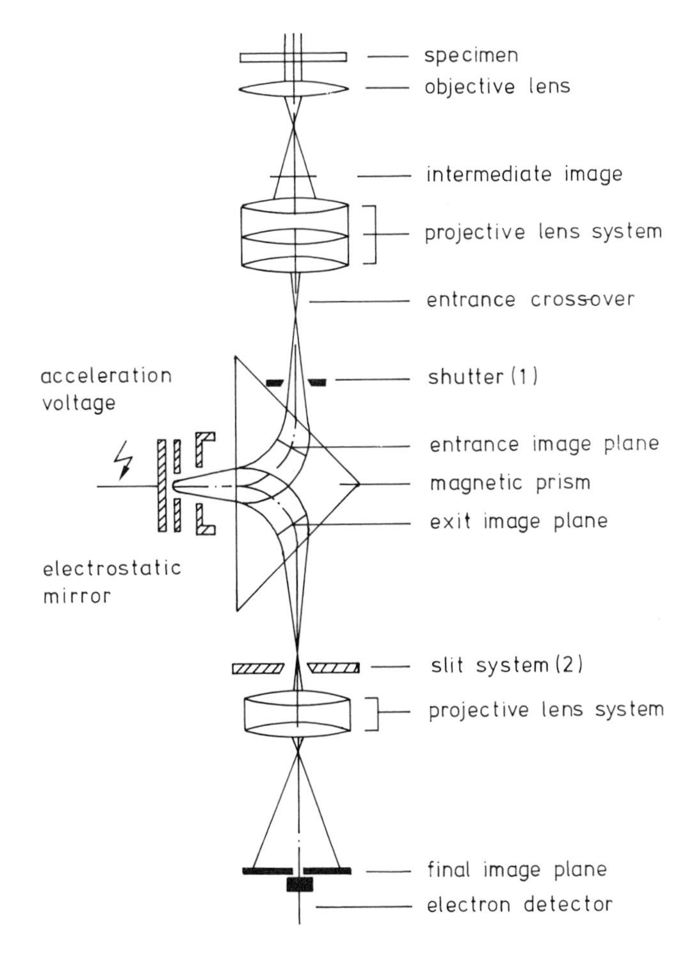

Fig. 2. Imaging and spectrometer system of the TEM902 ZEISS (after Bauer, 1985).

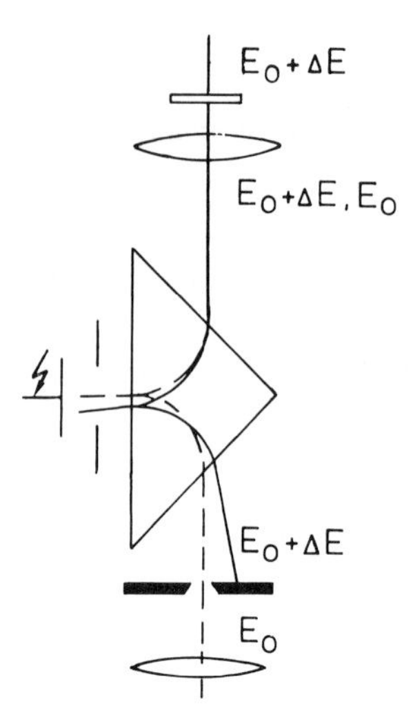

Fig. 3. Imaging by inelastically scattered electrons (after Bauer, 1985).

$\Delta U_0 = \Delta E \times e_0^{-1}$ (Fig. 3; for further details see Bauer, 1985, 1988). A magnetic sector energy analyser is placed below the final image plane (Fig. 2). Imaging is assisted by use of a photomultiplier (MIT camera). In this way contrasts of low intensity are enhanced on a monitor. The system may be supplemented by an optical disc and a videoprinter to allow the elemented maps to be stored and printed.

For our purposes there are three main ways in which the EELS and ESI data can be used. These are: (1) to identify the elements present; (2) to provide an elemental distribution map; and (3) to provide an energy-filtered image with improved contrast of very thin, unstained sections.

1. EEL spectra

EEL spectra of all stable elements have been obtained from thin films or foils and can be examined in the *EELS Atlas* (Ahn *et al.*, 1983), where the atomic absorption edges optimally used in the microanalysis of each element are depicted graphically. The threshold energy usually agrees to within a few eV with the known ionization energy for the appropriate shell of the atom. In general there is at least one strong edge in the usable energy region of 100–2000 eV for every element. For many elements there are two or more edges in the optimum EELS energy region of 200–1000 eV.

Energy resolution in the spectrum mode for small specimen areas is 0.8 eV for low beam current and 2.5 eV for higher beam current. The energy window is delineated by the variable slit (position 2 in Fig. 2). The area which can be simultaneously analysed by EELS is restricted by a 100 μm shutter at position (1) in Fig. 2, thus the area depends on the magnification (diameter of area: 12 000×, 1.2 μm; 30 000×, 0.5 μm; 85 000×, 0.18 μm; 140 000×, 0.1 μm). Best results are obtained when the analysed particle fills the opening of the shutter entirely, as background is then reduced to a minimum. Background noise can also be reduced by prolonging the exposure time, although greater damage to the specimen must then be expected. Minimal detectable mass is in the order of 2×10^{-21} g for iron. For other elements and especially in cell material real minimal detectable mass has still to be determined (see Section II.C for standards). Identification of an element not only depends on its concentration in the area under investigation, but is also influenced by the steepness of the threshold at the absorption edge of the atom, the region of energy loss, the bonding of the element, and sometimes by overlapping of the edges of two elements (Fig. 4). In case of smooth thresholds, "stripping" of the edge can assist presentation (Fig. 5a,b).

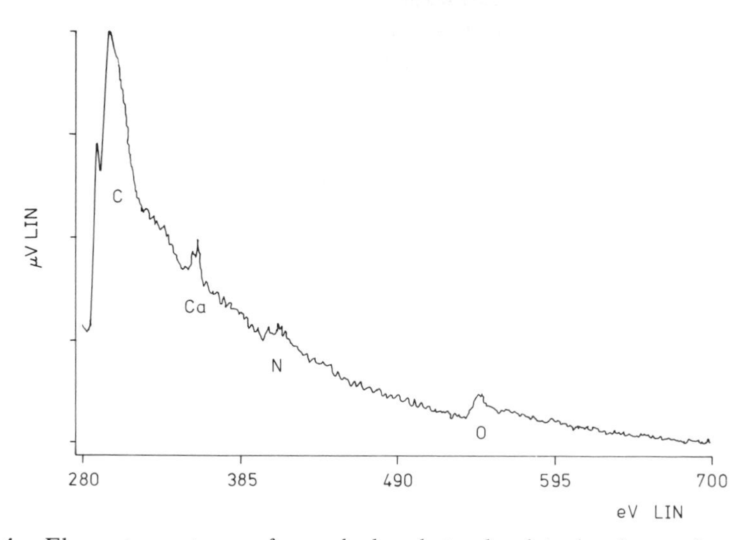

Fig. 4. Element spectrum of a polyphosphate droplet showing carbon (C), calcium (Ca), Nitrogen (N) and oxygen (O); acquired at 20 000× magnification.

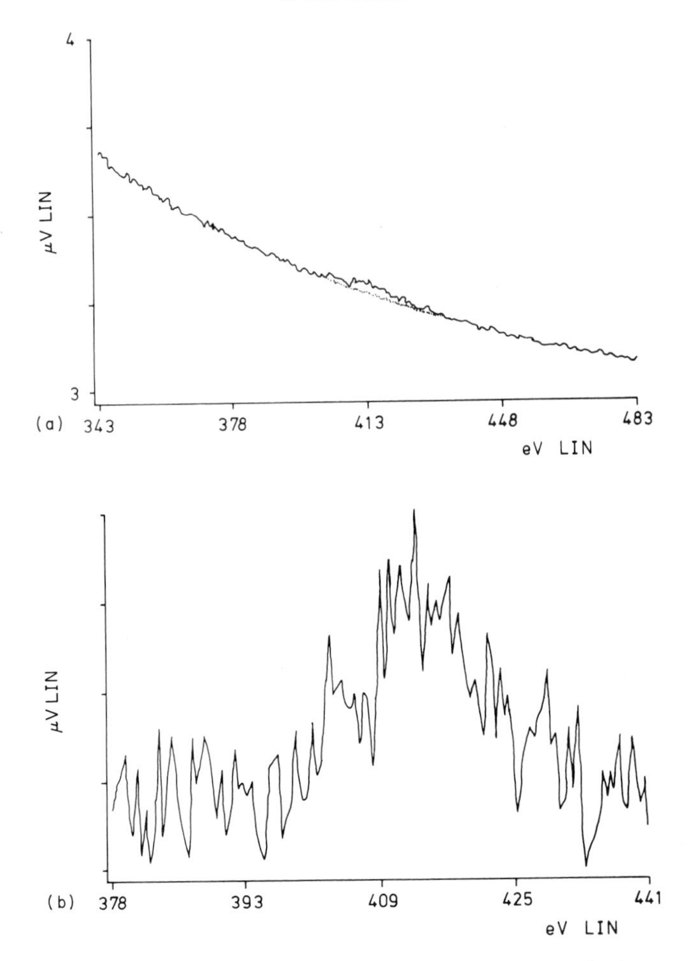

Fig. 5. Spectrum (a) and "stripping" (b) of low amount of nitrogen in the specimen; by "stripping" the element-specific threshold can be depicted more precisely.

2. *Elemental distribution maps*

Elemental maps are produced by a computer-assisted image-processing system using a photomultiplier MIT camera. Programs used on TEM902 are derived from IBAS (the Kenton Electronics software used for the analysis). The elemental map results from the increase of the energy intensity at the electron absorption edge which is superimposed on the background continuum (Fig. 6). To obtain the distribution of an element

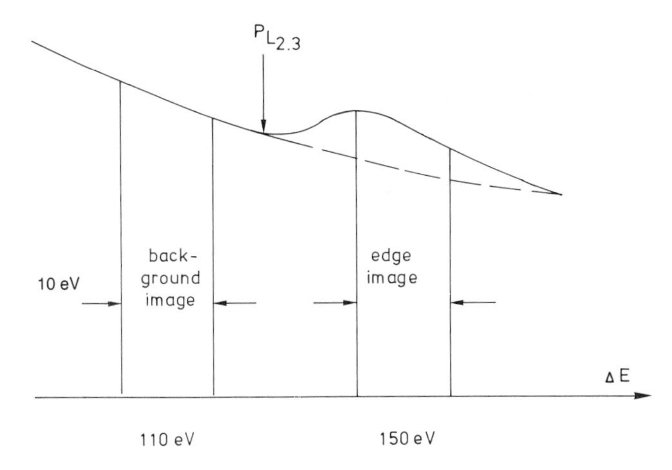

Fig. 6. Scheme illustrating aquisition of elemental map: the "edge image" is acquired close behind and the background image before the element edge (here P at 132 eV) (after Bauer, 1985).

at least two images have to be acquired one above the edge, carrying the information about the spatial distribution of the element, and a second to act as a reference image just below the chosen absorption edge to obtain a measure of the background continuum. The differences between the ESI above and below the atomic absorption edge, which are calculated with help of computer programs, constitute a map of the spatial distribution of the chosen element (Fig. 7). Where there is a steep decrease in energy (e.g. between 100 and 200 eV)—or where elements are superimposed on thresholds of other elements (e.g. Ca on C)—processing of a second background image at further distance is recommended. Special computer programs are available for this purpose. Energy resolution in the image mode is 20 eV or 10 eV, depending on procedure, and minimum spatial resolution is about 0.5 nm. Magnification recommended for elemental mapping is 20 000× or higher. Thus the area under simultaneous investigation, restricted by a 650 μm shutter (position 1 in Fig. 2), is small. To screen a larger region to obtain preliminary superficial information a "display element" program can be used, in which the ΔE edge and ΔE background searches switch over within seconds.

Elemental maps can also be recorded photographically (Fig. 8). The quality of such micrographs is much better than that achieved with a videoprinter. However, the net distribution has then also to be acquired photographically, a method difficult to handle (Lehmann *et al.*, 1990).

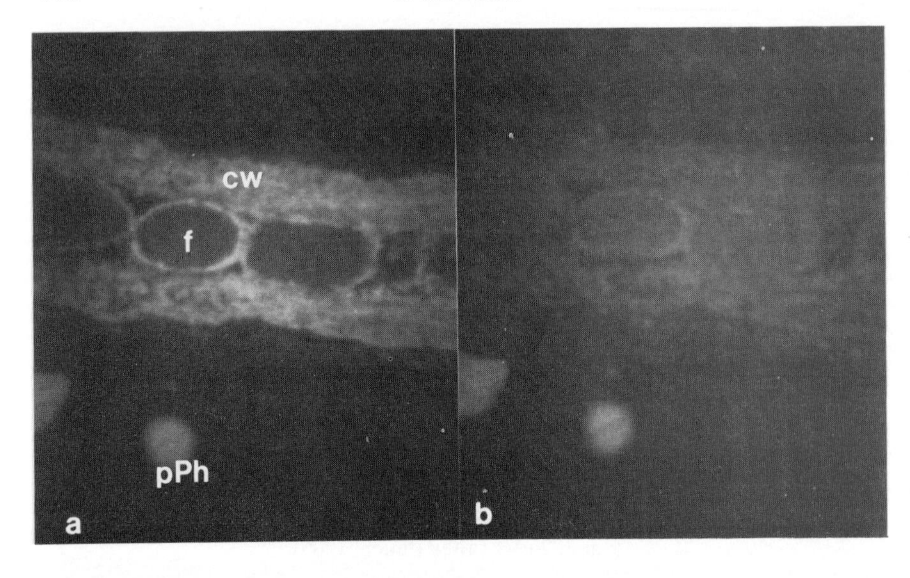

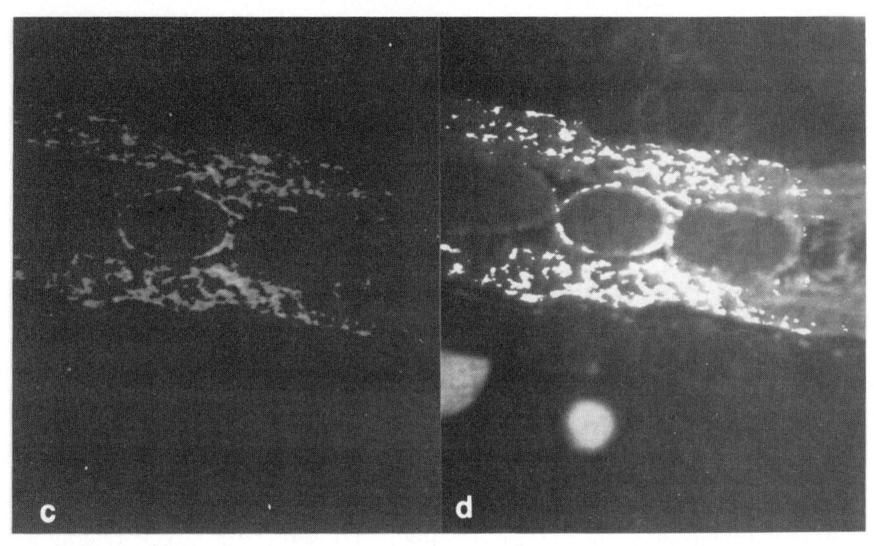

Fig. 7. Elemental map showing cerium in the fungal (f) and root cell walls (cw) of the Hartig net after exposing the mycorrhiza to a cerium solution; images reproduced by videoprinter. (a) Edge image at 900 eV; (b) background image at 860 eV; (c) net distribution computer calculated; (d) net distribution of cerium superimposed on the 250 eV image. Note that the polyphenol droplet (pPh) disappears in the net distribution image, indicating that even strong background is eliminated.

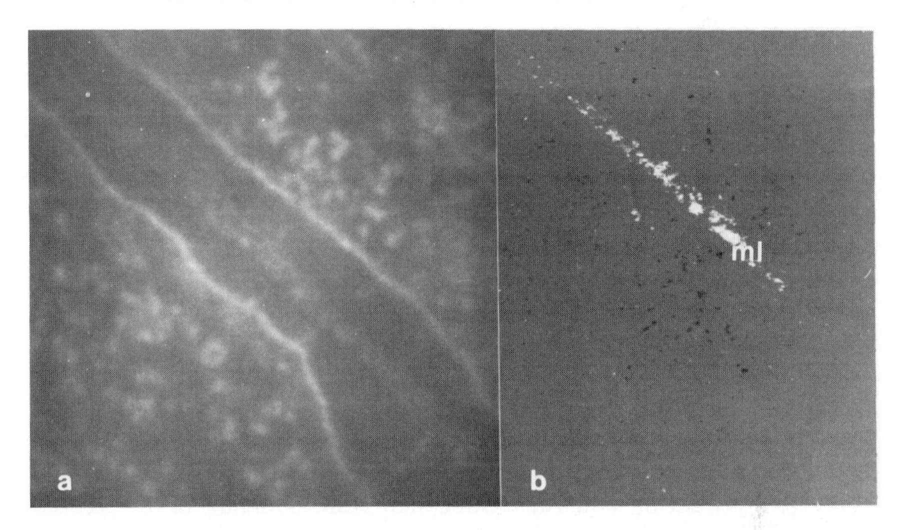

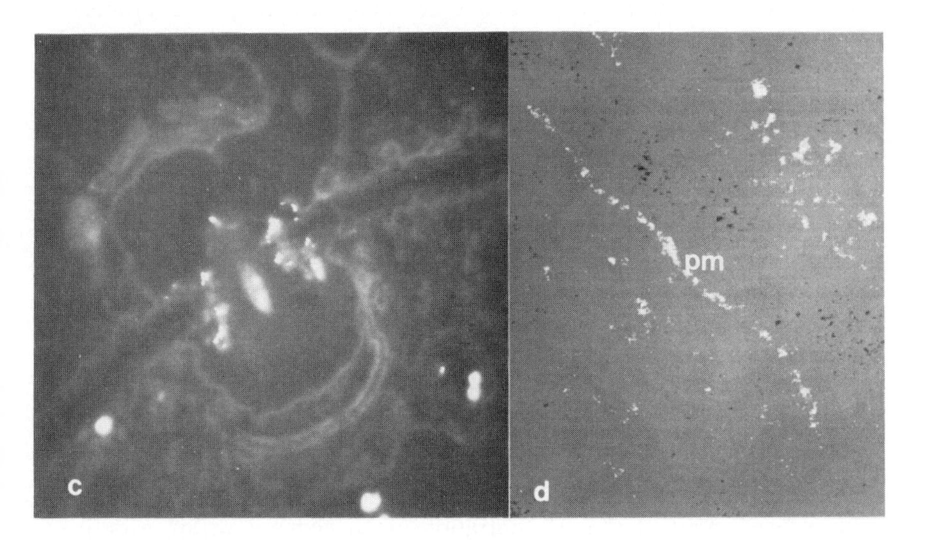

Fig. 8. Cell wall and elemental map of a meristematic cell of the root tip of *Picea abies*; videoprinter images acquired with two background images. (a) 250 eV image, middle lamella and plasma membranes visible, 30 000×; (b) net distribution of calcium mainly in the middle lamella (ml), 30 000×; (c) net distribution of phosphorus mainly on the plasma membrane (pm) and ribosomes, 30 000×; (d) net distribution of lanthanum superimposed on the 250 eV image, showing lanthanum in fungal vacuoles and the septal porus, 50 000×.

3. *High contrast images*

ESIs of high contrast can be processed on the monitor or photographically recorded at 200–250 eV, the carbon edge (Fig. 8a). High contrast and resolution can thus be attained in very thin, unstained sections. The elemental map can be superimposed in a chosen colour on the carbon edge image by computer program and displayed on the monitor. Thus both the ultrastructure and the chosen element are visible simultaneously. They can be printed by videoprinter (Figs. 7d, 9c) or photographed from the monitor for slides.

B. Sample preparation and sectioning

Only embedded specimens can be cut sufficiently thinly for EELS and ESI from mycorrhizal material. Samples must therefore be fixed chemically or by freezing and then embedded conventionally or by freeze-substitution (Lehmann *et al.*, 1990). From personal experience, only pre-fixation in glutaraldehyde–formaldehyde (Karnovsky, 1965), post-fixation in osmium tetroxide and embedding using Spurr's procedure (1969) can be recommended for ectomycorrhiza (Kottke and Oberwinkler, 1988). *En bloc* staining with uranyl acetate must be avoided otherwise high background noise will occur in EELS and ESI. Good cryofixation of mycorrhiza is extremely difficult to achieve and results in tremendous structural disturbance. It is therefore not advisable to use freeze-substitution followed by high resolution analysis in the TEM902.

The thin sections required for EELS and ESI can best be cut using a diamond knife. Because of their fragility these sections cannot be mounted on copper grids after dry sectioning. Sections have to be removed from the water trough and mounted on copper grids (600–700 mesh) immediately after sectioning to limit leaching of ions. They are analysed without any additional staining.

C. Preparation of standards

The measurement of the concentration of an element within a defined area, while being theoretically possible with EELs and ESI, has yet to be achieved. Besides the hazard of molecular damage of the specimen during exposure to the electron beam, the evaluation of the thickness of the section is the most serious difficulty on the TEM902. Therefore, co-embedding of ion-exchange beads of Chelex^{100}-type prepared as standards has been recommended (Verbueken *et al.*, 1984; Sorber *et al.*, 1991). The beads are exposed to a solution of defined concentration of a chosen element for about 24 h, oven-dried and embedded directly in ERL (Spurr's resin) together with the sample. The embedded bead is analysed by EELs in the same section as the mycorrhizal tissue. Ion concentration of the salt solution as well as of the beads can be analysed with atomic emission spectroscopy (AES) or voltametrically and the concentration of the element in the beads calculated. This method yields at least some rough estimation of the element concentration in the specimen.

III. Selected examples of analysis of ectomycorrhiza

Few investigations have so far been carried out on mycorrhiza using EELS and ESI, as the TEM902 has only been available for a few years—and is expensive. Examples shown are therefore all taken from my own research, which itself has only just started. The method is still in a testing phase with regard to use on mycorrhizal problems.

Example 1
Constitutive elements in cells of fungi and roots, such as calcium, phosphorus or nitrogen, have been localized by ESI and verified by EELS. Figure 8a,b,d demonstrates the distribution of calcium in cell wall, preferentially in the middle lamella, and phosphorus in the plasmamembrane of the same area. Identification of the elements is additionally verified by EELs spectra.

Example 2
Polyphosphates in hyphal vacuoles accumulate various cations, as has already been shown by X-ray analysis (Ashford *et al.*, 1986). Using ESI and EELS it was demonstrated for the first time that aluminium is also precipitated in these polyphosphates (Kottke and Oberwinkler, 1990). Figure 9a–c shows the same hyphae at different energy loss levels. The micrograph at 250 eV displays high contrast and reveals structural details of the hypha. At 150 eV the polyphosphate droplet in the vacuole is very prominent, while at 100 eV the aluminium in the polyphosphate droplet is visible. EEL spectra of the droplet confirm the identification of the elements (Fig. 10a,b).

Example 3
Figure 9d illustrates the precipitation of aluminium on the surface of a non-mycorrhizal root tip of *Picea abies* following exposure of roots to the element. The localization of aluminium in the mucilage of root tips

confirms the findings of Marschner (1990), who brushed off the mucigel from non-mycorrhizal root tips of *P. abies* and analysed it by AES.

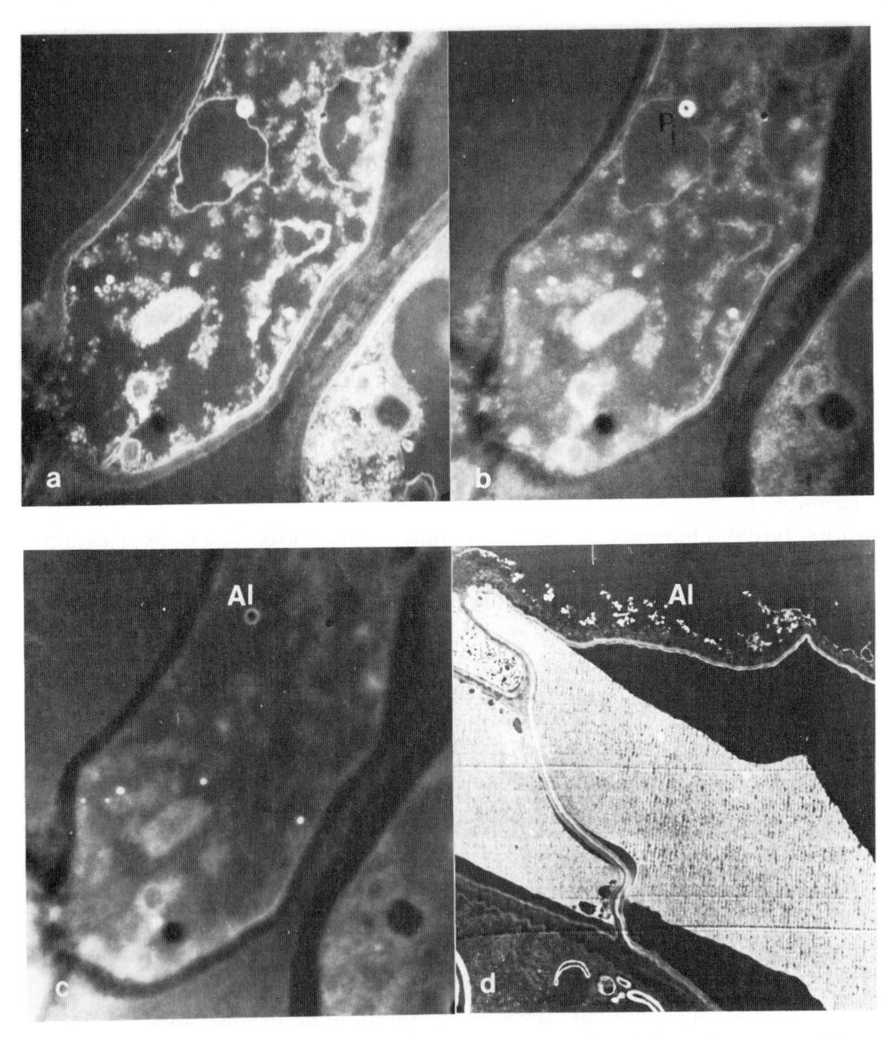

Fig. 9. Micrographs of a hypha of the outer mantle region of a mycorrhiza which has been exposed to aluminium solution: (a) photographed at 250 eV, revealing details of structure of a 30 nm thick, unstained section; (b) photographed at 150 eV P edge, polyphosphate droplet in the vacuole, confirmed by EELS (see Fig. 10a); (c) photographed by 100 eV Al–edge, aluminium in the polyphosphate droplet, confirmed by EELS (see Fig. 10b); (d) micrograph at 130 eV Al edge, showing aluminium precipitation on the surface of a non-mycorrhiza root tip of *Picea abies*.

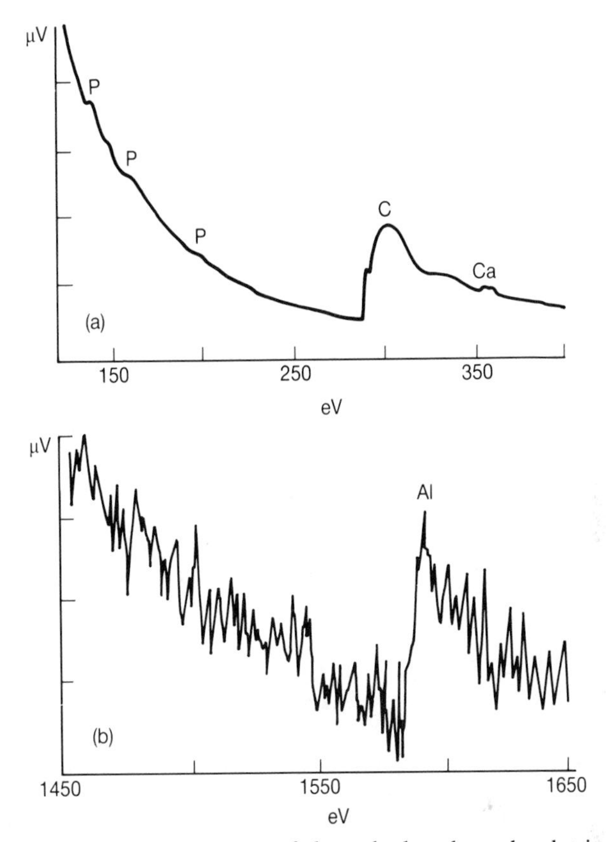

Fig. 10. (a) and (b) Element spectra of the polyphosphate droplet in the hypha (Fig. 9a–c), revealing phosphorus, calcium and aluminium (30 000×).

Example 4
EELS and ESI can be useful in detecting elements that have been used as tracers to search out the pathways of solutes. The lanthanides are especially suitable in the TEM902, as they mostly have steep thresholds between 600 and 1000 eV. Two examples for cerium and lanthanum are shown (Figs 7a–d, 8c).

IV. Advantages and limitations of the new technique

EELS and ESI on the TEM present a new dimension in electron microscopy. It is obvious that, especially for biological questions, the ability to reveal the association of an element with distinct cell compartments is important and the high resolution obtained is of great advantage compared with X-ray analysis or the use of radioactive markers. EELS and ESI results are very reliable, as element absorption edges are mostly well separated from each other; only a few overlap. One disadvantage is the need to embed material and by this to risk the loss of substances. Solutes in the apparent free space are probably lost totally; those of the Donnan free space are also leachable to some extent. Cells with damaged membranes may be leached more than intact cells. However, differences have still to be worked out for mycorrhiza as well as for other plant material. As yet experience with the technique is too short to enable us to recognize its full potential.

Acknowledgement

The author is indebted to the Projekt Europäisches Forschungszentrum (PEF), Karlsruhe for equipment and grants which made these investigations possible.

References

Ahn, C. C. Krivanek, O. L. Burgner, R. P. Disko, M. M. and Swann, P. R. (1983). *EELS Atlas*. Gatan Inc. Warrendale, USA.
Ashford, A. E., Peterson, R. L., Dwarte, D. and Chilvers, G. A. (1986). *Can. J. Bot.* **64**, 677–687.
Bauer, R. (1985). *Labor Praxis* **7–8**, 852–862.
Bauer, R. (1988). In *Methods in Microbiology* Vol. 20, pp. 113–146. Academic Press, London.
Karnovsky, M. J. (1965). *J. Cell Bio.* **27**, 137–138.
Kottke, I. and Oberwinkler, F. (1988). *Trees* **21**, 115–128.
Kottke, I. and Oberwinkler, F. (1990). *KfK-PEF* **61**, 65–73. Kernforschungszentrum, Karlsruhe.
Lehmann, H., Kramer, A. and Schulz, D. (1990). *Ultramicroscopy* **32**, 26–34.
Marschner, H. (1990). *International Congress in Forest Decline Research*, Friedrichshafen (B. Ulrich, ed.), Vol. 1, pp. 381–404. Kernforschungszentrum, Karlsruhe.
Sorber, C. W., Ketelaars, G. A. Gelsema, E. S. Jongkind, J. F. and de Bruijin, W. C. (1991). *Second European Workshop on Electron Spectroscopic Imaging and Analysis Techniques*, Dortmund 1990. *J. Microscopy* **162**, 43–54.
Spurr, A. R. (1969a), Abstracts. *J. Ultrastruct. Res.* **26**, 31–43.
Verbueken, A. H., van Grieken, R. E., Paulus, G. J. and Bruijin, W. C. (1984). *Anal. Chem.* **56**, 1362–1370.

17

Methods for Studying Nursery and Field Response of Trees to Specific Ectomycorrhiza

D. H. MARX and J. L. RUEHLE

US Department of Agriculture, Forest Sciences Laboratory, Athens, GA 30602, USA

C. E. CORDELL

Forest Pest Management, US Department of Agriculture, Forset Service, Southeastern Forest Experiment Station, Asheville, NC 28802, USA

I. Introduction

The use of specific fungi to form ectomycorrhiza on forest tree seedlings in the nursery can improve performance after outplanting. The dependence of forest trees on ectomycorrhiza has long been recognized (Hatch, 1937; Harley and Smith, 1983). Trees of many species, including pines, will not grow and develop normally without them. Reforestation efforts with species that require ectomycorrhiza inevitably fail in the absence of suitable ectomycorrhiza. In the past three decades, the value of nursery inoculation with selected ectomycorrhizal fungi in improving field performance on routine and adverse sites has been clearly demonstrated (Bowen, 1965; Mikola, 1973; Trappe, 1977; Molina and Trappe, 1982; Marx and Cordell, 1988).

The primary purpose of inoculation is to provide seedlings with adequate ectomycorrhiza for planting in man-made forests. Many inoculations with ectomycorrhizal fungi have provided excellent results in temperate countries (Marx, 1980). Various inoculation methods have been used in these countries. However, because of ecological and economic constraints, many of these methods may be unsuitable for tropical reforestation programmes in developing countries. Inoculation procedures and formation of ectomycorrhiza on planting stock of exotic pines should be investigated at locations where major reforestation efforts are planned. Experimental variables should include suitability of specific symbiotic fungi, use of bare-root or container-grown seedlings, properties of growing medium or soil, soil sterilization, fertilization practices, disease control, tree species and seed source.

Most research on inoculation with ectomycorrhizal fungi has been based on two working premises (Marx, 1980). First, any ectomycorrhiza on roots of planting stock are far better than none. Second, some species of ectomycorrhizal fungi under certain environmental conditions are more beneficial to trees than others. Research on the latter premise has been aimed at selecting, propagating, manipulating, and managing the most desirable ectomycorrhizal fungi to improve tree survival and growth in plantations.

Trappe (1977), Mikola (1973), and others (Imshenetskii, 1955; Shemakhanova, 1962; Bowen, 1965) thoroughly reviewed past work on ectomycorrhizal fungus inoculations. More recent reviews describe inoculation of seedlings to improve field performance on routine planting sites and drastically disturbed lands in the eastern USA (Marx and Cordell, 1988), France (Le Tacon *et al.,* 1988), and Canada (Lalonde and Piche, 1988). We, therefore, feel free to discuss selected reports as they relate to specific procedures. We concentrate on recent published and unpublished work on inoculations with ectomycorrhizal fungi and the resulting improvements in field performance.

II. Results of nursery and field research

The concept of forming ectomycorrhiza on planting stock in nurseries, with ectomycorrhizal fungi ecologically adapted to the planting site, to improve field performance was apparently developed in Hungary by Bokor (1954). Procedures were further refined in Austria by Moser (1958), in Argentina by Takacs (1967), in Australia by Theodorou and Bowen (1973), and in the USA by Vozzo and Hacskaylo (1971). These research programmes showed that seedlings with specific ectomycorrhiza exceeded the field performance of those that lacked or had few native ectomycorrhiza at planting.

A. Results of nursery and field research in the USA

For the past two decades research has been done by the US Department of Agriculture, Forest Service, Institute for Mycorrhizal Research and Development (IMRD) in the southern USA to determine the significance of specific ectomycorrhiza for survival and growth of tree seedlings on a variety of outplanting sites. Most work has concentrated on *Pisolithus tinctorius* (Pers.) Coker and Couch (Pt). Inoculum of Pt based upon commercial formulations and inoculation techniques has been developed for use in seedling production systems. The interest in Pt was prompted by its relatively wide geographic range and tree host distribution, its tolerance to a wide variety of environmental stresses, and the ease with which it can be grown in pure culture and manipulated (Schramm, 1966; Marx, 1980, 1981; Marx *et al.*, 1984a). Pt is an early-stage component in the normal fungal symbiont succession in southeastern forests of the USA and is very effective on seedlings and young trees. It is also present in mature natural stands of oak and pine, but here is mixed with hundreds of other higher fungi.

Marx (1980) discussed the early testing and development of vegetative Pt inoculum used for research purposes. The initial procedure involved growing mycelium of the fungus in jars containing a mixture of vermiculite and peat moss (ratio adjusted to give pH of 4.8–5.5) moistened with liquid medium. The mycelium, which developed in the laminated structure of the vermiculite particles, maintained viability and was protected from environmental stress and saprophytic microbial competition in nursery soil. This form of inoculum has been used in fumigated nursery soil and other growing media to form ectomycorrhiza on species of *Abies, Carya, Picea, Pinus, Pseudotsuga*, and *Quercus* seedlings in bare-root and container seedling nurseries throughout the USA and on various *Pinus* spp. in Brazil, Canada, Congo, France, Ghana, Liberia, Nigeria, Mexico, South Korea and Thailand (Marx and Bryan, 1975; Marx, 1979a,b, 1980; Alvarez and Trappe, 1983; Marx *et al.*, 1984a). Many other tree species have been inoculated with this fungus and formed abundant ectomycorrhiza in various tests at the IMRD. These included species of *Eucalyptus, Castanea, Fagus* and *Salix*. It is very possible that Pt, under the proper test conditions, will form ectomycorrhiza on all tree species that form ectomycorrhiza under natural field conditions. This broad host range is probably characteristic of most early-stage fungi (Dighton and Mason, 1985).

Marx and Cordell (1989) reviewed efforts to develop commercial vegetative inoculum of Pt. Since 1976, the effectiveness of different vegetative inoculum formulations has been tested under diverse cultural regimes on various tree species in bare-root and container seedling nurseries in the USA and Canada. The most effective commercial formulations developed initially (Marx, 1980) had: (1) the most hyphae of Pt inside the vermiculite particles (inoculum niche); (2) pH between 4.5 and 6.0 controlled with a 5–10% volume of acid peat moss mixed with pH 7.0 vermiculite; (3) low levels of bacterial and fungal contaminants, which apparently colonized inoculum during leaching and drying; and (4) low amounts of residual glucose ($mg\,g^{-1}$ inoculum). Commercial vegetative inoculum available today is produced in sterile plastic bags (10 litres) containing vermiculite–peat moss (pH 4.5–5.0) and having a C:N ratio of 55 in the growth medium. After inoculation and incubation at room temperature for 5–7 weeks, most of the carbohydrates are utilized by Pt, thus eliminating the need to leach and dry the inoculum prior to use (Marx *et al.*, 1989a). Vegetative inocula of Pt and other fungi are available commercially from Mycorr Tech Inc., University of Pittsburgh Applied Research Center, Pittsburgh, Pennsylvania 15238, USA.

Basidiospore inoculum of Pt has been used on an experimental and pilot operational scale in the USA and elsewhere. Recent studies have proved that spores are effective in various inoculum forms (Marx *et al.*, 1989b). In most studies testing these inoculum forms, spores were collected from multiple fruit bodies and, therefore, tended to represent inocula more genetically diverse than current vegetative inoculum. These studies also found that Pt basidiospore inocula rarely produce as many ectomycorrhiza as does the most effective vegetative inoculum (Marx *et al.*, 1989b). Large numbers of Pt ectomycorrhiza must be produced consistently on seedling roots after inoculation because a threshold level of at least half of all ectomycorrhiza must be those of Pt at planting time if maximum promotion of growth of southern pine seedlings planted on reforestation sites is to be achieved (Marx *et al.*, 1976, 1988). On routine reforestation sites, pine seedlings with less than half of all ectomycorrhiza formed by Pt often grow at the same rate as seedlings with the same amount of *Thelephora terrestris* (Ehrh.) Fr. (Tt).

The value of Pt ectomycorrhiza in tree regeneration has been demonstrated under diverse field conditions in the USA and in other countries. In the eastern USA, numerous studies on mined lands and on reforestation sites revealed significant and often dramatic enhancement of survival and growth of pine seedlings produced in bare-root or container seedling nurseries with abundant Pt ectomycorrhiza compared to seedlings with only naturally occurring ectomycorrhiza (i.e. Tt). These dramatic growth results have been observed in field studies on acid coal spoils, kaolin mining spoils, severely eroded sites, borrow pits, midwestern grassland prairies, urban sites and boreal forest sites. Similar improvements in pine and oak seedling performance were reported on routine reforestation sites in the USA. These field results were the subject of a recent review (Marx and Cordell, 1988).

Some of these reports noted only a minimal effect of Pt ectomycorrhiza on survival and growth of seedlings, while others reported increases of over 250% in volume yield per hectare. In those reporting large differences, control seedlings with only naturally occurring ectomycorrhiza (mostly Tt) at planting survived and grew poorly, whereas in those reporting small differences, the control seedlings survived and grew considerably better. These findings suggest that seedlings with abundant Pt ectomycorrhiza tolerate some environmental stress factors (e.g. soil water deficits, high temperatures) better than control seedlings.

Commercial vegetative and spore inocula of Pt are now used successfully in tree nurseries in the eastern USA. Vegetative inoculum is incorporated into nursery soil at 0.33 litre m^2 of soil surface or into container medium in the ratio of 1:10 (v/v) with rooting medium. In all bare-root nurseries the soil is fumigated before inoculation. This procedure effectively controls weeds, soil-borne pathogens, and other symbiotic fungi that compete with the introduced Pt. A tractor-drawn machine has been developed to inoculate fumigated nursery soils with commercial vegetative inoculum (Cordell *et al.*, 1981). It applies inoculum in bands at 4 to 6 cm depths under seed rows, reducing the application rate, time, cost, and labour associated with earlier broadcast methods. Spores of Pt are applied as pellets (Marx and Bell, 1985) or encapsulated seed (Marx *et al.*, 1984b). In 1989, over 8 million tree seedlings were inoculated with vegetative inoculum and another million were inoculated with spore inoculum of Pt in the eastern USA. These seedlings will be outplanted on routine reforestation sites, on various mined-land sites, and in Christmas tree plantations.

Spore inoculation has been very successful in bare-root nurseries in the northwestern USA (Castellano and Trappe, 1985). Basidiospores of seven species of hypogeous, ectomycorrhizal fungi have been applied to four conifer species. *Rhizopogon vinicolor* and *R. colossus* formed numerous ectomycorrhiza on Douglas fir following spore inoculation. Two years after outplanting, Douglas fir seedlings with *R. vinicolor* ectomycorrhiza at lifting had significantly greater survival, stem height, root collar diameter and biomass than non-inoculated seedlings. Basidiospores of these *Rhizopogon* species on container-grown Douglas fir seedlings have also produced abundant ectomycorrhiza when appropriate fertilization was used (Castellano *et al.*, 1985). In 1989, some 20 million Douglas fir seedlings were inoculated with spores of *R. vinicolor* in container (85%) and bare-root (15%) nurseries in Oregon, Washington, Idaho, and northern California (M.A. Castellano, pers. commun.).

B. Results of nursery and field research in France

Inoculation with ectomycorrhizal fungi in France has two purposes: to improve field performance in reforestation and to enhance production of edible fungi (Le Tacon *et al.*, 1988). Nursery procedures share many common features with procedures in the southern USA. In both countries, 95% of the seedlings for reforestation are grown in bare-root nurseries, soil fumigation is common, and *Thelephora terrestris* occurs naturally. *Laccaria laccata, L. bicolor*, and *Hebeloma crustuliniforme* form abundant ectomycorrhiza with Douglas fir, Norway spruce, and Scots pine seedlings following successful nursery inoculation with vegetative inocula. Instead of the normal 3–4 years, plantable Douglas fir seedlings can be produced in fumigated soil after 2 years following inoculation with *L. laccata*. Efforts are underway to develop commercial inoculum sources for reforestation programmes in France.

Since 1973, *Quercus* spp. artificially inoculated with *Tuber melanosporum* or *T. uncinatum* have been commercially produced in France for truffle production (Le Tacon *et al.*, 1988). Truffle fruit bodies can be obtained 3–5 years after transplanting the seedlings on proper sites. More recently, *Pinus pinaster* seedlings with *Suillus granulatus* ectomycorrhiza have been produced and outplanted to produce edible fruit bodies of the fungus.

Industrial fermentation procedures have been successfully employed to produce pure culture inoculum. Ectomycorrhizal inoculum has been produced by industrial liquid fermentation. Mycelium of *L. laccata* and *H. crustuliniforme* is produced in fermentors, leached to remove excess nutrients, and entrapped in calcium alginate beads containing powdered peat moss. This bead inoculum is better protected, survives longer, and is more effective than vermiculite–peat moss formulations in trials with Douglas fir and Norway spruce seedings in bare-root nurseries (Le Tacon *et al.*, 1988).

The French programme on ectomycorrhizal inoculations and field results was recently reviewed (Le Tacon *et al.*, 1988). During the last decade, many examples are found in which ectomycorrhizal inoculation is highly beneficial to reforestation. Best results have been achieved in experiments with conifers, especially Douglas fir, outplanted on sites, such as old fields, containing low resident inoculum of other ectomycorrhizal fungi. The most effective ectomycorrhiza on such sites were those formed by strains of *Laccaria laccata* from the USA and by local French strains of *L. bicolor*. One of us (DHM) visited France in July 1989 and observed several outplantings of Douglas fir with *L. laccata* ectomycorrhiza. Results on certain old fields are dramatic. Two- to three-fold increases in above-ground tree weights and volumes after 4–6 years are attributable to *Laccaria* ectomycorrhiza. Outplanting experiments with *Suillus granulatus* and *Boletus edulis* ectomycorrhiza are also underway.

C. Results of nursery and field research in Canada

Most ectomycorrhizal inoculation has concentrated on species of *Pinus, Picea* and *Larix* grown in containers. Strains of *Laccaria laccata, Hebeloma cylindrosporum, Cenococcum geophilum*, Tt and Pt have been grown vegetatively in solid and liquid substrate inoculum (Langlois and Gagnon, 1988).

Ectomycorrhizal inoculation in Canada is still in the experimental or developmental stage (Le Tacon *et al.*, 1988). The following points emerge from container-grown seedling research: (1) when a mixture of fungi is applied in solid inoculum only one fungus successfully colonizes the seedlings' roots; (2) the fertility of the substrate and the nutritional regimes applied influence ectomycorrhiza formation; (3) solid or liquid inoculum can be used with success at sowing; and (4) liquid inoculum injected in container cavities of 6- and 10-week-old seedlings after sowing produces abundant ectomycorrhiza.

Through 1992 the plans are to field test planting stock inoculated with commercially produced liquid inoculum (Rhizotec Laboratories, Inc., Laval University, Quebec, Canada). The objectives are to install plantations on different sites having diverse ecological conditions in order to determine the profitability of producing and planting artificially inoculated planting stock.

D. Results of nursery and field research in the Philippines and Venezuela

In the Philippines, de la Cruz and co-workers (P. E. de la Cruz, pers. commun.) at the National Institute of Biotechnology and Applied Microbiology (BIOTECH) developed a clay tablet inoculum containing basidiospores of Pt and a species of *Scleroderma*. Both fungi occur naturally in the forests of the Philippines. Tablet inoculum added to potted seedlings of *Pinus caribaea, Eucalyptus camaldulensis* and *E. deglupta* result in the formation of abundant ectomycorrhiza after 3 months. The response of the seedlings outplanted on acid soils low in organic matter and fertility and high in aluminium has been dramatic. After only 16–18 months in the field, diameter increases over control seedlings of more than 75% are common. Inoculation of seedlings with the clay-spore tablet is now a routine nursery practice in the Philippines. The biologically interesting feature of the growth response following inoculation of *Eucalyptus* seedlings is that this tree, unlike exotic *Pinus* spp., has not required ectomycorrhizal fungus inoculation in nurseries prior to outplanting. The response in the Philippines suggests that nursery inoculation with ectomycorrhizal fungi should be considered for *Eucalyptus* elsewhere.

In the Llanos Orientales of Venezuela, non-fumigated nursery soil is routinely treated with soil inoculum from an established *Pinus caribaea* plantation. This plantation was established in 1966 using soil inoculum from under pine brought in from Trinidad. The major, if not the only, ectomycorrhizal fungus in this inoculum is *Thelephora terrestris*. Following inoculation, it forms various amounts of ectomycorrhiza on seedlings of *Pinus caribaea, P. oocarpa*, and *Eucalyptus* spp. (D. H. Marx, unpubl. data). Since the early 1980s, basidiospores of Pt, originally from Georgia, USA, have been used to mass inoculate these nurseries (C.S. Hodges, pers. commun.). New spore inoculum is collected each year from fruit bodies of Pt produced in nursery beds inoculated the previous year. This procedure ensures maximum ectomycorrhizal development which is assumed to improve seedling survival and growth on Venezuelan reforestation sites. Such double inoculation is applied in nurseries that produce approximately 100 million pine seedlings each year.

E. Use of other fungi in inoculation programmes

In addition to the aforementioned fungi, other species have been used on various tree species to form specific ectomycorrhiza in experimental inoculation programmes in various parts of the world (Lalonde and Piche, 1988; Le Tacon *et al.*, 1988). Vegetative inocula of 11 other fungal species have been developed and tested on different tree species under various experimental conditions (Marx and Ruehle, 1989). These fungi are *Amanita muscaria, Corticium bicolor, Hebeloma cylindrosporum, Lactarius rufus, Paxillus involutus, Rhizopogon roseolus, Scleroderma auranteum, Suillus cothurnatus, Suill. luteus, Suill. variegatus* and *Tricholoma albobrunneum*. Commercial formulations of vegetative inocula of some of these fungi are currently available from Mycorr Tech Inc. However, only a few commercial vegetative formulations of these fungi have been tested for effectiveness under operational conditions (Kidd *et al.*, 1985; Hung and Molina, 1986). In addition to Pt, spore inocula of 18 other species of fungi have been reported to form ectomycorrhiza on various tree species under experimental conditions in various parts of the world (Marx and Ruehle, 1989). These fungi are *Hebeloma crustuliniforme, Rhizopogon colossus, R. luteolus, R. nigrescens, R. roseolus, R. vinicolor, Scleroderma auranteum, S. dictyosporum, S. flavidum, S. texanse, Suillus acidus, Suill. granulatus, Suill. grevillei, Tuber aestivum, T. brumale, T. maculatum, T. magnatum* and *T. melanosporum*.

III. Methods for nursery research

A. Methods of soil treatment

In successful ectomycorrhizal inoculation programmes, soil is either chemically or physically treated to eliminate or significantly reduce population densities of fungi, bacteria, insects and nematodes that might compete with, and reduce the efficacy of, the introduced ectomycorrhizal fungus, or cause malfunction of feeder roots. In bare-root nurseries where straw, grain, or pine needle mulch is applied after inoculation and seeding, the mulch should also be treated to eliminate undesirable competing microorganisms.

Soil treatment with fumigants such as methyl bromide–chloropicrin mixtures, dichloropropenes, ethylene bromide, metam sodium (vapam), vorlex, and dazomet is an accepted practice in many agricultural and forestry areas. More recently, a soil solarization technique (described below) has been developed as an alternative to chemical sterilization, and has been independently tested in several different countries (Katan *et al.*, 1976; Horiuchi, 1984; Stapleton and DeVay, 1986).

Chemical soil fumigants are non-selective soil biocides which kill a variety of soil organisms, including pathogenic and symbiotic fungi, insects, nematodes and weed seeds. The most commonly used fumigant in forest nurseries in the USA is the methyl bromide (98%)–chloropicrin (2%) mixture (MC-2), a fumigant that does an effective job preparing soil for ectomycorrhizal fungus inoculation. A methyl bromide (67%)–chloropicrin (33%) fumigant formulation (MC-33) has been particularly effective in controlling the more persistent soil pathogenic fungi (Cordell, 1983). Other chemical soil fumigants have produced variable and often less than satisfactory results in controlling soil pests (Sinha *et al.*, 1979).

Effective soil fumigation either eliminates or significantly reduces all living organisms in soil or potting mixtures in the zone of effective fumigant concentration. Beneficial ectomycorrhizal and saprophytic fungi can then rapidly invade treated soils if inoculum sources are available and often build up to higher population densities than those found in untreated soils (Cordell, 1983). Soil should be fumigated just before seed sowing for maximum effectiveness of inoculation with ectomycorrhizal fungi. Timing of fumigation in relation to inoculation and seeding is different depending on the chemical used. When MC-2 is used, inoculation and seeding are often carried out 1 week after fumigation, but other chemicals such as MC-33, vapam and dazomet, require longer soil aeration periods (4–6 weeks) before inoculation and seeding can be undertaken (Cordell, 1983).

In North America soil fumigants such as methyl bromide–chloropicrin mixtures, vapam and vorlex are commonly applied with a tractor-drawn machine that injects the chemical beneath the soil surface with chisels (Cordell and Kelley, 1985). Chisels are adjusted to a width of not over 30 cm and to an optimum soil depth of 20–25 cm (Cordell, 1983). Dosage rates are based on the amount of active chemical ingredient needed per hectare and these rates vary according to the chemical, target organisms, application method, soil types and environmental conditions. To be effective, a fumigant must remain in contact with the target organisms for sufficient time and in sufficient concentrations to kill. Therefore, highly volatile fumigants, such as methyl bromide–chloropicrin mixtures, require plastic covers or surface sealers to maintain sufficient dosage concentrations for adequate time periods to achieve effective control. Polyethylene covers of sufficient thickness (minimum 2 mil) and strength to minimize fumigant escape from soil or mulch are most effective (Cordell and Kelley, 1985).

In some smaller nurseries in the USA and in countries lacking the mechanical equipment needed to inject fumigants, soil fumigation is carried out by hand. Beds are covered with polyethylene covers which are supported above the soil, thus leaving a shallow air space (10–20 cm) between the soil and plastic. The fumigant is released from cans with a special applicator into evaporator pans placed on the soil surface under the plastic. The rate of application is usually 0.5 kg $10\,m^{-2}$ of soil surface. This very labour-intensive method is rather simple and requires no specialized equipment, but has the disadvantage of requiring considerable time to treat large areas and close supervision during application of this highly toxic gas (Cordell and Kelley, 1985).

Another soil fumigant that might be effective in forest nursery inoculation programmes is metam sodium (vapam, 33% a.i.), a material proven to be effective in Israel to control pathogens in nursery soil (Widin and Kennedy, 1983; Ben-Yephet *et al.*, 1988). Application methods include broadcasting granules and mixing them into the soil or drenching the soil with a liquid formulation. Depending on the soil type and pest problems, effective vapam rates of 250 to 750 litre ha^{-1} are needed for satisfactory control. Both methods require that effective concentrations of vapam are placed into the target soil profile (10–40 cm soil depth) with adequate water to activate release of the toxic breakdown chemical methyisothiocyanate. Vapam has also been effectively applied to nursery soil through a nursery irrigation system. The chemical was metered through the sprinkler system during 7 h of irrigation, then left undisturbed for 7 days to allow it to be effective. The soil was then aerated (disked, tilled) for 2–3 days before sowing.

Soil solarization is a rather simple alternative soil treatment that might be used in ectomycorrhizal inoculation programmes. Moist soil is covered with clear plastic to create a greenhouse effect with sunlight, which can raise soil temperatures high enough (40–50 °C) to reduce significantly population densities of undesirable soil organisms. Initially developed and perfected in Israel to control soil pathogens (Katan *et al.*, 1976), this technique involves covering moist soil with clear plastic (1–2 mil thickness) for a period of 4–6 weeks during the hottest and sunniest months of the year. Adequate soil moisture, at or near field capacity, greatly enhances the effectiveness of soil solarization (Horiuchi, 1984). Higher soil moisture often raises the soil temperature and extends the effective high thermal period. The plastic should be left in place until the soil is inoculated or sowed. Attempts to control certain pathogenic fungi which have high temperature tolerances (> 50 °C) have been relatively ineffective (Katan *et al.*, 1976; Horiuchi, 1984; Mihail and Alcorn, 1984; Stapleton and De Vay, 1986), but their control can be achieved using a combination of solarization and chemicals (Ben-Yephet *et al.*, 1988). This combination treatment is more effective than either treatment used separately because the efficacy of the chemical is increased at the higher soil temperatures.

B. Methods of inoculation

To keep costs down, one should apply the least amount of inoculum that will provide an acceptable quantity of specific ectomycorrhiza on seedlings in the nursery phase before outplanting. Soil can be inoculated: (1) before seeds are sown; (2) when seeds are sown; or (3) shortly after seedlings emerge. Inoculation of seedlings either at the end of the nursery phase or at the time of field planting is time-consuming, often less effective, and requires more inoculum. Seedling performance will not be improved until ectomycorrhiza have formed some weeks later. Thus, little improvement in initial seedling survival can be expected using this method.

1. Vegetative inoculum

(a) The use of broadcast inoculation. This procedure, used in bare-root nurseries, involves spreading a measured quantity of inoculum over a given area of soil surface followed by mixing into the root zone. Duff, soil inoculum, fruit bodies and spores, as well as pure culture vegetative inoculum in vermiculite–peat moss and in alginate beads have all been broadcast to provide infection. These inocula have also been mixed successfully with growing media before cavities were filled in container seedling nurseries.

The broadcasting of duff or soil inoculum at a rate of 2–4 kg m^{-2} of seedbed soil surface, followed by mixing into the upper 10 cm of soil prior to sowing, provides a simple and effective inoculation procedure that has been used throughout the world. However, duff or soil inoculum poses many potential hazards, including the introduction of pathogens and noxious weed seeds (Marx, 1980).

Pure-culture vegetative inoculum in vermiculite–peat moss or in alginate beads has been used when a specific ectomycorrhizal association has been required or when endemic nursery fungi were not ecologically adapted to the planting sites. Both laboratory and commercial forms of vegetative Pt inoculum in vermiculite–peat moss have been broadcast and mixed into the upper 10 cm of fumigated soil at the rate of 1 litre m^{-2} of soil surface (Marx *et al.*, 1984a). Vegetative inocula mixed 1:10 (v/v) in various container growing media before sowing have also proved satisfactory for producing specific ectomycorrhiza on container-grown seedlings (Marx *et al.*, 1982). Mycelium of *L. laccata* entrapped in alginate beads can be broadcast at a rate of 2–5 g (dry weight equivalent of mycelium) per m^2 of soil surface, then mixed into the upper 10–15 cm of fumigated soil.

(b) The use of banding inoculum. A nursery seedbed inoculator that applies vegetative inoculum has been designed and used successfully in the USA (Fig. 1). The machine places the inoculum 4–6 cm deep in soil, in a band 4–5 cm wide directly below the seed drill, before sowing. By banding the inoculum in the initial root development zone, less than one-third of the amount of inoculum is required (330 ml m^{-2}) as with broadcast applications. Development of Pt ectomycorrhiza is still comparable with that found with broadcasting (Cordell *et al.*, 1988). This machine is now marketed for operational inoculation of nursery soil with commercial vegetative inoculum in vermiculite–peat moss formulations. The seedbed inoculator was recently modified to apply vegetative inoculum of Pt in a slit between rows of established nursery seedlings. The slit is made with a coulter 15 cm deep between rows of seedlings planted 15 cm apart. Inoculum is spread in the slit at the same time. Lateral roots severed by the coulter proliferate in the inoculum slit and, after a few weeks, abundant Pt ectomycorrhiza are formed. The same amount of vegetative inoculum, 330 ml m^{-2} of soil surface, is used for both band and slit applications. Making slits between seedling rows with hand tools and hand-applying inoculum is also effective.

Banding is also very effective for inoculation of container-grown seedlings. Banding of 10 ml of vegetative inoculum in the centre of a 65 ml container just before seeding has proved satisfactory (Riffle and Maronek, 1982). Banding also works well when polyethylene bags (23 cm × 10 cm diameter) are used. Vegetative inoculum is added to a bag two-thirds filled with the desired soil, as a 5–10 mm layer. The remainder of the bag is filled with soil and the seeds are sown.

(c) The use of slurry techniques. Mycelial slurries added to soil of bare-root nurseries have shown little promise. This inoculum form, however, has been very effective when applied to container-grown seedlings of various tree species in Canada. Vegetative slurries of *Laccaria bicolor* can be injected into the growing medium at sowing or

Fig. 1. Ectomycorrhizal fungus applicator for applying vegetative inoculum in bare-root forest tree nurseries.

9–10 weeks after germination, when root systems begin to develop, to form ectomycorrhiza on *Picea* seedlings (Langlois and Gagnon, 1988).

Vegetative inoculum of Pt in a vermiculite–peat moss formulation has also been effective in a water slurry (Marx and Cordell, 1987). Inoculum was mixed 2:1 with water and vigorously stirred to keep the inoculum particles suspended. Roots of young seedlings were dipped into the slurry for a few seconds; the seedlings were immediately transplanted. Within a few weeks abundant Pt ectomycorrhiza were formed. This method, originally developed to determine root susceptibility and inoculum viability and survival, can be used to establish Pt ectomycorrhiza on seedlings in a production nursery (Marx and Cordell, 1987).

2. *Basidiospore inoculations*

Spores of various fungi have been used as inoculum to form specific ectomycorrhiza. Basidiospores of Pt have been effective when used as dry spores, suspended in water and drenched, or coated on seeds or vermiculite particles. Dry spores can be applied more effectively by pre-mixing them with a carrier (vermiculite, kaolin, sand or hydro-mulch) before application.

In bare-root and container nurseries, basidiospores of Pt are effective when mixed with a physical carrier before soil application (Marx *et al.*, 1979), suspended in water and drenched (Marx *et al.*, 1979), sprayed on soil and seedlings or dusted onto soil (Lee and Koo, 1984), broadcast in pellets (Marx and Bell, 1985), or placed in a seed encapsulating mixture (Marx *et al.*, 1984b). Encapsulated pine seeds have 0.5–1 mg of Pt basidiospores in a clay adhesive coating. There are about 1 million spores of Pt per mg. Spore pellets are spores coated on 1–3 mm vermiculite particles. Each pellet has about 0.5 mg of spores. After sowing or broadcasting, daily watering causes the spore coating to dissolve from either seeds or pellets, releasing the spores into the soil and down to the root zone.

The simplest method of applying Pt spores is by spraying them in a water suspension containing a wetting agent. Dry spores collected from the upper part of the Pt fruit body should be filtered through a 50 μm screen, dried at room temperature with 35–40% relative humidity to approximately 18% moisture content, and stored at 5 °C until used. Research results on southern pines suggest that a spore dose of 0.5–1 $g\,m^{-2}$ of soil surface is effective. This does not mean that lower dosages are ineffective elsewhere.

A spore suspension of Pt may be prepared as follows: to initially wet and suspend the spores, two to three drops of Tween 20 surfactant (or other wetting agent) are added to 30 ml of water in a 50 ml vial containing 1 g of spores. The vial should be shaken vigorously for 1–2 min and the contents decanted into a 4-litre bottle containing 3 litres of water. After mixing these contents vigorously, the 3-litre water suspension is decanted into a 4-litre capacity hydraulic sprayer. The bottle is rinsed with another litre of water and this volume is added to the sprayer, making a 4-litre final volume.

This spore volume of suspension is sprayed evenly over 1 m^2 of soil surface. The sprayer is then filled with another 4-litre volume of water. The contents are shaken and sprayed over the same 1 m^2 area. The soil should be irrigated immediately. The first spray volume dispenses the majority of spores from the sprayer, but spore clumping and floating make it necessary to wash the remaining spores from the sprayer with the second volume. The second volume also leaches the spores into the soil and reduces the possibility of spores drying on the soil surface and being blown off the nursery beds. Large volume hydraulic sprayers are more efficient than small ones. Sprayer volume is unimportant as long as the ratios between spores:water:soil surface area are approximately the same as in the example.

C. Cultural practices

All procedures applied to soil or seedlings from soil fumigation through to seeding and seedling lifting are cultural practices. Inoculation with specific fungi is only one aspect of producing a seedling of sufficient quality for outplanting. Frequency of watering; bed density; type, amount, and timing of fertilization; use of pesticides; root pruning; etc. are important in producing quality seedlings. Soil fertility (especially its nitrogen content), pH, and pesticides have all been shown to cause serious problems in past research programmes. In fertility studies, however, the form, amount, and methods of applying fertilizers, as well as the number of applications and the composition of available nutrients, differ widely. Other variables that influence results of inoculation treatments include the length of the experiment, fungal and host species, growth medium (soil or artificial), environment (greenhouse, growth chamber, bare-root nursery) and the method of expressing ectomycorrhizal development. Some early researchers unknowingly used late-stage fungi on seedlings and obtained mixed results. After adding nitrogen, there have been reports that ectomycorrhizal development declines, remains stable, or temporarily declines then recovers. Marx (1990) recently studied the effects of soil pH (pH 4.8, 5.8 and 6.8) and nitrogen fertilization (3, 6 or 9 applications of 50 kg N ha^{-1} as NH_4NO_3) on Pt ectomycorrhizal development of loblolly pine. Vegetative inoculum of Pt buried in soil lost viability to a significant degree after 54 days at soil pH 6.8. Up to three applications of NH_4NO_3 at 50 kg N ha^{-1} each did not affect inoculum survival at this time. At the end of the growing season, total Pt ectomycorrhizal ratings (combining the number of ectomycorrhiza and proportion of different morphological types) on seedlings at pH 6.8 were about 25% of those in more acid soil conditions. The highest applications of NH_4NO_3 resulted in soil concentrations of NO_3-N ranging from 60–120 kg ha^{-1} and of NH_4-N ranging from 90–130 kg ha^{-1} at pH 4.8 and 5.8 and were associated with the most abundant Pt ectomycorrhizal development. Increased nitrogen applications increased Pt development on seedlings at pH 5.8 and 6.8 relative to that achieved on seedlings receiving fewer nitrogen applications.

The presence of pesticides, especially systemic fungicides, can also confound results in mycorrhizal studies (Trappe *et al.*, 1964). Triadimefon (Bayleton), a systemic fungicide used to control fusiform rust in southern pine nurseries, seriously suppresses Pt and natural ectomycorrhizal development, even at rates significantly lower than recommended (Marx *et al.*, 1986; Marx and Cordell, 1987). In contrast, another systemic fungicide, Benomyl, stimulates Pt and natural ectomycorrhizal development as does the broad-spectrum fungicide Captan (Marx and Rowan, 1981). Both of the latter fungicides are recommended for practical establishment of Pt in the southern USA. Bayleton cannot be used following inoculation.

D. Experimental design

Variability is inherent in most nursery studies of ectomycorrhiza. Factors creating variability that can often confound treatment results are: size of seedlings, number of roots, soil fertility, growing system (bare-root versus container grown), seed source, genetic variation of host, variability of fungus symbiont, variability of inoculum and cultural practices. Every attempt should be made to standardize as many factors as possible.

To reduce some of this variability the following is suggested:

- Use half-sib seedlots from selected superior mother trees instead of mixed seedlots.
- Obtain properly processed seedlots with high germination potential.
- Use inoculum with high viability that has been produced and stored properly.
- Standardize fertility and soil pH to permit appropriate root development linked with suitable top growth.
- If containers are used, select growing medium with adequate drainage and gas exchange; avoid waterlogged conditions, water only as needed.
- In bare-root nurseries, select and standardize seedbed density adequate for the test tree species.

- Avoid unnecessary use of pesticides, especially those that suppress ectomycorrhizal development or have unknown effects.

Most inoculation experiments with forest tree species involve seedlings, rather than vegetatively propagated plants. Variability in seedling parameters often creates problems in the interpretation of results of inoculation trials. Some tree species vary so much in seedling characteristics, even within single mother tree seedlots, that selection for uniformity after germination may be advisable. One approach is to plant extra seed, then later to rogue the fastest and slowest germinants to achieve a standardized bed density for all treatments.

Pre-trials with seedling culture are recommended to familiarize the researcher with seed germination behaviour, effects of seed size on seedling size, cultural practices, length of study, time needed for development of ectomycorrhiza, and length and testing of effects of inoculation. Seedlings grown too long in containers have restricted root development, which causes stress and makes it difficult to evaluate growth responses to ectomycorrhizal fungi.

Care should be taken to select uniform conditions in nursery soils for experiments with bare-root seedlings. Soil should be treated to reduce competing micro-organisms and weeds and to avoid disease problems. Placement of study plots in relation to irrigation riser lines is important to assure uniformity of water applications in all plots.

Pre-trials are also needed to determine types of controls needed for inoculation tests. Initially, three control types might be considered: (1) no inoculum; (2) killed inoculum; and (3) inoculum substrate (e.g. vermiculite–peat moss) lacking the ectomycorrhizal fungus. If it is determined that there is no difference among controls, then only one such treatment is needed in later experiments. In greenhouse or controlled environment chambers, frequent re-randomization of the pot or cultural unit locations is recommended because light, air movement and other factors are rarely uniform and are not randomly variable. Re-randomization spreads the influence of uncontrolled variables across the experimental plant population.

Because of the wide range of factors causing inter- and intra-treatment variation, sampling of subsets of plants within each treatment per replicate is recommended. Stem diameters and dry weight of biomass are excellent measures of plant response to inoculation. With either measure, no fewer than 10 seedlings per treatment per replicate must be examined to discriminate relatively small differences in both bare-root nursery plots and container-grown seedling units. The measurement of 20 is preferred. At least five replicates (blocks) per treatment should be used for these trials. More replicates may be better. Pre-trials help to define treatment variability, identify the degree of precision needed, and provide an estimate of the most effective and efficient number of replicates.

E. Measurements of young seedlings

Dry weight is the most sensitive indicator of seedling response to any treatment. Tops should be weighed separately because treatments often influence root and top development differently. However, when weights of tops and roots reflect treatments similarly, total weights and top:root weight ratios should be recorded and measures of variabillty are required.

Particularly with young seedlings, height (stem length) usually is not a sensitive measure of small differences among ectomycorrhizal treatments. Stem diameter at root collar is an excellent indicator of growth difference. It can be measured non-destructively. Since it is strongly correlated with dry weight, especially of roots, root-collar diameter serves as a surrogate for weight measurement in seedlings that are not sacrificed. Root dimensions are seldom used because measurement is cumbersome and requires destructive sampling. Total length of lateral roots on young seedlings, however, is sensitive to ectomycorrhizal treatment and can be measured directly as described below.

F. Quantitative assessment of ectomycorrhiza

The most important measurement in any ectomycorrhizal research study, especially seedling studies, is the quantitative assessment of the ectomycorrhiza. Reliable quantification of ectomycorrhiza is essential to determine inoculum efficacy, ectomycorrhizal contribution to seedling biomass, nutrient absorption, relationships to root pathogen infection, etc. There is no one assessment technique that is, or should be, used for all studies. The choice of technique should be determined by the objectives of the study. There are two general types of assessments: visual estimates and direct counts (Grand and Harvey, 1982). In many seedling studies, entire root systems are examined with the unaided eye or at 5× magnification and the percentages of all short roots with specific ectomycorrhiza are visually estimated. An experienced observer can separate the percentage of those formed by individual fungi to within 5%. After estimation, the numbers can be assigned to rating systems such as + (< 1%), ++ (< 10%), +++(< 30%), and ++++

(> 30%), or excellent (75–100%), good (50–74%), moderate (24–49%), or poor (1–24%). Marx (1981) developed an index for assessing seedlings with Pt ectomycorrhiza. The Pt index is computed using the formula $a \times b/c$ where a = percentage of seedlings with any percentage of Pt ectomycorrhizae, b = average percentage of feeder roots with Pt ectomycorrhiza (including those with 0%), and c = average percentage of total ectomycorrhiza formed by all fungi including Pt. The Pt index can range from 0 to 100 and it provides a single value which represents fungus aggressiveness, inoculum efficacy and field performance. Other researchers have used modifications of this index for other fungi (Le Tacon *et al.,* 1988).

Direct counts of entire root systems have been used when only a few seedlings are involved. Each short root and ectomycorrhiza is counted under magnification and lateral root lengths are measured. The data can be presented as total number of ectomycorrhiza per seedling, number of ectomycorrhiza cm^{-1} of lateral root, or percentage of short roots that are ectomycorrhizal. Ectomycorrhiza on randomly selected lateral roots (cut from the plant) have also been counted directly on larger seedlings. Ectomycorrhiza can also be quantified by counting tips, by computing the average surface area, and by weighing the ectomycorrhiza.

Recently, Marx (1990) developed an assessment procedure involving a photocopying machine to determine lateral root length, number and percentage of ectomycorrhiza, and morphology of ectomycorrhiza. This procedure is as follows. Entire root systems are washed thoroughly of all debris. Short roots are counted on 1-cm segments of 5–10 randomly selected lateral roots still intact on each root system. Percentages of short roots ectomycorrhizal with Pt or other fungi are visually estimated twice at 5× magnification for each entire root system. The two estimates are averaged if they do not vary by more than 10%. If variation is more than 10%, a third estimate is made and the three estimates are averaged.

In order to reflect relative size differences in morphology of the Pt ectomycorrhiza a size estimate between 1 and 10 (Table I) is assigned on the basis of visual estimates of the proportions of ectomycorrhiza by morphological type on each entire root system. These estimates are made twice and averaged for each seedling. This number (between 1 and 10) is multiplied by the number of ectomycorrhiza (determined later) to reflect more accurately the total rating of ectomycorrhiza on each root system. After ectomycorrhizal assessment, first-order lateral roots (all lateral roots attached to the taproot) are removed from each seedling, placed between two sheets of plexiglass (3 mm thick) and photocopied (Wilcox, 1982). For each seedling, the first-order lateral

TABLE I

Numerical values based on visual estimates (percentages) of the relative proportion of ectomycorrhiza of different morphological types formed by *Pisolithus tinctorius* on loblolly pine seedlings (Marx, 1990).

	Ectomycorrhizal morphology			
Visual estimate rating	Non-forked simple (%)	Coralloid <4 tips (%)	Coralloid 4–10 tips (%)	Multiple coralloid fan-like >10 tips (%)
1	100	0	0	0
2	50	50	0	0
3	25	75	0	0
4	25	50	25	0
5	25	25	25	25
6	0	25	50	25
7	0	25	25	50
8	0	0	50	50
9	0	0	25	75
10	0	0	0	100

roots are counted and the total length of lateral roots of all orders supporting short roots and ectomycorrhiza are then measured to the nearest 0.5 cm with a planimeter. From these and other measurements, the total number of short roots and ectomycorrhiza are derived for each seedling.

Photocopying of root systems, originally suggested by Wilcox (1982) not only provides excellent permanent records, it also permits accurate determination of lateral root numbers and their length. However, visual determinations of percentages of ectomycorrhiza and morphological types of ectomycorrhiza cannot be obtained from the photocopies. Many ectomycorrhiza are damaged or lost before photocopying when lateral roots are cut from the taproot, free water is removed from the lateral roots, and the lateral roots are arranged on the plastic sheets. Short roots, the number per cm of lateral root, are easily counted from a sample before photocopying and used to derive the total number of short roots per seedling after lateral root lengths are measured on photocopies. These data, along with visual estimates of percentage of ectomycorrhizal development, provide accurate estimates of the total number of ectomycorrhiza on each seedling. When differences in total lateral root length are small because of soil treatments, significant differences between visual percentage estimates and derived numbers are rarely found. Sometimes treatments affect Pt index differently from other measurements. The Pt index may be markedly different from other Pt measurements when root colonization by Pt and naturally occurring ectomycorrhiza are both very low (Table II). In this situation, Pt indices are usually high for the quantity of Pt ectomycorrhiza involved and, therefore, misleading. The total ectomycorrhizal rating, which combines numbers and relative size of ectomycorrhiza on a root system, is an excellent indicator of ectomycorrhizal fungus vigour or its amount on the root system. In past research, most ectomycorrhizal assessments dealt only with percentage of colonized short roots or bifurcated tips. Size of ectomycorrhiza was not measured or estimated. Total ectomycorrhizal rating, percentage of roots colonized, and the derived number of ectomycorrhiza can be used together to reflect more accurately ectomycorrhizal status of root systems.

G. Statistical methods

Important statistical issues for ectomycorrhizal experiments are: design and analysis; relevance; combining results across experiments; and statistical power. Experimental designs are generally variations of split-plots or randomized blocks. Data are analysed by analysis of variance (ANOVA), analysis of covariance (ANCOVA) or regression techniques. In the case of ANOVA and ANCOVA, statistical models should be completely specified, i.e. fixed and random effects should be identified.

If seedling studies are to be useful, they must provide information that will enable the evaluation of hypotheses. Changes in seedling

TABLE II

Selected data (Marx, 1990) from loblolly pine seedlings to illustrate effects of treatments on Pt index and other *Pisolithus* ectomycorrhizal measurements.[a]

Treatment	% Pt ecto-mycorrhiza	No. Pt ecto-mycorrhiza	Pt morphological types[b]	Total ectomycorrhiza rating[c] ($\times 10^2$)	Pt index
Soil pH 4.8	53a	2325a	6.2a	144a	91a
Soil pH 5.8	37b	1030b	6.5a	67b	87a
Soil pH 6.8	6c	315c	3.1c	10c	68b

[a]Within columns, means followed by a common letter are not significantly different at $P = 0.05$ according to Duncan's new multiple range test.

[b]Relative visual estimate between 1 and 10 based on proportion of morphological types of ectomycorrhiza (see Table I).

[c]Total ectomycorrhizal rating = number Pt ectomycorrhiza × morphological rating.

condition due to treatment must also be representative of seedling populations. Ideally, information may also suggest hypotheses applicable to the study of more mature trees, but direct extrapolation from seedlings to mature trees is not usually possible.

Pooling of results from individual studies may be achieved by combining probabilities or test statistics or may involve the use of regression techniques. However, many experiments with widely different variations in objectives and design may not be comparable. In these cases, results should not be combined. In combining results of separate studies, one must identify potential problems with treatments or experimental procedures that might hinder the combination of results and cloud the final interpretations.

The power of the statistical test, the probability of detecting a consequential change in condition, should be considered for all ectomycorrhizal studies at both the initiation and conclusion of the research. Power is a function of experimental design, including sample size and experimental (uncontrolled) error, where sample size refers to both the number of seedlings used per treatment combination and the number of replicates. Each scientist should compute both the confidence limits of differences attributed to ectomycorrhizal treatments and the realized power.

IV. Methods for field research

As in nursery studies, field studies involve many factors that tend to influence, and often confound, responses to ectomycorrhizal treatments. The major differences are that field studies are of much longer duration and seedlings are exposed to far more natural stress than in nursery settings. Often, uncontrolled factors are not properly measured or accounted for. Initial factors that should be considered are variation in degree of ectomycorrhizal development on test seedlings, genetic variation in tree host, variation in initial seedling size and site variability. Deliberate attempts must be made to standardize as many as possible of the factors that directly or indirectly influence treatment effect and to standardize non-treatment growth response.

A. Selection of test seedlings

Frequently, seedlings inoculated in the nursery will be larger than the non-inoculated control seedlings. If initial size differences are taken to the field, they tend to confound interpretation of results when subsequent growth differences appear. This happens because big seedlings usually grow faster than small ones. Regardless of nursery treatment, grading of seedlings to a common size eliminates the bias of size differences attributable to treatment in the nursery. However, grading may introduce a genetic bias, especially when mixed seedlots are used. Selection of the larger control seedlings and the smaller inoculated seedlings may tend to select different genotypes from the initial mixed population. If inoculation produces larger seedlings, the use of similar sized or even larger control seedlings from a higher fertility treatment have been used in field experiments (Marx *et al.,* 1985). For interpreting field performance data, this treatment of controls eliminates any bias created by initial differences in seedling size at outplanting, but does create differences in initial seedling nutritional status.

B. Experimental field design

Selecting test sites with moderate variation in soil type, slope, etc., controlling competing vegetation, grading seedlings to equal size, and standardizing planting techniques all help to reduce unwanted variation and allow experimental treatment effects to be expressed and measured. However, many factors cannot be easily controlled, and, in some cases, excess experimental variation confounds treatment effects. Berry and Marx (1978) planted pines with different ectomycorrhizal treatments on severely eroded sites in the Copper Basin of Tennessee. The slope and soil type were quite variable and there was a problem of properly placing plots. After 2 years, with 24 trees per treatment in each of three replicate blocks, differences in seedling volumes of as much as 46% between some treatments were not statistically significant due to excessive variability caused by the site. It was obvious that site variability was not standardized with this design of only three blocks. More blocks were needed to accommodate the site variability. In another study involving 8-year performance of loblolly pine with Pt ectomycorrhiza outplanted on a good quality site (old agricultural field) in Georgia, on-site variation confounded treatment effects (Marx *et al.,* 1988). In this study there were 25 trees per plot in each of five blocks. Yet, because of undetected site variation, 18% differences between certain Pt treatments and controls in volume or weight per acre were not statistically significant. These differences might have been significant if six or seven blocks had been used to accommodate site variation.

Many options in field design are available. In our programme in the southern USA, we have standardized our design for low-variability reforestation sites. We use five replicate blocks and each treatment is represented by 49 seedlings (seven rows of seven seedlings each) in a plot within each block. For the first 2 years, all 49 seedlings are measured and evaluated. After that the interior 25 seedlings are used as measurement trees and the 24 border trees act as buffers against the edge effect. In our large-scale operational programmes, three rows of Pt seedlings and three rows of nursery-run seedlings are planted in repeating sequence over a planting site. The site may vary from 25 to 50 acres, and each row may contain up to 300 seedlings. Trees in the interior row of each three-row set are evaluated. On certain sites, each seedling treatment row is replicated in three-row sets up to 25 times. Statistical analyses of data can be done as with nursery tests.

C. Field measurements

Some measurements of trees in the field can be made directly; others must be derived. Survival, height, and either root-collar diameter (RCD) for small trees or diameter-breast height (DBH) for larger trees are common. Seedling volume using $(\mathrm{RCD})^2 \times$ height, total seedling volumes for a plot (plot volume index), which incorporates survival and growth into a single value, are useful derived values. Once trees are large enough, diameter at breast height is a preferred measurement, because it can be used in published tree volume tables. If the trees can be sacrificed, the best measurement is total above-ground dry weight. This value very accurately reflects plant response, at least above ground, but it is very labour-intensive to measure.

Accurate assessment of ectomycorrhiza in the field is difficult and often damages the test trees. To determine persistence and spread of test symbionts, extra test seedlings should be planted and totally excavated as desired. Ectomycorrhizal assessments can then be made as described in the nursery research section. Occurrence of fruit bodies of the test fungi in field plots has been used successfully by researchers as a substitute for fungus persistence on root systems.

D. Presentation of results

It is often necessary to examine data in different ways to understand fully the effects of ectomycorrhizal treatments. Sometimes standard statistical procedures on selected parameters (i.e. height, DBH, tree weight, tree volume, etc.) show no significant differences between treatments. Novel approaches for examining data have been employed to provide important information on effects of ectomycorrhizal treatments. For example, in a recent paper evaluating 8-year field performance of loblolly pine with initially different amounts of Pt ectomycorrhiza on a good-quality forest site, mean DBH values for trees with an initial Pt index 68 at planting was 5.3 in. and the mean values for control trees with Pt index 0 was 5.1 in. (Marx *et al.,* 1988). These means were statistically different, but the magnitude of difference was not of biological or practical significance. When trees were separated into size classes, however, a much larger proportion of the trees in the Pt index 68 treatment occurred in the two largest size classes (Fig. 2). Within the two largest DBH classes, there were 40% more trees from Pt index 68 treatment than controls. This type of data display was useful because it showed that 75% of the trees in the Pt index 68 treatment occurred in the two largest size classes. Canopy closure was complete in these plots after 8 years. Since more trees in the Pt treatment were in the dominant and co-dominant crown classes, they should maintain their growth rate superiority over the more suppressed and intermediate control trees for the life of the stand.

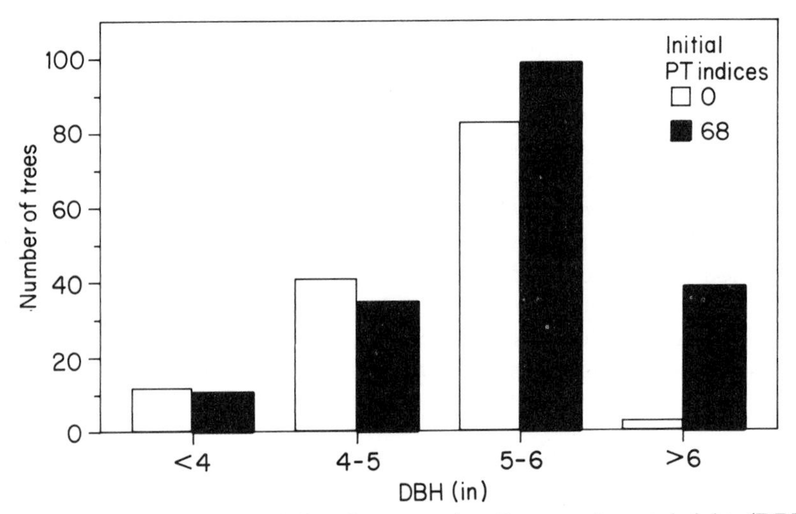

Fig. 2. Distribution of loblolly pine trees by diameter breast height (DBH) classes as affected by initial Pt indices after 8 years on a good forest site in Georgia, USA.

References

Alvarez, I. F. and Trappe, J. M. (1983). *Can. J. For. Res.* **13**, 533–537.

Ben-Yephet, Y., Melero-Vera, J. M. and DeVay, J. E. (1988). *Crop Prot.* **7**, 327–331.

Berry, C. R. and Marx, D. H. (1978). *Effects of* Pisolithus tinctorius *ectomycorrhizae on growth of loblolly and Virginia pines in Tennessee Copper Basin.* USDA Forestry Service Research Note SE-264, 6 pp.

Bokor, R. (1954). *Erdesz Kutatas, Budapest* **4**, 27–45.

Bowen, G. D. (1965). *Austral. For.* **29**, 231–237.

Castellano, M. A. and Trappe, J. M. (1985). *Can. J. For. Res.* **15**, 613–617.

Castellano, M. A., Trappe, J. M. and Molina, R. (1985). *Can. J. For. Res.* **15**, 10–13.

Cordell, C.E. (1983). In *Proceedings of the 1982 Southern Nursery Conferences*, USDA Forestry Service Technical Publication R8-TP4, pp. 196–201.

Cordell, C. E. and Kelley, W. D. (1985). In *Proceedings of the International Symposium on Nursery Management Practices for Southern Pines* (D. B. South, ed), pp. 496–504. Auburn University, Montgomery, AL.

Cordell, C. E., Marx, D. H., Lott, J. R. and Kenney, D. S. (1981). In Forest Regeneration, pp. 38–42. American Society for Agricultural Engineering, St Joseph, MI.

Cordell, C. E., Caldwell, C., Marx, D. H. and Farley, M. E. (1988). In *Proceedings of the 1988 on Mining, Hydrology, Sedimentology, and Reclamation*, pp. 229–235. University of Kentucky, Lexington, KY.

Dighton, J. and Mason, P. A. (1985). In *Developmental Biology of Higher Fungi* (D. Moore, L. A. Casselton, D. A. Wood and J. C. Frankland, eds), pp. 117–139. Cambridge University Press, Cambridge.

Grand, L. F. and Harvey, A. E. (1982). In *Methods and Principles of Mycorrhizal Research* (N. C. Schenck, ed), pp. 157–164. The American Phytopathological Society, St. Paul, MN.

Harley, J. L. and Smith, S. E. (1983). *Mycorrhizal Symbiosis*. Academic Press, New York. 483 pp.

Hatch, A. B. (1937). *Black Rock For. Bull.* No. 6, 168 pp.

Horiuchi, S. (1984). In *Soilborne Crop Diseases in Asia*, pp. 215–227. Food and Fertilizer Technology Center, Taipei, Taiwan.

Hung, L. L. and Molina, R. (1986). *Can. J. For. Res.* **16**, 802–806.

Imshenetskii, A. A. (1955). *Mycotrophy in Plants*, US Department of Commerce Translation TT67-51290 (1967), Washington, DC.

Katan, J., Greenberger, A. and Grinstein, A. (1976). *Phytopathology* **66**, 683–688.

Kidd, F., Brewer, D. and Miller, D. (1985). In *Proceedings of the 6th North American Conference on Mycorrhizae* (R. Molina, ed.), p. 218. Forest Research Laboratory, Oregon State University, Corvallis, OR.

Lalonde, M. and Piche, Y. (eds) (1988). *Canadian Workshop on Mycorrhizae in Forestry*. CRBF Faculté de Foresterie et de Geodesie, Université Laval, Ste-Foy, Quebec. 172 pp.

Langlois, C. G. and Gagnon, J. (1988). In *Canadian Workshop on Mycorrhizae in Forestry* (M. Lalonde and Y. Piche, eds), pp. 9–13. CRBF, Faculté de Foresterie et de Geodesie, Université Laval, Ste-Foy, Quebec.

Lee, K. J. and Koo, C. D. (1984). *J. Korean For. Soc.* **65**, 43–47.

Le Tacon, F., Garbaye, J., Bouchard, D. *et al.* (1988). In *Canadian Workshop on Mycorrhizae in Forestry* (M. Lalonde and Y. Piche, eds), pp. 51–74. CRBF, Faculté de Foresterie et de Geodesi, Université Laval, Ste-Foy, Quebec.

Marx, D. H. (1979a). *Synthesis of* Pisolithus *ectomycorrhizae on white oak seedlings in fumigated nursery soil*. USDA Forestry Service Research Note SE-280, 4 pp.

Marx, D. H. (1979b). *Synthesis of* Pisolithus *ectomycorrhizae on pecan seedlings in fumigated soil*. USDA Forestry Service Research Note SE-283, 4 pp.

Marx, D. H. (1980). In *Tropical Mycorrhiza Research* (P. Mikola, ed.), pp. 13–71. Clarendon Press, Oxford.

Marx, D. H. (1981). *Can. J. Res.* **11**, 168–174.

Marx, D. H. (1990). *For. Sci.* **2**, 224–245.

Marx, D. H. and Bell, W. (1985). *Formation of* Pisolithus *ectomycorrhizae on loblolly pine seedlings with spore pellet inoculum applied at different times*. USDA Forestry Service Research Paper SE-249, 6 pp.

Marx, D. H. and Bryan, W. C. (1975) *For. Sci.* **21**, 245–254.

Marx, D. H. and Cordell, C. E. (1987). Triadimefon affects *Pisolithus* ectomycorrhizal development, fusiform rust, and growth of loblolly and slash pines in nurseries. USDA Forestry Service Research Paper SE-267, 14 pp.

Marx, D. H. and Cordell, C. E. (1988). In *Canadian Workshop on Mycorrhizae in Forestry* (M. Lalonde and Y. Piche, eds), pp. 75–86 CRBF, Faculté de Forestrie et de Geodesie, Université Laval, Ste-Foy, Quebec.

Marx, D. H. and Cordell, C. E. (1989). In *Biotechnology of Fungi for Improving Plant Growth* (J. M. Whipps and R. D. Lumsden, eds), pp. 1–29. British Mycology Society. Cambridge University Press, Cambridge.

Marx, D. H. and Rowan, S. J. (1981). *For. Sci.* **27**, 167–176.

Marx, D. H. and Ruehle, J. L. (1989). In *Mycorrhizae for Green Asia*, 1st Asian Conference on Mycorrhizae (A. Mahadevan, N. Raman and K. Natarajan, eds). Sivakami Publications, Madras, India.

Marx, D. H., Bryan, W. C. and Cordell, C. E. (1976). *For. Sci.* **22**, 91–100.

Marx, D. H., Mexal, J. G. and Morris, W. G. (1979) *South. J. Appl. For.* **3**, 175–178.

Marx, D. H., Ruehle, J. L., Kenney, D. S. *et al.* (1982). *For. Sci.* **28**, 373–400.

Marx, D. H., Cordell, C. E., Kenney, D. S. *et al.* (1984a). *For. Sci. Monogr.* **25**, 101 pp.

Marx, D. H., Jarl, K., Ruehle, J. L. and Bell, W. (1984b). *For. Sci.* **30**, 897–907.

Marx, D. H., Hedin, A. and Toe, S. F. P. IV (1985). *For. Ecol. Management* **13**, 1–25.

Marx, D. H., Cordell, C. E. and France, R. C. (1986). *Phytopathology* **76**, 824–831.

Marx, D. H., Cordell, C. E. and Clark, A. (1988). *South. J. Appl. For.* **12**, 275–280.

Marx, D. H., Cordell, C. E., Maul, S. B. and Ruehle, J. L. (1989a) *New Forests* **3**, 45–56.

Marx, D. H., Cordell, C. E., Maul, S. B. and Ruehle, J. L. (1989b). *New Forests* **3**, 57–66.

Mihail, J. D. and Alcorn, S. M. (1984). *Plant Dis.* **68**, 156–159.

Mikola, P. (1973). In *Ectomycorrhizae: Their Ecology and Physiology* (G. C.

Marks and T. T. Kozlowski, eds) pp. 383–411. Academic Press, New York.
Molina, R. and Trappe, J. M. (1982). *For. Sci.* **28**, 423–458.
Moser, M. (1958). *Forstw. Cbl.* **77**, 32–40.
Riffle, J. W. and Maronek, D. M. (1982). In *Methods and Principles of Mycorrhizal Research* (N. C. Schenck, ed.), pp. 147–155. American Phytopathological Society, St. Paul, MN.
Schramm, J. R. (1966). *Trans. Am. Phil. Soc.* **56**, 1–194.
Shemakhanova, N. M. (1962). *Mycotrophy of Woody Plants*, US Department of Commerce Translation TT66-51073 (1967), Washington, DC.
Sinha, A. P., Agnihotri, V. P. and Kishan, S. (1979). *Plant Soil* **53**, 89–98.
Stapleton, J. J. and DeVay, J. E. (1986). *Crop Prot.* **5**, 190–198.
Takacs, E. A. (1967). *India Supplemento Forestal* **4**, 83–87.
Theodorou, C. and Bowen, G. D. (1973). *Soil Biol. Biochem.* **5**, 765–771.
Trappe, J. M. (1977). *Ann. Rev. Phytopathol.* **15**, 203–222.
Trappe, J. M., Molina, R. and Castellano, M. A. (1984). *Ann. Rev. Phytopathol.* **22**, 331–339.
Vozzo, J. A. and Hacskaylo, E. (1971). *For. Sci.* **17**, 239–245.
Widin, K. D. and Kennedy, B. W. (1983). *Phytopathology* **73**, 429–434.
Wilcox, H. E. (1982). In *Methods and Principles of Mycorrhizal Research* (N. C. Schenck, ed.), pp. 103–113. American Phytopathological Society, St. Paul, MN.

18

Maintenance of Ectomycorrhizal Fungi

HELVI HEINONEN-TANSKI and TOINI HOLOPAINEN

Department of Environmental Sciences, University of Kuopio, PO Box 6, SF 70211 Kuopio, Finland

I. Introduction

Detailed accounts of the isolation, cultivation and maintenance of ectomycorrhizal fungi have been given in the article by Molina and Palmer (1982) and many of their procedures have been adopted in our work. In addition, some other methods used in studies of ectomycorrhizal fungi of coniferous trees of Scandinavia, e.g. Scots pine (*Pinus sylvestris* L.) and to a lesser degree on Norway spruce (*Picea abies* L.) are described.

METHODS IN MICROBIOLOGY
VOLUME 23 ISBN 0-12-521523-1

Copyright © 1991 by Academic Press Limited
All rights of reproduction in any form reserved

II. Cultivation of ectomycorrhizal fungi

A. Establishment of a collection of mycorrhizal fungi

When starting a collection of ectomycorrhizal fungi it is desirable to obtain cultures from well-known collections or from other ectomycorrhiza research workers. Cultures are in most cases obtained as agar slants which can be easily sent by post or carried on the person. In cold climates special care must be taken to ensure that mycorrhizal cultures are not frozen during transport.

The isolation and cultivation of ectomycorrhizal fungi is greatly facilitated by the use of a laminar flow-chamber. Apparatus and instruments useful for isolation purposes are Petri dishes containing suitable nutrient agar (see below), sterile scalpels, thick double or triple cultivation loops (Schaffer, 1982) and metal forceps which can be sterilized by flaming in ethanol before each isolation or subculture.

Suitable isolation media include modified Hagem's agar (Modess, 1941) and modified Melin–Norkrans' agar (Marx, 1969). The main difference between these media is that Melin–Norkrans' agar contains more thiamine while Hagem's agar contains more ammonium salt and is nutritionally more versatile than the synthetic modified Melin–Norkrans' agar. Half of the carbon source in the former medium comes from malt extract, the content of which varies but usually contains more than 50% maltose and approximately 40% other carbohydrates, some protein and ash (Oxoid, 1982). Both of these media are recommended when isolating new strains of ectomycorrhizal fungi. The content of modified Hagem's agar is: KH_2PO_4 0.5 g, NH_4Cl 0.5 g, $MgSO_4 \cdot 7H_2O$ 0.5 g, $FeCl_3$ (1%) 0.5 ml, glucose 5.0 g, malt extract 5.0 g, thiamine HCl 50 μg, aureomycin 35 mg, agar 15 g and water 1000 ml. The content of the modified Melin–Norkrans' medium is KH_2PO_4 0.5 g, NH_4Cl 0.25 g, $MgSO_4 \cdot 7H_2O$ 0.15 g, $FeCl_3$ (1%) 0.5 ml, $CaCl_2$ 0.05 g, NaCl 0.025 g, glucose 10.0 g, thiamine HCl 100 μg, aureomycin 35 mg, agar 15 g and water 1000 ml. The media are autoclaved for 15 min at 121 °C and the pH should be 4.5–5.0 for Hagem's agar and 5.7–6.2 for Melin–Norkrans' agar. If the agar medium is too soft, the amount of agar can be increased to 20–25 g $litre^{-1}$. Because cultivation times are often quite long, it is important that the agar does not dry out and 20–25 ml of media per 9 cm Petri dish is often suitable.

The antibiotic aureomycin can be added to the media at a concentration of 35 mg $litre^{-1}$ but is omitted when pure cultures are being maintained. Benomyl (1 mg $litre^{-1}$) is sometimes used against air-borne fungal contaminants (Marx, 1969), with or without aureomycin.

1. *Isolation from sporophores*

Fresh isolates can be obtained from mycorrhizal roots or from sclerotia in boreal coniferous forests during the growing season. Many fungi are easily isolated in late summer or autumn when ectomycorrhizal fungi appear on the soil as sporophores.

The isolation of mycorrhizal fungi is easiest from a sporophore, which of course must be identified correctly. Palmer (1971) pointed out that it is important to make careful descriptions of the growth site, host plants and if possible to take colour photographs of the site and the sporophores present in addition to the preservation of reference specimens, which can be dried (Molina and Palmer, 1982). Kendrick and Berch (1985) advise depositing the reference specimen in an internationally recognized herbarium listed by Holmgren *et al.* (1981).

The sporophores collected should be young, healthy and undamaged. Material containing mites and insects should be rejected. The sporophores should be wrapped in paper and rapidly transported to the laboratory, ideally during the same day without additional damage. Isolations should be made soon after arrival in the laboratory. If the rapid transport of sporophores is not possible, the use of a portable field cultivation chamber (Molina and Palmer, 1982) is recommended. The chamber used by Molina and Palmer consisted of a 15 cm × 40 cm chamber of open frame, with a plexiglas lid and sides and bottom of plywood.

The sporophores should be pre-cleaned before being moved to the inoculation laboratory. In most cases tissue taken from the inner parts at the junction between the stipe and the cap of the sporophore has been found to be most successful. The inoculum can be removed after breaking the sporophore carefully by hand to expose inner surfaces which are cut as small pieces using a sterile scalpel. The inner surfaces must not be touched during the preparation and contamination from the outer surfaces must be avoided. Three or four cubic pieces with dimensions of approximately 1–3 mm are cut with a sterile scalpel and placed on agar in each Petri dish. Normally several inoculum pieces can be readily obtained from the same sporophore.

2. *Isolation from basidiospores*

The mature hymenal surface can sometimes be taken for isolation of ectomycorrhizal fungi using the basidiospore-drop technique of Palmer (1971). A piece of hymenium is attached using agar or petroleum jelly to the surface of a glass tube so that spores can fall readily onto the agar slant placed just below, but not in contact with, the hymenium. The lid of a Petri dish can also be used, the hymenium again being arranged so that spores fall onto agar in the lower half of the Petri dish.

3. *Isolation from mycorrhizal rootlets*

Ectomycorrhizal fungi can be isolated from mycorrhizal rootlets. This is the only available technique in the case of those fungi that do not produce sporophores. Fresh roots are collected and washed to remove adherent soil. Lateral, young, fine root organs containing mycorrhiza are surface sterilized for 15–30 s with hydrogen peroxide (30%, v/v) or for 3–6 min with sodium hypochlorite (with 6–8% active Cl) and washed repeatedly in sterile water to remove all traces of the sterilant. The material should not be moved to the cultivation laboratory until it has been through the sterilization procedure. The organs, cut into smaller pieces, are then placed onto an appropriate agar medium (see above) for incubation. Mercuric chloride used as sterilant and recommended in the older literature should be avoided because of its toxicity. Contaminants from soil are common in spite of careful precautions. In many cases also the mycorrhizal fungus does not grow readily on artificial medium or where growth does occur, it is too slow for experimental purposes.

4. *Isolation from sclerotia*

Some fungi, for example *Cenococcum geophilum*, can also be isolated from sclerotia. Sclerotia are sieved from soil, washed, surface sterilized and carefully rinsed (Molina and Palmer, 1982). Similar procedures can be applied to structures such as rhizomorphs produced by some ectomycorrhizal fungi (Kendrick and Berch, 1985).

5. *Incubation of primary isolations*

In most cases hyphae develop during an incubation time of 1–7 weeks at 17–25 °C. Drying and contamination can be minimized by sealing the rims of the closed Petri dishes with Parafilm. Growth can be followed by eye or by stereo-microscope after 2–3 days and then at weekly intervals. If contaminants are found, the desired fungus can sometimes be saved with a serial transfer to a fresh medium. When making these transfers a thick cultivation loop or a sterile, sharp scalpel are useful.

III. Maintenance methods

The traditional maintenance methods for ectomycorrhizal fungi are agar slant cultures and the symbiotic synthesis of ectomycorrhizal fungi with the macrosymbiotic plant. There are also other more or less successful methods that will be described here.

A. Agar slant cultures

Amongst the agar media which can be used for maintenance of slant cultures are modified Hagem's and modified Melin–Norkrans' agar (described in Section II.A) with or without the addition of aureomycin. Good growth has also been obtained using the low pH mycological agar, which can be obtained commercially (Difco Laboratories, 1984). It contains soytone 10.0 g, glucose 10.0 g, agar 15.0 g and water 1000 ml and pH adjusted to 4.8 ± 0.2. We have adjusted the pH with HCl instead of lactic or acetic acid. Many fungi grow well on the medium of Marx–Melin–Norkrans (Marx, 1969) containing $CaCl_2$ 0.05 g, NaCl 0.025 g, KH_2PO_4 0.5 g, $(NH_4)_2HPO_4$ 0.25 g, $MgSO_4 \cdot 7H_2O$ 0.15 g, $FeCl_3$ (1%) 1.2 ml, thiamine HCl 100 μg, malt extract 3 g, sucrose 10 g, agar 15 g and 1000 ml distilled water.

Test tube cultures save space and can be held for 3–6 months in a cold room (3–4 °C). Molina and Palmer (1982) have used cultivation tubes containing 3 ml of agar medium per tube but we use 5–10 ml medium per tube in order to minimize the effects of drying. Standard test tubes (16 mm × 160 mm or 20 mm × 150 mm) with straight rims and aluminium Cap-O-Test or Labocap caps are useful. Molina and Palmer (1982) recommended serial transfers every third or fourth month but Marx and Daniel (1976) and M. Lindeberg (pers. commun.) transfer them at 5- or 6-monthly intervals. A minimum of four tubes should be inoculated and stored in at least two refrigerators at 3–4 °C.

Ectomycorrhizal cultures have been saved on agar slants for years or even decades, but some of these cultures have altered characters after long preservation in the laboratory. Most importantly, they may be unable to form mycorrhiza with plant symbionts, although there might still be significant stimulation of the growth of host plants (Sen, 1990). After many serial transfers some strains will inevitably be lost.

B. Sterile water

The use of sterile water for the maintenance of ectomycorrhizal fungi, originally described by Marx and Daniel (1976), has been found in our laboratory to be a very valuable method that saves both labour and materials (Heinonen-Tanski, 1989). The maintenance times tested by us varied from 6 to 24 months. This procedure preserved the growth of all strains (over 50 tested, half from old collections and half recently isolated from Finnish forests, mainly *Cenococcum geophilum*, *Pisolithus tinctorius* and different *Suillus* spp.) with the exception of one old collection strain of *Paxillus involutus*, which was unable to grow further on agar slants, and two *Suillus* strains that gave modest growth (Table I). For this procedure the only requirements are young cultures, sterile water in screw cap bottles and a refrigerator.

The bottles used are autoclavable reagent bottles (50–150 ml) containing respectively 30 and 100 ml water sterilized for 15 min at 121 °C. Smaller bottles are more useful, because it is easier to take slippery growth pieces from them for further studies.

Pieces of inocula from the edges of young ectomycorrhizal colonies growing on Petri dishes containing modified Hagem's or modified Melin–Norkrans' agar are removed with a thick loop, scalpel or sterile cork borer. The size of a piece is approximately 5 mm × 5 mm. Between 4 and 6 pieces can be added to the same bottle, which is placed in a refrigerator (3–4 °C).

Recently Richter and Bruhn (1989) using similar methods reported that only about a half of their tested mycorrhizal strains survived after

TABLE I
Viability of ectomycorrhizal fungi after different maintenance methods

Method	Number of strains (and percentage in parenthesis)		
	Good growth	Modest growth	No growth
Sterile water at least 24 months	41 (93)	3 (7)	0 (0)
Sterile water at least 6 months[a]	3 (50)	2 (33)	1 (17)
Silica gel	15 (42)	1 (3)	19 (54)
Sand	29 (66)	4 (9)	11 (25)
Garden soil	2 (7)	0 (0)	24 (92)
Slants at −20°C without protectant	4 (16)	5 (20)	16 (64)
Slants at −20°C with 20% glycerol	4 (17)	1 (4)	18 (78)
Slants at −20°C with 20% mannitol	3 (13)	1 (4)	19 (83)

[a]No tests in sterile water during 24 months.

2–4 years in sterile water. It is important that research is done to discover the reasons for these contradictory results.

C. Maintenance at −20 °C, in silica gel, or in sand or soil

The maintenance methods described for fungi by Smith (1984) have also been tested on ectomycorrhizal fungi (Heinonen-Tanski, 1989). These methods were: (a) agar slants at −20 °C with or without addition of a protection agent (20% glycerol or mannitol); (b) maintenance in sterile, dried silica gel granules, where mycelium suspension in a minimal volume of milk was added to bottles containing sterile, frozen silica gel at 4 °C and shaken daily in closed bottles; (c) maintenance in thrice heat-sterilized sand or garden soil moistened with sterile water to 20%, where mycelium was added to the medium and allowed to grow for one week at room temperature before storage at 5 °C (see Table I). Growth was monitored after one month.

As can be seen from Table I, the maintenance methods involving deep-freezing, silica gel and garden soil were unsuccessful, giving either very weak or no growth. Only some *Cenococcum geophilum* strains survived. These methods therefore cannot be recommended. Sand preservation gave better results (29 strains of 44 survived), but was not efficient enough to be recommended for general use.

D. Maintenance as a grain culture

If storage for a relative short period (less than one year) is required, the grain culture technique may be helpful. Park (1971) and Göbl (1975) have found this method suitable for the preservation of some ectomycorrhizal fungi for up to 9 months when making inoculum preparations. Liquid medium inoculum is introduced into a bottle or into autoclavable plastic bags containing cooked and sterilized wheat or other cereal grains with the addition of some calcium carbonate or calcium sulphate. The fungi are allowed to grow and the containers are shaken gently once a week. The cultures developed can then be stored in a cold room for up to nine months. The value of this simple and cheap method as a maintenance method should be carefully determined for different types of ectomycorrhizal fungi.

Grain cultures as forest plantation inoculants have also been used by other research workers as reviewed by Marx (1980) although these were not always found to give good results.

E. Maintenance at extra-low temperatures

Cell material has in many cases been preserved at −75 to −80 °C or in liquid nitrogen. Experience of such methods for maintaining ectomycorrhizal fungal strains is as yet insufficient for any recommendations to be given.

F. Maintenance of ectomycorrhizal fungi in symbiosis

The maintenance of ectomycorrhizal fungi with their macrosymbiotic plant partners is the most natural way and is to be recommended, at least for the most valuable strains. These synthesized cultures can be either true diaxenic or only semi-aseptic. Synthesis is also useful in rejuvenating strains. The fungi can be reisolated from mycorrhiza or from sporophores (Marx, 1980). Synthesis methods are described in detail by Peterson and Chakravarty (Chapter 3, this volume).

IV. Conclusions

To reduce labour and to allow comparisons between old cultures and fresh isolates, safe, easy, economical and ecologically acceptable methods are needed for maintaining ectomycorrhizal fungi. The physiological diversity of these fungi means that they can adapt themselves to very different environments. Thus, for example, *Cenococcum geophilum* was resistant to most of the maintenance methods tested by us and was also found to be relatively resistant to industrial air pollutants (Holopainen and Vaittinen, 1988). Conversely *Suillus variegatus* was quite sensitive to most of the maintenance methods tested and also to air pollutant exposure (Ohtonen *et al.*, 1990).

A single method may not always be the most suitable for maintaining many different ectomycorrhizal strains. The method selected will depend on the facilities available in a particular laboratory. For maintenance of a small collection serial transfer would be the simplest solution, although the strains are susceptible to genetic changes; a risk which could be reduced by making regular plant–mycorrhiza syntheses. In larger collections if there is a shortage of laboratory technicians the suitability of sterile water or maybe oil as maintenance methods should be investigated.

There is a pressing need for further development of safe and reliable maintenance methods for ectomycorrhizal fungi. Reports of successful and unsuccessful methods are welcome to help us in developing the utilization of ectomycorrhizal fungi in tree cultivation and for protecting forests against environmental stresses.

Acknowledgements

We thank Mr Robin Sen (MSc) for his critical comments and for correcting the English text and Mrs Mirja Korhonen for careful technical assistance.

References

Difco Laboratories (1984). Difco Manual: *Dehydrated Culture Media and Reagents for Microbiology*, 10th edn. Difco Laboratories, Detroit, MI.
Göbl, F. (1975). *Centralbl. Ges. Forstwesen* **92,** 225–237.
Heinonen-Tanski, H. (1989). *Agric. Ecosyst. Environ.* **28,** 171–174.
Holmgren, P. K., Keuken, W. and Schofied, E. K. (1981). *Index Herbariorum, Part I: The Herbaria of the World*, 7th edn. *Regnum Vegetabile*, Vol. 106. W. Junk, The Hague.
Holopainen, T. and Vaittinen, S. (1988). *Karstenia* **28,** 35–39.
Kendrick, B. and Berch, S. (1985). In *Comprehensive Biotechnology. The Principles, Applications and Regulations of Biotechnology in Industry, Agriculture and Medicine* (M. Moo-Young, ed.), Vol. 4, pp. 109–152. Pergamon Press, Oxford.
Marx, D. H. (1969). *Phytopathology* **59,** 153–163.
Marx, D. H. (1980). In *Tropical Mycorrhiza Research* (P. Mikola, ed.), pp. 13–71. Oxford University Press, Oxford.
Marx, D. H. and Daniel, W. J. (1976). *Can. J. Microbiol.* **22,** 338–341.
Modess, O. (1941). *Symb. Bot. Upsaliensis* **5,** 1–147.
Molina, R. and Palmer, J. G. (1982). In *Methods and Principles of Mycorrhizal Research* (N. C. Schenck, ed.), pp. 115–129. The American Phytopathological Society, St Paul, MN.
Ohtonen, R., Markkola, A. M., Heinonen-Tanski, H. and Fritze, H. (1990). In *Acidification in Finland* (P. Kauppi, P. Antila and K. Kenttämies, eds), pp. 373–393. Springer-Verlag, Berlin.
Oxoid (1982). Oxoid Manual: *Culture Media, Ingredients and Other Laboratory Services*, 5th edn. Oxoid Limited, Basingstoke.
Palmer, J. G. (1971). In *Mycorrhizae. Proceedings of the First North American Conference on Mycorrhizae* (E. Hacskaylo, ed.), pp. 132–144. USDA, Forest Service Laboratory,
Park, J. Y. (1971). In *Mycorrhizae. Proceedings of the First North American Conference on Mycorrhizae* (E. Hacskaylo, ed.), pp. 239–240. USDA, Forest Service Laboratory,
Richter, D. L. and Bruhn, J. (1989). *Can. J. Microbiol.* **35,** 1055–1060.
Schaffer, A. G. (1982). In *Sourcebook of Experiments for the Teaching of Microbiology* (S. B. Primrose and A. C. Wardlaw, eds), pp. 16–20. Academic Press, London.

Sen, R. (1990). *New Phytol.* **114,** 617–629.
Smith, D. (1984). In *Maintenance of Microorganisms. A Manual of Laboratory Methods* (B. E. Kirsop and J. J. Snell, eds), pp. 83–197. Academic Press, London.

19

Root Window Technique for *in vivo* Observation of Ectomycorrhiza on Forest Trees

SIMON EGLI and IGNAZ KÄLIN

Swiss Federal Institute of Forest, Snow and Landscape Research, CH-8903 Birmensdorf, Switzerland

I. Introduction

Our knowledge of the condition and behaviour of ectomycorrhiza in their natural surroundings is very limited, both in terms of physiology and morphology. Studies on the decline of forests in particular have revealed how little we really know in this field. We cannot even define the threshold above which fine roots and mycorrhiza should be classified as "damaged", because we do not know what their "normal" state is. We know practically nothing about the range of morphological characteristics of any one particular mycorrhizal type, its growth dynamics, seasonal variation, or life-span. The main reason for this lack of knowledge is that the observation of mycorrhiza in their natural surroundings is very difficult.

METHODS IN MICROBIOLOGY
VOLUME 23 ISBN 0-12-521523-1

Copyright © 1991 by Academic Press Limited
All rights of reproduction in any form reserved

Soil core sampling, the most common method for studying root systems and mycorrhiza, is certainly useful for observing conditions at any given time, but has the great disadvantage of being destructive and thus precluding continuous observation of particular microsites, so that it does not allow conclusions about developments or changes in the mycorrhiza. These problems can be largely overcome through the installation of root windows, which allow continuous, long-term observation with minimum disturbance. Photographs can be used to register morphological changes.

The root window is not a new concept. As early as the beginning of the 1950s, glass plates laid flat on the soil were employed to observe root growth in forest trees (Orlov, 1957, 1960). At East Malling Research Station (Maidstone, Kent, UK), underground tunnels were installed in the mid-1960s for the observation of the root system of fruit trees under natural and experimental conditions. Göttsche (1972) and Turner and Streule (1983) used vertical root windows to measure root growth and growth periodicity. It is surprising that such methods have practically never been employed in the study of mycorrhiza.

In developing the method described in this paper, special emphasis was laid on using basic laboratory equipment so as to keep the method as simple as possible.

II. Methods

A. Technical details

1. Installation

The root window is installed by ramming a sharp-edged 40 cm steel plate about 20 cm into the soil tangentially to, and about 2 m from, the tree trunk; this distance was chosen because preliminary studies have shown that, at least for mature spruces, the density of the fine roots is at its greatest there. Behind the plate a space of about 50 cm square and 40 cm deep is excavated. The steel plate is then carefully removed and a glass plate (35 cm × 20 cm × 6 mm) is inserted in its place and fixed with plastic nails directly against the soil face. In an alternative method of installation, a gap of 1 cm is left between the glass plate and the soil face. The gap is filled with dry, sieved soil, following the horizons of the profile in such a way that natural cavities are filled.

In either case it is important that glass with no optical defects is used and that the profile is covered with black, lightproof foil clamped to the upper edge of the glass sheet (Fig. 1). This is important to prevent light

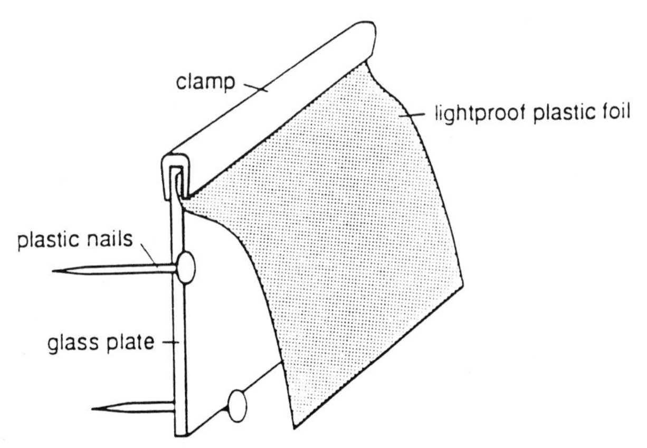

Fig. 1. Root window with lightproof plastic foil.

from entering and disturbing root growth. The rest of the excavated cavity is filled with styrofoam chips sewn into nylon curtain netting and covered with a wooden frame, so as to ensure conditions of insulation and water permeability as nearly natural as possible.

2. *Photography*

A Wild M8 binocular microscope is used for observation and photography. The observation distance can be extended by means of an extra lens (e.g. magnification 0.4×). To allow observation of the vertical root window, the image is turned through 90° by a surface-coated deflection mirror. A halogen lamp is used for illumination, the light rays being directed onto the object through optical fibre tubes (Fig. 2).

A first photograph is taken to show the immediate vicinity of the subject (magnification on the slide: 0.9×). Then individual subjects are photographed in such a way that the magnification on the slide is 4.2× (an example is shown in Fig. 3). A Kodak colour slide film is used (EPY, Artificial Light, ASA 50). The exposure time is automatically controlled by a light meter, and with a medium diaphragm aperture ranges from 2 to 60 s, depending on the illumination and the subject.

B. Special problems

When the temperature conditions are unfavourable, that is, when the gradient between soil temperature and air temperature is too steep, condensation on the glass plate may cause problems. This is a particular problem when the air temperature is lower than the soil temperature

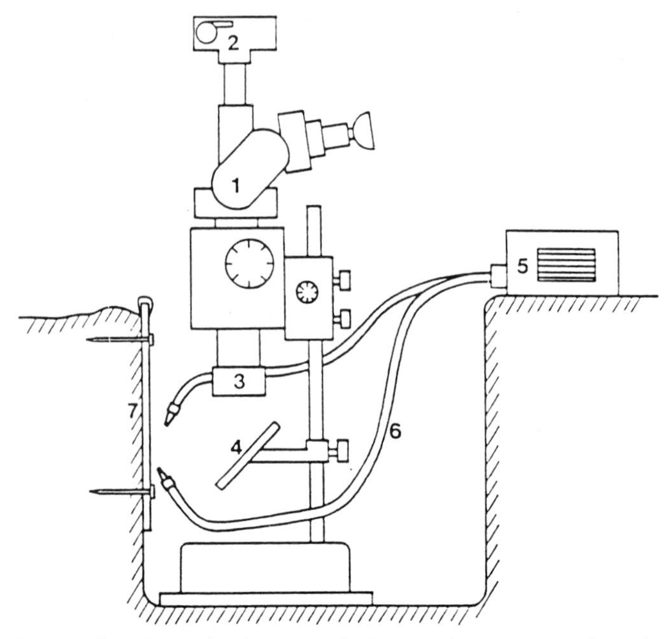

Fig. 2. Schematic diagram of root window and apparatus for observation. Key: 1, binocular microscope; 2, camera; 3, reducing lens; 4, deflection mirror; 5, halogen lamp; 6, optical fibre tubes; 7, glass plate.

and condensation forms on the inner side (next to the soil) of the plate. This mainly occurs in autumn. In such cases, the glass plate can be warmed with the palms of the hands or by directing the warmth from the lamp onto the glass plate through a tube, so that the plate is carefully warmed and only as much as necessary.

Root windows which remain in position for long periods may become obscured by a deposit of fine soil particles, especially in soils containing much clay, so that they may have to be cleaned. In such cases the glass plate is very carefully loosened from the soil profile, cleaned with water and then with alcohol, and equally carefully replaced. This, however, is only done when absolutely necessary.

III. Applications

The root window has many applications. Experience so far shows that it is particularly suitable for the observation of various qualitative and quantitative changes in fine roots and mycorrhiza, for instance, seasonal patterns of growth and formation, life-span of different mycorrhizal types, growth disturbances, formation of subsidiary roots, dying-off processes, etc. This allows synchronization of observations on changes and developments in the underground and aerial parts of the tree. Furthermore, the root window provides a means of determining whether particular morphological features used in the description of mycorrhiza, such as colour or mode of branching, remain constant or change under the influence of growth and ageing. Photography offers many possibilities for quantitative observation: for example, by means of calibrated magnification, the growth of fine tree roots and mycorrhiza in one particular area can be recorded over a fairly long period of time. With transparent foil, the growth of long roots can be traced directly onto the glass plate. The technique described here was primarily designed for field studies, but, with a few modifications, may also be employed in the greenhouse.

IV. Experience to date

As part of the Swiss National Research Programme NFP 14+ root windows were installed on various sites to determine the relationships between the development of mycorrhiza and the state of health of the tree (Egli and Kälin, 1990). Regular observations were made every 6 weeks over a period of 3 years. The windows had to be cleaned each spring. The problem of condensation was overcome by directing the warmth from a lamp onto the glass plate. No other problems arose during these 3 years. The root windows with glass plates laid directly against the soil were found to give significantly better results than those filled in behind. The majority of mycorrhiza are found in the natural soil cavities, which are preserved through this type of installation. In this case, mycorrhiza can be observed and photographed from the very beginning. Where the window is filled in from the rear, the tree roots must grow through the soil before observation is possible. In such cases, only the growth of long roots could be observed during the first of the 3 years.

V. Advantages over other techniques

In the literature various techniques are described for the *in vivo* observation of roots using an "intrascope" or "rhizoscope" (endoscope),

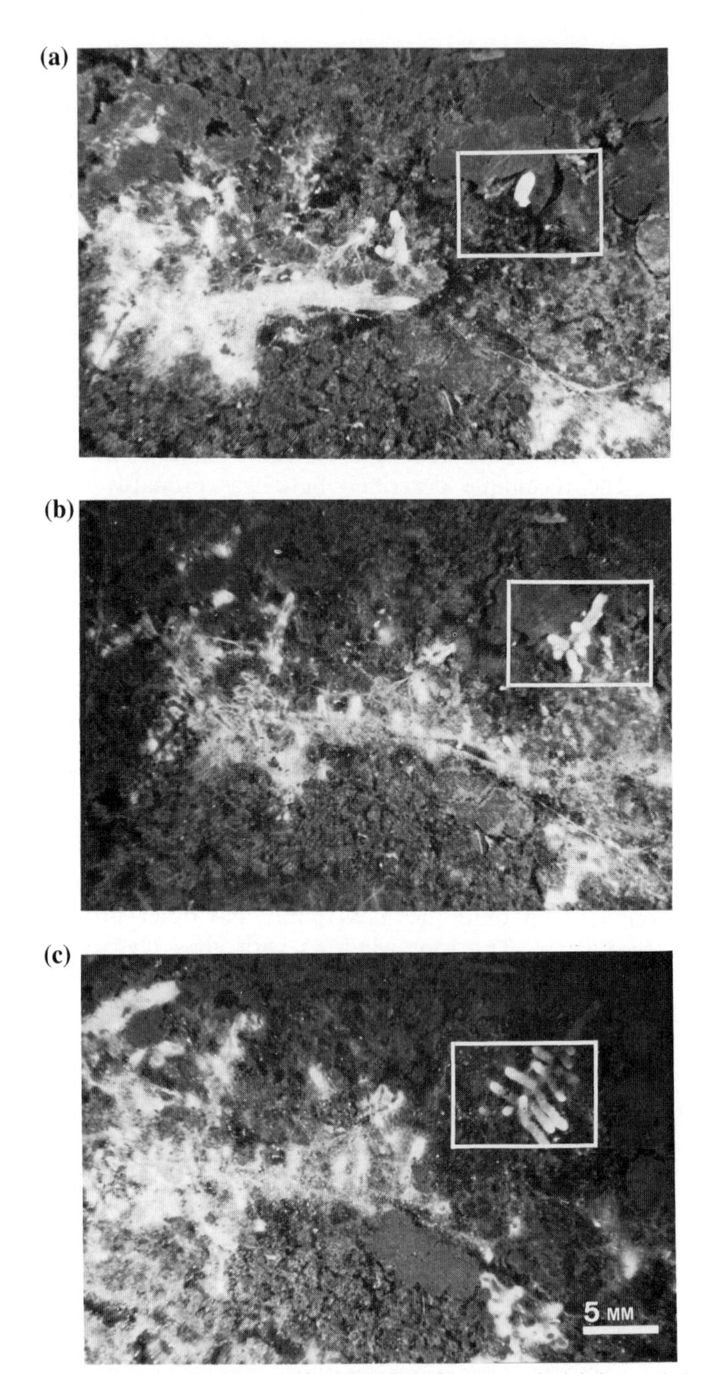

Fig. 3(a)-(f). Example of photo series showing the general vicinity and details of spruce mycorrhiza.

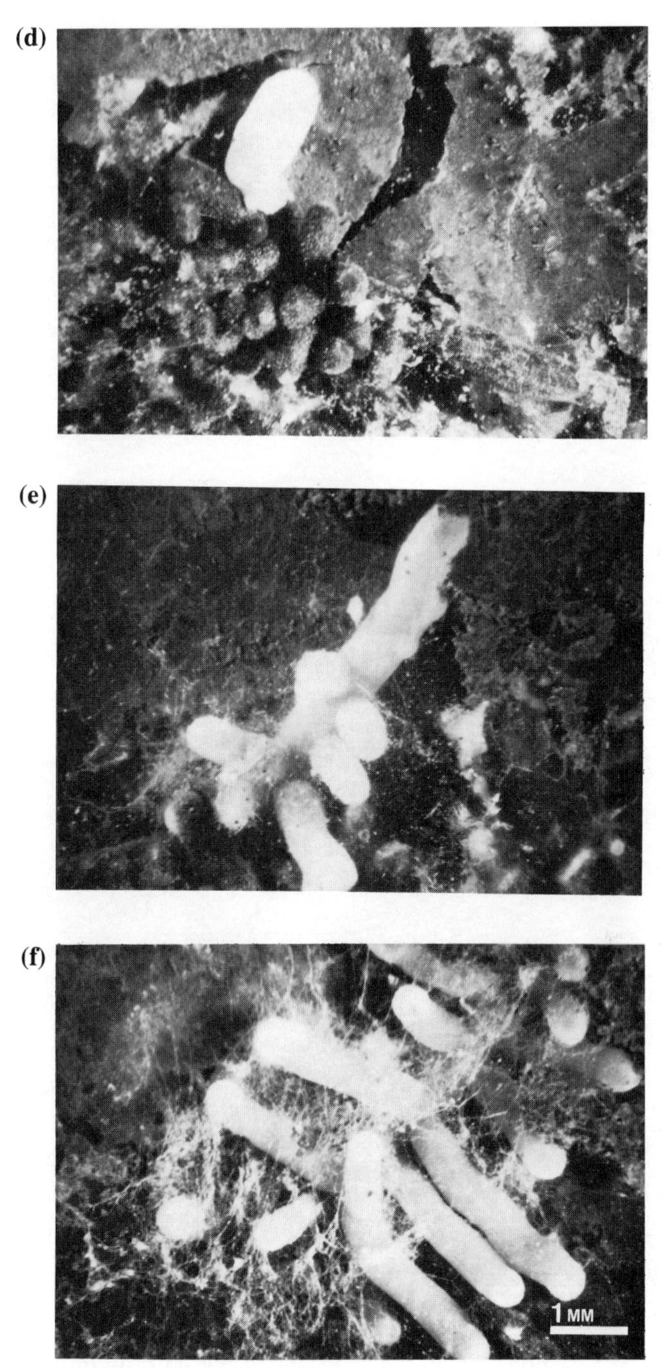

(a, d) Time zero, (b, e) 4 weeks later, (c, f) 12 weeks later.

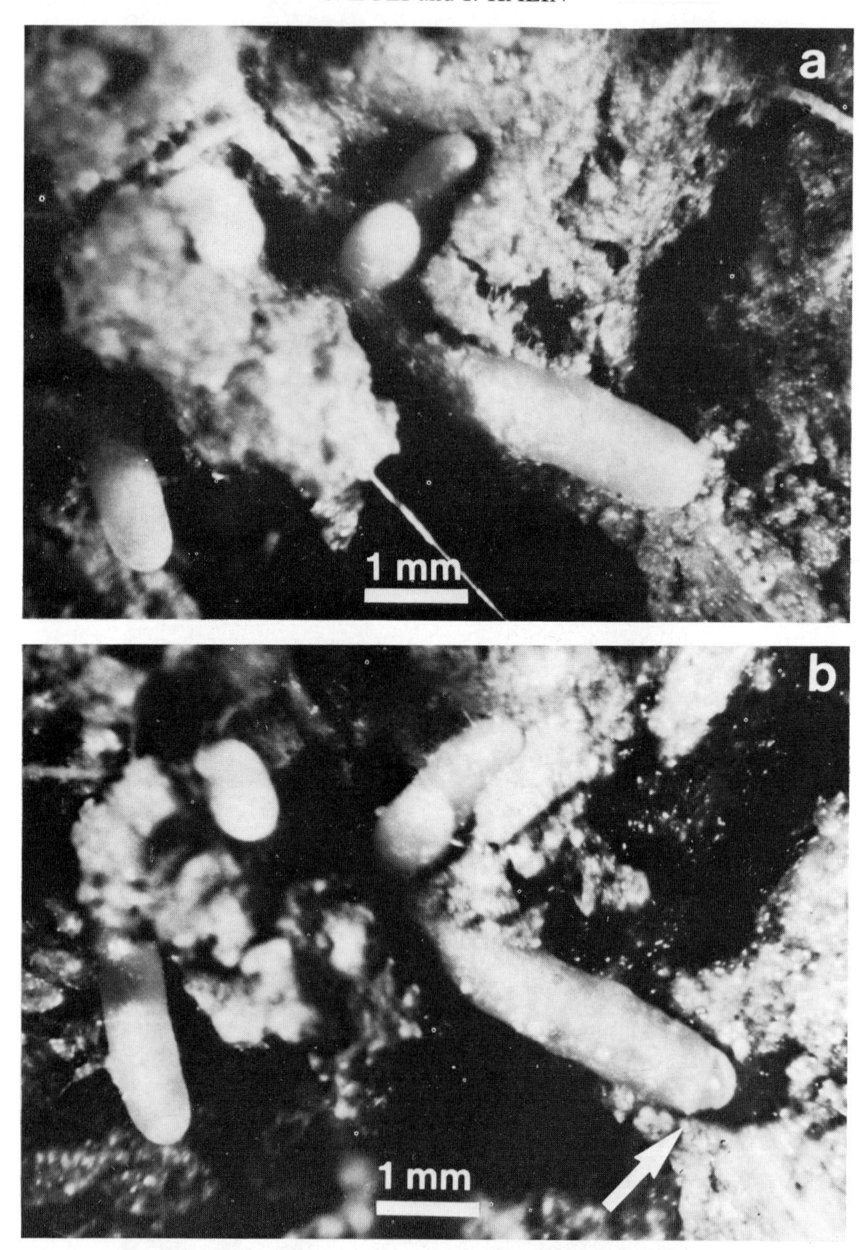

Fig. 4. Development of a spruce mycorrhiza infected by a pathogen, as observed through the root window. (a) December 1986: appearance normal. (b) April 1987: splitting of the fungal mantle at the root cap (arrow), probably due to rapid root elongation.

Fig. 4(c). July 1987: a pathogen has entered through the split in the mantle and the subseguent decay of the root tip is visible, while subsidiary roots (arrow) are already under formation.

a device usually employed in medicine for the examination of internal organs. Tweel and Schalk (1981) and Limonard and Smits (1985) describe such techniques which were developed for greenhouse studies and require special, perforated containers. However, this technique can also be employed in the field; a glass tube is permanently fixed in the soil, and an endoscope used to observe its immediate vicinity. Details are described by Sanders and Brown (1978), Seufert *et al.* (1986) and Taylor *et al.* (1990).

In comparison, the root window has the advantage that it allows the observation of a larger area for a longer period, since the mycorrhiza seldom grow beyond the observation field (or the glass plate) during the observation time. Furthermore, if necessary the window can be removed, cleaned and replaced without great disturbance, whereas with the rhizoscope that is not possible, as the extraction of the glass tube damages the root network. This possibility of cleaning has proved very useful, particularly in certain soil types where fine particles are rapidly washed out during heavy rainfall and deposited on the glass plate.

Another technique for in vivo observation—though only applicable to seedlings—involves the use of special folding containers, for instance the "root-trainers" developed by Spencer Le Maire. Our own experience has shown that it is often difficult to loosen the folding walls of the containers from the soil, so that the soil and root structures are greatly disturbed.

Yet another technique is to lay root windows horizontally on the soil surface, but this has the disadvantage that it hinders precipitation from penetrating into the soil. Nevertheless, this technique is an extremely interesting variant as a very simple means of observing the growth of fine tree roots and mycorrhiza within very small areas (i.e. only a few cm^2) as Orlov (1957, 1960) reports.

VI. Conclusions

Any serious effort to understand the complex relationships and interactions of the forest ecosystem with all its different environmental factors must include the observation of mycorrhiza in their natural habitat. Only through such observations is it possible to learn about the whole range of variability over time and space and to deduce criteria for assessment and the interpretation of changes. Here, the root window has proved a very useful technique.

Research nowadays seems to neglect pure observation as a means of study. Perhaps it is a sign of the times that we no longer have time or patience for such things. Again, simple observation is often rejected as a scientific research method, or at least is not recognized as such, mainly because it is difficult to verify the findings statistically. Yet such judgements are not justified if we look at the past findings of famous scientists: for instance, Frank's descriptions of mycorrhiza, dating from 1885, are astonishingly precise and still furnish a great deal of information. Observation must naturally go hand in hand with experimentation, and the two approaches must be combined if we are to gain more knowledge in this field.

Acknowledgement

The development of this technique was supported by the Swiss National Science Foundation (SNCF) as part of the Project No. 4.824.0.85.14.

References

Egli, S. and Kälin, I. (1990). *Agric. Ecosyst. Environ.* **28**, 107–110.

Goettsche, D. (1972). *Verteilung von Feinwurzeln und Mykorrhizen im Bodenprofil eines Buchen- und Fichtenbestandes im Solling.* Mitt. Bundesforschungsanstalt für Forst- und Holzwirtschaft, Nr. 88.

Limonard, T. and Smits, W. (1985). In *Proceedings of the 6th North American Conference on Mycorrhizae* (R. Molina, ed.), p. 254. Forest Research Laboratory, Oregon State University, Corvallus, OR.

Orlov, A. Y. (1957). *Botaniceskij Z. SSSR* **42**, 1172–1181.

Orlov, A. Y. (1960). *Botaniceskij Z. SSSR* **45**, 888–896.

Sanders, J. L. and Brown, D. A. (1978). *Agron. J.* **70**, 1073–1076.

Seufert, G., Woellmer, H., Arndt, U. and Babel, U. (1986). *Allg. Forstzeit.* **20**, 493–496.

Taylor, H. M., Upchurch, D. R. and McMichael, B. L. (1990). *Plant Soil* **129**, 29–35.

Turner, H. and Streule, A. (1983). *Wurzelökologie und ihre Nutzanwendung* International Symposium Gumpenstein, 1982 Bundesanstalt Gumpenstein, Irdning, pp. 617–635.

Van den Tweel, P. A. and Schalk, B. (1981). *Plant Soil* **59**, 163–165.

20

Experiments with Ericoid Mycorrhiza

J. R. LEAKE and D. J. READ

Department of Animal and Plant Sciences, University of Sheffield, Sheffield, S10 2UQ, UK

I. Introduction

Early work on the biology of mycorrhiza in the Ericales (Rayner, 1915; Doak, 1928; Friesleben, 1933; Bain, 1937; Burgeff, 1961; McNabb,

METHODS IN MICROBIOLOGY
VOLUME 23 ISBN 0-12-521523-1

Copyright © 1991 by Academic Press Limited
All rights of reproduction in any form reserved

1961) established that the distal portions of the root systems of most members of the order consisted of fine hair-like structures, the epidermal cells of which were occupied by "coils" of fungal hyphae. Harley (1969) called these distinctive structures "ericoid" mycorrhiza. These early studies revealed much about the infection processes, something of the nature of the fungi involved in formation of the structure, but relatively little of its functional attributes.

It was later established (Read, 1974) that the most commonly isolated fungal endophyte of ericoid roots was an ascomycete *Hymenoscyphus* (*Pezizella*) *ericae* (Read) Korf and Kernan which, when re-inoculated into aseptically grown seedlings, produced typical ericoid mycorrhiza and so satisfied the simplest requirements of Koch's (1882) postulates. However, in experiments with mycorrhiza the isolation and culturing of a fungus and the demonstration that it forms typical "structures" is only the first stage in the determination of the nature of the relationship between heterotroph and autotroph. Since many fungi have the ability to invade roots, evidence that under conditions approaching those prevailing in nature the relationship between host and fungus is of a mutualistic kind is required before the association can legitimately be described as being mycorrhizal. The "occurrence" of a given fungus is clearly not synonymous with its being "mycorrhizal", although it is widely interpreted as being so.

Those examining the so-called "mycorrhiza" of ericaceous plants have often failed to satisfy even the basic requirements of Koch's postulates by isolating the fungus observed and re-inoculating it to form the same structure. Thus, the early erroneous assumption that a species of *Phoma* (isolated from the shoots of *Calluna*!), was its mycorrhizal partner (Rayner, 1915, 1925) has been followed more recently by claims based upon "occurrence" that ectomycorrhizal (Largent *et al.*, 1980) and vesicular-arbuscular (Koske *et al.*, 1990) fungi all form "mycorrhizas" with ericaceous plants.

When re-inoculation has been attempted, the identification of "mycorrhizal infection" has often been based upon morphological rather than functional criteria. It is suggested here that the following requirements, which are essentially an extension of those of Koch, must be met before any root–fungus association can be described as being mycorrhizal.

(1) The putative mycorrhizal fungus must be isolated and maintained in pure form either as a mycelial or spore culture.
(2) The fungus must be grown with its putative host plant under semi-natural conditions, i.e. on an ecologically meaningful substrate *without* the addition of exogenous carbon supplies, and the

essential structural features of the association observed. Whereas in the case of vesicular-arbuscular mycorrhizal infection arbuscules or hyphal coils must be formed and in ectomycorrhizal associations the formation of a Hartig net is required, the key structural features of the ericoid infection are the compact intracellular hyphal complexes formed in the epidermal cells of the "hair-roots" of the hosts (Fig. 1).

(3) There must be experimental evidence that infection by the fungus leads to enhancement of growth or nutrient capture by the host, and knowledge of the extent of the "dependence" of the heterotroph upon its host for carbon supplies should be obtained.

In this chapter the critical experimental procedures involved in satisfying all of the requirements outlined above are described as they relate to the study of ericoid mycorrhiza. Procedures for isolation, culture and re-inoculation of the mycorrhizal endophyte are first provided in brief. The extent to which systematic experimental analysis of the response of host plants to infection has enabled us justifiably to apply the term "mycorrhizal" to the relationship between *H. ericae* and its host plants is then examined.

II. Isolation of endophytes

A modification of the Harley and Waid (1955) serial washing procedure was employed for isolation of the endophyte by Pearson and Read (1973) and is still in routine use. When delicate tissues such as ericoid "hair-roots" are involved it has the advantage that no chemical sterilants are involved.

Entire root systems of seedlings or distal portions of root systems of mature plants are excavated in the field, transferred to the laboratory. Soil is carefully removed, initially with the aid of a water-jet and, in the final stages, with the aid of forceps under a dissecting microscope. A small portion of the "hair-root" system, from which all microscopic debris has been removed, is transferred to a Universal vial containing approximately 10 ml of sterile distilled water. The capped vial is agitated vigorously for a period of 2–3 min before the root sample is transferred to a further vial of sterile water. This procedure is repeated to provide a series of washes. There is no hard and fast rule as to the number of washes but it is obvious that the greater the number the better the chance of removing surface contaminants. Using freshly collected actively growing "hair-roots", 30 washes are normally sufficient, but 50

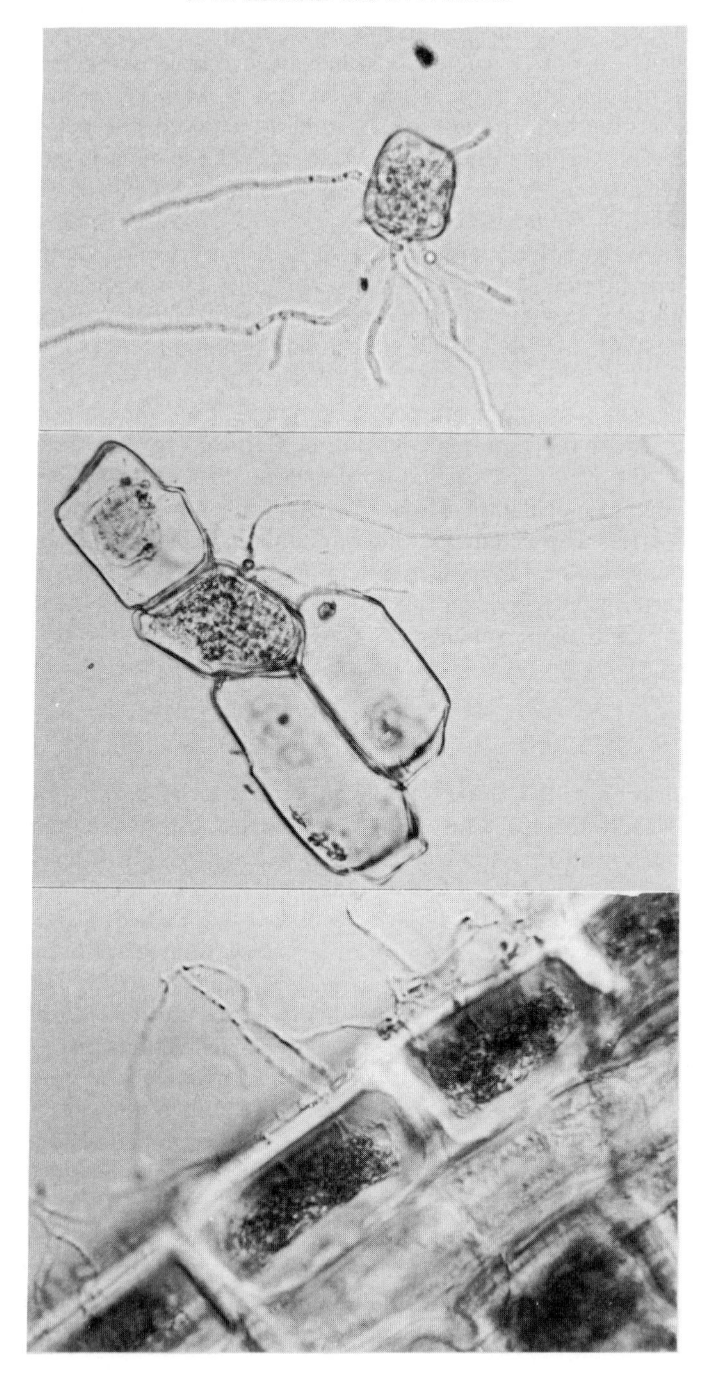

may be necessary under less favourable circumstances. A significant increase in overall efficiency of the procedure can be obtained if the agitations are achieved using a "Whirlimix" instrument, the rubber cups of which will grip a universal vial sufficiently firmly to remove the need for manual assistance.

A supplementary procedure which can be interpolated between the initial cleaning and the serial washing stages and which is designed to reduce the numbers of sterile washes required, is to expose the selected root sample in a tied muslin bag to running tap water for up to 2 h. Tap water is normally chlorinated and so itself acts as a very mild sterilant.

After serial washing the root sample can be further manipulated to produce entire segments, groups of epidermal cells or individual cells. If segments are required the sample is transferred aseptically to a Petri dish containing 10 ml of sterile water and pieces of root, normally between 5 and 10 mm in length, are aseptically cut with a scalpel. Groups of epidermal cells can be stripped from the root axis using a fine sharp needle. Large numbers of epidermal cells, both as groups and individuals, can be obtained by transferring the washed sample to a "macerator" vial which can be operated at high speed using an MSE "Top-drive" macerator. More laborious manual maceration is equally effective.

Whether root segments or suspensions of cells are produced, both are transferred to dilute (*c.* 0.5–1.0%) agar containing no added nutrients. In the case of suspensions a small volume can be transferred to the surface of the agar on the dish, and distributed evenly to disperse the cells. The advantage of the latter procedure is that it is subsequently possible, using a microscope at 100× magnification, to scan the plate and to make the critical determination as to the origin of any emerging hyphae. Such determinations are more difficult to make using root segments.

The upper surface of this nutrient-poor agar can normally be scanned on the laboratory bench without fear of aerial contamination but a Laminar flow bench can be used if available. Hyphae begin to emerge from infected cells within one to two days of incubation at room temperature (Fig. 1). At this stage the agar supporting the entire

---

Fig. 1. Top: *H. ericae*, emerging from single epidermal cell dispensed as part of a suspension onto water agar after serial washing of root system. Middle: As above, but showing growth from a single infected cell amongst a group that are uninfected. Bottom: Phase contrast view of endophyte emerging from epidermal cells of a serially washed root segment.

infected cells or the emerging hyphae can be removed to a further dish of water or nutrient agar. Malt agar (2%) is a convenient medium upon which to maintain cultures.

Even after the extensive washing procedures described above, bacterial contamination may be a problem. The addition of streptomycin sulphate (20 mg litre^{-1}) to the agar medium can help to alleviate this problem. If contamination does occur it can often be detected during scanning of water agar, but will certainly be apparent on transferring to nutrient agar. Because serial washing leads to a dilution of such contaminants it is frequently the case that within a batch of dishes from a single maceration some are free of contamination.

III. The fungi isolated

A. *Hymenoscyphus ericae*

The earliest systematic attempts to isolate the endophyte from ericoid mycorrhizal cells (Friesleben, 1936; Bain, 1937; Burgeff, 1961) showed that most of the fungi obtained when grown on nutrient agar, were dark, slow-growing and sterile. This has been repeatedly confirmed in later studies using the procedures described above (Pearson and Read, 1973; Vegh *et al.*, 1979; Read, 1983). Old cultures of some of these isolates growing on a wide range of nutrient media, solid and liquid, with and without the presence of a host plant, occasionally produce discocarps enabling confirmation of the identity of the fungus as *Hymenoscyphus ericae*. Not only is this the only fungus to be consistently isolated from the epidermal cells, but it appears to be the only one in which all three criteria for confirmation of mycorrhizal relationship, namely isolation, re-inoculation *and* demonstration of a positive response in the host, have been obtained (Read and Stribley, 1973).

The mycelial characteristics of *Scytalidium vaccinii*, a dematiaceous hyphomycete recently isolated from *V. angustifolium* and described as being a new species of ericoid endophyte (Dalpé *et al.*, 1989), are so similar to those of *H. ericae* as to suggest that they are closely related fungi.

B. Other fungi isolated from ericoid mycorrhizal roots

Other fungi have been isolated from, or found to be closely associated with, ericoid mycorrhizal roots.

Burgeff (1961) reported that one of his many fungal isolates of ericoid roots was an *Oidiodendron* species. The occurrence of *O. griseum* as an endophyte of ericoid mycorrhiza has since been confirmed (Couture *et al.*, 1983; Dalpé, 1986.) Douglas *et al.* (1989) isolated *O. maius* from roots of *Rhododendron* and the ultrastructural features of the infection appear to be comparable with those seen when hosts of this genus are infected with *H. ericae*. (Peterson *et al.*, 1980; Duddridge and Read, 1982). The occurrence of *Oidiodendron* species in close association with and even inside some cells of ericaceous plants is perhaps not surprising. They are amongst the commonest of all fungal species isolated from boreal forest soils (Söderström and Bååth, 1978) and are readily isolated also from the ectomycorrhizal sheaths of roots of coniferous species which often co-exist with ericaceous plants in these environments (Schild *et al.*, 1988; Summerbell, 1988). Evidence is emerging that when grown in pure culture with a host plant, be it ectomycorrhizal (Abuzinadah and Read, 1989) or ericoid (see below), the isolate of *O. griseum* obtained by Dalpé (1986) can have the beneficial effects required to establish its mycorrhizal credentials. However, the critical question, addressed again later, concerns the ability of such fungi, when present in mixed natural populations of putative mycorrhizal fungi, to compete for infection sites, establish infection and produce the same beneficial effects.

The ability of known *Oidiodendron* teleomorph species to form ericoid mycorrhiza was also explored by Dalpé (1989). She found that *Myxotrichum setosum* and *Gymnascella dankaliensis* produced ericoid infection in *Vaccinium angustifolium*, as did a species of *Pseudogymnoascus*, *P. ruseus*, which has a *Geomyces* anomorph.

Stoyke and Currah (1991), in a study of endophytic fungi of ericaceous roots in the Canadian Rocky Mountains, observed the widespread occurrence of a dark-pigmented simple-septate fungus, some strains of which were identified as being *Phialocephala fortinii*. Infections of this kind are also widespread in Europe and are particularly prominent on old roots at high altitude (Haselwandter and Read, 1980). However, in view of the failure of *P. fortinii* to form defined mycorrhizal structures, and of the absence of any evidence that putative hosts in the Ericaceae respond positively to inoculation by this or related fungi, there would appear to be no justification at this stage for describing these infections as being mycorrhizal.

C. Other fungi associated with ericaceous roots

The frequent occurrence of basidiomes of *Clavaria* species, particularly *C. argillacea*, with roots of ericaceous plants growing in heathland (Gimingham, 1960), in horticultural nurseries (Moore-Parkhurst and Englander, 1982) and in pots (Seviour *et al.*, 1973) has led to the belief that basidiomycetous fungi may also form ericoid mycorrhiza. Indirect evidence of a close association between *Clavaria* and ericaceous roots has been obtained using serological (Seviour *et al.*, 1973) and nutrient transfer (Englander and Hull, 1980) techniques. Using immunogold labelling techniques (see *Methods in Microbiology* Vol. 24, Chapter 8) Mueller *et al.* (1986) were able to identify hyphae of *Clavaria*, as well as those of *H. ericae*, in an intracellular position in a root of *Rhododendron*. This observation was rightly interpreted to indicate that more than one fungus can be "associated" with ericaceous roots, but was not considered to provide evidence that *Clavaria* contributed to the mycorrhizal association. Since such a fungus appears never to have been isolated from ericoid epidermal cells there remains no evidence that the association is anything but casual. There are some indications that fruit body production is most prolific after ericoid roots have been damaged by natural events such as drought or by cutting. To this extent the relationship between fungus and "host" is analogous to that seen between *Morchella* and species of elm (*Ulmus*) where, again, there is a prolific production of fruit bodies associated with death of the host but no evidence, either structural or functional, of mycorrhiza formation.

There are other reports of the occurrence of hyphae with dolipore septa, and hence of basidiomycetous affinities, inside epidermal cells with ericoid infection (Bonfante-Fasolo, 1980; Peterson *et al.*, 1980) but, again, this is not evidence of mycorrhizal infection. Indeed, bearing in mind the very large number of sections of ericoid roots examined in recent years, by a wide range of authors, it would be surprising if such occurrences had not been recorded.

The reports of the occurrence of vesicular-arbuscular mycorrhizal infections in ericoid roots (Johnson *et al.*, 1980; Koske *et al.*, 1990) are also unconvincing, in particular because they provide no indication of the presence of arbuscules in the "host" root. It is widely recognized that infection of true host plants can provide vesicular-arbuscular mycorrhizal fungi with sufficient inoculum potential to enable them to invade the tissues of "non-host" species (Read *et al.*, 1976), but such invasion in the absence of arbuscule production satisfies none of the three criteria, established above, for mycorrhizal infection. In the same way, *H. ericae* can "infect" non-host plants (Bonfante-Fasolo *et al.*, 1984) without producing any of the structures typical of ericoid infection.

IV. Aseptic culture of seedlings

Confirmation that any isolate obtained from infected epidermal cells is a candidate mycorrhizal organism can be achieved only by re-inoculation into aseptically grown seedlings.

Seeds of the required host plant can conveniently be surface sterilized by shaking them for up to 20 min in the filtrate derived from a 10/140 (w/v) solution of calcium hypochlorite (bleaching powder). The excess sterilant is removed by washing the seeds in sterile water before they are individually transferred to a Petri dish of water agar. Plates are scanned regularly over the few days after sowing and any contaminated seeds are removed. Germination of most ericaceous species in the genera *Calluna*, *Erica*, *Vaccinium* and *Rhododendron* will take place within three weeks of incubation in light at room temperature.

V. Synthesis of mycorrhiza

A number of methods for synthesis of mycorrhiza have been used successfully but perhaps the simplest involves the use of water agar. Although not a natural substrate, water agar has colloidal and organic properties comparable with those of the peats which so often support ericaceous plants. A closer approach to the natural situation can be obtained by dispersing small quantities of sterilized heathland soil through the agar. This can provide increases of amounts and of uniformity of infection (Pearson and Read, 1973). If plants are to be maintained on the synthesis medium for a lengthy period it is advisable that mineral nutrients be added to the agar. A suitable mineral nutrient medium contains $CaSO_4 \cdot 2H_2O$, 69 mg; $(NH_4)_2SO_4$, 53 mg; $MgSO_4 \cdot 7H_2O$, 48.7 mg; K_2HPO_4, 69.5 mg; $FeCl_3 \cdot 6H_2O$, 2.9 mg; $MnSO_4 \cdot 4H_2O$, 2.03 mg; H_3BO_3, 2.9 mg; $(NH_4)_6MO_7O_{24} \cdot 4H_2O$, 1.3 mg; $ZnSO_4 \cdot 7H_2O$, 0.4 mg; $CuSO_4 \cdot 5H_2O$, 0.4 mg; agar, 10 g; water 1 litre, at pH 4.0. The addition of small quantities of activated charcoal (1.0 g $litre^{-1}$) and of glucose (0.5 g $litre^{-1}$) was shown by Duclos and Fortin (1983) to enhance mycorrhiza formation. While it is perhaps legitimate to add glucose at this concentration when dealing with a fungus which has been previously established to be mycorrhizal, higher concentrations should not be used when assaying for mycorrhizal potential because, as Duclos and Fortin (1983) demonstrated, an abundance of exogenous carbon strongly influences the vigour and inoculum potential of ericoid fungi, which can become pathogenic under these circumstances.

Interestingly, the incorporation of activated charcoal has been found to be useful in quite a different context from that originally envisaged by Duclos and Fortin (1983). A major stumbling block in production of uniform material for analysis of responses of ericaceous plants to mycorrhizal infection has been that in the *absence* of fungi, root development can be completely inhibited. Addition of charcoal to the medium supporting aseptically grown seedlings enables their root systems to develop in a manner comparable with that seen in cultures grown with the fungus. This observation is itself of biological importance. It suggests that facilitation of development of the delicate and sensitive "hair-root" system may be one of the most fundamental functions of the mycorrhizal fungus in nature. Increasingly experimental evidence, reported below, confirms that detoxification is one of the primary functions of ericoid mycorrhizal infection.

The procedures described above enable two of the basic requirements for verification of mycorrhizal status to be achieved, namely isolation of the fungus and reproduction of essential structural features of the association. Experimental evidence in favour of the third requirement, that the infection leads to enhancement of nutrient capture or growth, is now examined. It has become increasingly clear that the ericoid mycorrhizal fungus plays a key role both in the nutrition and in the "detoxification" of the environment of its host. These roles are first considered separately and their ability, in combination, to contribute to the "fitness" of ericaceous plants in the heathland environment, is then assessed.

VI. The role of mycorrhiza in the nutrition of plants with ericoid infection

In heaths the hair-roots of dominant plants such as *Calluna vulgaris*, *Erica* spp. and *Vaccinium* spp. are characteristically confined to the top 10 cm, or less, of the soil profile where they are closely associated with the litter (Reiners, 1965; Gimingham, 1972; Persson, 1980). Interestingly, when herbaceous species such as *Deschampsia flexuosa*, *Molinia caerulea*, *Eriophorum vaginatum* and *Carex* spp. coexist with ericaceous shrubs their roots are concentrated at lower levels of the soil profile (Gimingham, 1972), so the two groups of plants are not competing for the same resources.

The use of the isotopes ^{32}P, applied to the soil as phosphate (Boggie *et al.*, 1958), and ^{14}C, fed to the shoots as CO_2 (Wallen, 1983), has confirmed that even when ericaceous plants root to considerable depth, the active roots are confined to the top few centimetres of the soil profile. In this position the substrates available for attack are largely those produced by the ericaceous plants themselves and by their fungal associates.

Recent studies have indicated that the ericoid mycorrhizal fungus has the ability to attack most of these substrates despite the chemical complexity of some of them. Thus it can cleave the phenolic ring structures of the lignin polymer (Haselwandter *et al.*, 1990) and can break down pure chitin (Leake and Read, 1990c) and tannic acid (Leake and Read, 1989a). Such attributes will serve to facilitate the enzymic degradation of organic nitrogen sources which are otherwise inaccessible. The question of the extent to which the nitrogen can then be mobilized by ericoid mycorrhizal fungi is crucial to an understanding of the success of these plants in heathland environments.

An early suggestion that the mycorrhiza might be involved in the release of nitrogen from organic complexes in soil came from a study (Stribley and Read, 1974b) in which ammonium, which was then thought to be the only significant source of nitrogen for such plants, was added as $^{15}NH_4$ to heathland soil containing infected or uninfected plants. It was revealed in this experiment that despite their increased total yields and nitrogen contents, the mycorrhizal plants contained less ^{15}N than their uninfected counterparts. The inescapable conclusion was that the mycorrhizal fungus provided the plant with access to a nitrogen source other than ammonium, and that in the absence of nitrate in these soils, the only alternative source was organic nitrogen. Subsequent work has confirmed that *H. ericae* has the ability to use amino-acids (Bajwa and Read, 1986), peptides (Bajwa and Read, 1985), pure protein (Bajwa *et al.*, 1985; Leake and Read, 1989b) and even protein co-precipitated with tannin (Leake and Read, 1989a) as nitrogen sources. Most importantly, the nitrogen assimilated by the fungus is readily passed to the host plant (Bajwa *et al.*, 1985).

The ability to cleave protein into its constituent amino-acids is attributable to the release by the fungus of a carboxy-proteinase enzyme, the optimal pH for both the activity and production of which is between 2.0 and 4.0, a range coinciding almost exactly with that typical of soils dominated by ericaceous plants (Leake and Read, 1990a). It is likely to be of considerable significance for the nitrogen nutrition of ericaceous plants that the proteolysis achieved by *H. ericae* in the presence of exogenous carbon does not involve decarboxylation. The products are assimilated as amino acids and ammonium is released only under conditions of carbon starvation (Read *et al.*, 1989).

Induction of proteinase production is not dependent upon the presence of exogenous protein, activity of the enzyme being expressed in the presence of mineral nitrogen (Leake and Read, 1990b) and of mixtures of amino-acids (Leake and Read, 1991). It has been proposed (Leake and Read, 1990b) that regulation of enzyme production is achieved by the "Noah's Ark" strategy (Burns, 1986) in which the release of small quantities of enzyme molecules as "emissaries", in the absence of proteinaceous substrates, is followed by rapid increase of production of enzyme once a "reporter" molecule in the form of a protein hydrolysis product is detected. This ability to respond rapidly to the availability of substrate would enable the fungus to be highly effective in competition with other micro-organisms for complex organic sources of nitrogen.

Phosphorus is thought to be the major growth-limiting nutrient in some heathlands, particularly in warmer climates where more active mineralization of nitrogen occurs (Groves, 1981). However, the demands of the plant for phosphorus are considerably lower than those for nitrogen and this, together with the fact that heathland plants are active with remarkably low tissue phosphorus concentrations, probably reduces the extent to which phosphorus stress is experienced. In the rooting environment of ericaceous plants phosphorus, like nitrogen, is present primarily in organic residues. It is known that *H. ericae* produces acid phosphatase enzymes (Pearson and Read, 1975; Straker and Mitchell, 1986) and that it can release and assimilate phosphorus from calcium, iron and aluminium phytates (Mitchell and Read, 1981). A recently revealed attribute of this enzyme, which is likely to be of ecological significance, is its resistance to inhibition by the metal ions Al^{3+} and Fe^{2+}, both of which occur in solution at the pH prevailing in "mor" humus (Shaw and Read, 1989).

Studies of iron uptake by mycorrhizal and non-mycorrhizal ericaceous plants have shown that infection provides for significantly increased absorption of the element over the range of concentration (0–18 mg $litre^{-1}$ = 0–360 μM) typically found in soil solution in heathlands (Shaw *et al.*, 1990). Significantly greater root yields and root iron concentration were found over this range in mycorrhizal than in non-mycorrhizal plants grown in sand culture. In short-term studies comparing the rates of uptake of ^{59}Fe by infected and uninfected plants from solution culture, it was shown that the mean uptake rates of 3.9 nmol Fe (mg $root)^{-1}$ h^{-1} achieved by mycorrhizal roots of *Calluna* were more than double those of non-mycorrhizal plants. In contrast to the situation found for the other metals examined, significant proportions of the absorbed iron are rapidly transferred from root to shoot. However, differences in transfer patterns between infected and uninfected plants became less marked with increasing concentration of exogenous iron, suggesting that the rate of translocation of iron is under metabolic control (Shaw *et al.*, 1990).

In the presence of calcium salts, which are likely to decrease the availability of iron in nature, mycorrhizal infection gave increases of shoot iron concentrations of up to 234% (Leake *et al.*, 1990).

Schüler and Haselwandter (1988) have reported the production of hydroxamate siderophores in cultures of *H. ericae* and it is likely that such compounds play a key role in the mechanism of iron capture. That iron-scavenging mechanisms are effective in the field is suggested by data of Woodwell *et al.* (1975), who determined mineral element concentration in an oak–pine woodland with an understorey of ericaceous species. Of the eight elements examined the contrasts between the patterns of iron accumulation in the overstorey and understorey components was most marked. The ericoid shrubs contained consistently high and the ectomycorrhizal trees consistently low concentration of the element.

Iron-scavenging mechanisms would be expected to become progressively more important in soils of higher pH in which the availability of the element is reduced. They may therefore be a vital attribute of the mycorrhiza of that relatively small number of ericaceous species such as *Erica carnea*, *Rhododendron hirsutum* and *Rhodothamnus chamaecistus* which are important components of heaths on calcareous soils. Interestingly, recent results of Haselwandter suggest that the pH optima for activity of siderophores produced by endophytes of the calcicolous species *R. chamaecistus* are somewhat higher than those of fungi isolated from the typically calcifuge genera.

In contrast to iron, the concentrations of most base cations, and in particular of calcium, are characteristically depleted in acidic heathland soils, where the presence of ericaceous plants even accelerates the processes of leaching and cation loss (Grubb and Suter, 1971). Mycorrhizal infection appears to provide an effective mechanism for capture of calcium ions. In experiments comparing calcium capture by mycorrhizal and non-mycorrhizal plants of *Calluna* which were supplied with the element in three different salt forms (Leake and Read, 1989c), shoots of infected plants were found to contain 60, 32 and 23% higher calcium concentrations than uninfected controls when grown on $CaCO_3$, $CaCl_2$ and $CaSO_4$, respectively. These observations confirm an earlier report (Powell, 1982) that infection caused a particularly pronounced enhancement of calcium concentration in blueberry (*Vaccinium corymbosum*).

VII. The extent of "dependence" of the heterotrophic partner upon its autotropic host

Since the term "mutualism" implies benefits to both partners of the association it is necessary, especially where, as in ericoid mycorrhiza, there is a large fungal biomass, to determine the nature of the carbon economies of the association. It can readily be shown that carbon fed as $^{14}CO_2$ to the host plant is transferred to the endophyte where host assimilate is converted to "fungal" sugars (Stribley and Read, 1974a; see also Chapter 6, this volume). Indeed, there is a suggestion, arising from the previously reported observation that addition of simple carbon compounds exogenously to ericoid mycorrhizal systems leads to increased virulence of the fungus, that control of the heterotroph by the host may be achieved by restricting the release to it of assimilates.

In the case of fungi such as *H. ericae* which have a proven ability to attack polymeric sources of carbon in soil, the possibility must be considered that these can supplement the carbon budget of the association. Recent laboratory studies in which carbon was supplied heterotrophically to ericoid mycorrhizal plants in the form of amino acids and proteins, have indicated that assimilation of carbon from these sources can account for 10% of that achieved autotrophically (Davies and Read, in press).

VIII. Screening of putative mycorrhizal isolates using an assay based upon nutrient release to the host plant

As indicated earlier, claims, based upon the detection of hyphal complexes in epidermal cells of hair-roots, that a number of fungi other than *H. ericae* are mycorrhizal with ericaceous hosts, have not so far been verified by demonstration that infection by the re-introduced fungus is followed by nutritional or growth response. There is thus an urgent need for an ecologically meaningful screening procedure which facilitates the analysis of growth or nutritional responses of host plants in the presence of putative mycorrhizal fungi. Using one such procedure (Leake and Read, in press), *Calluna vulgaris* was grown aseptically with a range of putative mycorrizal associates on nutrient agar in which the nitrogen was supplied in a form, protein, known to be accessible to the host only through the activities of compatible fungi.

The fungi employed were *H. ericae*, ericoid (i), ericoid (ii), which had originally been isolated from *Calluna*; isolates from other ericoid hosts of acid soils, *Erica scoparia* (*E. scop.*), *E. hispidula* (*E. hisp.*), *E. mauritanica* (*E. maur.*), *Vaccinium* (*O. gris.* and *Scyt. vacc.*), and an isolate from a host *Rhodothamnus chamaecistus* (*Rhodo.*) which normally occurs on relatively base-rich soils (Haselwandter and Read, 1983).

The results (Fig. 2) demonstrate a very wide range of growth responses in the host. While *H. ericae* produced the most effective positive response in terms of enhancement of yield, thus confirming its status as a truly mycorrhizal associate, some of the other isolates,

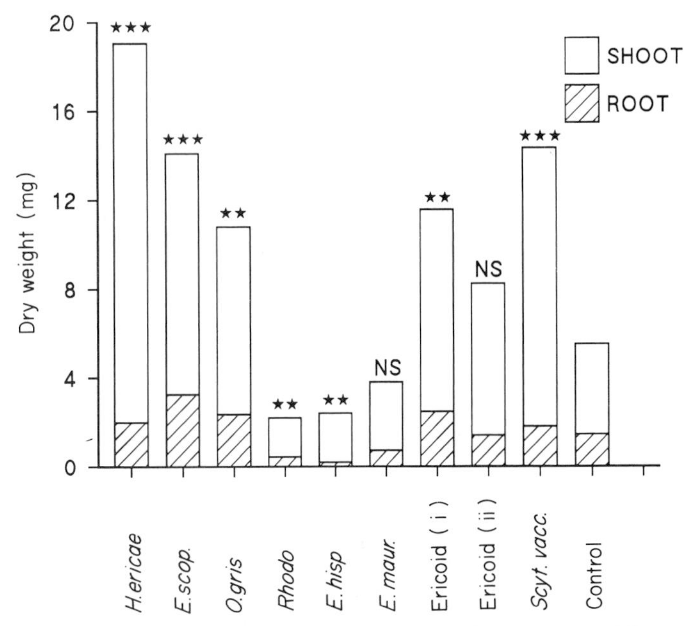

Fig. 2. Dry weight yields of *Calluna* seedlings when grown with a range of fungal isolates obtained from ericoid mycorrhizal roots and in the absence of a fungus (control). See text for details. Key: ★★★ represents statistically significant increase of yield over control at $p < 0.01$; ★★ represents increase or decrease of yield at $p < 0.01$; NS, no significant difference.

notably those from *E. scoparia* and *Vaccinium* spp. (*O. gris.* and *Scyt. vacc.*) also yielded significant growth increases over the controls, indicating that they also are truly mycorrhizal with this host. The isolates of *E. mauritanica* and even one of the *Calluna* isolates *ericoid* (ii) failed to produce a significant positive growth response while two fungi, those from *E. hispidula* and *R. chamaecistus*, reduced the yield of *Calluna* and could thus even be seen as pathogens.

While this assay procedure represents an advance, it too provides only a restricted view of the true situation. At the nutritional level it indicates responses of the host–fungus combination to challenge by the single constraint imposed by availability of organic nitrogen. In ecological terms it would be appropriate, indeed ultimately necessary, to examine the responses of each combination when grown under the influence of the particular constraint, be it nutritional or toxicological, predominating in the natural environment from which the putative partners were taken.

At the biological level it is necessary to broaden the base of such screening procedures to employ a wider range of potential host plants as well as other fungal isolates, including saprotrophs. A further highly desirable refinement of such an assay procedure would involve the use of mixed inocula, the responses of the hosts being subsequently assessed not only in terms of their growth, but also in terms of the extent and nature of the infection achieved by individual fungi within the mixture. Such an analysis might be achieved by re-isolation from large numbers of individual cells of the host but this would not only be time-consuming but could yield misleading results if more than one fungus occupied an individual cell. This is clearly an area in which the systematic application of immunocytochemical techniques could yield important advances in our understanding.

IX. Detoxification of the soil environment

Both laboratory studies referred to earlier and analyses of growth in heathland soil indicate that the roots of ericoid plants such as *Calluna* are extremely sensitive to the presence of low levels of the organic and metallic compounds which are widely distributed in acid heathland soils. In the absence of mycorrhizal infection, complete inhibition of root growth can occur. Under these circumstances, which have been described as a "toxicity syndrome" (Read, 1984), it is evident that physico-chemical rather than nutritional constraints are of primary importance in determining survival, and hence "fitness", in the environment.

As a first step towards the elucidation of the role of mycorrhiza in the toxicity syndrome, the effects of some of those phenolic and aliphatic compounds identified by Jalal and Read, (1983a,b) in extracts of heathland soil have been examined (Table I). The compounds have been added to culture media supporting the ericoid endophyte as well as to seeds and seedlings of *Calluna* growing in the presence and absence of the endophyte.

A. Growth of *H. ericae* on media containing phenolic acids

Only two of the phenolic compounds, *p*-methoxybenzoic and benzoic acid, produced severe inhibition of growth of the endophyte, while a third, salicylic acid, produced some inhibition up to 10 days (Fig. 3). Analysis of ultraviolet (UV) absorption patterns suggests that the endophyte was assimilating or degrading all those compounds which were not inhibitory to its growth. Even those compounds which were found to inhibit the endophyte when added individually had no inhibit-

TABLE I

The phenolic acid composition of soil under *Calluna* and in the two mixtures used in the seedling bioassays. Values expressed as mg kg^{-1} fresh soil, assuming water content of 80%, and as mg $litre^{-1}$ H_2O in the mixtures (after Jalal and Read, 1983b)

	Maximum annual level in *Calluna* soil (mg $litre^{-1}$)	May level (mg $litre^{-1}$)	Concentration in field mixture (mg $litre^{-1}$)	Concentration in equal weight mixture (mg $litre^{-1}$)
Benzoic	1.87	1.51	1.5	0.25
p-Methoxybenzoic	1.08	1.02	1.0	0.25
Salicylic	2.36	1.70	1.7	0.25
p-Hydroxybenzoic	0.70	0.70	0.7	0.25
Vanillic	1.75	1.75	1.8	0.25
Syringic	1.04	1.04	1.0	0.25
p-Coumaric	0.55	0.74	0.4	0.25
Ferulic	1.58	1.58	1.0	0.25

ory effect when supplied at ecologically meaningful concentrations similar to those observed in the field.

B. Effects of phenolic acids upon germination of *Calluna* seed

Since seed of *Calluna* normally falls directly onto soil enriched in phenolic compounds, phytotoxicity can be experienced from the germination phase onwards. Incorporation of these compounds into agar on which *Calluna* seed was sown, showed that some can indeed inhibit the germination process (Fig. 4). Salicylic and *p*-methoxybenzoic acids were the most toxic individual acids but, more importantly, the field mixture also produced a 17% inhibition. In the presence of the fungus, however, such an inhibition was almost entirely removed. Thus even before infection takes place, the presence of endophyte mycelium in the environment around the root can lead to detoxification.

C. Effects of phenolic acids upon growth of mycorrhizal and non-mycorrhizal seedlings of *Calluna*

The two phenolic acid mixtures (Table I) were added to agar cultures on which mycorrhizal (M) and non-mycorrhizal (NM) seedlings were grown.

Responses of *Calluna* were assessed in terms of dry weight yield (Fig. 5a) and extension growth (Fig. 5b). As frequently observed when plants

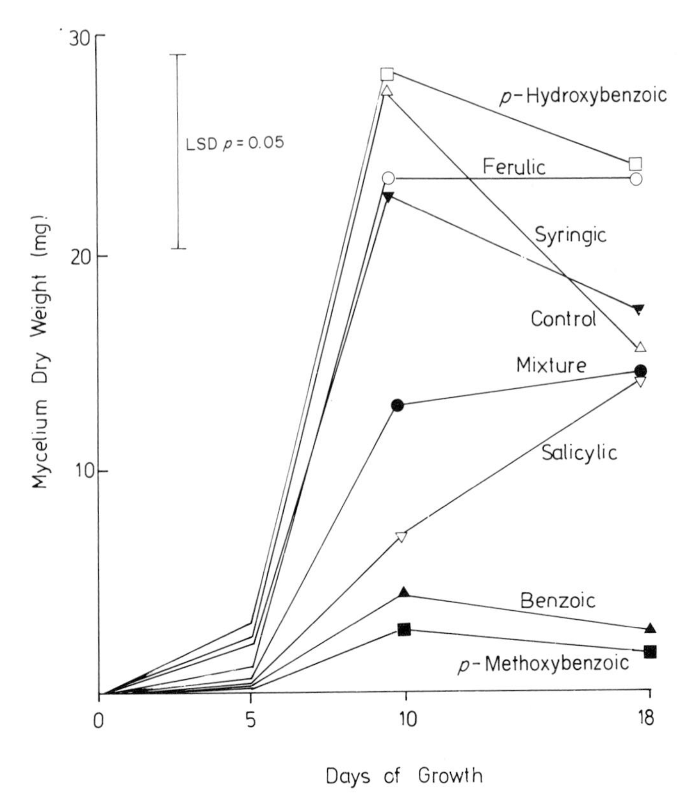

Fig. 3. The mycelial yield of *H. ericae* when grown on basal medium amended with individual phenolic acids at a concentration of 0.3 mM and with an equimolar mixture containing each acid at a concentration of 0.037 mM. Values are means of five replicates. Vertical bar gives LSD $p = 0.05$.

are grown aseptically on semi-solid media there was significant inhibition of root development even in the control treatments when the fungus was not present. Infection led to stimulation of root growth and a shift towards a higher root/shoot ratio. The field mix of phenolics produced an even greater inhibition of root yield and extension growth, an effect which was eliminated by the endophyte, which provided a seven-fold increase of root yield and a four-fold increase of length in the longest root compared to those in the NM plants.

D. Effects of acetic acid upon *H. ericae* and upon mycorrhizal and non-mycorrhizal plants

A feature of organically enriched soils, particularly under water-logged

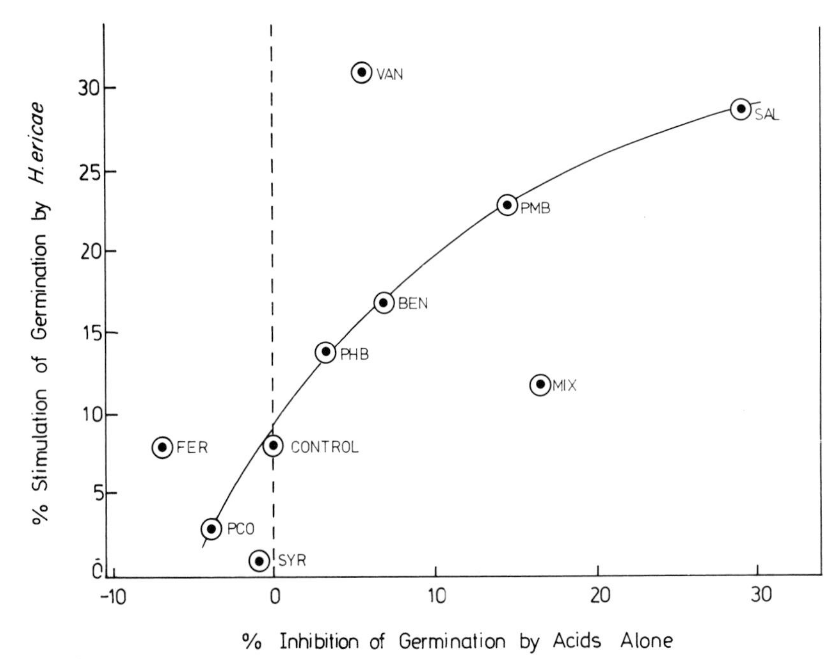

Fig. 4. Plot of inhibitory effect of phenolic acids on the aseptic germination of *Calluna* against the stimulation of germination following inoculation with the endophyte, both expressed as percentages. Each acid was added at a concentration of 0.3 mM and at 0.037 mM in the mixture. Values are means of 10 replicates. Key: MIX, equimolar mixture; PHB, *p*-hydroxybenzoic; SAL, salicylic; FER, ferulic; PMB, *p*-methoxybenzoic; PCO, *p*-coumaric; VAN, vanillic; SYR, syringic; BEN, Benzoic.

conditions, is the accumulation of short-chain fatty acids, the toxicity of which is greatly exacerbated when they are in the undissociated condition at low pH (Jackson and Taylor, 1970), While the volatile fatty acids (VFSs) have been shown to be generally less phytotoxic than the aromatic compounds (Prill *et al.*, 1949; Lee, 1977), this situation can be counteracted in some circumstances by the fact that the VFAs, in particular acetic acid, can accumulate to high levels (Lynch, 1977, 1980, 1985). In order to examine the effect of acetic acid on the growth of *H. ericae* and upon M and NM seedlings, the acid was added to liquid mineral media over a concentration range of 0–100 mg litre^{-1} which was designed to cover that likely to be encountered in soil. No inhibition of yield was observed over this range. Growth of the ectomycorrhizal fungus *Suillus bovinus* was, in contrast, strongly inhibited at the higher concentrations. The effects of the VFA upon the development of non-mycorrhizal roots and the response to infection were similar to

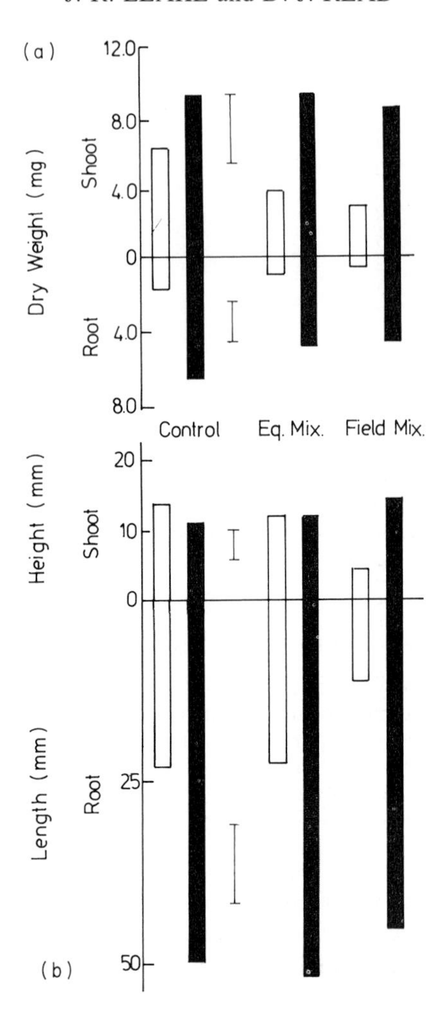

Fig. 5. (a) Dry weights of shoots and roots of non-mycorrhizal (open bars) and mycorrhizal (shaded bars) plants of *Calluna* grown on agar in the presence of equal weight (Eq. MIX.) and field mixture (Field Mix) of phenolic acids and without phenolics (control). Vertical bar gives LSD $p = 0.05$. (b) As (a) but showing mean shoot heights and lengths of the longest root.

those seen in the phenolic acid treatments, the presence of *H. ericae* providing release from the inhibitory effects of the acid (Fig. 6).

The role of the hydroxy-alkanoic acids identified by Jalal and Read (1983a,b) in the toxicity syndrome remain to be investigated. It has been shown, however, that the longer chain fatty acids, 18:1, 18:2 and 18:3,

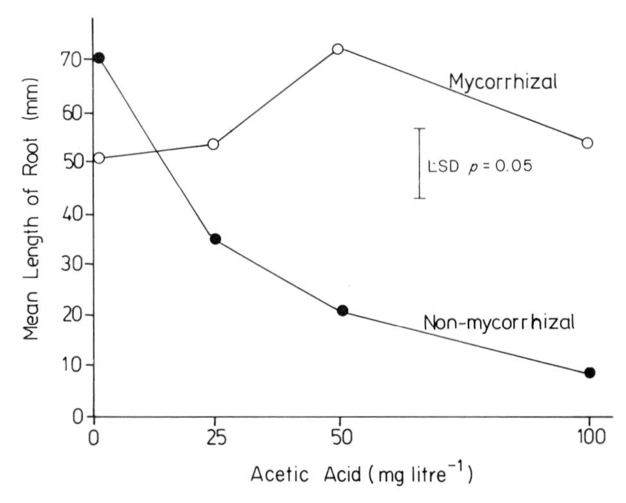

Fig. 6. Effects of concentration of acetic acid upon mean length of the longest root of *Calluna* plants grown in mycorrhizal and non-mycorrhizal conditions.

which are characteristically present in high concentrations as storage products of ericaceous shrubs (Tschager *et al.*, 1982), are readily used as carbon sources by the endophyte.

E. Ericoid mycorrhizal infection and metal toxicity

As with organic acids, the potential for metallic ions to exert toxic effects is greater at low soil pH, in their case because of increased solubility under acid conditions. It has been shown previously that ericoid infection can confer resistance to copper and zinc toxicity by reducing inflow of the metals to the shoots (Bradley *et al.*, 1981, 1982). Of greater importance as potential toxins in natural environments are the metals aluminium and iron. The inherent toxicity of the former element has long been recognized as a factor involved in the exclusion of some species from acid soils, while the role of the latter is complicated by the fact that it is an essential element which can be toxic at high concentrations.

Recent studies have shown that the ericoid endophyte has remarkable resistance to aluminium, no inhibition of yield being found even in media containing 800 mg litre^{-1} Al^{3+} (Burt *et al.*, 1986).

In solutions containing in excess of 400 mg litre^{-1} the metal strongly inhibits root development in non-mycorrhizal plants, characteristic "stilt" roots being formed. While these plants survive under protected growing conditions, it is unlikely that such root systems would be adequate to sustain them in the field. Mycorrhizal infection enables roots to develop normally and leads to a reduction of inflow of the metal to the shoots. A similar enhancement of resistance to high levels of iron has been shown to be provided by infection (Shaw *et al.*, 1990).

X. Discussion and conclusion

The experiments with ericoid mycorrhiza described here provide evidence to support the view that the relationship between *H. ericae* and its host plants is truly mycorrhizal, and go some way towards identification of the physiological and ecological basis of the mutualism. It may be the very wide range of beneficial attributes demonstrated by this fungus that has led it to be so widely selected as the dominant mycorrhizal associate of ericoid roots.

Many questions, however, remain to be answered. Of particular importance is the fact that in the ericoid, as indeed in the other types of mycorrhiza discussed in these volumes, the function of the mutualism in truly natural situations involving soils and mixed populations of heterotrophic and autotrophic organisms remains largely unexplored.

While there are strong indications that infection provides the host with the wherewithal to mobilize the nutrients, and to assimilate and hence de-toxify some of the organic acids contained in natural substrates, these experiments have generally employed pure compounds or mixtures which at best provide simulations of possible field conditions rather than the conditions themselves. If experiments with mycorrhiza are to have any relevance to the real world they should be designed and executed, wherever possible, to take account of circumstances prevailing in nature.

The reductionist approach to analysis of mycorrhizal function, exemplified most widely by experiments in which the processes of uptake of a single mineral ion are studied under controlled conditions, has provided valuable insights into some of the specific mechanisms which may be in operation in natural communities. The challenge now is to determine the true nature of the resources being exploited by mycorrhizal roots in the far more complex soil environment, and to evaluate the extent to which interactions within and between heterotrophic and autotrophic populations influence the expression of mycorrhizal potential.

The notion that enhancement of nutrient supply or of growth over a short part of the life-cycle of a plant provides an adequate criterion with which to identify a mycorrhizal relationship, may itself come to be called into question. In natural ecosystems it is likely to be at least as important to consider reproductive capability or survivorship as indicators of "fitness" and there is little doubt that, in future, experiments with mycorrhiza will turn increasingly to investigating the hypothesis that the infection is involved at these deeper levels, to ensure survival of the individual components of the community, and hence to sustain the ecosystem itself.

References

Abuzinadah, R. A. and Read, D. J. (1989). *New Phytol.* **112**, 61–68.

Bain, H. F. (1937). *J. Agric. Res.* **55**, 811–835.

Bajwa, R. and Read, D. J. (1986). *Trans. Br. Mycol. Soc.* **87**, 269–277.

Bajwa, R., Abuarghub, S. and Read, D. J. (1985). *New Phytol.* **101**, 469–486.

Boggie, R., Hunter, R. F. and Knight, A. H. (1958). *J. Ecol.* **46**, 621–239.

Bonfante-Fasolo, P. (1980). *Trans. Br. Mycol. Soc.* **75**, 320–325.

Bonfante-Fasolo, P., Gianinazzi-Pearson, V. and Martinengo, L. (1984). *New Phytol.* **98**, 329–334.

Bradley, R., Burt, A. J. and Read, D. J. (1981). *Nature (Lond.)* **292**, 335–337.

Bradley, R., Burt, A. J. and Read, D. J. (1982). *New Phytol.* **91**, 197–209.

Burgeff, H. (1961). *Mikrobiologie des Hochmores.* Gustav Fischer Verlag, Stuttgart.

Burns, R. G. (1986). In *Interactions of Soil Minerals with Natural Organics and Microbes*, pp. 429–451. Soil Scince Society of America Special Publication 17.

Burt, A. J., Hashem, A. R., Shaw, G. and Read, D. J. (1986). In *Proceedings of the First European Symposium on Mycorrhizas* (V. Gianinazzi-Pearson and S. Gianinazzi, eds), pp. 683–687. INRA Press, Paris.

Couture, M., Fortin, J. A., and Dalpé, Y. (1983). *New Phytol.* **95**, 375–380.

Dalpé, Y. (1986). *New Phytol.* **103**, 391–396.

Dalpé, Y. (1989). *New Phytol.* **113**, 523–527.

Dalpé, Y., Litten, W. and Sigler (1989). *Mycotaxon* **35**, 371–378.

Doak, K. D. (1928). *Phytopathology* **18**, 101–108.

Douglas, C. G., Heslin, M. C. and Reid, C. (1989). *Can. J. Bot.* **67**, 2206–2212.

Duclos, J. L. and Fortin, J. A. (1983). *New Phytol.* **94**, 95–102.

Duddridge, J. A. and Read, D. J. (1982). *Can. J. Bot.* **60**, 2345–2356.

Englander, L. and Hull, R. J. (1980). *New Phytol.* **84**, 661–667.

Friesleben, R. (1933). *Ber. Dtsch. Bot. Ges.* **51**, 351–356.

Friesleben, R. (1936). *Jahrb. Wiss. Bot.* **82**, 413–459.

Gimingham, C. H. (1960). *J. Ecol.* **48**, 455–483.

Gimingham, C. H. (1972). *Ecology of heathlands.* Chapman & Hall, London.

Groves, (1981). In *Ecosystems of the World*, Vol. 9B: *Heathlands and Related Shrublands* (R. Specht, ed.), pp. 151–163. Elsevier, Amsterdam.

Grubb, P. J. and Suter, M. B. (1971). In *The Scientific Management of Animal and Plant Communities for Conservation* (E. Duffey and A. S. Watt, eds), pp. 115–133. Blackwell, Oxford.

Harley, J. L. (1969). *The Biology of Mycorrhiza*. Leonard Hill, London.
Harley, J. L. and Waid, J. S. (1955). *Trans. Br. Mycol. Soc.* **38**, 104–118.
Haselwandter, K. and Read, D. J. (1980). *Oecologeca (Berlin)* **45**, 57–62.
Haselwandter, K. and Read, D. J. (1983). *Sydowia Ann. Mycol. Ser. II* **36**, 75–77.
Haselwandter, K., Bobleter, O. and Read, D. J. (1990). *Arch. Mikrobiol.* **153**, 352–354.
Jackson, P. C. and Taylor, J. M. (1970). *Plant Physiol.* **46**, 538–542.
Jalal, M. A. F. and Read, D. J. (1983a). *Plant Soil* **70**, 257–272.
Jalal, M. A. F. and Read, D. J. (1983b). *Plant Soil*. **70**, 273–286.
Johnson, C. R., Joiner, J. N. and Crews, C. E. (1980). *J. Am. Soc. Hort. Sci.* **105**, 286–288.
Koch, R. (1882). *Über die Milzbrandimpfung, eine Entgegnung auf den von Pasteur in Genf gehaltenen Vortrag*. Berlin: Fischer.
Koske, R. E., Gemma, J. N. and Englander, L. (1990). *Am. J. Bot.* **77**, 64–68.
Largent, D. L., Sugihara, N. and Wishner, C. (1980). *Can. J. Bot.* **58**, 2274–2279.
Leake, J. R. and Read, D. J. (1989a). *Agric. Ecosyst. Environ.* **29**, 225–236.
Leake, J. R. and Read, D. J. (1989b). *New Phytol.* **112**, 69–76.
Leake, J. R. and Read, D. J. (1989c). *New Phybl.* **113**, 535–544.
Leake, J. R. and Read, D. J. (1990a). *New Phytol*. **115**, 243–250.
Leake, J. R. and Read, D. J. (1990b). *New Phytol*. **116**, 123–128.
Leake, J. R. and Read, D. J. (1990c). *Mycol. Res.* **94**, 993–995.
Leake, J. R., Shaw, G. and Read, D. J. (1990). *New Phytol*. **114**, 651–657.
Lee, R. B. (1977). *J. Exp. Bot.* **28**, 578–587.
Lynch, J. M. (1977). *J. Appl. Bacteriol.* **42,** 81–87.
Lynch, J. M. (1980). *Plant, Cell Environ.* **3**, 255–259.
Lynch, J. M. (1985). In *Soil Organic and Biological Activity* (D. Vaughn and R. E. Malcolm, eds), Developments in Plant and Soil Science, Vol. 16, pp. 152–174. Martinus Nijhoff/Dr W Junk, Dordrecht.
Nabb, R. F. R. (1961). *Austral. J. Bot.* **9**, 57–61.
Mitchell, D. T. and Read, D. J. (1981). *Trans. Br. Mycol. Soc.* **76**, 255–260.
Moore-Parkhurst, S. and Englander, L. (1982). *Can. J. Bot.* **60**, 2342–2344.
Mueller, W. C., Tessier, B. J. and Englander, L. (1986). *Can. J. Bot.* **64**, 718–725.
Pearson, H. (1980). *Oikos* **34**, 77–87.
Pearson, V. and Read, D. J. (1973). *New Phytol*. **72**, 371–379.
Pearson, V. and Read, D. J. (1975). *Trans. Br. Mycol. Soc.* **64**, 1–7.
Peterson, T. A., Mueller, W. C. and Englander, L. (1980). *Can. J. Bot.* **58**, 2421–2433.
Powell, C. L. (1982). *J. Am. Soc. Hort. Sci.* **107**, 1012–1015.
Prill, E. A., Barton, L. V. and Solt, M. L. (1949). *Contrib. Boyce Thompson Inst.* **15**, 429–435.
Rayner, M. C. (1915). *Ann. Bot. (Lond.)* **29**, 97–133.
Rayner, M. C. (1925). *Br. J. Exp. Biol.* **2**, 265–291.
Read, D. J. (1974). *Trans. Br. Mycol. Soc.* **63**, 381–383.
Read, D. J. (1983). *Can. J. Bot.* **61**, 985–1004.
Read, D. J. (1984). *Aspects Appl. Biol.* **5**, 195–209.
Read, D. J. and Stribley, D. P. (1973). *Nature (Lond.), New Biol.* **244**, 81–82.
Read, D. J., Koucheki, H. K. and Hodgson, J. R. (1976). *New Phytol*. **77**, 641–651.

Read, D. J., Leake, J. R. and Langdale, A. R. (1989). In *Nitrogen, Phosphorus and Sulphur Utilization by Fungi* (L. Boddy, R. Marchant and D. J. Read, eds), pp. 181–204. Cambridge University Press, Cambridge.
Reiners, W. A. (1965). *Bull. Torrey Bot. Club* **92**, 448–464.
Schild, D. E., Kennedy, A. and Stuart, M. R. (1988). *Eur. J. For. Pathol.* **18**, 51–61.
Schüler, R. and Haselwandter, K. (1988). *J. Plant Nutr.* **11**, 907–913.
Seviour, R. J., Willing, R. R. and Chilvers, G. A. (1973). *New Phytol.* **72**, 381–385.
Shaw, G. and Read, D. J. (1989). *New Phytol.* **113**, 529–533.
Shaw, G., Leake, J. R., Baker, A. J. M. and Read, D. J. (1990). *New Phytol.* **115**, 251–259.
Söderström, B. and Bååth. E. (1978). *Hol. Ecol.* **1**, 62–72.
Stoyke, G. and Currah, R. S. (1991). *Can. J. Bot.* **69**, 347–352.
Straker, C. J. and Mitchell, D. T. (1986). *New Phytol.*, **104**, 243–256.
Stribley, D. P. and Read, D. J. (1974a). *New Phytol.* **73**, 731–741.
Stribley, D. P. and Read, D. J. (1974b). *New Phytol.* **73**, 1149–1155.
Summerbell, R. C. (1988). *Can. J. Bot.* **66**, 553–557.
Tschager, A., Hilscher, H., Franz, S., Kull, U. and Larcher, W. (1982). *Acta Oecologia* **3**, 119–130.
Vegh, I., Fahre, E. and Gianinazzi-Pearson, V. (1979). *Phytopathol. Z.* **96**, 231–243.
Wallen, B. (1983). *Oikos* **40**, 241–248.
Woodwell, G. M., Whittaker, R. H. and Houghton, R. A. (1975). *Ecology* **56**, 318–322.

Index

D

F

T

U

V

S·ience Library